WAVE PROPAGATION FOR TRAIN-INDUCED VIBRATIONS

A Finite/Infinite Element Approach

WAVE PROPAGATION FOR TRAIN-INDUCED VIBRATIONS

A Finite/Infinite Element Approach

Y B Yang

National Taiwan University, Taiwan

H H Hung

National Center for Research on Earthquake Engineering, Taiwan

World Scientific

NEW JERSEY · LONDON · SINGAPORE · BEIJING · SHANGHAI · HONG KONG · TAIPEI · CHENNAI

Published by

World Scientific Publishing Co. Pte. Ltd.
5 Toh Tuck Link, Singapore 596224
USA office: 27 Warren Street, Suite 401-402, Hackensack, NJ 07601
UK office: 57 Shelton Street, Covent Garden, London WC2H 9HE

British Library Cataloguing-in-Publication Data
A catalogue record for this book is available from the British Library.

WAVE PROPAGATION FOR TRAIN-INDUCED VIBRATIONS
A Finite/Infinite Element Approach

ISBN-13 978-981-283-582-6
ISBN-10 981-283-582-2

Printed in Singapore by Mainland Press Pte Ltd.

To our families

Foreword

In the four years since the senior author published his first book on Vehicle-Bridge Interaction Dynamics (World Scientific 2004), interest in this subject has spread over the whole globe, especially in Taiwan and Mainland China on the planning, design and construction of high speed railways. Since the completion of the high speed railways in Taiwan in 2007, our theoretical and our practical knowledge of the subject have expanded rapidly. The senior author with his many students has contributed fruitful ideas and computational techniques to advance the field.

The first book was focused on super structure vibration involving vehicle-bridge resonance caused by the moving trains at high speeds. This book deals exclusively with the theoretical principles and numerical techniques on infra-structural vibrations caused by the wave propagation of the ground and the vibration of buildings located along side the railways.

Here, as in every other field of engineering, the first theoretical principles are developed on the basis of a highly idealized condition with radical simplification of material model for soils and structures. Hence it is advisable and realistic that the authors began their new book exclusively with the theoretical treatment of the problem separated from practical applications. The magnitude of the difference between the performances predicted on the basis of the theory can only be ascertained by field experience. Under certain conditions and restrictions, some of the theories have stood the test of experience which is applicable to the approximate solution of practical problems.

It is a great pleasure to commend to you, the reader, this remarkably comprehensive book combining fundamental theory and numerical techniques on wave propagation and train-induced vibrations. The authors have brought together in a unified manner for the first time so much of what until now was available in journal papers authored mostly by the senior author and his former students which are known only to a few in the field. Students, researchers and practitioner will all benefit much from reading this book and having it for reference in the years to come.

W. F. Chen
Honolulu, Hawaii
March 2008

Preface

The commercial operation of the first high-speed train, i.e., the bullet train, in 1964 with a speed of 210 km/hr in the Japanese railways connecting Tokyo and Osaka marked the advent of a new era for railway engineering. Since then, high-speed trains with speeds over 200 km/hr have emerged as an effective tool for intercity transportation in several countries in Europe and Asia, including Japan, Germany, France, Italy, Spain, United Kingdom, Korea, Taiwan, China, etc. The trend of constructing new high-speed railways or upgrading existing railways to raise the train speeds is expected to remain upward for some years.

In order to provide unobstructed right of way, especially in densely populated areas, high-speed railway tracks are often carried by multi-unit elevated bridges. By doing so, the railway tracks can be maintained in an exclusive way and alleviated significantly from the settlement caused by the adjacent sinking ground. This has been the philosophy behind the construction of high-speed railways in Taiwan. Of the total length of 345 km high-speed railways in Taiwan, it is amazing to see that 73 percent of the railway track runs through the multi-unit elevated bridges, 18 percent runs through the tunnels, and only 9 percent runs through the traditional embankments.

The vibrations caused by moving trains over multi-unit elevated bridges may be classified into two categories related to the super- and infra-structural vibrations. As far as the *super-structural vibrations* are concerned, it is essential that the vibrations of the bridge and vehicles under the moving loads be kept within the design tolerance limits, as they both relate to the safety and maneuverability of the moving trains at high speeds. One key phenomenon in this regard is the vehicle-bridge resonance caused by trains moving at some critical speeds. Such a subject has been addressed by the senior author and co-workers in their book entitled *Vehicle-Bridge Interaction Dynamics – with Applications to High-Speed Railways*, World Scientific, 2004. It is totally excluded from the present coverage.

As for the *infra-structural vibrations*, one is concerned with the wave propagation of the ground and the vibration of buildings caused by trains moving on the ground surface or through the underground tunnels. This is exactly the subject to be covered in this book. Emphasis will be placed on the development of simple and accurate analysis methods for use by engineers in simulating the ground-borne vibrations encountered in practice.

The vibration of an elastic half-space problem is not a new problem. Research on this subject dates back to the classic work of Lamb in 1904. However, the analytical approaches developed in early days apply only to some ideal problems, e.g., a uniform elastic half-space with point or line loads. As far as numerical modeling is concerned, the half-space can be divided into to a *near field* and a *far field*. The near field, composed of the source, foundations, buildings, and underlying soils, which is generally irregular in nature, has been the focus of interest of design engineers. In contrast, the far field is defined as the soil domain excluding the near field, which is unbounded in nature. One key concern with the far field is how to model its effect of geometric attenuation or radiation damping due to the unbounded domain.

The boundary element method offers us a means to solve a wider class of half-space problems than the analytical approaches. However, this method is not as handy a tool as the finite element method in coping with the geometric and material variations of the near field. To overcome such a drawback, the near field will be modeled by finite elements in this book, for their general versatility in dealing with the irregularities in geometry and materials. Meanwhile, the far field will be modeled by the infinite elements derived, for their capability in simulating the radiation damping associated with unbounded domains.

Two special features exist with the infinite elements presented in this book. First, both the amplitude decay factor and wave number involved in the shape function for the direction leading to infinity are determined in a rational way. Second, a dynamic condensation procedure is presented for computing the far-field impedance for waves of lower frequencies consecutively from the one established for waves of the highest frequency. By such a procedure, the effort required in preparing the finite/infinite element mesh to meet the demands of various frequencies is greatly reduced.

Starting from a general review of related previous works in Chapter 1, the fundamental theory for elastic waves in the elastic half-space is summarized in Chapter 2, in which the loading functions for moving loads of various forms are presented.

Based on the plane strain assumption, the finite/infinite element approach is first presented for modeling the 2D profile perpendicular to the railway track. The basic derivation of the 2D approach will be presented in Chapter 3. Such an approach will be employed to study the characteristics of foundation vibrations in Chapter 4, the reduction efficiency of three wave barriers in Chapter 5, and the vibration reduction of buildings located alongside the railways in Chapter 6.

The 2D finite/infinite element approach presented above suffers from the drawback that the Mach radiation along the load-moving direction was ignored. To remedy such a problem, a third degree of freedom is introduced to each node of the original 2D elements to account for the out-of-plane wave transmission, assuming the material properties of the half-space to be uniform along the load-moving direction. The 3D wave propagation behavior caused by the moving trains along the railway line can be simulated using basically the 2D finite/infinite element mesh established for the profile considered. Because of its elegant feature, such an approach has been termed the 2.5D approach.

The basic theory of the 2.5D finite/infinite element approach is presented in Chapter 7. With such an approach, a parametric study for the key parameters involved in the ground vibrations will be presented in Chapter 8. In Chapter 9, such an approach will be adopted to study the efficiency of three wave barriers in reducing the ground vibrations caused by moving trains. One area that is particularly suitable for application of the 2.5D approach is the subway-related soil vibrations, since the geometric and material properties along the tunnel direction can be reasonably assumed to be uniform. This is exactly the topic to be presented in Chapter 10.

For most of the vibration problems studied, three speed ranges are considered for the moving loads, i.e., sub-, trans-, and super-critical speeds with respect to the Rayleigh wave speed of the ground surface. The maximum operating speed for some high-speed railways is known to be in the range of 300 to 350 km/hr. Recently, some railway companies have demonstrated that their test trains can easily exceed the speed of 570 km/hr. Since these speeds are generally higher than the Rayleigh wave speed of the ground, there is an urgent need to understand the vibration behaviors of trains, tracks, rails, and soils for trains moving in the super-critical range.

Parts of the materials presented in this book have been revised from the papers published by the authors and their co-workers in a number of technical journals, as well as the theses by the second author. Efforts have been undertaken to update, digest, and rewrite the materials acquired from different sources, such that a unified style of presentation can be maintained throughout the book. In particular, the authors would like to acknowledge the use of materials from the following papers and express their thanks the respective copyright holders: Hung (1995), Yang *et al.* (1996), Yang and Hung (1997), Hung (2000), Yang and Hung (2001), Hung and Yang (2001), Hung *et al.* (2001), Yang *et al.* (2003), Hung *et al.* (2004), and Yang and Hung (2008).

This book has been prepared as part of the results of research carried out by the senior author at the National Taiwan University. Many of the former graduate students have contributed directly or indirectly to the success of this work. Finally, a book can never be completed without the continuous support and expectation from the families of the authors, colleagues, friends, and the society in which they live.

Y. B. Yang
H. H. Hung
Taipei, Taiwan

Contents

Chapter 1

Introduction

The basic features of wave propagation in the half-space with soil medium are discussed. A review is presented of the various methods used in investigating the soil vibrations, including the analytical methods, field measurements, empirical prediction models, and numerical methods of simulation. Particular emphasis is placed on the vibrations induced by trains moving on the ground or through underground tunnels. Also summarized are the methods of isolation for ground-borne vibrations and the evaluation criteria adopted by different countries.

1.1 Ground-Borne Vibrations

Railway trains have been a major form of mass public transportation in the world for more than one and half centuries. There exists a wide variety of railway trains, ranging from the traditional freight and passenger trains to subway trains and high-speed trains. Different types of railway trains should meet different service, safety and environmental considerations. Even though great progress has been made in air transportation in the past century for long-distance, international, and trans-ocean travels, the status of railways as a key transportation tool for medium and short-distance travels remains the same. As a matter of fact, almost all major cities in the world have built their own subway systems, while high-speed railways have become increasingly popular in Asian and European countries following the launch of the *bullet train* in Japan in 1964.

Owing to the popularity of subways and high speed railways worldwide, most major cities have encountered the problem that railway lines have inevitably come close to some vibration-sensitive residential areas, laboratories, hospitals, high-precision science parks or telecommunication buildings. Although the vibrations induced by passing trains may not result in structural damages on adjacent buildings, they are known to cause the malfunctioning of some high-precision instruments or facilities housed in the buildings and to result in higher imperfection rates for integrated circuit production lines, while becoming a source of continuous annoyance to the occupants of buildings located alongside the railways. It should be mentioned that the effect of vibration on human comfort and annoyance is a very complex problem, which cannot be specified solely by the magnitude of monitored vibrations alone.[1] Recently, more and more vibration data have been collected from existing railway lines. Partly due to stricter environmental considerations, the problem of train-induced vibrations and their influence on human comfort and operation of sensitive equipment has received increasing attention from engineers, researchers and urban transportation planners in recent years.

The ground vibration induced by moving trains is a complicated dynamic problem. Vibrations of various sorts can be generated by the passage of trains due to the surface irregularities of wheels and rails, the rise and fall of the axles over sleepers. They can be transmitted through the track structure, including the rails, sleepers, ballast and sub-layers, propagate as waves through the soil medium, and then reach the buildings located nearby, creating a sense of discomfort to the occupants there. It should be noted that even for a train with perfect wheels moving over smooth rails, i.e., with no imperfections or unevenness in components such as wheels and rails, vibrations can still be generated by the regular repetitive action of the moving loads of the train.

Four major phases can be identified for the transmission of vibrations from the moving train through the railway, ties and subsoils to the neighboring structures: (a) *Generation*, i.e., the excitation caused by the regular repetitive action of moving wheel loads on the rails, plus the impact caused by the rotating wheels over the rails due to surface irregularities; (b) *Transmission*, i.e., the propagation of waves through

[1] Both the vibration and noise induced by passing trains may be of concern in this regard.

the surrounding soils; (c) *Reception*, i.e., the vibrations received by nearby buildings; and (d) *Interception*, i.e., the reduction in vibrations through implementation of wave barriers, such as piles, trenches, isolation pads, etc.

In each phase, various factors may affect the levels of vibrations to certain extents. The primary factors entering into consideration include the train type, train speed, track design, embankment design, ground condition, building foundation, building type, and the distance between the railway and buildings. The lack of an in-depth understanding of all these factors makes it difficult to simulate the problem in an accurate manner. Under certain circumstances, however, it is possible to estimate the levels of ground vibrations transmitted from the railways or traffic roads using a combination of empirical and theoretical formulas that have been made available.

Previously, the problem of ground-borne vibrations has been dealt with using mainly four different approaches, i.e., the *analytical approaches*, *field measurements*, *empirical prediction models*, and *numerical simulation*. In this chapter, a general survey will be given of each of the four approaches, followed by two separate sections each on the isolation of traffic-induced vibrations using some control devices and the evaluation criteria of vibrations adopted in different countries. All in all, we realize that research on train-induced vibrations has been voluminous and will grow continuously. It is almost impossible to come up with a compressive listing of all the relevant papers. We also realize that some of the research works were not available in English, especially those from Europe and Japan, where high-speed railways and subways have been put into operation for several decades. Under such a constraint, only papers that are readily available to the writers and have been written in English or Chinese will be cited in this chapter.

1.2 Analytical Approaches

By an analytical approach, one uses the theoretical models to describe the wave propagation characteristics of the source-path-receiver system. Because of the simplifications inevitably made in the modeling, exact closed-form solutions for most practical problems are at present not available. However, even for the limited number of ideal cases studied, the solutions obtained by previous researchers did provide us with a general picture of the key parameters involved. Solutions such as these are useful references for validating the results obtained by other numerical approaches.

1.2.1 *Classical theory of wave propagation*

The pioneering work of Lamb (1904) contained most of the elements that are essential to analytical studies of the sources and transmission paths in soils. In this work, Lamb investigated the disturbance generated in an elastic medium due to an impulsive force applied along a line or at a point on the semi-infinite surface or inside an unbounded full space. These solutions can be extended to yield the steady-state solutions for the cases with moving loads at constant speeds, if a new coordinate system moving synchronously with the loads is adopted. In reality, Lamb's solutions were also used by researchers as the basis in developing empirical prediction models. For the reasons stated, major features of the elastic half-space problem subjected to a point or line load, as was studied by Lamb, should be further explained. In this regard, it is realized that the same problems were analyzed subsequently by a number of researchers at different times, including, in particular, Ewing *et al.* (1957), Fung (1965), Graff (1973), and Achenbach (1976), among others. In the two somewhat tutorial papers presented by Gutowski and Dym (1976) and Dawn and Stanworth (1979), some major features of the elastic half-space problem were thoroughly discussed.

The governing equations for a homogenous isotropic solid can be written in terms of displacements **u** as

$$(\lambda+\mu)\nabla\nabla\cdot\mathbf{u}+\mu\nabla^2\mathbf{u}+\rho\,\mathbf{f}=\rho\,\ddot{\mathbf{u}}, \tag{1.1}$$

where λ and μ, the *Lamé constants*, are the elastic constants for the material; the latter is also known as the *shear modulus* and denoted as G in later chapters. Both constants can be expressed in terms of other elastic constants, such as *Young's modulus E*, *Poisson's ratio* ν, and the bulk modulus $\boldsymbol{K}$ (Graff, 1973):

$$E = \frac{\mu(3\lambda + 2\mu)}{\lambda + \mu}, \tag{1.2a}$$

$$\nu = \frac{\lambda}{2(\lambda + \mu)}, \tag{1.2b}$$

$$K = \lambda + \frac{2}{3}\mu. \tag{1.2c}$$

In Eq. (1.1), ρ is the mass density per unit volume of the material and $\mathbf{f}$ is the body force per unit mass of the material. Consider the governing equations in the absence of body forces. By performing the vector operation of divergence, one obtains

$$(\lambda + \mu)\nabla \cdot (\nabla\nabla \cdot \mathbf{u}) + \mu\nabla \cdot (\nabla^2 \mathbf{u}) = \rho\nabla \cdot \ddot{\mathbf{u}}. \tag{1.3}$$

Since $\nabla \cdot \nabla = \nabla^2$ and $\nabla \cdot (\nabla^2 \mathbf{u}) = \nabla^2 (\nabla \cdot \mathbf{u})$, the preceding equation can be reduced to

$$(\lambda + 2\mu)\nabla^2 \mathbf{\Delta} = \rho \frac{\partial^2 \mathbf{\Delta}}{\partial t^2}, \tag{1.4}$$

where $\mathbf{\Delta} = \nabla \cdot \mathbf{u}$ is the *dilation* of the material. Equation (1.4) can be recognized as the *wave equation*, expressible in the following form:

$$\nabla^2 \mathbf{\Delta} = \frac{1}{c_P^2} \frac{\partial^2 \mathbf{\Delta}}{\partial t^2}, \tag{1.5}$$

where the *propagation velocity* c_P is given by

$$c_P = \sqrt{\frac{\lambda + 2\mu}{\rho}}. \tag{1.6}$$

We thus conclude that *dilatational waves* will propagate at the velocity c_P in the solid.

We now perform the operation of curl on the governing equation in Eq. (1.1) neglecting the body forces. Since the curl of the gradient of a scalar is zero, we obtain

$$\mu \nabla^2 \boldsymbol{\omega} = \rho \frac{\partial^2 \boldsymbol{\omega}}{\partial t^2}, \tag{1.7}$$

where $\boldsymbol{\omega} = \nabla \times \mathbf{u} / 2$ is the rotation vector. The preceding equation can also be expressed in the form of *vector wave equation*, i.e.,

$$\nabla^2 \boldsymbol{\omega} = \frac{1}{c_S^2} \frac{\partial^2 \boldsymbol{\omega}}{\partial t^2}, \tag{1.8}$$

where the propagation velocity c_S is given by

$$c_S = \sqrt{\frac{\mu}{\rho}}. \tag{1.9}$$

Thus, *rotational waves* will propagate at velocity c_S in the medium.

We have found that waves may propagate through the interior of an elastic solid at two different speeds c_P and c_S. Dilatational waves, involving no rotation, propagate at the speed c_P, while rotational waves, involving no volume changes, propagate at the speed c_S. In general, the speed for the dilatational waves, c_P, is greater than that for the rotational waves, c_S, as can be observed by comparing Eqs. (1.5) with (1.8). A variety of terminology exists for the two types of waves. Dilatational waves are also known as *irrotational* or *primary* (*P*) *waves*, and the rotational waves as *equi-voluminal*, *distortional*, or *secondary* (*S*) *waves*. The P and S wave designations have arisen mainly from the seismological society. Other designations frequently used for the P waves are *longitudinal* or *compressional waves* and for the S waves are *transverse* or *shear waves*.

When an elastic wave encounters a boundary between two media, energy is reflected from and refracted across the boundary. If the boundary is a free surface, no refraction can occur. A major feature of the wave-boundary interaction process is mode conversion.

Except the two types of waves mentioned above, a third type of waves may exist whose effects are confined closely to the surface. Such waves were called *Rayleigh* (*R*) *waves*, as they were first investigated by Lord Rayleigh, who showed that their effect decreases rapidly with depth and their velocity of propagation is somewhat less than that of shear waves. The Rayleigh wave velocity c_R can be approximately related to the shear wave velocity c_S as

$$\frac{c_R}{c_S} = \frac{(0.87 + 1.12\nu)}{(1+\nu)}. \tag{1.10}$$

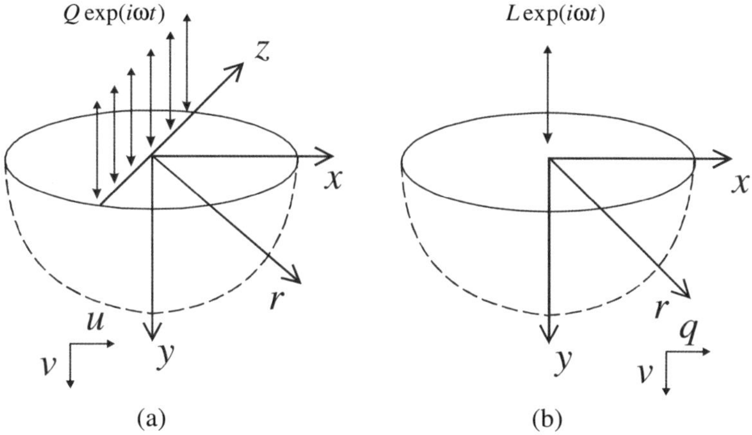

Fig. 1.1 Classical Lamb's problems with harmonic: (a) line load; (b) point load.

To honor the contribution of Lamb to the classical theory of wave propagation, the problems studied by Lamb have been named after him and grossly referred to as *Lamb's problems*.

Let us consider two typical Lamb's problems that are central to the present study in Fig. 1.1, in which parts (a) and (b) show a uniform elastic half-space subjected to a harmonic line load and an oscillating point load, respectively. Since the early work of Lamb, the same problems have been studied again by many researchers, including Ewing *et al.* (1957), Graff (1973), and Achenbach (1976), among others.

For the case of a harmonic line load $Q\exp(i\omega t)$ applied on the surface of the half-space, i.e., case (a) in Fig. 1.1, where Q is the magnitude of the applied load and ω the frequency of excitation, the horizontal displacement u and vertical displacement v on the surface ($y = 0$) of the half-space can be given as follows (Ewing *et al.*, 1957):

$$u = (Q/\mu)\left\{-H\exp[i(\omega t - k_R x)] + C(k_P x)^{-3/2}\exp\left[i(\omega t - k_P x)\right] + D(k_S x)^{-3/2}\exp\left[i(\omega t - k_S x)\right] + \cdots\right\}, \tag{1.11}$$

$$v = (Q/\mu) = \left\{-iK\exp[i(\omega t - k_R x)] + C_1(k_P x)^{-3/2}\exp\left[i(\omega t - k_P x)\right] + D_1(k_S x)^{-3/2}\exp\left[i(\omega t - k_S x)\right] + \cdots\right\}, \tag{1.12}$$

where $\cdots$ represents the higher order terms of the solutions, which can be neglected for farther distance x. The factors C, D, C_1, D_1, H and K depend on the wave numbers $k_S = \omega / c_S$, $k_P = \omega / c_P$ and $k_R = \omega / c_R$, but not on the distance x from the source:

$$C = -i\sqrt{\frac{2}{\pi}}\frac{k_P^3 k_S^2 \left(k_S^2 - k_P^2\right)^{1/2}}{\left(k_S^2 - 2k_P^2\right)^3}\exp\left(-i\frac{\pi}{4}\right), \tag{1.13a}$$

$$D = \sqrt{\frac{2}{\pi}}\sqrt{1 - \frac{k_P^2}{k_S^2}}\exp\left(-i\frac{\pi}{4}\right), \tag{1.13b}$$

$$C_1 = \frac{i}{2}\sqrt{\frac{2}{\pi}}\frac{k_P^2 k_S^2}{\left(k_S^2 - 2k_P^2\right)^2}\exp\left(-i\frac{\pi}{4}\right), \tag{1.13c}$$

$$D_1 = 2\sqrt{\frac{2}{\pi}}\left(1 - \frac{k_P^2}{k_S^2}\right)\exp\left(-i\frac{\pi}{4}\right), \tag{1.13d}$$

$$H = -\frac{k_R\left(2k_R^2 - k_S^2 - 2\sqrt{k_R^2 - k_P^2}\sqrt{k_R^2 - k_S^2}\right)}{F'(k_R)}, \tag{1.13e}$$

$$K = -\frac{k_S^2\sqrt{k_R^2 - k_P^2}}{F'(k_R)}. \tag{1.13f}$$

In Eqs. (1.13e) and (1.13f),

$$F'(k_R) = \left.\frac{dF(k)}{dk}\right|_{k=k_R}, \tag{1.14}$$

where $F(k)$ is the *Rayleigh function*,

$$F(k) = \left(2k^2 - k_S^2\right)^2 - 4k^2\sqrt{k^2 - k_P^2}\sqrt{k^2 - k_S^2}. \tag{1.15}$$

In Eqs. (1.11) and (1.12) for the displacement responses, the first, second, and third terms represent the contribution of the R-, P-, and S-waves, respectively. Clearly, for the case of harmonic line load considered, the R-waves do not suffer from any geometric attenuation on the ground surface, while both the P- and S-waves show a rate of attenuation proportional to $x^{-3/2}$.

Now, let us consider the case of a point load $L\exp(i\omega t)$ applied on the surface of the half-space, as shown in Fig. 1.1(b), where L denotes the magnitude of the applied load. The displacements of the half-space can be derived in the cylindrical coordinates, since they are symmetrical about the y-axis penetrating into the body. According to Ewing *et al.* (1957), the radial displacement q and vertical displacement v on the ground surface ($y = 0$) are

$$q = \frac{L}{\mu}\left\{-ik_R H\sqrt{\frac{1}{2\pi k_R x}}\exp\left[i\left(\omega t - k_R x - \frac{\pi}{4}\right)\right] + \frac{M}{(k_P x)^2}\exp[i(\omega t - k_P x)] + \frac{N}{(k_S x)^2}\exp[i(\omega t - k_S x)] + \cdots\right\}, \tag{1.16}$$

$$v = \frac{L}{\mu}\left\{k_R K\sqrt{\frac{1}{2\pi k_R x}}\exp\left[i\left(\omega t - k_R x - \frac{\pi}{4}\right)\right] + \frac{M_1}{(k_P x)^2}\exp[i(\omega t - k_P x)] + \frac{N_1}{(k_S x)^2}\exp[i(\omega t - k_S x)] + \cdots\right\}, \tag{1.17}$$

where x denotes the radial distance from the point load to the point of concern. Again, the first terms of Eqs. (1.16) and (1.17) represent the contribution of the R-waves. However, for the case of a point load applied on the half-space, the R-waves attenuate along the surface inversely proportional to the square root of the distance from the source, i.e., with a decaying rate proportional to $x^{-1/2}$. The remaining terms in the preceding two equations represent the contribution of the P- and S-waves, where M, N, M_1 and N_1 depend on the wave numbers k_S and k_P, but not on the distance x. As can be seen, the amplitudes of P- and S-waves diminish with the distance as a function of x^{-2} on the ground surface. Thus, the *geometric attenuation* of the P- and S-waves on the surface is more severe than that of the R-waves.

Concerning the interior of the half-space, Graff (1973) showed that for the case with a point load applied on the half-space, both the P- and S-waves decay at a rate proportional to r^{-1}, where r is the distance from the source to the point of concern in the interior of the half-space. In contrast, the response for the case of a line load is typical of cylindrical energy spreading, for which the attenuation rate for the P- and S-waves is proportional to $r^{-1/2}$ in the interior of the half-space.

As can be seen from the above discussions, the R-waves have a relatively good capability in travelling for a long distance on the surface of the half-space, but the same is not true for the P- and S-waves. In other words, the R-waves exist primarily near the surface, but the P- and S-waves have a better capability in penetrating through the interior of the half-space. For this reason, the R-waves have also been referred to as the *surface waves*, and the P- and S-waves as the *body waves*.

1.2.2 *Elastic medium subjected to moving loads*

With the continuous increase in the moving speed of passenger trains worldwide, the effect of speed of the moving loads has drawn much more attention from researchers than ever. One of the major concerns on the moving speed of passenger trains is the shock waves that may be produced as the train passes through some elastic barriers. It is well known that when an airplane passes through the sound barrier, the so-called *Mach radiation* of shock waves will occur. Likewise, when a moving train surpasses the *characteristic speed* of the waves of the soil medium, significant radiation effect can be expected for the ground motions. Obviously, the classical theory of wave propagation becomes insufficient, as no account has been taken of the effect of speed of the moving objects with respect to the soil medium.

After the classical questions associated with wave propagation had been answered to a certain level of satisfaction, scientists working on soil dynamics began to extend the framework established primarily by Lamb to the analysis of moving load problems. Consider an elastic medium subjected to a load moving at speed c. The solution for such a problem can be divided into three speed ranges:

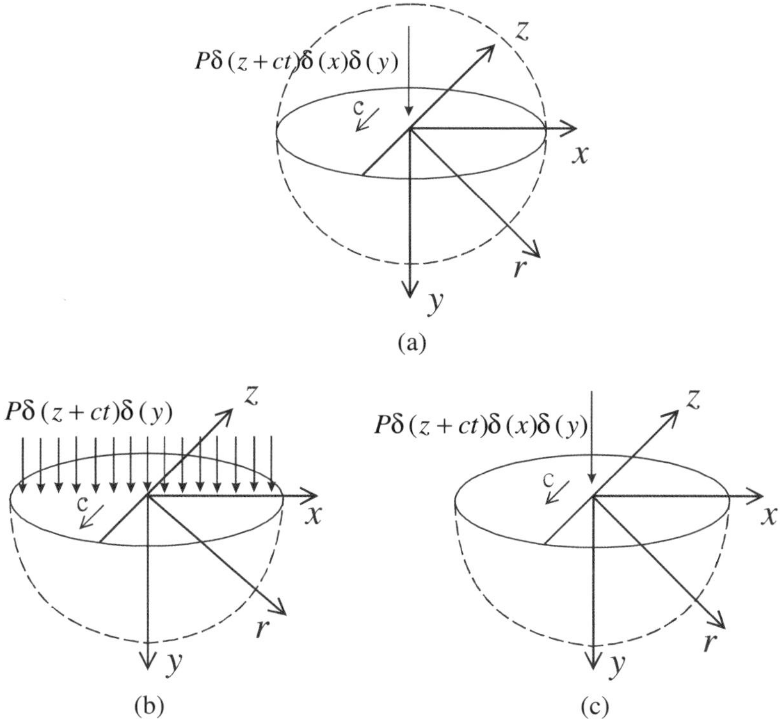

Fig. 1.2 Elastic body subjected to a moving load: (a) unbounded elastic body with point load; (b) elastic half-space with line load; (c) elastic half-space with point load.

(a) *Sub-critical speed* ($c < c_S$): The load is moving at a speed less than the S-wave speed of the elastic medium;
(b) *Tran-critical speed* ($c_S < c < c_P$): The load moves at a speed greater than the S-wave speed, but smaller than the P-wave speed; and
(c) *Super-critical speed* ($c_P < c$): The moving speed of the load is greater than the P-wave speed.

In the literature, terms such as *sub-*, *trans-*, and *super-sonic speeds* have also been used. However, due to the fact that the original meaning of sonic speed refers to the speed of sound or air, which is not the case encountered herein, we prefer to use the terms *critical speeds* to refer to the speeds of soils or the ground in this book. Besides, for problems where the surface waves play a much more important role than the body waves, e.g., those studied in Chapters 8 and 9, the Rayleigh wave speed may be chosen as the critical speed instead.

With regard to the effect of speed of the moving loads, three problems have been studied by researchers, as depicted in Figs. 1.2(a)-(c), in which part (a) shows an *elastic unbounded body* subjected to a moving point load, and parts (b) and (c) show an elastic half-space subjected to a moving line load and a moving point load, respectively. All the three cases are not purely of mathematical interest, but may have some implications in reality. For instance, the problem in Fig. 1.2(a) may find applications in computation of the response of soils around a tunnel through which the train passes. The problem in Fig. 1.2(c) represents the effect of an at-grade moving train. As far as the three ranges of moving speeds are concerned, there is a total of nine solutions for the three problems considered. However, only the three solutions for problem (a) are available in closed form. The solutions to problems (b) and (c) have to be computed by numerical procedures. Major features for the three problems will be briefly summarized in the following.

1.2.2.1 *Elastic unbounded body subjected to a moving point load*

Frýba (1972) analyzed the response of the unbounded elastic body in Fig. 1.2(a) to a moving point load by the technique of triple Fourier integral transformation. The solution obtained by Frýba (1972) for the vertical displacement v in the elastic body at $t = 0$ subjected to a point load P with speed c moving in the negative z-direction is

- Sub-critical speed:

$$v = \frac{P}{4\pi\mu M_2^{\,2}}\left[\frac{M_2^{\,2}}{R_2} + \frac{x^2}{r^4}(R_2 - R_1) - \frac{y^2 z^2}{r^4}\left(\frac{1}{R_2} - \frac{1}{R_1}\right)\right], \tag{1.18a}$$

- Trans-critical speed:

$$v = \frac{P}{4\pi\mu M_2^{\,2}}\left\{\frac{M_2^{\,2}}{R_2} H(z - a_2 r) + \frac{x^2}{r^4}\left[R_2 H(z - a_2 r) - R_1\right]\right.$$
$$\left. - \frac{y^2 z^2}{r^4}\left[\frac{1}{R_2} H(z - a_2 r) - \frac{1}{R_1} + \frac{a_2 r R_2}{z^2}\delta(z - a_2 r)\right]\right\}, \tag{1.18b}$$

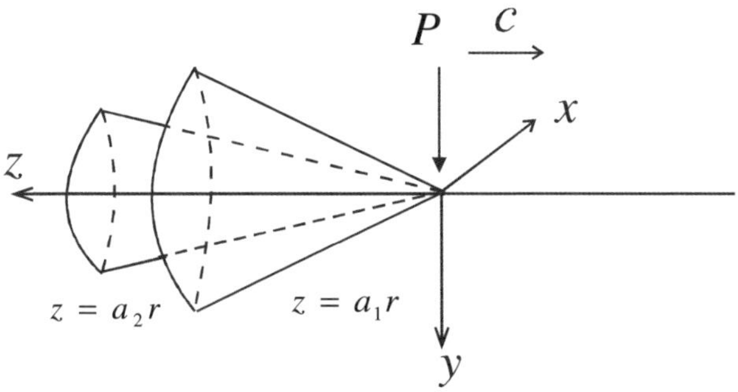

Fig. 1.3 Two Mach cones existing at super-critical speeds.

- Super-critical speed:

$$v = \frac{P}{4\pi\mu M_2^{\ 2}}\left\{\frac{M_2^{\ 2}}{R_2}H\left(z-a_2r\right)+\frac{x^2}{r^4}\left[R_2H\left(z-a_2r\right)\right.\right.$$
$$\left.-R_1H\left(z-a_1R\right)\right]-\frac{y^2z^2}{r^4}\left[\frac{1}{R_2}H\left(z-a_2r\right)-\frac{1}{R_1}H(z-a_1r)\right.$$
$$\left.\left.+\frac{r}{z^2}\left(a_2R_2\delta(z-a_2r)-a_1R_1\delta(z-a_1r)\right)\right]\right\}, \qquad (1.18c)$$

where

$$a_i^2 = \left|1-M_i^2\right|, \quad r^2 = x^2+y^2, \quad R_i^2 = z^2+(1-M_i^2)r^2, \qquad (1.19)$$

for i = 1,2. Here, $M_1 = c/c_P$ and $M_2 = c/c_S$ denote the *Mach numbers* related to the P- and S-waves, respectively, μ is the shear modulus for the elastic body, $H(x)$ the Heaviside function, and $\delta(x)$ the Dirac function.

As can be seen from Eq. (1.18), the vertical displacement is symmetrical about the x axis for the sub-critical speed range. But when the load speed exceeds the speeds of the P- or S-waves, the effects of the corresponding waves are confined to a region of the solid bounded by a trailing *Mach cone* with apex at the loading point and moving with it. The P-wave Mach cone can be written as $z = a_1r$ and the S-wave Mach cone as $z = a_2r$ (see Fig. 1.3). From Eq. (1.18c), one observes that no vibrations will be induced ahead of the P-wave front.

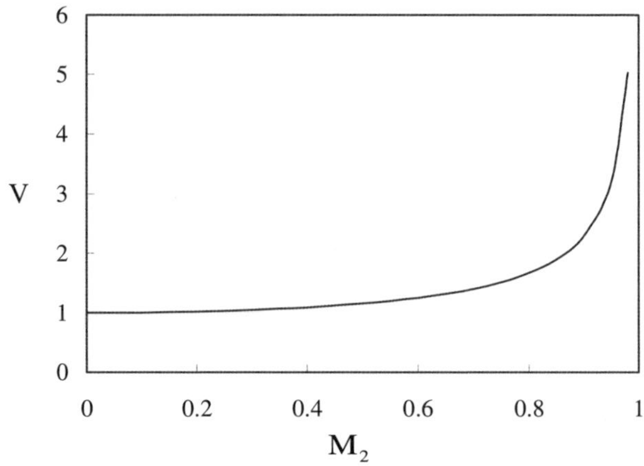

Fig. 1.4 Maximum vertical displacement versus the S-wave Mach number for an unbounded elastic body subjected to a moving point load.

Based on Eq. (1.18a) for the sub-critical speed range, the maximum vertical displacement v at the point (x = 0 m, y = 1 m, z = 0 m) was plotted with respect to the S-wave Mach number M_2 in Fig. 1.4, where the displacements have been given in a normalized form, i.e., $V = (4\pi\mu/P)v$. As can be seen, the displacement increases with the speed of the moving load. The variation appears to be gradual in the range with $M_2 < 0.6$, but for the range $M_2 > 0.6$, the displacement increases dramatically following the increase of M_2. Also, there exists a tendency that as the moving speed of the load approaches the S-wave speed, i.e., as M_2 approaches unity, the displacement becomes infinite.

1.2.2.2 *Elastic half-space subjected to a moving line load*

A general integral solution was given by Sneddon (1951) for the two-dimensional problem of a line load moving with a uniform sub-critical speed over the surface of a uniform elastic half-space. Cole and Huth (1958), Fung (1965), and Frýba (1972) considered the same problem for a normal line load and obtained solutions for the sub-, trans-, and super-critical speeds. The transient problem for a line load that suddenly appears on the surface of an elastic half-space and then moves with constant speed was considered by Payton (1967). From the aforementioned works, it is observed that if the load moves steadily at a speed equal to the R-wave speed, the response will become infinitely large. For the load moving at a super-critical speed, two *Mach planes* ($z = a_1 y$ and $z = a_2 y$) with singularities in displacement, instead of the two Mach cones, will occur.

1.2.2.3 *Elastic half-space subjected to a moving point load*

Eason (1965) studied the three-dimensional steady-state problem for a uniform half-space subjected to a point load moving at constant speeds. Besides the point load, Eason also considered the case of moving loads distributed over a circular or rectangular area. The governing equations were solved by means of integral transform, with the resulting multiple integrals reduced to single finite integrals for the sub-critical speed case. Gakenheimer and Miklowitz (1969) derived the transient displacements for the interior of an elastic half-space under a normal point load that is suddenly applied and then moves at a constant speed on the free surface. All the sub-, trans-, and super-critical speed cases were studied, while the inverse transform is evaluated by the Carniard-de Hoop technique. The steady-state response for the same problem was also given by Frýba (1972) in integral form. Using a method similar to Eason's (1965), Alabi (1992) studied the response due to an oblique moving point load on the free surface. By numerical integration, a parametric study was performed to investigate the effects of the load speed, distance and ground depth for the sub-critical speed case.

Following generally the procedure proposed by Luco and Aspel (1983), the steady-state displacements and stresses were solved by de Barrors and Luco (1994) for a multi-layered viscoelastic half-space due to a buried or surface point load moving along a horizontal straight line with sub-, trans-, or super-critical speed. The effect of layering was considered by using an exact factorization for the displacement and stress fields in terms of the generalized transmission and reflection coefficients. By the Fourier transform and a special integration scheme, Yeh *et al.* (1997) obtained the response of an elastic half-space to a moving point load for the sub-critical speed case. In the study by Lieb and Sudret (1998), the inverse transformation was performed by a decomposition in wavelets and the layered half-space was modeled by one-dimensional finite elements for the vertical direction in the transformed domain. Grundmann *et al.* (1999) studied the response of a layered half-space to a single moving periodic load as well as a simplified train load.

By the Fourier transform and a special integration technique, Hung and Yang (2001) proposed an analytical procedure for studying the mechanism of wave propagation for a uniform elastic half-space under the moving loads with static and dynamic components, considering four different types of moving load patterns. Also presented is a parametric study to investigate the effect of moving loads of the sub-, trans-, and super-critical speed ranges on the response of the underlying soils. Sheng *et al.* (2004) proposed a ground vibration model comprising the vehicles, track and ground for investigating the ground vibration in the presence of rail irregularities.

1.2.3 *Beam on elastic half-space subjected to moving loads*

As mentioned above, when an object moves at a speed greater than the wave speed of the surrounding medium, a *Mach cone* that moves with the object will be generated. For the case of moving trains, the moving load is first acting on the rails and then transmitted via the track and foundations to the underlying half-space. Obviously, the *characteristic speed* of the rails and foundations should be taken into account. In the literature, the supporting railroad track has been modeled as a beam resting on a *Winkler foundation* by a number of researchers. For instance, Frýba (1972) presented a detailed solution for the problem of a constant load moving along an *infinite beam* on an elastic foundation, considering all possible speed ranges and values of viscous damping. By the concept of *equivalent stiffness* for the supporting structure, a *critical speed* was identified for the moving load, at which the response of the beam becomes infinite. Such a speed corresponds exactly to the propagation speed of waves in the beam. For a load with speeds smaller than the critical speed, the largest amplitude of waves occurs near the point of loading. On the other hand, for a load with speeds greater than the critical speed, the waves moving ahead of the load are of small wavelengths and amplitudes compared with those behind the load.

The critical speed for a *Bernoulli-Euler beam* is the lowest bending wave speed, which can be given as

$$c_{cr} = \sqrt[4]{\frac{4sEI}{m^2}} \tag{1.20}$$

in which m is the mass per unit length, E the elastic modulus, I the moment of inertia of the beam, and s the coefficient of the Winkler foundation, usually assumed to be a constant. Similar results were obtained by Duffy (1990) in his study for the vibrations arising when a moving, vibrating mass passes over an infinite railroad track lying on a Winkler foundation.

By substituting the material properties for typical railroads into Eq. (1.20), researchers concluded that coincidence of the train speed with the critical speed is extremely unlikely (Frýba, 1972; Heckl *et al.*, 1996). However, the accuracy of these results can be largely influenced by the value used for the foundation coefficient s, which in practice is difficult to determine. Dieterman and Metrikine (1996, 1997) and Metrikine and Dieterman (1997) conducted a series of analysis to derive the *equivalent stiffness* of an elastic half-space interacting with a Bernoulli-Euler beam of finite width. They found that the equivalent stiffness depends mainly on the frequency and wave number of the beam. With this equivalent stiffness taken into account, the analysis indicated that there exist two *critical speeds*. One corresponds to the R-wave speed and the other is somewhat smaller than the R-wave speed. Both speeds can result in severe amplification of the displacement of the beam. Later, Lieb and Sudret (1998) performed a similar analysis and found that severe displacements can be observed on the half-space underlying the rails at the critical speeds as well. Metrikine *et al.* (2001) used a similar track model overlying a visco-elastic half-space to theoretically investigate the phenomenon of *visco-elastic drag* associated with the excitation of ground waves by a high-speed train.

Suiker *et al.* (1998) further studied the critical behavior of a Timoshenko-beam-half-space system under the moving load. If the highway traffic, instead of the railroad traffic, is considered, then it is more proper to use a model with a plate on elastic foundations. Kim and Roësset (1998) investigated the dynamic response of an *infinite plate* on an elastic foundation subjected to the moving load. The *critical speed* observed for such a case is

$$c_{cr} = \sqrt[4]{\frac{4sD}{m^2}}, \tag{1.21}$$

where D is the flexural rigidity of the plate, and m and s are the mass and stiffness of the foundation per unit area, respectively. Chen and Huang (2000) derived the critical velocities for both the Bernoulli-Euler and *Timoshenko beams* on *Winkler foundations*. It was found that the lowest bending wave speed of the rails and the Rayleigh wave speed of the ground are both over 500 km/hr for real situations.

1.2.4 *Tunnel structure subjected to moving loads*

Considering the interaction of a tunnel-soil-building system due to passing trains, Balendra *et al.* (1991) proposed a simple semi-analytical approach for predicting the level of ground-borne vibrations using a *substructure method*. The plane strain model they used comprises a rigid tunnel in the elastic half-space with a rigid embedded footing for supporting the building that was modeled as a lumped mass. The entire problem was decomposed into a foundation radiation boundary-value problem and a tunnel radiation boundary-value problem for the purpose of computing the impedance matrix for the entire system. By using the substructure method, the response of the building to the train loading was evaluated and compared with the allowable vibration limits.

To investigate the level of ground vibrations due to trains moving in a tunnel, Metrikine and Vrouwenvelder (2000a) proposed an analytical approach with a simple two-dimensional model consisting of a visco-elastic layer and a Bernoulli-Euler beam located inside the layer. Assuming the layer and the beam to be infinitely long in the longitudinal direction, they analyzed the surface vibration under three types of loadings moving along the beam, namely, constant, harmonically varying, and stationary random loads. Later, Metrikine and Vrouwenvelder (2000b) improved their procedure by using two identical Bernoulli-Euler beams connected by distributed springs, instead of a single beam. They recognized that the results obtained using such a two-dimensional model can be regarded as an upper estimate of the level of ground vibrations, as the practical situations may be quite different. Recently, Forrest and Hunt (2006a,b) proposed a three-dimensional analytical model for studying the train-induced ground vibration from a deep underground railway tunnel of circular cross section. The tunnel is assumed to be an infinitely long, thin cylindrical shell, whereas the surrounding soil is modeled by means of wave equations for an elastic continuum.

1.2.5 *Load generation mechanism*

For continuously welded rails and perfect wheels, the most important mechanism of excitation for ground vibrations by the moving trains is the *quasi-static pressure* exerted by the wheel axles onto the track. Such a pressure with certain patterns will move with the wheels. Krylov and Ferguson (1994) studied the ground vibration associated with railways by the *Green function* formalism, in which the deflection curve of a beam lying on a *Winkler foundation* and subjected to a stationary point load was adopted as the shape of the pressure generated by each wheel axle on the rails. The pressure generated by the wheel axle is distributed and radiated to the ground through the sleepers. By superposition of the elastic waves radiated by the sleepers caused by the passage of all the wheel axles and by taking into account the time lag between the forces and their locations in space, a load generating mechanism that is capable of simulating the influence of sleeper spacing, train length and train speed was constructed. As for the effect of subsoils, they utilized the results of the axisymmetric Lamb's problem for the half-space subjected to a vertical harmonic point load to determine the Green function, but considered only the contribution of the R-waves. Krylov (1995) further extended this analysis to studying the response caused by superfast trains, from which the *Mach radiation* can be observed as the train moves at a speed faster than the R-wave speed of the subsoil. In Takemiya's (1997) study, the same deflection curve was adopted to account for the quasi-static pressure generated by the wheel axles onto the ground. In this study, however, the sleeper spacing was not taken into account.

A *random vibration method* was used by Hunt (1991) to model the road traffic-induced ground vibration. In his study, vehicles were modeled as two-axle systems, each with four degrees of freedom, and the ground as a uniform elastic half-space with viscous damping. Based on Lamb's (1904) solution of the half-space response generated by a harmonic load on the surface, he derived the *frequency response function* for an elastic isotropic half-space. Later, Hunt (1996) extended the above approach to computation of the vibration transmission from railways into buildings using the random process. A similar method was adopted by Hao and Ang (1998) to estimate the *power spectral densities* of traffic-induced ground vibrations. In order to circumvent the difficulties associated with numerical integrations, they considered the contribution of the R-waves only and derived an approximate closed-form solution accordingly.

As far as the ground vibrations due to trains moving over multi-unit elevated bridges are concerned, a semi-analytical approach was proposed by Wu *et al.* (2002) with the following two features. First, the analytical solution of an elastically supported beam travelled by the moving loads is used to simulate the load-transmitting mechanism from the superstructure of the bridge. Second, the *Green's function* for an elastic half space under a point load is adopted to simulate the wave propagation behavior of the soil. For the kind of structures considered, such a semi-analytical approach is much more efficient for studying the ground-borne vibrations than those based on full numerical modeling. This approach was later extended by Wu and Yang (2004a,b) and Yang and Wu (2007) to the analysis of ground and building vibrations due to high speed trains of different types moving over multi-unit elevated bridges, in which the effect of elastic bearings, bridges piers, pile foundations, and foundation-soil interactions are all taken into account.

1.3 Field Measurement

By field measurement, the response of existing structures to various disturbances can be obtained directly. When sufficient measurements are taken, a data base can be compiled, by which the results can be analyzed using statistical approaches. These results can then be used as a basis for predicting the vibration levels of structures under similar conditions.

However, only after a statistically meaningful number of results have been made available for each case, can the effect of various parameter changes and corresponding vibration control measures be reliably predicted (Melke and Kraemer, 1983). Besides, successful site measurements of structures and ground vibrations require sophisticated electronic equipment. It is essential that a sufficient number of vibration sensors be deployed at different control points such that the representative response of wave transmission can be simultaneously measured. This will call for a number of high-quality long cables and remote amplifiers, if the data are to be transmitted to a central data logger without picking up extraneous noises (Newland and Hunt, 1991). Obviously, a thorough field measurement is the most time-consuming and expensive way among the four approaches mentioned above. Given below is a partial review of the experimental results obtained by some previous researchers for the railroads.

Dawn and Stanworth (1979) presented a few of the experimental studies conducted on the *British Railways*. Melke and Kraemer (1983) proposed an approach for analyzing the field measurement data, by which information was extracted and used in establishing a prediction model. From a *one-third octave band* analysis of the experimental data obtained, they observed the existence of two peak levels in the frequency domain. One has a fixed value corresponding to the tunnel/soil natural frequency, and the other has a value that increases with the train speed corresponding to the *sleeper passing frequency* f_s, that is,

$$f_s = \frac{c}{l_s}, \tag{1.22}$$

where c is the train speed and l_s the spacing between sleepers. Whenever the two peak frequencies coincide at certain train speed, the vibration level will increase drastically due to the effect of *resonance*.

From the analysis of some measurement data, remarks were made by Heckl *et al.* (1996) concerning the mechanism of vibrations excited by the moving trains. They found that besides the sleeper passing frequency f_s, the *wheel passing frequency* f_a is another crucial parameter for the vibrations induced by moving trains. Here, the wheel passing frequency f_a is defined as

$$f_a = \frac{c}{l_a}, \tag{1.23}$$

in which l_a is the distance between two consecutive wheels. Since the distance between two consecutive wheels is generally not constant for a commercial train, the wheel passing frequency is less apparent than the sleeper passing frequency. Other possible excitation mechanisms of ground vibrations by moving trains include the *quasi-static pressure* generated by the wheel axles onto the track, the effects of joints in unwelded rails, unevenness of wheels or rails, and the effects of carriage- and wheel-axle bending vibrations associated with their natural frequencies (Krylov and Ferguson, 1994).

Volberg (1983) carried out measurements of vibration propagation induced by passing trains at three different sites with different ground properties. The measured data appear to be rather independent of the sites investigated. A calculation scheme involving a simple power law was proposed for predicting the train-induced vibrations in the vicinity of planned railroad tracks. Noise and vibration measurements have been conducted for the rapid rail transit system in Calcutta, India by Mohanan *et al.* (1989). The results of measurement showed that both the noise and vibration levels were higher than the recommended values. Based on these data, the factors influencing the results were discussed and the methods for reducing the noise and vibration levels were proposed. With reference primarily to the *German standards*, Kurze (1996) reviewed various measurement procedures and prediction schemes in use for determining the environmental impact of railway noise and vibration. Common methods for noise and vibration control at sources and in the propagation paths were also discussed in this paper.

In China, experimental measurements were carried out for a tunnel in *Beijing subways* by Pan and Xie (1990). Two in-situ experiments were performed for a bridge site and buildings near the railway lines by Xia *et al.* (2005) concerning the vibrations induced by running trains. In Japan, Okumura and Kuno (1991) studied the effects of various factors on the railway noise and vibration through a regression analysis of the field data collected for 79 sites along 8 traditional railway lines in an urban area. Among the six factors they used to explain the vibration peak level, i.e., the distance, railway structure, train type, train speed, train length and background vibration, they found the influence of distance to be most crucial. The second prominent factor is background vibration, which is considered to be characteristic of the soil properties at each site. They also found that the influence of train speed is not so obvious. Such an observation can be attributed to the fact that the field data is collected from traditional railways, whose running speed is generally below 100 km/hr (27.77 m/s), far smaller than the R-wave speed for usual soil conditions. Regarding the influence of railway structures, the vibration levels for the concrete bridges and retaining walls are lower than those for the at-grade structures.

Takemiya (1998a) analyzed the measured field data alongside one *Shinkansen railway* during the train passage, which has an average speed of around 240 km/hr. He concluded that the high-speed train generates rather impulsive ground motions of short duration corresponding well with the wheel distance. Consequently, the vibration property can be modeled quite well, given the information of wheel distance and the number of carriages connected. From his observation, the response features are significantly different for different types of supporting structures. For instance, much more waves are reflected through the layered soils for the at-grade track, while for railways of the viaduct type, the frequency contents of the structure-borne vibration are closely related to the soil-structure system.

Lang (1988) performed some experiments to test the effectiveness of floating concrete slabs and trenches in isolating the vibrations of buildings located near the railway track. The results indicate that both methods are effective for reducing the vibrations, but the barriers appear to be more effective for reducing the vibrations if they are installed at places closer to the track.

Recently, a project named *CONVURT* (Clouteau *et al.*, 2005) was conducted by the European Union, aimed at controlling the vibrations generated by underground rail traffic through a multi-national effort. Within the framework of this project, Chatterjee *et al.* (2003) and Degrande *et al.* (2006a) performed in-situ vibration measurements in Paris and London, respectively.

1.4 Empirical Prediction Models

Due to the lack of a proper understanding of the excitation generation mechanism from railway trains and the difficulty involved in determining the soil properties, to precisely model the soil-structure system of concern using existing numerical methods is not an easy task. In this regard, one feasible, but approximate, approach is to construct a simplified but reasonable model for predicting the responses based on the empirical and theoretical results available. By and large, most prediction models existing in the literature are composed of several separable independent formulas, each of which contains a control parameter and can affect to a certain extent the final response. A simple prediction model such as this can be used to provide tentative estimates, when extensive measurements or investigations are not affordable or cannot be achieved in a short time.

Gutowski and Dym (1976) and Verhas (1979) combined the essence of measurement and theory into a predictive model, which is given in a simplified form of *attenuation function*, taking into account the effects of material attenuation and geometrical attenuation. Kurzweil (1979) presented a model for predicting the vibration of buildings caused by the trains passing nearby, in which the vibration attenuation due to propagation through the ground, the ground-building interaction, and the propagation characteristics of the building were all taken into account. Melke (1988) proposed a procedure for predicting the structure-borne noise and vibration from underground railway lines. Based on the analytical techniques and laboratory measurements, a chain of *transmission losses*, including the track transmission loss, tunnel transmission loss, ground transmission loss, and building transmission loss, was proposed to predict the final velocity level of the building. Trochides (1991) presented a simple method for predicting the excitation levels due to ground-borne vibrations in buildings located near the subways. This model is based on approximate impedance formulas for the tunnel and structure, as well as simple energy considerations. Comparisons between the calculations and measurements on scaled models showed that the predictions were generally acceptable for design purposes.

By the use of a statistical formulation, Madshus *et al.* (1996) proposed a semi-empirical model for predicting the low frequency vibrations, based on a large number of vibration measurements in Norway and Sweden. To make possible a unified and systematic handling of the empirical data, a database was established, too. This model includes five separable statistically independent factors, i.e., the train type specific vibration level, speed factor, distance factor, track quality factor, and building amplification factor. In order to minimize the negative influence of vibrations in buildings located near railways, *Swiss Federal Railways* developed a three-part computer program for predicting the emission of vibrations and structure-borne noises for every newly constructed or extended railway track (Kuppelwieser and Ziegler, 1996). Recently, a simple prediction model specially tailored for the *Italian high-speed railway* was proposed by Rossi and Nicolini (2003).

1.5 Numerical Simulation

Concerning the literature on ground-borne vibrations, most of the early researches were conducted by analytical or experimental approaches. Whenever an analytical approach was adopted, however, restrictions were often imposed on the geometry and material properties of the problem considered, as closed-form solutions cannot be easily made available for most practical situations. On the other hand, although the results obtained by the experimental approaches appear to be most reliable and close to real situations, an exhausted field test may cost a lot. Starting from the mid 1970s and enhanced by the advent of high-performance computers, various numerical methods emerged as effective tools for solving the wave propagation problems, including in particular the finite element method, boundary element method, and their variants.

In the past three decades, a great portion of the studies on wave propagation problems were performed by the *boundary element method*. A significant amount of the relevant works can be found in the review papers by Beskos (1987, 1997). Using the boundary element method, the *radiation damping* can be accurately taken into account through use of suitable fundamental solutions. However, the irregularities in geometry

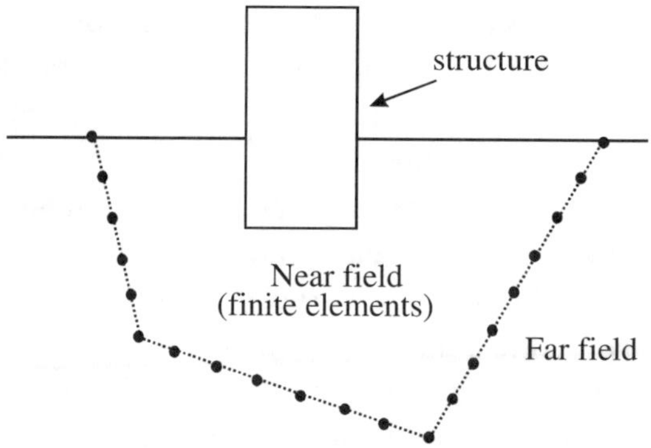

Fig. 1.5 Schematic diagram of the hybrid method.

and materials of the structure and underlying soils, as may be encountered in practice, cannot be dealt with in an easy way. It is true that some modern versions of the boundary element method have also been equipped with the capability to deal with the inhomogeneity in geometry. Nevertheless, this has been achieved at the expense of using a much more complicated *Green's function* or a finer subdivision of the interior domain considered.

In contrast, the finite element method appears to be more versatile in applications, with which various irregularities in geometry, including the embedded structures and multi soil layers, can be simulated with no difficulty. Thus, as far as the vibration of structures and surrounding soils is concerned, a finite element modeling remains the most favorable choice. However, the finite element method suffers from the drawback that the soil, which is semi-infinite by nature, can only be modeled by elements of finite size. Consequently, the *radiation damping* that accounts for the loss of energy due to waves traveling to infinity cannot be accurately modeled.

To overcome this drawback, other auxiliary methods are often called for to model the infinite region, which leads to the so-called *hybrid method*. By this method, the domain of a soil-structure system is divided into two sub-domains, i.e., the *near field* and *far field* (Fig. 1.5). The near field consisting of the structure and the region of the soil of interest, as enclosed by the dotted line in the figure, is modeled by the finite elements as conventional. The far field is a semi-infinite domain excluding the near field. The dotted line in the figure can be regarded as the interface between the near and far fields. In a finite element analysis, the *impedance matrix* for the far field is established in terms of the nodal points at the interface, as indicated by the dotted line, for relating the nodal forces to the nodal displacements.

In the literature, a number of methods exist for modeling the infinity property of the far field for use in the finite element simulation, which include, for instance, the traditional boundary element method, consistent boundary, transmitting boundary, viscous boundary, superposition boundary, paraxial boundary, double-asymptotic boundary, extrapolation boundary, multi-direction boundary, infinite element, and the so-called consistent infinitesimal finite-element cell method. A discussion of the advantages and disadvantages of each of these methods can be found in Wolf and Song (1996), which will not be recapitulated here.

Owing to its flexibility, the *hybrid method* has often been used in dealing with problems involving the wave barriers, buildings, embankment, layered soils, as well as rails and tracks. In general, there are three approaches for modeling the half-space problems: two-dimensional (2D) modeling, three-dimensional (3D) modeling, and two-point-five-dimensional (2.5D) modeling. By the coupled finite element-boundary element method, Andersen and Jones (2006) investigated the quality of the results obtained from the 2D model of a railway tunnel through comparison with those obtained from a corresponding 3D model. They concluded that 3D models are required for absolute predictions. However, the 2D model provides results that agree qualitatively with those of the 3D model at most frequencies. Consequently, for problems of which the qualitative behavior, rather than the quantitative behavior, is of primary concern, a 2D model is considered sufficient.

Intuitively, the results obtained by 3D modeling are believed to be most trustworthy. Following general finite element analysis procedures, one can establish the 3D model for simulating the structure of concern and surrounding soils in a straightforward manner, and then use such a model to analyze the dynamic response of the structure-soil system. A major concern in this regard is the large effort required in establishing the three-dimensional analysis model and the huge amount of computation required for the frequency-domain analysis, which may involve the operations of complex numbers.

Zhao and Valliappan (1993) presented a dynamic infinite element for 3D infinite-domain wave problems. Park *et al.* (2004) developed 3D elastodynamic infinite elements for soil-structure interaction problems. As far as the practical applications are concerned, a 3D finite element analysis was conducted by Ju (2002) for the ground vibration due to trains moving over a seven-span bridges, using the absorbing boundary conditions to simulate the infinite boundary of the soils. The dynamic 3D finite element program ABAQUS was adopted by Hall (2003) to simulate the train-induced ground vibrations with the boundary simulated by a number of dashpots. However, as was pointed out by both authors, a full 3D dynamic finite element analysis of the half-space problem is extremely time-consuming. For this reason, only a limited amount of research works have been carried out along these lines of research for ground-borne vibrations. It is realized that for problems with great variations in the geometric and material properties of the soil-structure system, a full 3D finite element modeling may still be necessary, in order to capture some of the local effects that may be hidden by the 2D or other simplified models.

Fortunately, due to the periodic nature of the loading and geometry of the half-space along the direction of the moving loads, a third type of modeling called the *2.5D modeling* has been made possible for simulating the 3D problem. By the 2.5D approach, one uses a finite element mesh that is basically of the 2D nature, but with due account taken of the load-moving effect in the third dimension, to simulate the 3D dynamic behavior of the half-space. Based on the above discussions, only a brief review of the 2D and 2.5D modeling will be given in the following.

1.5.1 *Two-dimensional modeling*

A survey of the literature indicates that most early researches on the ground-borne vibrations were based on the two-dimensional modeling with *plane strain* assumption. Under the condition that the external loading can be regarded as an *infinite line load*, and that the material and geometric properties of the system are identical along the direction of the line load, the assumption of plane strain applies and therefore the two-dimensional modeling can be adopted. According to Gutowski and Dym (1976), the passage of train loads can be reasonably simulated as a moving line load, provided that the receiver from the track is approximately less than $1/\pi$ times the length of the train.

Segol *et al.* (1978) used finite elements along with special non-reflecting boundaries to investigate the isolation efficiency of open and in-filled trenches in layered soils. Balendra *et al.* (1989) used finite elements along with the viscous boundary to investigate the vibration of a subway-soil-building system in Singapore. Thiede and Natke (1991) adopted a similar method to study the influence of thickness variation of subway walls. Laghrouche and Le Houedec (1994) used finite elements along with consistent boundaries to study the effectiveness of an elastic mattress lying under a railway in reducing the traffic-induced ground vibrations. Chua *et al.* (1995) analyzed a subway-soil-building system using a two-dimensional finite-element idealization, in conjunction with an analytical derivation of the train-loading spectrum at the tunnel invert. Yang *et al.* (1996) and Yang and Hung (1997) combined the finite and infinite elements to investigate the effect of trenches and elastic foundation in reducing the ground vibrations induced by moving trains. Hung *et al.* (2001) used the same procedure to study the vibration of the building alongside the railway. Analytical frequency-dependent infinite elements were presented by Yun *et al.* (2000) and Kim and Yun (2000) for analysis of 2D soil-structure systems.

1.5.2 *2.5-dimensional modeling*

For practical reasons, one may assume that the material and geometric properties are identical along the direction of a railway track. Consider a 2D profile perpendicular to the track, which consists of the cross section of the railway, surrounding soils and even the bedrock. If the load-moving effect in the third dimension is not of concern, then the use of the 2D profile including the geometric and material variations of the half-space is generally sufficient.

However, if the effect of load-moving in the third dimension is to be considered, then the use of the 2D profile alone is not sufficient. This is especially true when the train speed increases and approaches the critical speed of the soil, as the *Mach radiation* effect of the soil cannot be ignored. In reality, such a problem is two-dimensional in geometry, but three-dimensional in wave propagation. Strictly speaking, it can be analyzed only using a 3D model. However, for problems where the geometry and material properties are uniform along the railway direction, the use of a 3D model to simulate a problem that is 2D in nature is not computationally efficient.

By taking into account the relation of displacements between two nodes on the neighboring finite elements along the direction of wave-traveling, a solid element was reduced to a plane element with three degrees of freedom per node by Hwang and Lysmer (1981). This element was used to study the response of underground structures under the traveling seismic waves. A similar idea called the *dimensionality reduction* was later employed by Luco and de Barros (1994, 1995) to study the seismic response of a cylindrical shell and a layered cylindrical valley embedded in a layered half-space, and by Stamos and Beskos (1996) to study the seismic response of long lined tunnels excavated in a half-space.

As an extension of the work by Hwang and Lysmer (1981), the three-dimensional wave propagation behavior of traffic-induced vibrations was analyzed by Hanazato *et al.* (1991), in which the weights, speed, intervals, and vibrating conditions of the vehicles were all taken into account. The near field was modeled by finite elements, and the far field by thin-layered elements. Later, Takemiya (1997) used similar finite elements to model the embankment, while adopting the boundary element procedure to derive Green's function for the underlying layered soils by discretization along the depth.

As was stated previously, the problem of train-induced ground vibrations is two-dimensional in geometry, but three-dimensional in wave propagation. Thus, if the original 2D formulation can be modified to include the load-moving effect in the third dimension, then we can use basically the same 2D mesh to generate the 3D response of the problem considered. This has been the idea behind the 2.5D approach proposed by Hung (2000) and Yang and Hung (2001), which can also be regarded as an extension of the original 2D approach by Yang *et al.* (1996) for modeling the soil-structure system in the wave-number and frequency domain using the finite/infinite elements.

By the 2.5D approach, the geometry and material properties of the half-space along the load-moving direction are assumed to be invariant. An extra degree of freedom is introduced at each node to account for the out-of-plane wave transmission, in addition to the two in-plane degrees of freedom conventionally used for the plane strain element. The profile of the half-space is divided into a *near field* and a semi-infinite *far field*. The near field containing the acting loads, structures, and soil region of concern is simulated by finite elements, while the far field containing infinite soil domains by infinite elements. By first transforming the system equations to the frequency domain and then back to the time domain, the 2.5D finite/infinite element method can be used to simulate the three-dimensional wave traveling behavior of the soil-structure system due to the moving loads for all ranges of speeds considered. Later, the 2.5D approach was adopted by Yang *et al.* (2003) to study the wave propagation behavior of layered soils due to surface moving trains. The results from this study allow us to visually apprehend how the Mach cones are formed along the railway track as the train speed increases from the sub- to the super-critical speed range. The 2.5D approach was also adopted to study the reduction efficiency of various wave barriers by Hung (2000) and the ground vibrations caused by underground moving trains by Yang and Hung (2008).

Similarly, but not based on the finite/infinite element approach, Sheng *et al.* (2006) used the boundary element method incorporating the wave number in the track direction to predict ground vibrations from trains running on the ground surface and in tunnels. Recently, by the hypothesis that the tunnel and soil are periodic in the longitudinal direction of the tunnel, Degrande *et al.* (2006b) proposed a periodic coupled finite element-boundary element formulation for predicting the free-field vibrations caused by metro trains moving through the tunnels. By the periodicity assumption for the geometry, the discretization of the soil-tunnel system in the direction of the tunnel is limited to a single-bounded reference cell.

1.6 Isolation of Ground Vibrations

There have been a number of methods developed for the control of ground-borne vibrations due to moving trains. The most popular countermeasures include the installation of trenches, wave impeding barriers, and floating slab tracks. Depending on whether the isolation device is installed near the source of excitation or near the structure to be protected, the method of isolation can be classified as *active isolation* or *passive isolation*, respectively. In what follows, the major features and literature associated with each type of wave barriers will be discussed. Other possible methods of railway vibration reduction include the installation of very thick tunnel walls, resilient mount under buildings (Newland and Hunt, 1991), an increase in tunnel depth, rail grinding and wheel truing, or using rail pads, under-sleeper pads, ballast mats, etc. (Wilson *et al.*, 1983).

1.6.1 *Trenches*

The trenches, including open and in-filled ones, have been used as wave barriers for isolating the vibration of machine foundations for years. Relevant literature on this subject has been abundant. An experimental investigation on the screening effect of open trenches was performed by Woods (1968). By the lumped mass method, Lysmer and Waas (1972) studied the effectiveness of a trench in reducing the horizontal shear wave motion induced by a harmonic load acting on the rigid footing lying over a horizontal soil layer. Segol *et al.* (1978) used the finite elements, along with special *non-reflecting boundary*, to investigate the isolation efficiency of open and bentonite-slurry-filled trenches in layered soils. Yang and Hung (1997), Hung (2000), and Hung *et al.* (2004) used the 2D and 2.5D finite/infinite elements to parametrically analyze the isolation effect of open trenches, in-filled trenches, and elastic foundations. Other related works that should be cited here include those of Aboudi (1973), Emad and Manolis (1985), Beskos *et al.* (1986), Beskos *et al.* (1990), Ahmad and Al-Hussaini (1991), Ahmad *et al.* (1996), Ni *et al.* (1994), Al-Hussaini and Ahmad (1991, 1996), Yeh *et al.* (1997), and Ni and Hung (1998).

As indicated by the aforementioned works, the most important requirement for the trench to achieve a good effect of isolation is that the trench should have a depth of an order of the surface wave length. Primarily for this reason, the isolation of ground-borne vibrations by trenches is effective only for moderate to high frequency vibrations.

1.6.2 *Wave impeding block*

Because of the presence of a rigid rock base, a soil stratum has some intrinsic eigenmodes for the waves to transverse, according to Wolf (1985). No vibration eigenmodes can be induced below the *cut-off frequency* of the soil stratum, which equals $c_P/(4H)$ for the vertical injected longitudinal waves, and equals $c_S/(4H)$ for the shear waves, with H denoting the depth of the soil stratum. It is therefore possible to take advantage of this vibration transmission property of the soil layer over the bedrock to impede the spreading of vibrations, say, by installing an artificial stiff plate at a certain depth below the source. Such an idea has led to invention of the so-called *wave impedance barrier* (WIB) for vibration reduction. Among the works conducted on the subject, the following should be cited: Schmid *et al.* (1991), Antes and von Estorff (1994) and Takemiya and Fujiwara (1994). All of these studies show that the WIB can effectively reduce the ground-borne vibrations. If an *artificial bedrock* is used, the foundation and soil vibrations can be significantly reduced, but the propagation of waves into the surrounding area cannot be totally prevented (commonly known as the *leaking problem*) for two reasons. First, the artificial bedrock is limited in length. Second, the artificial bedrock may vibrate by itself, in violation of its role as a rigid base. The effectiveness of the artificial bedrock can be improved by enlarging its length and stiffness. Shielding of the building from soil vibrations can also be achieved by installing an artificial bedrock directly beneath the building.

From the construction point of view, a WIB with a rectangular shape requires a substantial amount of excavation of the soils before the concrete block can be poured and cast on site. To overcome this drawback, the rectangular WIB was later modified by Takemiya (1998b) to be of the X shape, and referred to as the X-WIB. Such a device can be constructed by the conventional soil improvement procedure on site through mixing and injecting the cement paste directly into the soils. More recently, another WIB named honeycomb WIB was presented by Takemiya (2004) for mitigating the vibration induced from a high-speed train viaduct with pile foundations.

1.6.3 *Floating slab track*

The floating slab tracks, which consist basically of the concrete slab track supported by resilient elements, have been widely used on modern rail transit systems (Wilson *et al.*, 1983). It is well known that greater effectiveness can be achieved for reducing the ground-borne vibration and noise at frequencies above $\sqrt{2}$ times the vertical *resonant frequency* of the floating slab system. However, as the frequency is close or equal to the resonant frequency, the vibration will be greatly amplified. The design of floating slab tracks is based on the assumption of a single-degree-of-freedom system, with the lumped mass determined as the summation of the mass of the floating slab and the unsprung mass of the train, and the spring stiffness determined solely from the supporting resilient pads. In order to raise the effectiveness of the floating slab track, namely, to lower the resonant frequency, the mass of the floating slab should be enlarged as much as possible, because the resilient pads should not be too soft to ensure rail stability under full axle loads. Such highly resilient elements can be incorporated in different places of the transmission path to reduce the level of vibrations. Many different devices can be used as the resilient elements, including the rubber springs under the rails, Cologne eggs (Esveld, 1989), ballast, resilient devices under sleepers, plates under the rails, foam rubber mats under the ballast, etc. (Heckl *et al.*, 1996).

Balendra *et al.* (1989) used a two-dimensional finite element model to compare the effects of two different supporting systems, the direct fixation and the one with a floating slab. It was found that the vibration levels for the *floating slab track system* exceed those of the *direct fixation track system* in the low frequency range. However, in the high frequency range, the floating slab track system behaves as an effective vibration isolator. Grootenhuis (1977) introduced several types of floating track slabs that were already used in engineering applications, while proposing a new design that can be constructed inside a bored tunnel without increase in the tunnel diameter. Wilson *et al.* (1983) also studied the effectiveness of a floating slab trackbed for a rapid transit system in Washington, D.C. Laghrouche and Le Houedec (1994) and Yang and Hung (1997) investigated the isolation efficiency of an *elastic foundation* in reducing the train-induced vibrations. In general, the function of an elastic foundation constructed right underneath the track is similar to that of the floating slab track. Nelson (1996) discussed some of the developments and applications of vibration mitigating measures taken at the vibration sources in the U.S.A. and Canada. According to this study, the best performance in terms of the vibration attenuation can be expected from the floating slab system, among those techniques implemented for the source.

1.7 Evaluation Criteria of Vibration

Many design guides and standards have offered methods for assessing or reducing human exposure to vibrations in buildings. The effect of vibration on comfort and annoyance, however, is a very complex issue and cannot be specified solely by the magnitude of monitored vibrations alone. In other words, vibration associated phenomena, such as structure-borne noise, airborne noise, rattling, movement of furniture and other objects, as well as visual effects, may relate to the degree of complaints. Some studies, including the works done by Howarth and Griffin (1991), Paulsen and Kastka (1995), and Knall (1996), have been conducted to predict the subjective response of human beings to simultaneous noise and vibration produced in buildings located alongside the railways. It was concluded that for a proper evaluation of annoyance, the combined effects of the noise and vibration should be taken into account, rather than either the noise or vibration alone. However, researches related to the combined effect of disturbances by noise and vibration are still insufficient to form a valid basis for implementation of design standards. Further investigations with field experiments are required to establish appropriate criteria for evaluation of human response to train-induced vibrations in buildings.

Table 1.1 Standards for evaluation of human exposure to vibration in buildings.

Nation	Name of Standard	Standard Number
International Organization for Standardization	International Standards	ISO 2631-1 ISO 2631-2
United States of America	American National Standards Institute	ANSI S3.29
United Kingdom	British Standards	BS 6472
Germany	Deutsches Institut für Normung	DIN 4150-2
Norway	Norwegian Standard	NS 8176
Japan	Japanese Industrial Standards	JIS C 1510 JIS Z 8735

Experience in many countries has indicated that the occupants of residential buildings are likely to complain even if the vibration levels only slightly exceed the perception threshold, which for instance may range from about 0.01 to 0.02 m/s^2 peak, according to the *International Standards* ISO 2631-1:1997 (1997). Therefore, most standards provide values representing approximately the same human response with respect to annoyance of various frequencies, but no acceptable magnitudes on the building vibrations alone. Table 1.1 lists the standards used in some countries. Most standards for the evaluation criteria of vibrations, for example, the *Norwegian Standard* NS 8176 (Turunen-Rise *et al.*, 2003), contain two main objectives. The first is to define a unified method for measuring and quantifying vibrations, and the second to give some limit criteria for vibrations.

The International Standards ISO 2631-2 is the most commonly used standards and has often been regarded as the basis of other standards for development of related criteria for evaluating the human exposure to vibrations in buildings. A brief conceptual review of such a standard will be given in the following. For those who are interested in applications of the vibration criteria for buildings, this standard should be consulted for more details. As the part of the standards to be summarized below is related to assessment of public vibration nuisance, it should find applications to ground-borne vibrations induced by the moving trains as well.

The International Standards ISO 2631-2 is a part of ISO 2631, which offers guidance on the evaluation of human exposure to whole-body vibrations, especially for vibrations in buildings from 1 to 80 Hz. The measurement of vibrations should follow the methods given in ISO 2631-1. As human sensitivity to vibration is highly frequency-dependant, the summation effects should be considered for vibrations of different frequencies. Thus, overall weighted vibration values in terms of acceleration are often used in the evaluation. The *frequency-weighted acceleration* a_ω is determined by appropriate weighting and addition of one-third octave band data as follows:

$$a_\omega = \left[\sum_i (\omega_i a_i)^2 \right]^{1/2}, \tag{1.24}$$

where ω_i is the weighting factor for the ith one-third octave band and a_i is the root-mean-square (r.m.s.) acceleration for the ith one-third octave. The frequency weighting is normally incorporated in the design of measuring equipment with built-in weighting filters and band-limiting filters. Most modern vibration meters give an overall level of frequency-weighted acceleration on the measured axis a_ω. For brevity, the values of the frequency weighting factors will not be listed here. Those who are interested in calculation of the frequency-weighted acceleration should refer to the standard ISO 2631-1:1997 (1997) for further details.

The basic evaluation parameter given in ISO 2631-1:1997 (1997) is the *weighted r.m.s. acceleration* a_ω (in m/s^2 or rad/s^2), defined as

$$a_\omega = \left[\frac{1}{T} \int_{t=0}^{T} a_\omega^2(t)dt \right]^{1/2}, \tag{1.25}$$

where T is the duration of measurement (s). The weighted r.m.s. acceleration a_ω should be determined for each axis (x, y and z) of the principal surface of the floor supporting the human body. For undefined axis of human vibration exposure, the combined effects of vibrations in buildings are also taken into account by the combined standard base curve shown in Fig. 1.6.

According to the ISO 2631-2:1989 (1989), satisfactory vibration magnitudes for rooms of various functions should be specified in multiples of the base curve magnitudes. The ranges of multiplying

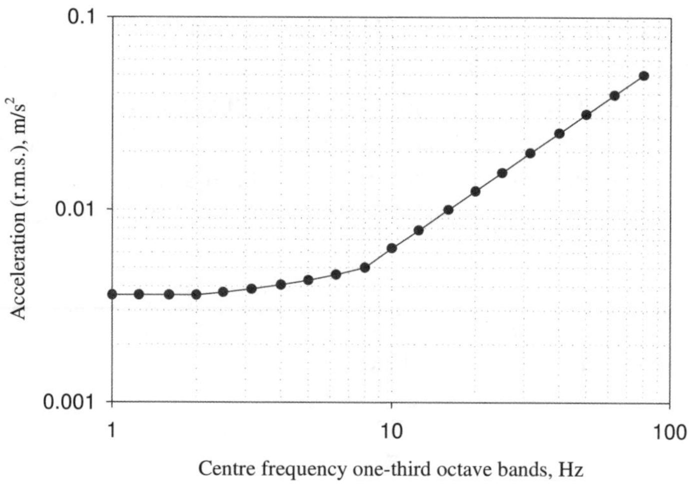

Fig. 1.6 Combined direction (*x*-, *y*-, *z*-axis) acceleration base curve for building vibrations: ISO 2631-2:1989.

Table 1.2 Multiplying factors given for vibration magnitudes below which the probability of adverse human reaction is low (ISO 2631-2:1989).

Place	Time	Continuous or intermittent vibration	Transient vibration excitation with several occurrences per day
Critical working areas (e.g. some hospital, operating theatres, some precision laboratories, etc.)	Day Night	1	1
Residential	Day Night	2 to 4 1.4	30 to 90 1.4 to 20
Office	Day Night	4	60 to 128
Workshop	Day Night	8	90 to 128

factors used in several countries were listed in Table 1.2. Complaints are likely to arise from the occupants of buildings when the vibration magnitudes, i.e., the weighted r.m.s. accelerations, exceed the value represented by the corresponding curve related to each axis. This does not necessarily mean that the values above this curve will give rise to

Table 1.3 VDV suggested above which adverse reactions may be expected from residential building occupants (unit: $m/s^{1.75}$) (BS 6472:1992).

Place	Low probability of adverse comment	Adverse comment possible	Adverse comment probable
Residential buildings 16 hours (Day)	0.2 to 0.4	0.4 to 0.8	0.8 to 1.6
Residential buildings 8 hours (Night)	0.13	0.26	0.51

adverse reactions, as the magnitude which is considered to be satisfactory depends on the real circumstance.

According to Griffin's (1996) comprehensive handbook for human vibration, the *Vibration Dose Value* (VDV) is a preferred measurement unit for assessment of human exposure to railway vibrations, which is evaluated at the center of the floor of interest during the measurement period. The VDV is used as a measure of the cumulative exposure to vibrations from a passing train, and also as one of the defined means for assessing the vibration severity in ISO 2631-1:1997 (1997). It is defined as the fourth root of the integral of the fourth power of the frequency-weighted acceleration a_ω over a period T:

$$\text{VDV} = \left[\int_{t=0}^{T} a_\omega^4(t)dt \right]^{1/4}, \tag{1.26}$$

where $a_\omega(t)$ is the *instantaneous frequency-weighted acceleration* (m/s^2) or (rad/s^2) and T is the duration of measurement (s). The SI unit of VDV is $m/s^{1.75}$.

In general, the *British Standards* BS 6472:1992 (1992) is quite similar to ISO 2631-2:1989 (1989). Besides the frequency-weighted r.m.s. acceleration, the VDV is used by the BS 6472:1992 (1992) as another measurement for evaluation of the human response to vibrations in buildings. Examples for calculating the VDV were given in this standard. Table 1.3 lists the VDV suggested by BS 6472:1992 (1992), in which the ranges for adverse reactions expected from residential building occupants were also listed.

In the newest edition of ISO 2631-2:2003 (2003), the baseline curves are not used any more. This edition gives only measurement methods for vibration, while the same guidance as in ISO 2631-1:1997 (1997) is adopted for evaluating the annoyance of human beings. In this standard, it was pointed out that complaints of occupants of residential buildings are likely to arise if the vibration magnitudes evaluated in terms of the frequency-weighted r.m.s. acceleration slightly exceed the perception threshold extended from about 0.01 to 0.02 m/s^2 peak. Due to the lack of comprehensive understanding of vibration associated annoyance, collecting data for evaluation of the human response to building vibrations is encouraged for updating the future version of ISO 2631-2.

Another frequently used measurement of vibration is the *vibration acceleration level* L_{va} with unit decibel (dB), which is defined as

$$L_{va} = 20 \log \frac{a}{a_0}, \tag{1.27}$$

where a is the r.m.s. value of the vibration acceleration and a_0 is the reference vibration acceleration, which is taken as 10^{-5} m/s^2 by the *Japanese Industrial Standards* (JIS). If the perception threshold of human being is 0.01 m/s^2, the vibration acceleration level of the perception threshold is 60 dB. When the r.m.s. vibration acceleration is weighted, another definition of vibration measurement often used is the *vibration level* L_v with unit decibel (dB), which is defined as

$$L_v = 20 \log \frac{a_c}{a_0}, \tag{1.28}$$

where a_c is the r.m.s. value of the vibration acceleration weighted by the vertical or horizontal characteristics, and a_0 is the reference vibration acceleration (10^{-5} m/s^2 for JIS).

In Japan, the methods of measurement for vibration levels, especially for the ground vibrations due to public vibration nuisance, were standardized in JIS Z 8735 (1981) and JIS C 1510 (1995) for vibration level meters. The ground vibration caused by road traffic, factory facilities and construction work have been regulated by law so to protect the quality of life environment. The *Vibration Regulation Law* issued by the Ministry of the Environment (1976), Japan, applies to vibrations measured on the ground surface. Owing to the fact that people are more sensitive to vertical than horizontal vibrations in the frequency range of

Table 1.4 Vibration criteria regulated by the Vibration Regulation Law.

		Day	Night
Type I	Residential area	65 dB	60 dB
Type II	Commercial area Industrial area	70 dB	65 dB

Note: The criteria for the area within 50 m away from schools, hospitals, libraries and sanatoria are obtained with a reduction of 5 dB from the values listed above.

vibration nuisances and that the vertical ground vibration is usually more serious than the horizontal ground vibration, the focus of vibration impact assessment is placed mainly on the vertical vibration. The criteria of vibrations listed in the vibration regulation law have been reproduced in Table 1.4. The magnitude of vibration on the floor of a house is usually estimated by adding a value of 5 dB to the one measured on the nearby ground surface (Yokota, 1996). However, this correction value was obtained 20 years ago when most of the houses were made of wood. Nowadays, further researches on this subject are conducted to achieve a more reasonable value for modern buildings in Japan, which are made mainly of steel or reinforced concrete.

1.8 Concluding Remarks

An overall review arranged in an approach-oriented manner has been presented on the vibration issues associated with railways. Also commented are the countermeasures for vibration mitigation and evaluation criteria of vibration. Regardless of which approach was used, all the papers cited play a role in advancing the research on this subject.

The theoretical results serve as a useful reference for development of the other methods. By the analytical approaches, the major factors affecting each problem, such as the train speed, distance, and soil condition, can be identified, and guidelines for evaluating the relative influence of each of the factors can be drawn.

To perform a complete field measurement for the train-induced vibrations is always costly and labor-intensive, not only because the related equipment may not always be available, but because it is difficult to find a site with tracks and scheduled trains that is safe and convenient for testing. For the reasons stated, a database compiled from the field measurements is highly valuable. It can offer clues for evaluating the key factors involved in the overall dynamic response, such as the spacing of sleepers, spacing of wheels, unsprung masses, and the type of track supporting structures.

Empirical prediction models seem to be the roughest among the four approaches considered. But they offer a hands-on approach for engineers to draw a quick estimate when there is a lack of time for tedious numerical analysis and extensive field measurement.

With the rapid advancement of high-performance computers, numerical simulation emerges as a very effective tool for modeling the wave propagation problems. As a matter of fact, for many practical problems, the numerical approach remains the only approach that can be undertaken at a reasonably low cost and within a short period. Nevertheless, the reliability of numerical simulation in predicting the vibration levels depends largely on the accuracy of the input data and the choice of an appropriate theoretical framework, which can be evaluated using some benchmark problems through comparison with experimental or theoretical results previously made available.

The most complex process among the four processes mentioned in Sec. 1.1 for vibration transmission is the source generation mechanism. In the literature, most researchers considered only the effect of quasi-static pressure generated by the axle loads. But in reality, there may exist dynamic terms which may be generated by the unevenness of the wheels and rails, or associated with the sleeper passing frequency, rail passing frequency, and resonance in the vehicle suspension. All these factors should be taken into account in future studies.

From the analytical studies, we know that if a train travels at a speed greater than the propagation speed of the ground waves, a shock wave will be generated on the ground. Such a phenomenon should not be regarded merely as one of mathematical interest. It may arise in the real world due to the continuous rise in the operation speed of modern high-speed trains. For example, it was reported that train speeds over 500 km/hr have been achieved on an experimental track in France (Krylov, 1995). In May 1990, nine runs of TGV trains moving at speeds over 500 km/hr or 138.8 m/s were made by the *French Railway Company* (SNCF) on the section of track between Courtalain and Tours. More recently, according to a news released by SNCF on April 3, 2007, their new test train achieved a record-high speed of 574.8 km/hr in one of their eastern railway lines. These speeds have already surpassed the speed of Rayleigh waves of the sustaining soils. As a result, significant radiation effect on the ground vibrations became visible in these areas and has resulted in restriction of the speed for the TGV trains on that part of track (Dieterman and Metrikine, 1996). Measurements by the railway companies in Swiss (SBB), France (SNCF), Germany (DB), Holland (NS) and Great Britain (BR) have also confirmed the amplification of the vertical movement in the track when the train moves with a speed of the same order as that of the Rayleigh wave speed of the subsoil (Dieterman and Metrikine, 1997).

Of the previous works concerning the trans-Rayleigh wave behavior, most were conducted by theoretical investigations. However, as this phenomenon is becoming not merely as a theoretical issue, but can really take place in certain circumstances, much more realistic models should be adopted to thoroughly study such an effect, at least through in-depth numerical simulations. On the other hand, apart from passively setting a speed limit on the train and/or improving the supporting strength of the subsoil, few countermeasures have been proposed for vibration reduction of trains moving over critical speeds. As far as the trains moving at super-critical speeds is concerned, it is suggested that further research be conducted to investigate the stability of the track system, including the rails, and that the effectiveness of conventional wave barriers, including those mentioned in Sec. 1.6, be re-examined.

This review is conducted with the hope that it may provide useful information to engineers and researchers for evaluation of the environmental vibrations associated with high-speed railways and subways in different parts of the world. Besides, the papers cited herein serve as good references for further investigation.

Chapter 2

Elastic Waves in Half-Space Due to Vehicular Loads

In this chapter, the response of a visco-elastic half-space subjected to moving loads with static and dynamic components is investigated. Four types of vehicular loads are considered, including the moving point load, uniformly distributed wheel load, elastically distributed wheel load, and a train load simulated as a sequence of elastically distributed wheel loads. In each case, the influence of the moving speeds in the sub-, trans-, and super-critical ranges on the dynamic responses of the half-space is studied. The parametric study conducted herein enables us to grasp insight into the mechanism of wave propagation for a visco-elastic half-space under the moving loads. The results obtained also serve as benchmarks for verifying the accuracy of the numerical simulation technique to be presented in later chapters.

2.1 Introduction

The problem of ground-borne vibrations induced by moving vehicles has been one of increasing interest, partly enhanced by the construction of mass rapid transit systems and high speed railways worldwide. Eason (1965) studied the three-dimensional steady-state response for a uniform half-space subjected to loads moving at constant speeds, in which both point loads and loads distributed over a circular or rectangular area are considered. The governing equations were solved by means of integral transforms, with the resulting multiple integrals reduced to single finite integrals for the sub-critical speed case. Gakenheimer and Miklowitz (1969) derived the transient displacements for the interior of an elastic half-space under a suddenly applied point load moving at a constant speed on the free surface. All the sub-, trans-, and super-critical speed cases were studied, while the inverse transform is evaluated by the Carniard-de Hoop technique. The steady-state response for the same problem was also given by Frýba (1972) in integral form.

Using a method similar to Eason's (1965), Alabi (1992) studied the soil response to an oblique moving point load applied on the free surface. By numerical integration, a parametric study was performed to investigate the effects of the load speed, distance and ground depth for the sub-critical case. De Barrors and Luco (1994) obtained the steady-state displacements and stresses within a multi-layered visco-elastic half-space generated by a buried or surface point load moving along a horizontal straight line with sub-, trans-, or super-critical speeds, following generally the procedure proposed by Luco and Aspel (1983). Grundmann *et al.* (1999) studied the response of a layered half-space subjected to a single moving periodic load, as well as a simplified train load. The inverse transformation was performed by a decomposition in wavelets by Lieb and Sudret (1998), along with the layered half-space modeled by one-dimensional finite elements for the vertical direction in the transformed domain.

The objective of this chapter is not to give a comprehensive review of related previous works. Rather, efforts will be focused on study of the key parameters involved in the wave propagation of a half-space generated by moving vehicles. Four types of vehicular loads are considered, including a moving point load, a uniformly distributed wheel load, an elastically distributed wheel load, and a train load simulated as a sequence of elastically distributed wheel loads. In each case, the influence of the moving loads traveling in the sub-, trans-, and super-critical speed ranges on the dynamic responses of the half-space will be studied.

The elastically distributed wheel load was considered by Krylov and Ferguson (1994), Krylov (1995), and Takemiya (1997). However, they have not proceeded to investigate the effect of different vehicle speeds. The materials presented in this paper have been revised significantly from the paper by Hung and Yang (2001).

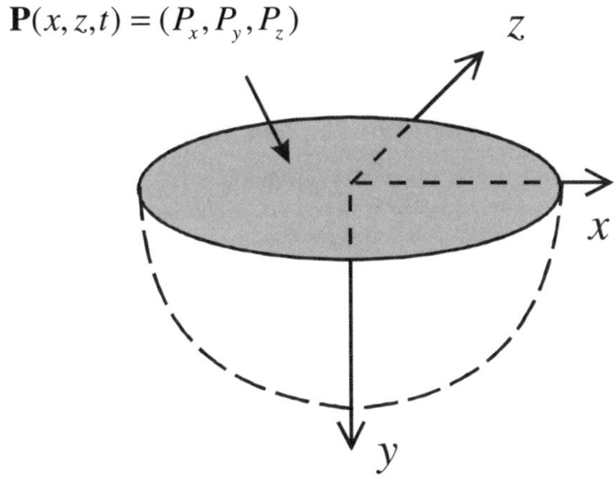

Fig. 2.1 Uniform elastic half-space subjected to a general load.

2.2 Fundamentals of the Problem

In this section, the basic equations to be used, along with the triple Fourier Transform will first be outlined.

2.2.1 *Equation of motion*

The governing equations in terms of the displacements for the homogenous isotropic elastic half-space shown in Fig. 2.1 can be written as follows:

$$(\lambda+\mu)\nabla\nabla\cdot\mathbf{u}+\mu\nabla^2\mathbf{u}+\rho\,\mathbf{f}=\rho\,\ddot{\mathbf{u}}, \tag{2.1}$$

where λ and μ are Lamé's constants, $\mathbf{u}$ and $\mathbf{f}$ denote the displacement and body force components, respectively, and ρ is the mass density of the elastic solid. The preceding equation can be reduced to a simple set of equations through use of the *Helmholtz potential*. Let the displacement field $\mathbf{u}$ be represented as

$$\mathbf{u}=\nabla\Phi+\nabla\times\mathbf{\Psi}, \tag{2.2}$$

where $\Phi(\mathbf{x},t)$ is a scalar function and $\mathbf{\Psi}(\mathbf{x},t)$ a vector-valued function. In the absence of body forces, substituting the displacements $\mathbf{u}$ in Eq. (2.2) into Eq. (2.1) yields

$$\nabla\left[(\lambda+2\mu)\nabla^2\Phi-\rho\ddot{\Phi}\right]+\nabla\times\left[\mu\nabla^2\mathbf{\Psi}-\rho\ddot{\mathbf{\Psi}}\right]=0. \tag{2.3}$$

Clearly, the equation of motion will be satisfied if the displacement components **u** can meet the following conditions:

$$\Delta\Phi-\frac{1}{c_P^2}\frac{\partial^2\Phi}{\partial t^2}=0, \tag{2.4a}$$

$$\Delta\mathbf{\Psi}-\frac{1}{c_S^2}\frac{\partial^2\mathbf{\Psi}}{\partial t^2}=0, \tag{2.4b}$$

in which the *compressional* and *shear wave speeds*, c_P and c_S, are defined as follows:

$$c_P=\sqrt{\frac{\lambda+2\mu}{\rho}}, \tag{2.5a}$$

$$c_S=\sqrt{\frac{\mu}{\rho}}. \tag{2.5b}$$

The compressional wave speed c_P can also be expressed in the following form based on Eq. (1.2b):

$$c_P=\sqrt{\frac{2\mu(1-\nu)}{\rho(1-2\nu)}}. \tag{2.6}$$

The first equation with the scalar potential Φ in Eq. (2.4) describes the propagation of the compressional waves, and the second one with the vectorial potential $\mathbf{\Psi}$ the shear waves. The implication from Eq. (2.4) is that the waves may propagate into the interior of an elastic solid at two different speeds, i.e., at c_P and c_S. From Eq. (2.2), the three components of the displacement **u** can be expressed as:

$$u=\frac{\partial}{\partial x}\Phi+\frac{\partial}{\partial y}\Psi_z-\frac{\partial}{\partial z}\Psi_y, \tag{2.7a}$$

$$v=\frac{\partial}{\partial y}\Phi+\frac{\partial}{\partial z}\Psi_x-\frac{\partial}{\partial x}\Psi_z, \tag{2.7b}$$

$$w=\frac{\partial}{\partial z}\Phi-\frac{\partial}{\partial y}\Psi_x+\frac{\partial}{\partial x}\Psi_y, \tag{2.7c}$$

where u, v, w are the displacement components in the time and space domains along the three directions x, y, z.

By Hooke's law and the strain-displacement relations, the stresses can be expressed in terms of $\mathbf{\Phi}$ and $\mathbf{\Psi}$ as:

$$\sigma_{yy} = 2\mu\left(\frac{\partial^2}{\partial y^2}\Phi + \frac{\partial^2}{\partial y \partial z}\Psi_x - \frac{\partial^2}{\partial y \partial x}\Psi_z\right) + \lambda\left(\frac{\partial^2}{\partial x^2} + \frac{\partial^2}{\partial y^2} + \frac{\partial^2}{\partial z^2}\right)\Phi, \tag{2.8a}$$

$$\tau_{xy} = \mu\left[2\frac{\partial^2}{\partial x \partial y}\Phi + \frac{\partial^2}{\partial x \partial z}\Psi_x - \frac{\partial^2}{\partial y \partial z}\Psi_y + \left(\frac{\partial^2}{\partial y^2} - \frac{\partial^2}{\partial x^2}\right)\Psi_z\right], \tag{2.8b}$$

$$\tau_{zy} = \mu\left[2\frac{\partial^2}{\partial z \partial y}\Phi + \frac{\partial^2}{\partial y \partial x}\Psi_y - \frac{\partial^2}{\partial x \partial z}\Psi_z - \left(\frac{\partial^2}{\partial y^2} - \frac{\partial^2}{\partial z^2}\right)\Psi_x\right], \tag{2.8c}$$

for an elastic solid.

2.2.2 *Triple Fourier transform*

In this chapter, the triple Fourier transform and its inverse transform that are adopted throughout are defined as follows:

$$\hat{f}(k_x, y, k_z, \omega) = \frac{1}{(2\pi)^3}\int_{-\infty}^{\infty}\int_{-\infty}^{\infty}\int_{-\infty}^{\infty} f(x, y, z, t) \times \exp(-ik_x x)\exp(-ik_z z)\exp(-i\omega t)dxdzdt, \tag{2.9a}$$

$$f(x, y, z, t) = \int_{-\infty}^{\infty}\int_{-\infty}^{\infty}\int_{-\infty}^{\infty} \hat{f}(k_x, y, k_z, \omega) \times \exp(ik_x x)\exp(ik_z z)\exp(i\omega t)dk_x dk_z d\omega, \tag{2.9b}$$

where k_x and k_z denote the *wave numbers* along the x- and z-axes, respectively.

By applying the triple Fourier transformation, one can transform the governing equation in Eq. (2.3) from partial differential equations into ordinary differential equations with the vertical coordinate y serving as the variable, that is,

$$\left[-k_x^2 - k_z^2 + \left(\frac{\omega}{c_P}\right)^2\right]\hat{\mathbf{\Phi}} + \frac{\partial^2}{\partial y^2}\hat{\mathbf{\Phi}} = 0, \tag{2.10a}$$

$$\left[-k_x^2 - k_z^2 + \left(\frac{\omega}{c_S}\right)^2\right]\hat{\mathbf{\Psi}} + \frac{\partial^2}{\partial y^2}\hat{\mathbf{\Psi}} = 0. \tag{2.10b}$$

By letting the wave numbers k_P and k_S for the compressional and shear waves as

$$k_P = \frac{\omega}{c_P}, \tag{2.11a}$$

$$k_S = \frac{\omega}{c_S}, \tag{2.11b}$$

and introducing the two parameters m_1 and m_2:

$$m_1^2 = k_x^2 + k_z^2 - k_P^2, \tag{2.12a}$$

$$m_2^2 = k_x^2 + k_z^2 - k_S^2, \tag{2.12b}$$

the partial differential equations in Eq. (2.10) can be rewritten in a more compact form as

$$\frac{\partial^2}{\partial y^2}\hat{\mathbf{\Phi}} - m_1^2\hat{\mathbf{\Phi}} = 0, \tag{2.13a}$$

$$\frac{\partial^2}{\partial y^2}\hat{\mathbf{\Psi}} - m_2^2\,\hat{\mathbf{\Psi}} = 0, \tag{2.13b}$$

both of which are of the same form.

2.3 Solution for the Soil Response

In this section, the boundary conditions for the problem will first be outlined, and then solution will be given for the steady state response in time domain.

2.3.1 *Boundary conditions*

The solution to the transformed differential equations in Eq. (2.13) can be given as

$$\hat{\Phi} = A\exp(-m_1 y), \tag{2.14}$$

and

$$\hat{\Psi}_x = B_1 \exp(-m_2 y), \tag{2.15a}$$

$$\hat{\Psi}_y = B_2 \exp(-m_2 y), \tag{2.15b}$$

$$\hat{\Psi}_z = B_3 \exp(-m_2 y), \tag{2.15c}$$

where A, B_1, B_2 and B_3 are the constants to be determined and the exponentially increasing terms have been discarded because of the radiation and finiteness conditions imposed for the half-space at infinity. The boundary conditions on the free surface of the half-space are

$$\begin{aligned} \hat{\sigma}_{yy}(y=0) &= -\hat{P}_y, \\ \hat{\tau}_{xy}(y=0) &= -\hat{P}_x, \\ \hat{\tau}_{zy}(y=0) &= -\hat{P}_z, \end{aligned} \tag{2.16}$$

where $\hat{P}_x$, $\hat{P}_y$, $\hat{P}_z$ denote the components of the applied load in the transformed domain.

It should be noted that Eq. (2.2) relates the three components of the displacement vector to four other functions: the scalar potential and the three components of the vectorial potential. This indicates that Φ and the components of $\mathbf{\Psi}$ should be subjected to an additional constraint condition. Generally, but not always, the following relation is taken at the additional constraint condition:

$$\nabla \cdot \mathbf{\Psi} = 0. \tag{2.17}$$

By substituting Eqs. (2.14) and (2.15) into the boundary conditions in Eq. (2.16) and considering the constraint condition in Eq. (2.17), the four constants A, B_1, B_2 and B_3 can be solved in terms of the transformed load. Then, by substituting these constants into Eqs. (2.14) and (2.15), and in turn into the transform of Eq. (2.7), one can obtain the displacements in the transformed domain, also in terms of the transformed load, as

$$\begin{Bmatrix} \hat{u} \\ \hat{v} \\ \hat{w} \end{Bmatrix} = -\frac{1}{2\mu Q}[\mathbf{D}][\mathbf{H}][\mathbf{G}]\begin{Bmatrix} \hat{P}_y \\ \hat{P}_x \\ \hat{P}_z \end{Bmatrix}, \tag{2.18}$$

where

$$[\mathbf{D}] = \begin{bmatrix} ik_x & -m_2 & 0 \\ -m_1 & -ik_x & ik_z \\ ik_x & 0 & m_2 \end{bmatrix}, \tag{2.19a}$$

$$[\mathbf{H}] = \begin{bmatrix} e^{-m_1 y} & 0 & 0 \\ 0 & e^{-m_2 y} & 0 \\ 0 & 0 & e^{-m_2 y} \end{bmatrix}, \tag{2.19b}$$

and

$$Q = \left(k_x^2 + k_z^2 - \tfrac{1}{2}k_s^2\right)^2 - m_1 m_2 \left(k_x^2 + k_z^2\right). \tag{2.20}$$

The matrix [**G**] is given as follows:

$$[\mathbf{G}] = \begin{bmatrix} g_{11} & g_{12} & g_{13} \\ g_{21} & g_{22} & g_{23} \\ g_{31} & g_{32} & g_{33} \end{bmatrix}, \tag{2.21}$$

where

$$g_{11} = k_x^2 + k_z^2 - \tfrac{1}{2}k_s^2, \tag{2.22a}$$

$$g_{12} = -ik_x m_2, \tag{2.22b}$$

$$g_{13} = -ik_z m_2, \tag{2.22c}$$

$$g_{21} = -ik_x m_1, \tag{2.22d}$$

$$g_{22} = \frac{k_z^2}{m_2^2}\left(k_x^2 + k_z^2 - \tfrac{1}{2}k_s^2 - 2m_1 m_2\right) + \left(k_x^2 + k_z^2 - \tfrac{1}{2}k_s^2\right), \tag{2.22e}$$

$$g_{23} = k_x k_z \left[2\frac{m_1}{m_2} - \frac{1}{m_2^2}\left(k_x^2 + k_z^2 - \tfrac{1}{2}k_s^2\right)\right], \tag{2.22f}$$

$$g_{31} = -ik_z m_1, \tag{2.22g}$$

$$g_{32} = -k_x k_z \left[2\frac{m_1}{m_2} - \frac{1}{m_2^2}\left(k_x^2 + k_z^2 - \tfrac{1}{2}k_s^2\right) \right], \tag{2.22h}$$

$$g_{33} = -\left[\frac{k_z^2}{m_2^2}\left(k_x^2 + k_z^2 - \tfrac{1}{2}k_s^2 - 2m_1 m_2\right) + k_x^2 + k_z^2 - \tfrac{1}{2}k_s^2\left(k_x^2 + k_z^2 - \tfrac{1}{2}k_s^2\right) \right]. \tag{2.22i}$$

2.3.2 *Steady state response in time domain*

The final expression of the displacements of the half-space in time domain can be obtained by employing the inverse Fourier transformation to Eq. (2.18), that is,

$$\begin{Bmatrix} u \\ v \\ w \end{Bmatrix} = -\int_{-\infty}^{\infty}\int_{-\infty}^{\infty}\int_{-\infty}^{\infty} \frac{1}{2\mu Q}[\mathbf{D}][\mathbf{H}][\mathbf{G}] \begin{Bmatrix} \hat{P}_y \\ \hat{P}_x \\ \hat{P}_z \end{Bmatrix} \times \exp(ik_x x)\exp(ik_z z)\exp(i\omega t)dk_x dk_z d\omega. \tag{2.23}$$

Similarly, the velocities and accelerations in time domain can be written as follows:

$$\begin{Bmatrix} \dot{u} \\ \dot{v} \\ \dot{w} \end{Bmatrix} = -\int_{-\infty}^{\infty}\int_{-\infty}^{\infty}\int_{-\infty}^{\infty} \frac{i\omega}{2\mu Q}[\mathbf{D}][\mathbf{H}][\mathbf{G}] \begin{Bmatrix} \hat{P}_y \\ \hat{P}_x \\ \hat{P}_z \end{Bmatrix} \times \exp(ik_x x)\exp(ik_z z)\exp(i\omega t)dk_x dk_z d\omega, \tag{2.24}$$

$$\begin{Bmatrix} \ddot{u} \\ \ddot{v} \\ \ddot{w} \end{Bmatrix} = \int_{-\infty}^{\infty}\int_{-\infty}^{\infty}\int_{-\infty}^{\infty} \frac{\omega^2}{2\mu Q}[\mathbf{D}][\mathbf{H}][\mathbf{G}] \begin{Bmatrix} \hat{P}_y \\ \hat{P}_x \\ \hat{P}_z \end{Bmatrix} \times \exp(ik_x x)\exp(ik_z z)\exp(i\omega t)dk_x dk_z d\omega, \tag{2.25}$$

which are all functions of the three loading components $\hat{P}_x$, $\hat{P}_y$, and $\hat{P}_z$.

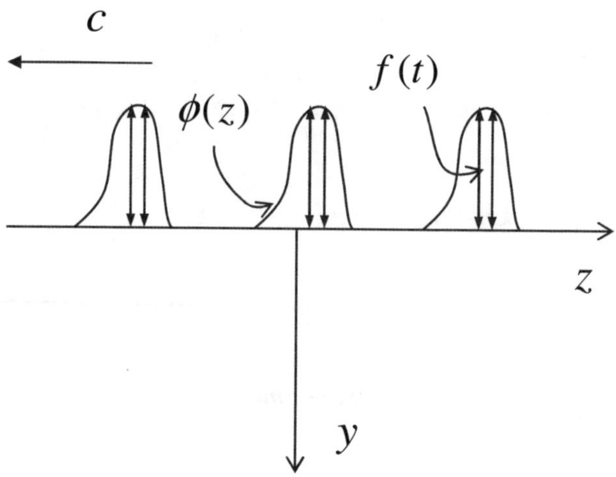

Fig. 2.2 Schematic diagram of a train-induced general moving load.

2.4 Loading Functions for Moving Loads of Different Forms

In this section, the loading functions for moving loads of different forms will be derived. The train loads will be simulated by a sequence of wheel loads, of which each is simulated as a moving load acting on an *infinite beam* resting on an elastic foundation.

2.4.1 *General loading function of a moving train*

Concerning the train-induced vibrations on the soils, the loadings are transmitted to the track and underlying soils through the contact points existing between the wheels and rails. Depending on the nature of loadings, they can be spilt up into two parts as depicted in Fig. 2.2. The first part relates to the distribution of the axle loads passing a fixed point given as $\phi(z-ct)$, where c is the speed of the moving loads and $\phi(z)$ the *distribution function* of each axle load. This part leads to excitation at the vehicles' passing frequencies and is the major source for the train speed-dependent components of the low frequency vibration spectra (Krylov 1995).

The second part is generated by the interaction between the wheels and rails, which, as indicated by $f(t)$, moves with the wheels and is independent of the moving direction z. The following is a general expression for the moving load $\mathbf{P}(x, z, t)$ moving along the z-direction:

$$
\begin{aligned}
P_x &= \delta(x)\phi_x(z-ct)f_x(t),\\
P_y &= \delta(x)\phi_y(z-ct)f_y(t),\\
P_z &= \delta(x)\phi_y(z-ct)f_z(t),
\end{aligned}
\tag{2.26}
$$

where the subscripts x, y, and z denote the acting directions of the load. Both the functions $\phi(z)$ and $f(t)$ will be determined in the following sections.

By applying the Fourier transformation to Eq. (2.26), the moving load can be expressed as

$$
\begin{aligned}
\hat{P}_x &= \frac{1}{2\pi}\tilde{\phi}_x(k_z)\tilde{f}_x(\omega+k_z c),\\
\hat{P}_y &= \frac{1}{2\pi}\tilde{\phi}_y(k_z)\tilde{f}_y(\omega+k_z c),\\
\hat{P}_z &= \frac{1}{2\pi}\tilde{\phi}_z(k_z)\tilde{f}_z(\omega+k_z c),
\end{aligned}
\tag{2.27}
$$

in which $\tilde{\phi}(k_z)$ and $\tilde{f}(\omega)$ are the one-dimensional Fourier transforms of $\phi(z)$ and $f(t)$, respectively, with respect to the variables z and t. In this chapter, the symbols $\hat{}$ and $\tilde{}$ are used to denote a triple and one-dimensional Fourier transform, respectively.

2.4.2 *Distribution function $\phi(z)$ of the loading*

Theoretically, the distribution function $\phi(z)$ of the loading should be determined based on the field data collected for the wheel loads of the train. However, using simple models enables us to grasp the fundamental features of ground vibrations induced by the moving loads. Moreover, most related previous works that can be used to verify the accuracy of the present approach in load-modeling deal only with the simplest case of moving point loads. Based on these considerations, four different forms of distribution function $\phi(z)$ will be considered in the following.

The model for the train loads will be conceived as a sequence of moving wheel loads, with each treated as an elastically distributed wheel load. The Fourier transform $\tilde{\phi}(k_z)$ for each of the four distribution functions will be given in an analytical form.

2.4.2.1 *Single point load*

For a point load, the distribution function along the z-axis can be written in terms of Dirac's delta function as:

$$\phi(z) = \delta(z). \tag{2.28}$$

Correspondingly, the Fourier transform is

$$\tilde{\phi}(k_z) = \frac{1}{2\pi}, \tag{2.29}$$

which is a constant regardless of the value of k_z.

2.4.2.2 *A uniformly distributed wheel load*

In the real situation, the contact point existing between the wheel and rail is not a "point" but an "area". Thus, a better representation for the wheel load is a uniformly distributed load given as:

$$\phi(z) = \begin{cases} \dfrac{1}{2a} & \text{for } -a \le z \le a, \\ 0 & \text{otherwise,} \end{cases} \tag{2.30}$$

where a is a constant, representing half of the width of the distributed load. Based on the definition of the distribution function $\phi(z)$ in Eq. (2.30), the integration of $\phi(z)$ from $-\infty$ to ∞ is 1, implying that the total contribution of the loading function equals a unit value, same as that implied by Dirac's delta function. By applying the Fourier transformation to Eq. (2.30), the transformed load $\tilde{\phi}(k_z)$ can be obtained as follows:

$$\tilde{\phi}(k_z) = \begin{cases} \dfrac{\sin(ak_z)}{\pi k_z} & \text{for } k_z \ne 0, \\ \dfrac{a}{\pi} & \text{for } k_z = 0. \end{cases} \tag{2.31}$$

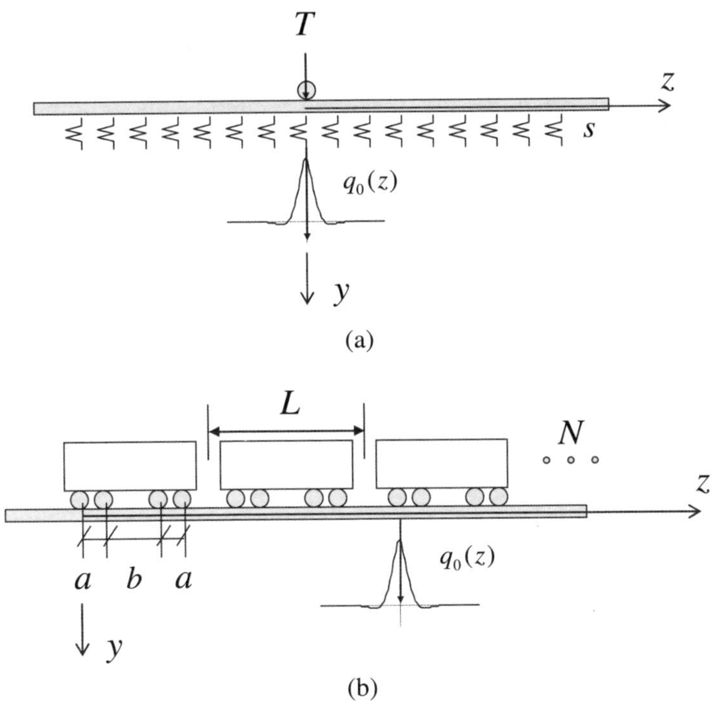

Fig. 2.3 Train-induced loadings between the rails and soil: (a) single wheel load; (b) a sequence of wheel loads.

As can be seen, the transformed function $\tilde{\phi}(k_z)$ is no longer a constant, but tends to decay with the increase of k_z.

2.4.2.3 *An elastically distributed wheel load*

If the wheel load is regarded as the force exerted from the track onto the underlying soils, rather than the one from the wheels onto the track, one may use the deflection curve of the track to simulate the distribution of the wheel load (Krylov and Ferguson 1994; Krylov 1995; Takemiya 1997). In this connection, the track is treated as an infinite Bernoulli-Euler beam supported by an *elastic foundation* of stiffness s as shown in Fig. 2.3(a). Let EI denote the bending stiffness of the beam. For an elastically supported beam with an axle load T acting at z = 0, the vertical displacement v is (Esveld 1989):

$$v(z)=\frac{T}{2s\alpha}\exp\left(\tfrac{-|z|}{\alpha}\right)\left[\cos\left(\tfrac{|z|}{\alpha}\right)+\sin\left(\tfrac{|z|}{\alpha}\right)\right], \tag{2.32}$$

where the *characteristic length* α can be related to the bending stiffness EI as

$$\alpha=\sqrt[4]{\frac{4EI}{s}}\ \text{(m)}, \tag{2.33}$$

with s (N/m^2) denoting the spring coefficient of the foundation.

Consequently, the load distribution function can be written as

$$\phi(z)=q_0(z)=\frac{T}{2\alpha}\exp\left(\tfrac{-|z|}{\alpha}\right)\left[\cos\left(\tfrac{|z|}{\alpha}\right)+\sin\left(\tfrac{|z|}{\alpha}\right)\right]. \tag{2.34}$$

Note that the integration of $\phi(z)$ from $z = -\infty$ to ∞ equals the axle load T. Obviously, the total force transmitted through the rails remains unchanged, although the pattern of distribution has been changed. By applying the Fourier transformation to Eq. (2.34), the load distribution function can be transformed as

$$\tilde{\phi}(k_z)=\tilde{q}_0(k_z)=\frac{4T}{4+k_z^4\alpha^4}, \tag{2.35}$$

which is a function of the characteristic length α.

2.4.2.4 *A sequence of wheel loads*

Let us extend the single wheel load case to the case of a train consisting of N carriages of equal length L in Fig. 2.3(b). Here, each carriage is assumed to have two bogies separated by distance b, each of which in turn comprises two axles, i.e., two sets of wheels, separated by distance a. Suppose that each set of wheels has the same load distribution function $q_0(z)$ as the one given in Eq. (2.34). The total distribution function of loading for the present case can be written as:

$$\begin{aligned}\phi(z)=\sum_{n=0}^{N-1}\big[&q_0(z-nL)+q_0(z-nL-a)\\&+q_0(z-nL-a-b)+q_0(z-nL-2a-b)\big],\end{aligned} \tag{2.36}$$

along with its Fourier transform as

$$\tilde{\phi}(k_z) = \tilde{q}_0(k_z)\sum_{n=0}^{N-1}\exp(-ik_z nL)\{1+\exp(-ik_z a)$$
$$+\exp[-ik_z(a+b)]+\exp[-ik_z(2a+b)]\}, \qquad (2.37)$$

where $\tilde{q}_0(k_z)$ is the same as the one given in Eq. (2.35).

2.4.3 *Interaction forces between wheels and rails*

In reality, the three components of the interaction force, i.e., $\mathbf{f}(t) = (f_x(t), f_y(t), f_z(t))$, between the wheels and rails may be simulated by a quasi-static term of constant value plus a dynamic term that varies with time t (Esveld 1989). Take the interaction force $f_y(t)$ along the vertical direction as an example. The static term is contributed mainly by the wheel weight, whereas the dynamic term by the track irregularities and vehicle defects, such as wheel flats, natural vibrations and hunting. The dynamic term is extremely complex and is by no means generally accessible, due to the fact that the train always interacts with the track, and in turn, with the soils, during its movement. Unless all the carriages of a train, together with track and underground soils, are accurately included in an analytical model, the actual interaction force can never be found. For simplicity, in the present analysis, the dynamic term $f(t)$ is assumed to depend only on a single frequency ω_0, i.e., $f(t) = \exp(i\omega_0 t)$. By so setting, it is easy to see that when $\omega_0 = 0$, $f(t) = 1$, implying that a moving load with no oscillation is considered. For the case with $\omega_0 \neq 0$, the moving load oscillates by itself at a constant frequency $f_0 = \omega_0/(2\pi)$ Hz. The following are the ranges of frequencies that may be induced by a moving train: (a) sprung mass: 0 - 20 Hz; (b) unsprung mass: 0 - 125 Hz; (c) corrugations, welds, and wheel flats: 0 - 2,000 Hz (Esveld 1989).

By setting $f(t) = \exp(i\omega_0 t)$, the dynamic function $\tilde{f}(\omega + k_z c)$ in Eq. (2.27) can be expressed as

$$\tilde{f}(\omega + k_z c) = \frac{1}{c}\delta\left(\frac{\omega - \omega_0}{c} + k_z\right), \qquad (2.38)$$

which has been given in analytical form.

2.4.4 *Calculation of inverse Fourier transform*

At this stage, both the functions $\tilde{\phi}(k_z)$ and $\tilde{f}(\omega+k_z c)$ have been made available. By substituting these two functions into Eq. (2.27) and, in turn, into Eqs. (2.23)-(2.25), the final responses of the soils in time domain can be obtained. Note that the inversion of the Fourier transform with respect to z in Eqs. (2.23)-(2.25) can be done analytically by only replacing k_z with $-(\omega-\omega_0)/c$ because of the involvement of Dirac's delta function in Eq. (2.38). It follows that the original triple integral is reduced to a double integral with respect to frequency ω and wave number k_x only.

By carefully examining Eqs. (2.23)-(2.25), we find that there exist two radicals (as represented by $m_1 = 0$ and $m_2 = 0$) and a singularity (as represented by $Q = 0$) in the integrands. This makes the evaluation of these integrals a formidable task. Because of this, equations similar to Eqs. (2.23)-(2.25) were frequently presented in integral form in most previous works. However, over the last decade, more and more researchers tend to evaluate integrals of this sort by numerical methods that can automatically skip the singularities and yield the desired results. In this study, appropriate quadrature routines available in IMSL (Piessens *et al.* 1983) will be used to perform the inverse transform with respect to k_x and the fast Fourier transform with respect to ω. For the cases of trans- and super-critical speeds, the pole of the integrands can be shifted off the real axis k_x by introducing material attenuation. The material damping can be taken into account through use of *complex Lamé's constants*, i.e., $\mu^* = \mu(1+2i\beta)$ and $\lambda^* = \lambda(1+2i\beta)$, where β denotes the *hysteretic damping ratio*.

2.5 Numerical Studies and Discussions

The moving point load was the model frequently adopted by researchers in their study of vehicle-induced vibrations. For the purpose of verification, we shall adopt here the same model to study the response of an elastic half-space, assuming the load to move at a sub-critical speed. The steady-state response for the same problem was presented by Eason (1965) by reducing the resulting multiple integrals to single finite integrals, as given in Appendix. The Eason results presented in the following as the reference have been obtained by using appropriate subroutines available in IMSL (Piessens *et al.*, 1983) to evaluate the finite integrals.

2.5.1 *Verification of the present approach*

In this example, the load is assumed to move at c = 90 m/s. The elastic half-space considered has a shear wave (S-wave) speed of $c_S = 100$ m/s, Poisson's ratio $\upsilon = 0.25$ and mass density $\rho = 2{,}000$ kg/m^3. The corresponding compressional wave (P-wave) speed c_P and Rayleigh wave (R-wave) speed c_R are 173.2 and 92 m/s, respectively. In Fig. 2.4, the displacements at the observation point (x, y, z) = (0, y_0,0), with y_0 set to be 1 m, have been presented in a non-dimensional normalized form $\mathbf{U}_i = 2\pi\mu y_0 \mathbf{u}_i / P_i$, where the subscript $i = x$, y or z represents the direction along which the loading is applied. Each component of the vectors **u** and **U**, i.e., u, v, w and U, V, W, denote the displacement and its normalized value along each of the three directions. As can be seen, the results obtained by the present approach are in good agreement with those based on Eason (1965).

To verify the accuracy of the present procedure on application to the super-critical speed range, the displacements at the observation point (0 m, 1 m, 0 m) have been computed for a point load moving at speed c = 200 m/s (greater than those of the Rayleigh, shear and compressional waves) in Fig. 2.5 in terms of the normalized displacement $\mathbf{U}_i^* = \mu y_0 \mathbf{u}_i / P_i$ and normalized time $t^* = c_S t / y_0$. The underlying soil has the same properties as those of the preceding example, except that a material damping of β = 0.01 is considered for the present case. As can be seen, the present results are in good agreement with those obtained by de Barros and Luco (1994).

In the preceding two examples, the present approach has been demonstrated to be capable of computing the soil displacements for both the sub- and super-critical speed ranges. In what follows, the same procedure will be adopted to study the influence of moving speeds for different load distribution functions.

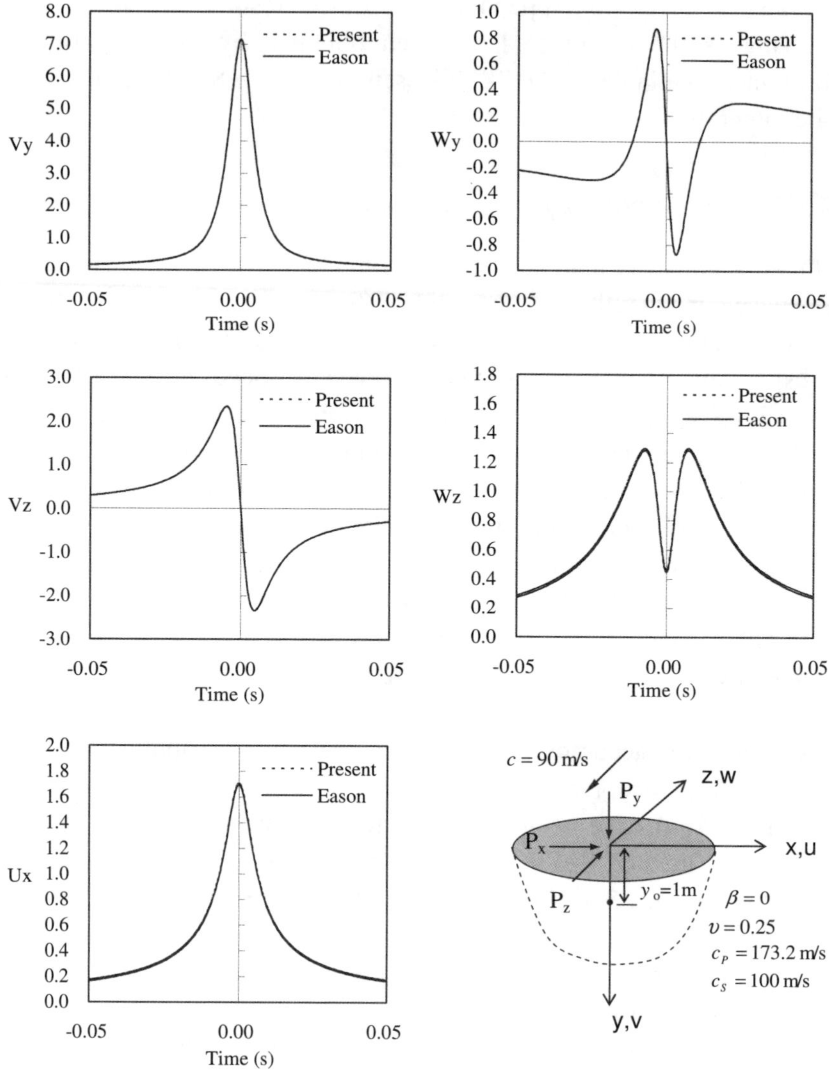

Fig. 2.4 Comparison of present results with those of Eason (1965).

Throughout the following analyses, the visco-elastic half-space considered is assumed to have an S-wave speed $c_S = 100$ m/s, P-wave speed $c_P = 173.2$ m/s, R-wave speed $c_R = 92$ m/s, Poisson's ratio $\upsilon = 0.25$, mass density $\rho = 2{,}000$ kg/m^3, and damping ratio $\beta = 0.02$.

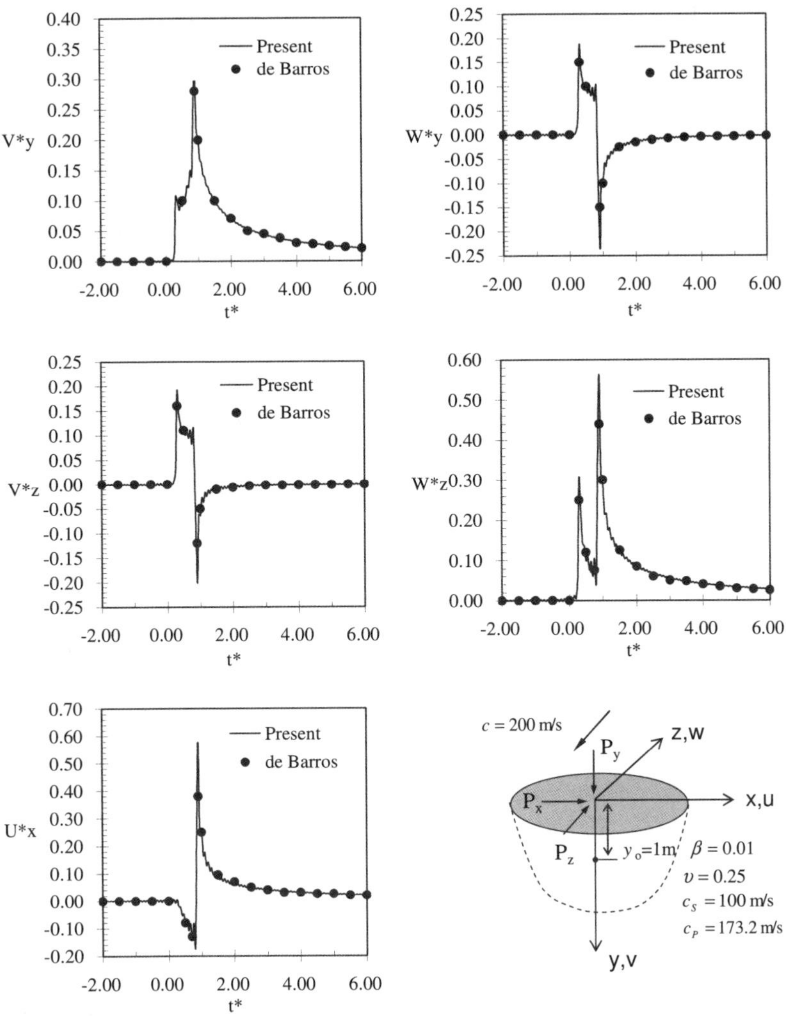

Fig. 2.5 Comparison of present results with those of de Barros and Luco (1994).

2.5.2 *Single moving point load*

To evaluate the effect of the load-moving speeds, the normalized displacements $\mathbf{U}_i$ have been computed for the soils subjected to loads moving at the sub-critical (c < 100 m/s), trans-critical (100 m/s < c < 173.2 m/s), and super-critical (c > 173.2 m/s) speed ranges. The vertical displacement V and longitudinal displacement W respectively computed at the observation point (0 m, 1 m, 0 m) were plotted in Figs. 2.6 and 2.7 for loads applied along the vertical y-direction, and in Figs. 2.8 and 2.9 for loads applied along the z-direction, i.e., the load-moving direction. In Fig. 2.10, the transverse displacements U_x at the position (0 m, 1 m, 0 m) were plotted for a point load applied along the x-axis. The displacement components U_y, U_z, V_x and W_x were not shown because they are all equal to zero. In each figure, the time $t = 0$ corresponds to the moment at which the point load passes through the origin ($z = 0$).

The following observations can be drawn from Figs. 2.6-2.10: The shapes of the responses for load-moving speeds in the sub-critical speed region ($c < 100$ m/s) are generally different, depending on whether the speed is lower or greater than the R-wave speed $c_R = 92$ m/s. The response is almost symmetric (for V_y, W_z, U_x) or anti-symmetric (for V_z, W_y) with respect to $t = 0$ for load-moving speeds lower than the R-wave speed. For the case of zero damping, it becomes entirely symmetric or anti-symmetric, as shown in Fig. 2.4. As the speed increases and becomes larger than the R-wave speed, the response becomes more asymmetrical as a result of the *Mach radiation effect*.

For the trans-critical speed case (100 m/s $< c <$ 173.2 m/s), the first peak of the response occurring after time $t = 0$ corresponds to the arrival of the S-wave front, followed immediately by the R-wave front. However, the S- and R-wave fronts are so close to each other that they cannot be clearly distinguished. For the super-critical speed case ($c = 200$ m/s), the first peak of the response immediately after time $t = 0$ represents the arrival of the P-wave front, followed by the arrival of the S- and R-wave fronts, which are very close to each other.

The maximum displacements computed at the observation point (0 m, 1 m, 0 m) for loadings applied along the y-, z-, and x-directions have been plotted in Figs. 2.11(a)-(c), respectively. Here, with the S-wave *Mach number* defined as $M_2 = c / c_S$, we may interpret $M_2 < 1.0$ as the *sub-critical speed range*, $1.0 < M_2 < 1.73$ as the *trans-critical speed range*, and $M_2 > 1.73$ as the *super-critical speed range*. As can be seen, for the cases with loadings applied along the y- and z-axes, the critical speed is equal to the R-wave speed, as indicated by $M_2 = 0.92$,

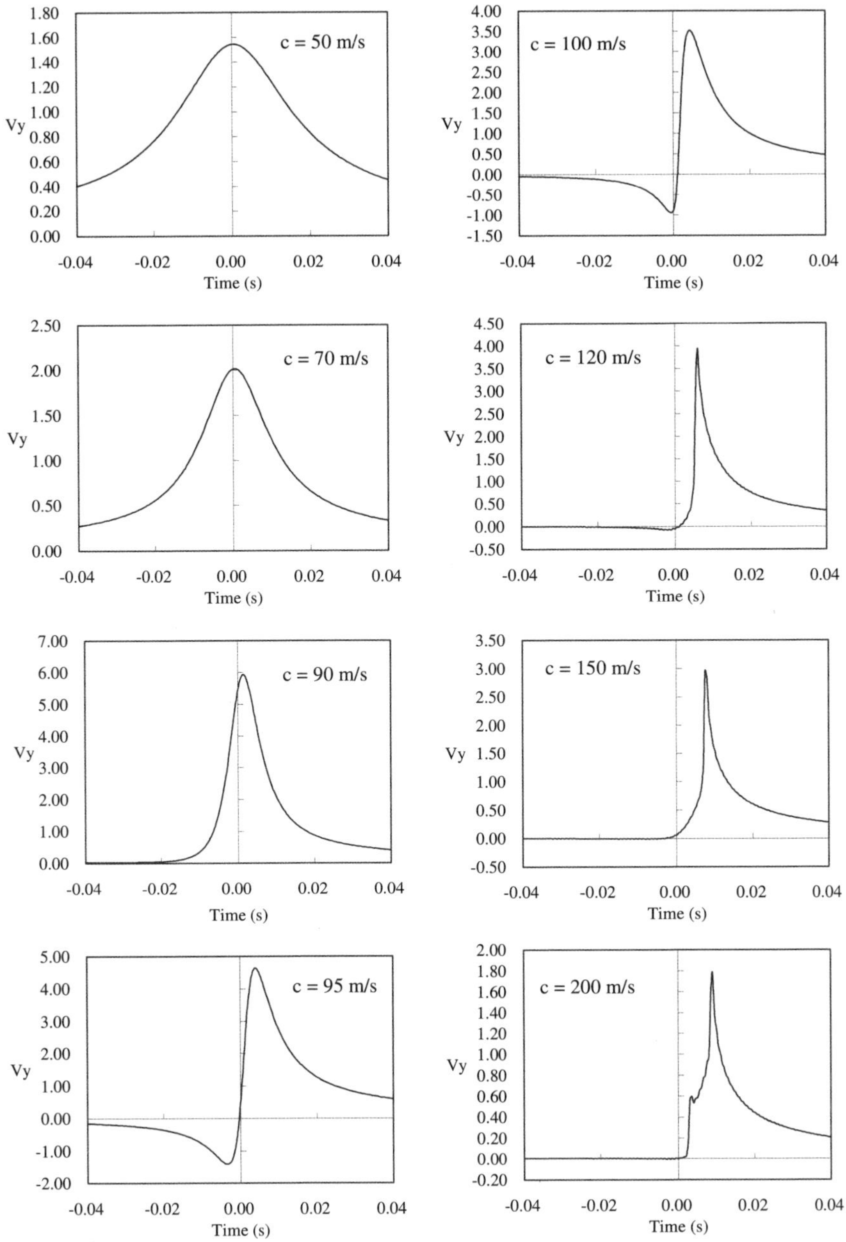

Fig. 2.6 Vertical displacement V_y for a point load directed along the y-axis.

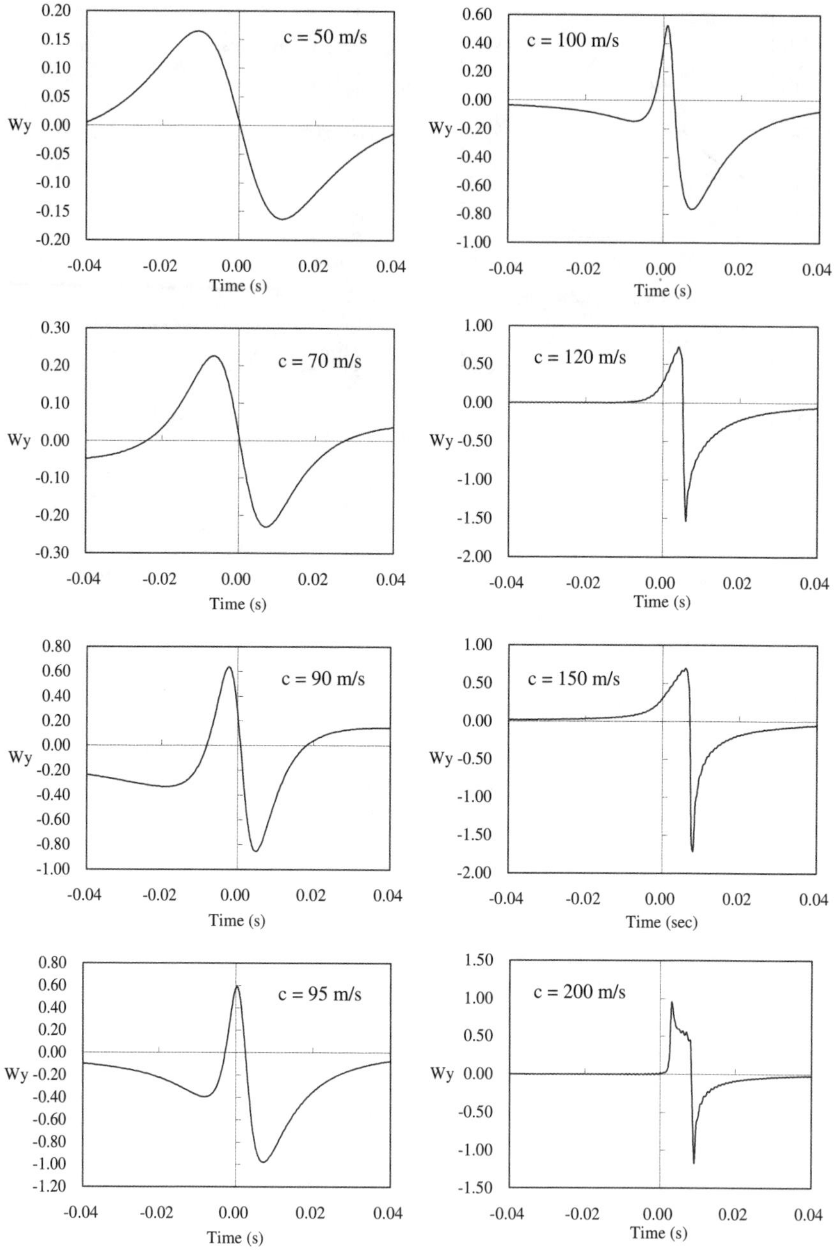

Fig. 2.7 Longitudinal displacement W_y for a point load directed along the y-axis.

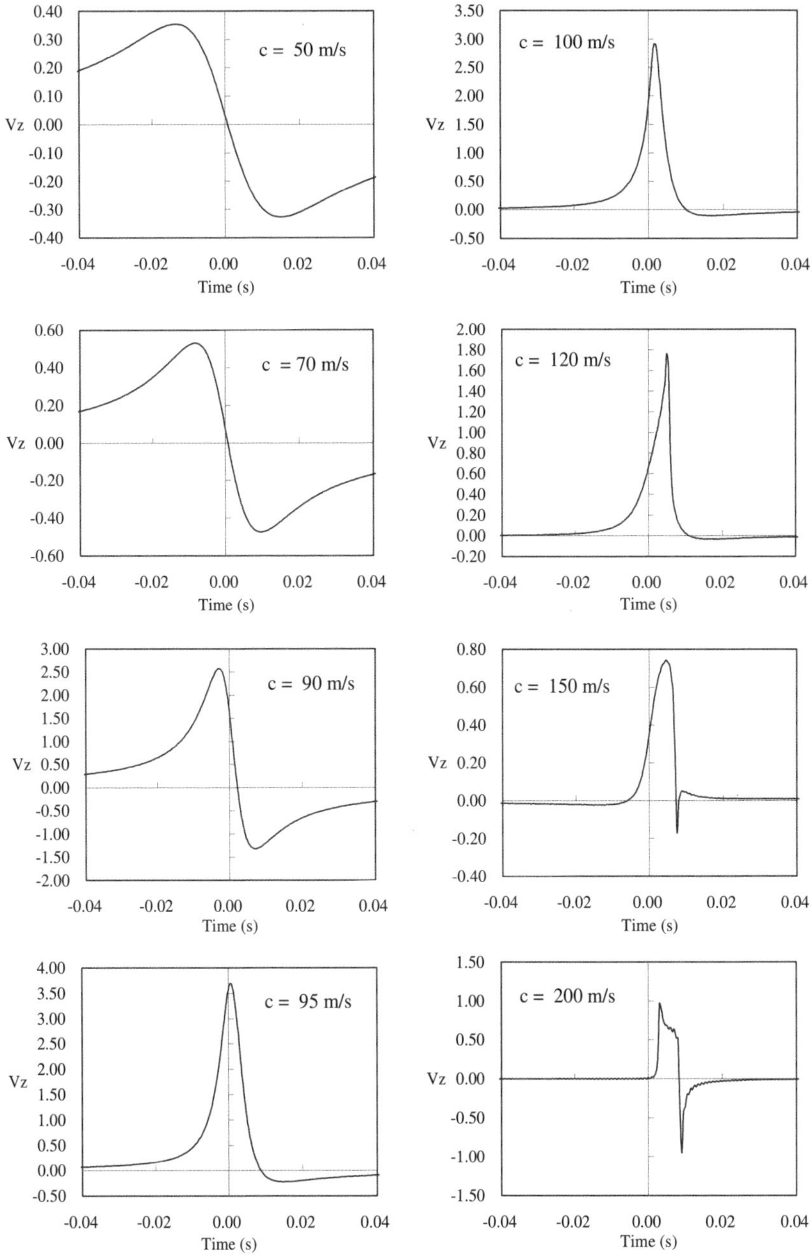

Fig. 2.8 Vertical displacement V_z for a point load directed along the z-axis.

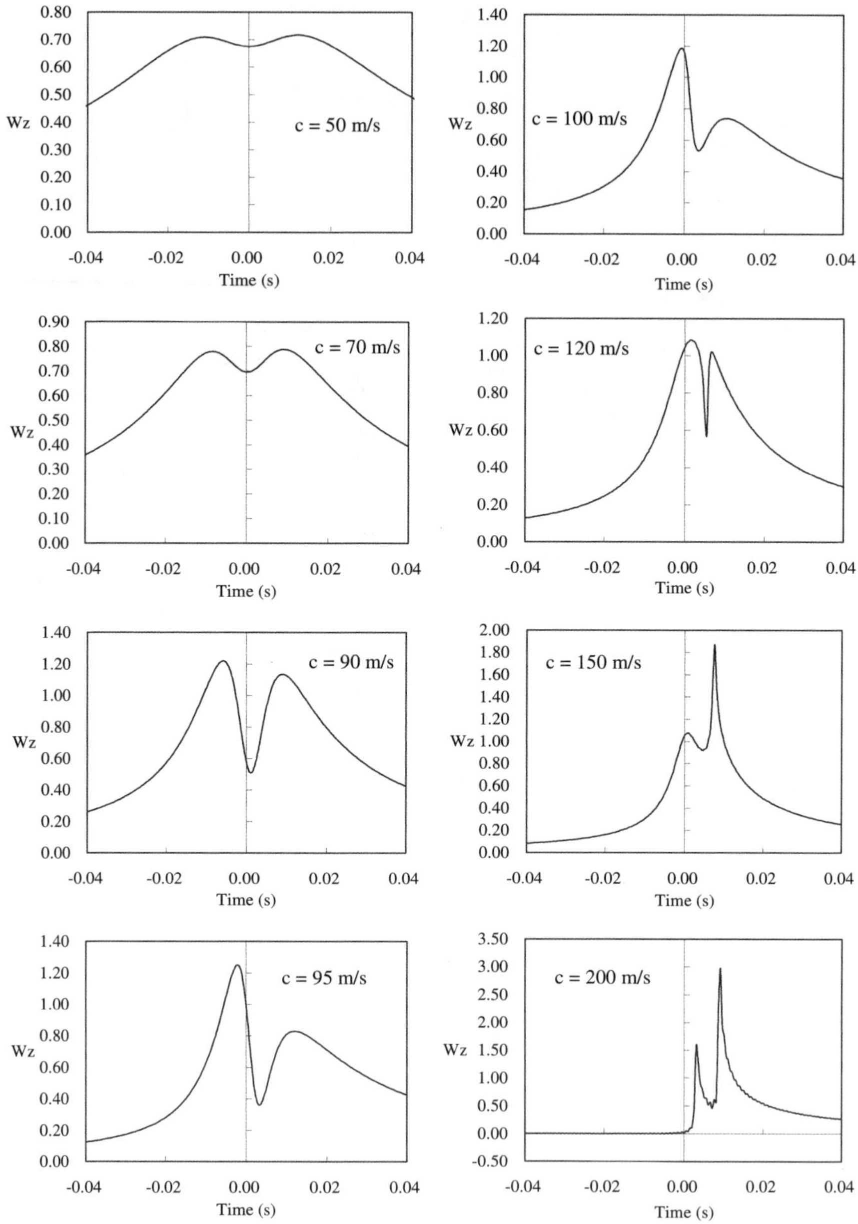

Fig. 2.9 Longitudinal displacement W_z for a point load directed along the z-axis.

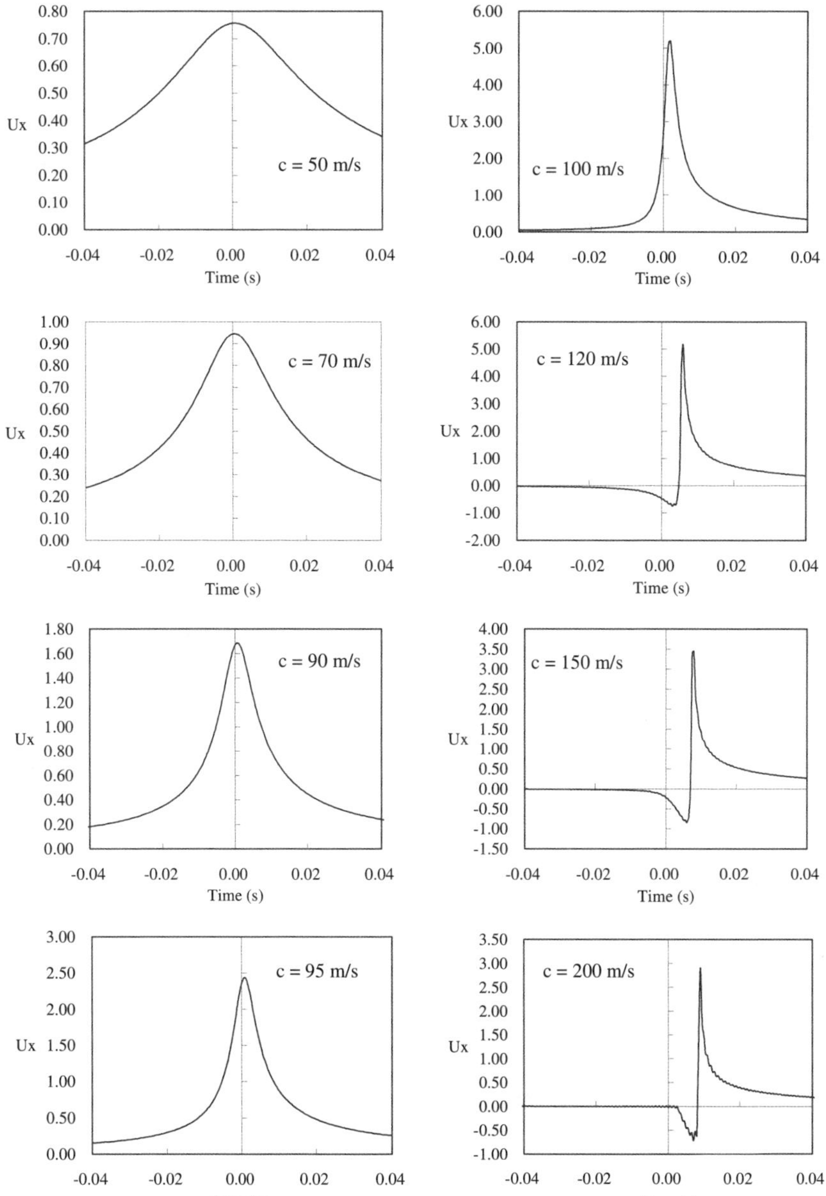

Fig. 2.10 Transverse displacement U_x for a point load directed along the x-axis.

while for the case with loadings directed along the x-axis, the critical speed is slightly larger than the S-wave speed, as indicated by $M_2 = 1.0$.

For all the displacements given in Fig. 2.11, the general trend is that as the moving speed increases, they all show a tendency to increase until the *critical (resonant) speed* is reached, and then they all decrease. However, the vertical displacements (Vy, Vz) increase at a rate faster than that of the longitudinal displacements (Wy, Wz) before the critical speed is reached. Besides, both the vertical displacements (Vy, Vz) attain their maximum at the first critical speed ($c = c_R$), while the longitudinal displacements (Wy, Wz) at a speed higher than the R-wave speed c_R. For the case with loadings directed along the load-moving (z) direction, as shown in Fig. 2.11(b), another critical speed, i.e., $M_2 = 1.73$, can be identified for both the displacement components Vz and Wz, which should be interpreted as the P-wave speed.

As shown in Fig. 2.11(a), for loadings applied along the vertical direction and for the special case of zero moving speed, the present problem reduces to that of a static point force acting at the origin, known as the classical *Boussinesq's problem*. According to Fung (1965), the vertical displacement v for a Boussinesq's problem with load P is

$$v = \frac{P}{4\pi\mu R}\left[2(1-\upsilon) + \frac{y^2}{R^2}\right], \tag{2.39}$$

with

$$R^2 = x^2 + y^2 + z^2, \tag{2.40}$$

which yields a value of 1.25 for the normalized displacement at the observation point for the case with $\upsilon = 0.25$, exactly the same as the one shown in the figure for $M_2 = 0$.

2.5.3 *A uniformly distributed moving wheel load*

Consider a uniformly distributed moving wheel load with $a = 0.5$ (see Eq. (2.30) for definition). For the present case, the load distribution function $\phi(z)$ and its Fourier transform $\tilde{\phi}(k_z)$ have been drawn in Figs. 2.12(a) and (b). The maximum displacements computed for the point (0 m, 1 m, 0 m) with respect to different S-wave Mach numbers

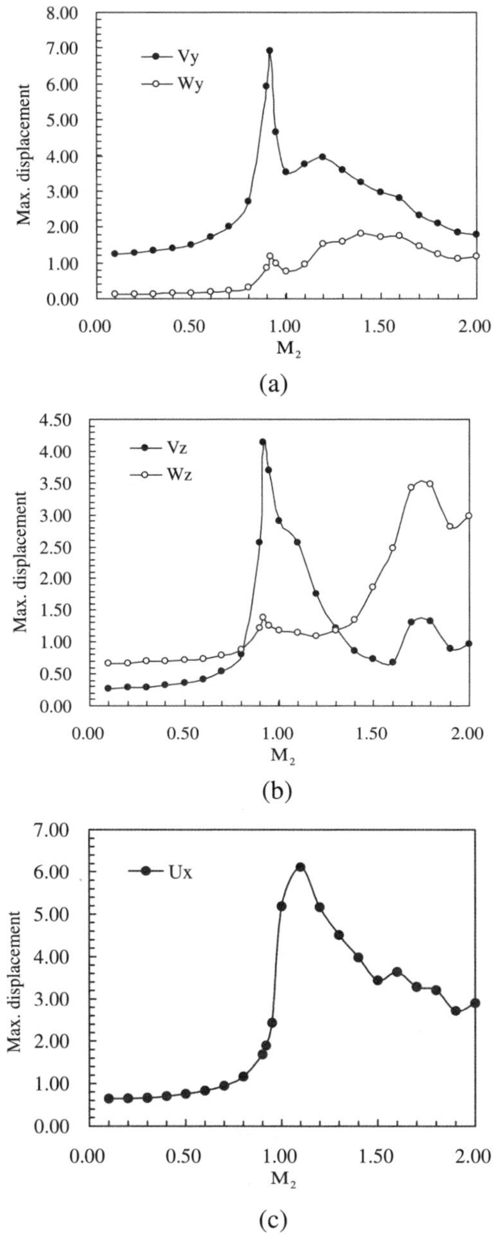

Fig. 2.11 Maximum displacements for a point load directed along: (a) *y*-axis; (b) *z*-axis; (c) *x*-axis.

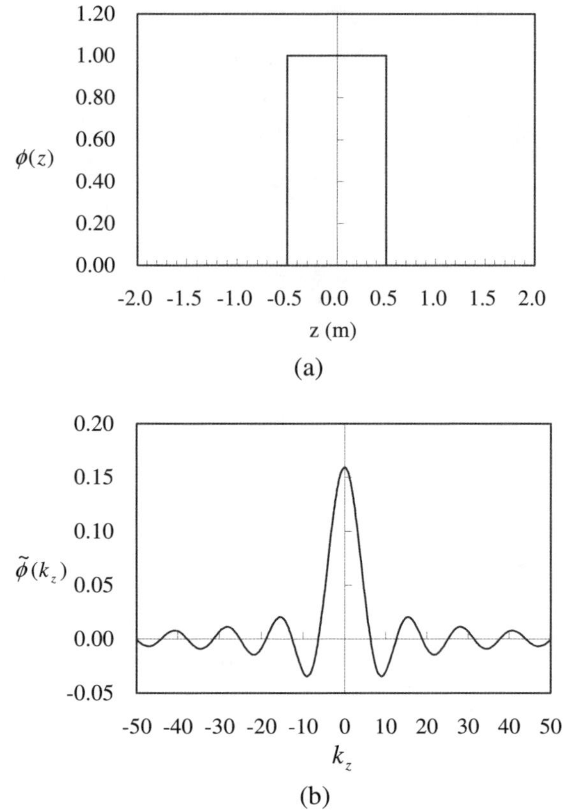

Fig. 2.12 A uniformly distributed moving load: (a) load distribution function; (b) Fourier transform.

M_2 have been plotted in Fig. 2.13, from which a trend similar to that of Fig. 2.11 can be observed, except that the response amplitudes are much lower throughout all the speed range. Such a result is consistent with the understanding that for the same load to be distributed in a wider area, the response induced should be smaller.

Besides, the maximum displacements in Fig. 2.13 show a trend much smoother than that of Fig. 2.11. One reason for this is that the contribution of $\tilde{\phi}(k_z)$ for a uniformly distributed load is mostly concentrated on the lower k_z, as can be seen from Fig. 2.12(b). Thus, only a small range of frequencies need be considered when performing

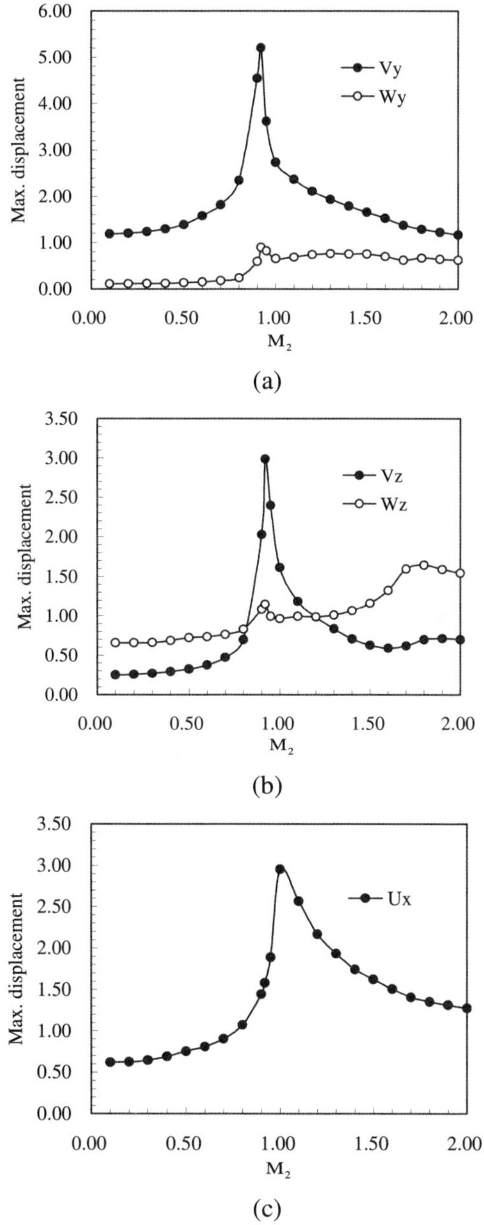

Fig. 2.13 Maximum displacements for a uniformly distributed load directed along: (a) *y*-axis; (b) *z*-axis; (c) *x*-axis.

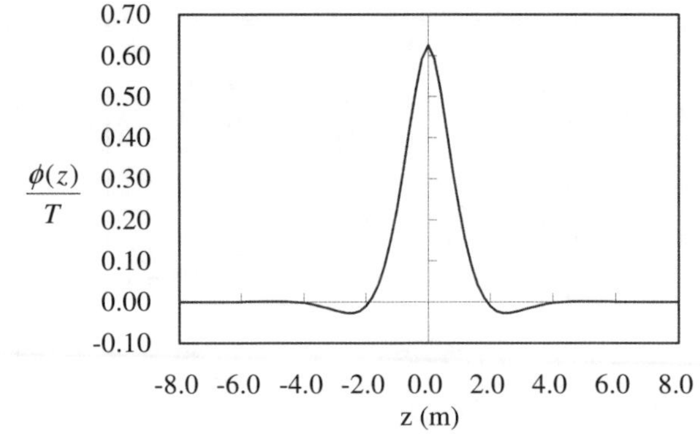

(a)

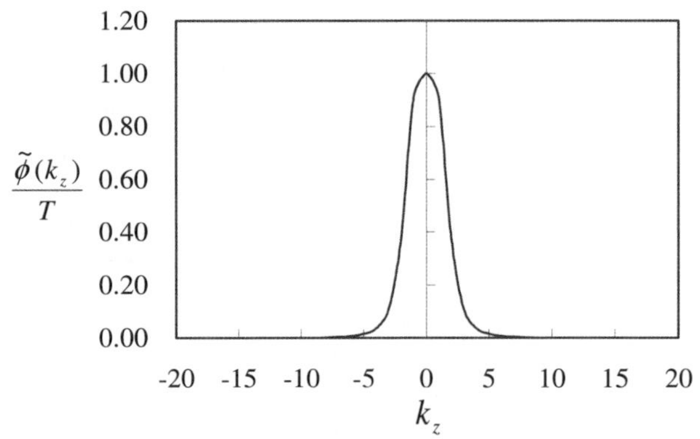

(b)

Fig. 2.14 An elastically distributed moving load: (a) load distribution function; (b) Fourier transform.

the inverse FFT. For the case of a point load, however, the function $\tilde{\phi}(k_z)$ is a constant, as indicated by Eq. (2.29), implying that "all" frequencies should be considered to guarantee the convergence of the maximum displacements to the exact ones. In this study, the range of frequencies considered is 0 - 1,000 Hz.

2.5.4 *An elastically distributed moving wheel load*

Consider an elastically distributed moving wheel load as defined in Eq. (2.34), assuming the wheel load $T = 10$ t and the characteristic length

α = 0.8 m. The load distribution function and its transform were shown in Figs. 2.14(a)-(b). Realizing that the vertical loadings are much larger than other loading components for a moving train, and so too are the vertical displacements, we will concentrate only on the vertical responses induced by the vertical applied loads in this subsection.

The vertical displacements, velocities and accelerations computed at the observation point (0 m, 1 m, 0 m) for a static wheel load moving at different speeds have been plotted in Figs. 2.15-2.17. In contrast, the real-part responses for the *self oscillation* effect of the wheel load with a frequency of f_0 = 10 Hz were included in the results shown in Figs. 2.18-2.20. These results can also be interpreted as the responses caused by a moving wheel load with self oscillation of the form:

$$f(t) = \cos(2\pi f_0 t). \tag{2.41}$$

Compare the time history responses for the case with a static wheel load and those for the case with a dynamic wheel load. We observe that for the former case, the responses are concentrated only in a very small duration, whereas for the latter case, the responses oscillate and propagate for a rather long duration, with a larger frequency of fluctuation observed for the waves ahead of the arrival of the moving load than that following in the sub-critical speed range. The two frequencies can be computed as

$$f_{cr} = \frac{f_0}{1 \pm c/c_R}, \tag{2.42}$$

where c_R = 92 m/s, according to the *Doppler effect*.

The effect of vibration frequency f_0 of the moving wheel load on the maximum (i.e., absolute) displacements, velocities and accelerations of the observation point was plotted in Figs. 2.21(a)-(c) with respect to the Mach number M_2. As can be seen, for the case with no self oscillation, i.e., with f_0 = 0 Hz, there exists a distinct critical speed at M_2 = 0.92, but for the case with fluctuating loads, i.e., with $f_0 \neq 0$ Hz, no distinct critical speeds can be observed. These results suggest that only when the wheel vibrations are neglected, can the critical region be determined for the train speed in relation to the R-wave speed of the soils. As a matter of fact, the critical speed for this case can be obtained from Eq. (2.42) as:

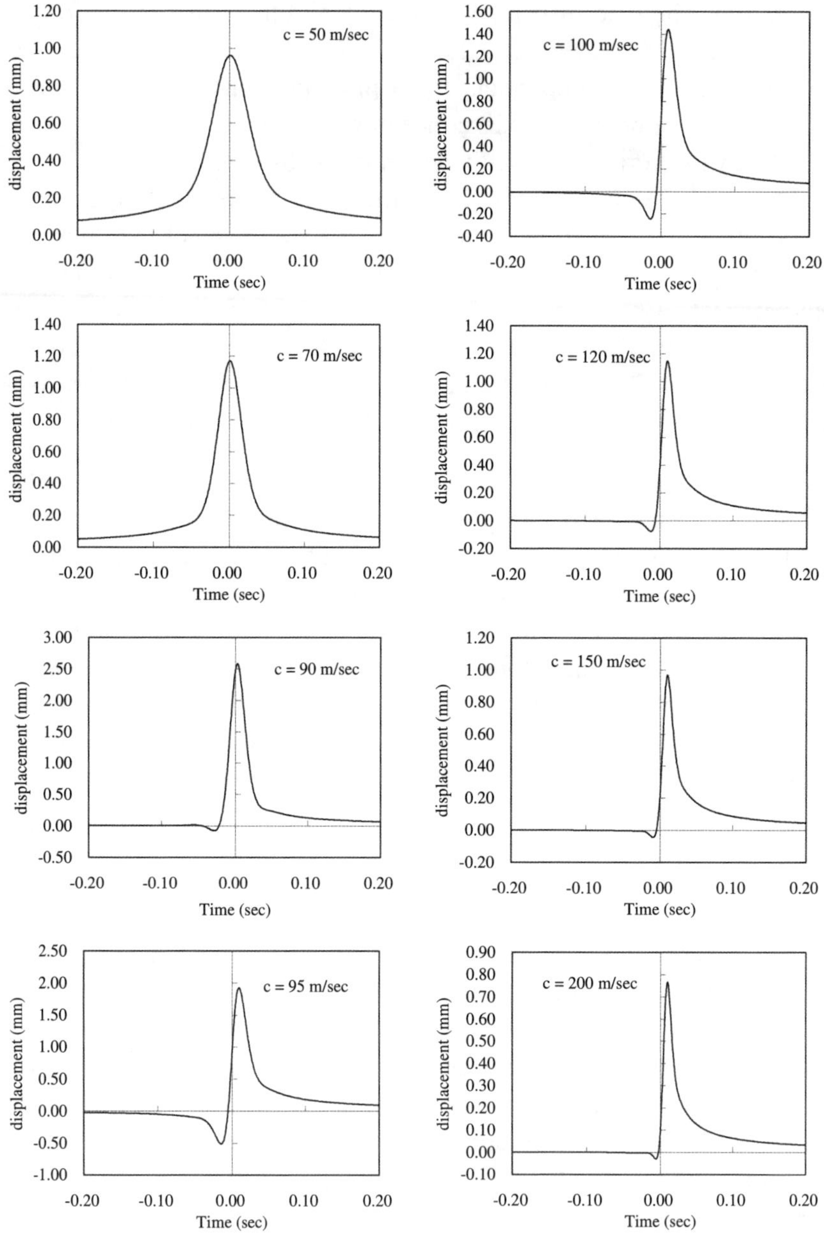

Fig. 2.15 Vertical displacement for an elastically distributed load with no self oscillation directed along the *y*-axis.

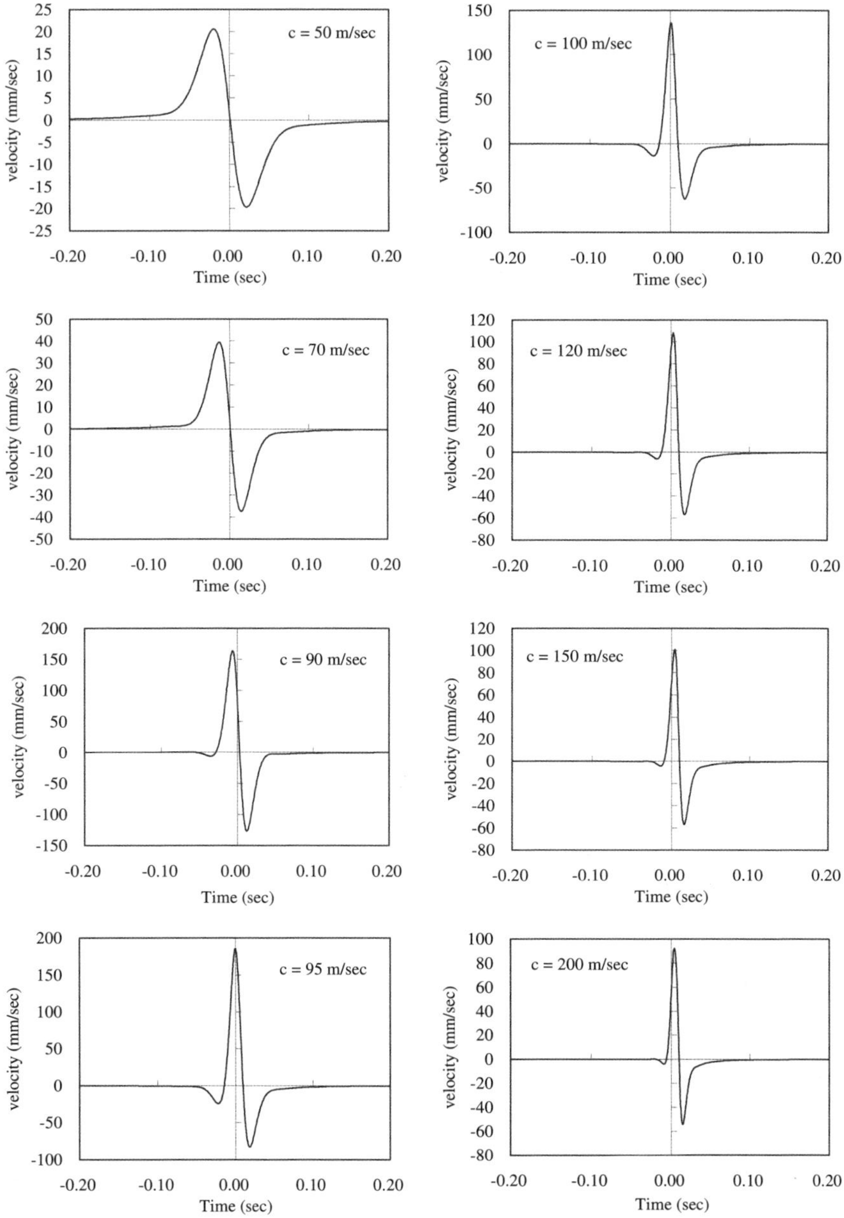

Fig. 2.16 Vertical velocity for an elastically distributed load with no self oscillation directed along the *y*-axis.

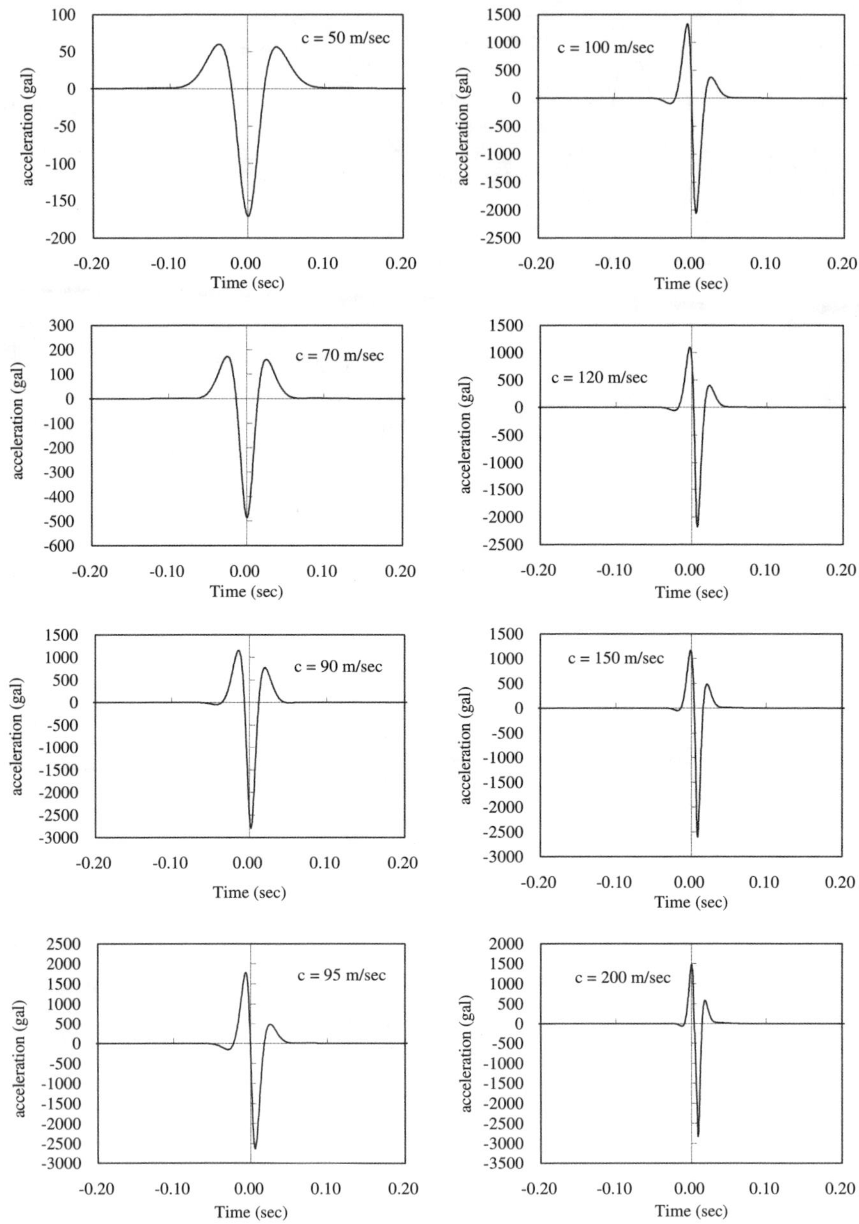

Fig. 2.17 Vertical acceleration for an elastically distributed load with no self oscillation directed along the *y*-axis.

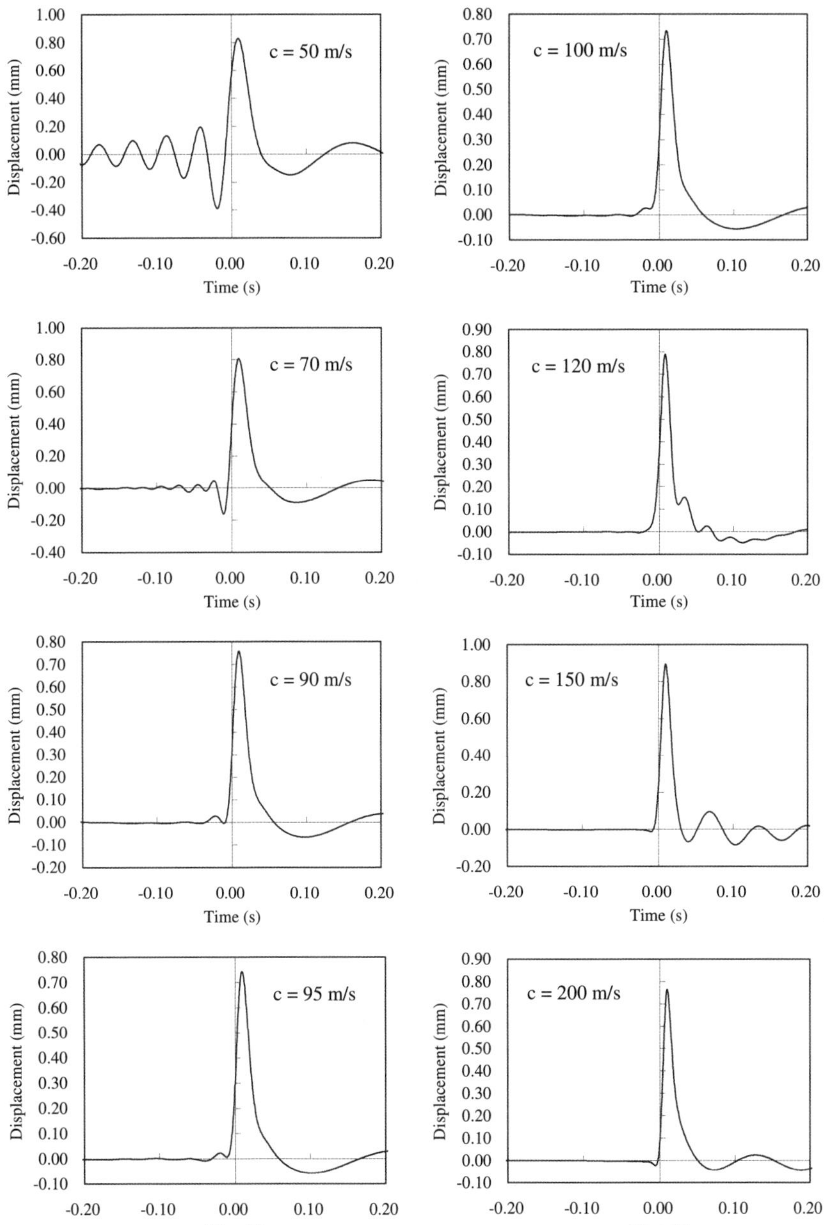

Fig. 2.18 Real-part vertical displacement for an elastically distributed load with frequency $f_0 = 10$ Hz directed along the y-axis.

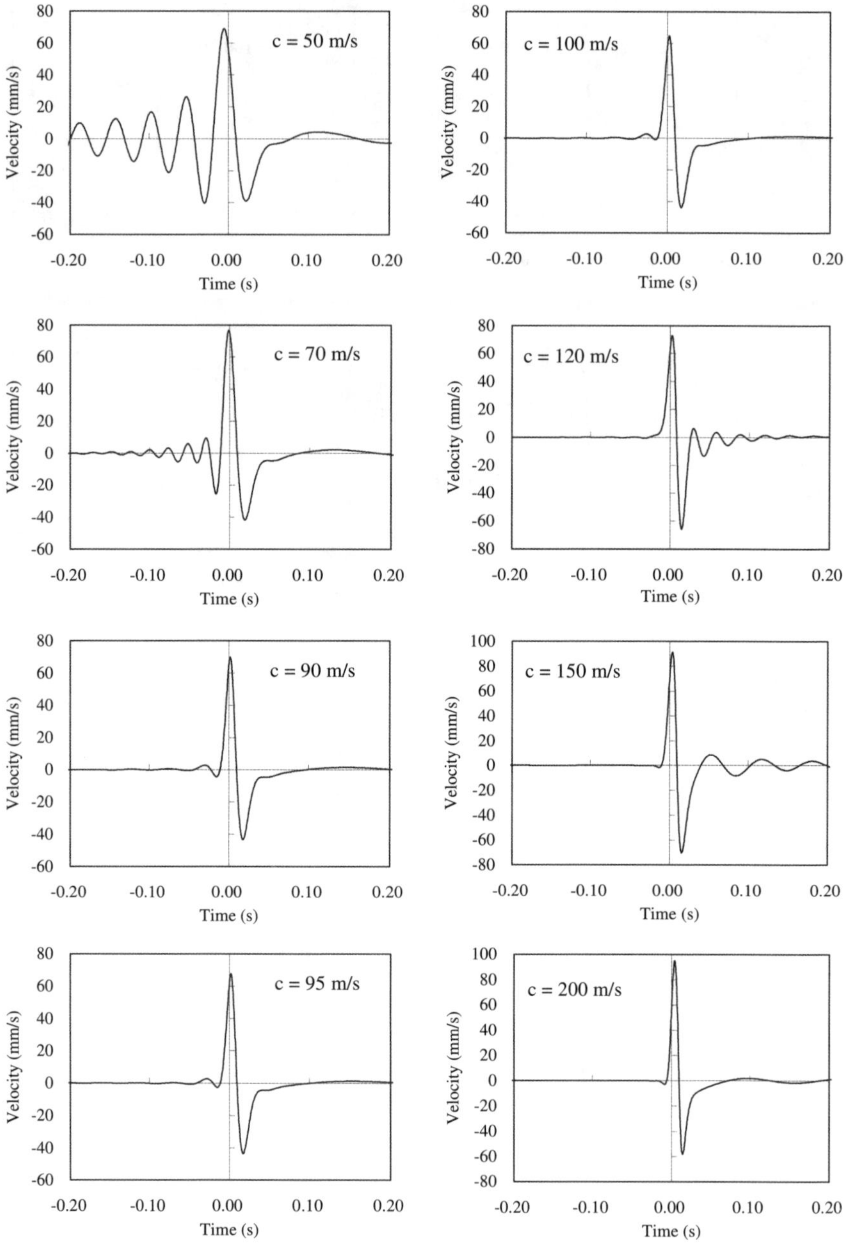

Fig. 2.19 Real-part vertical velocity for an elastically distributed load with frequency $f_0 = 10$ Hz directed along the y-axis.

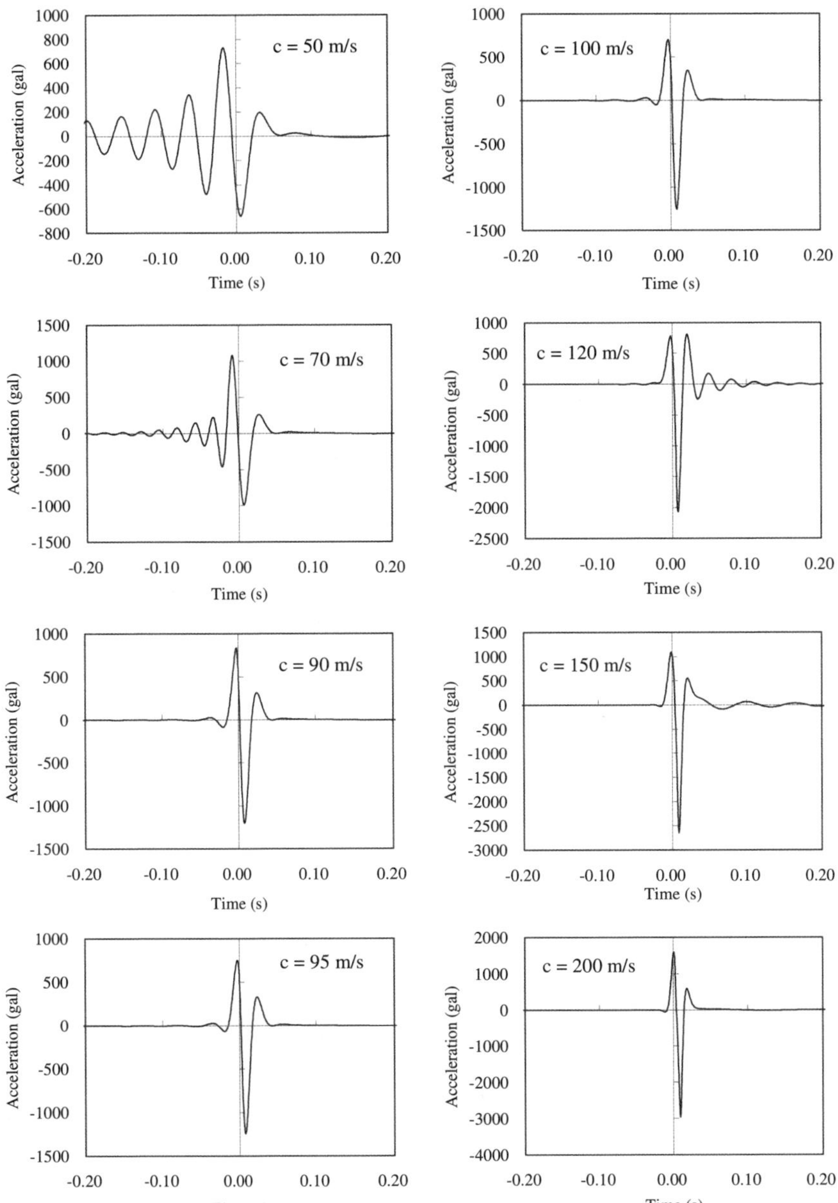

Fig. 2.20 Real-part vertical acceleration for an elastically distributed load with frequency f_0 = 10 Hz directed along the y-axis.

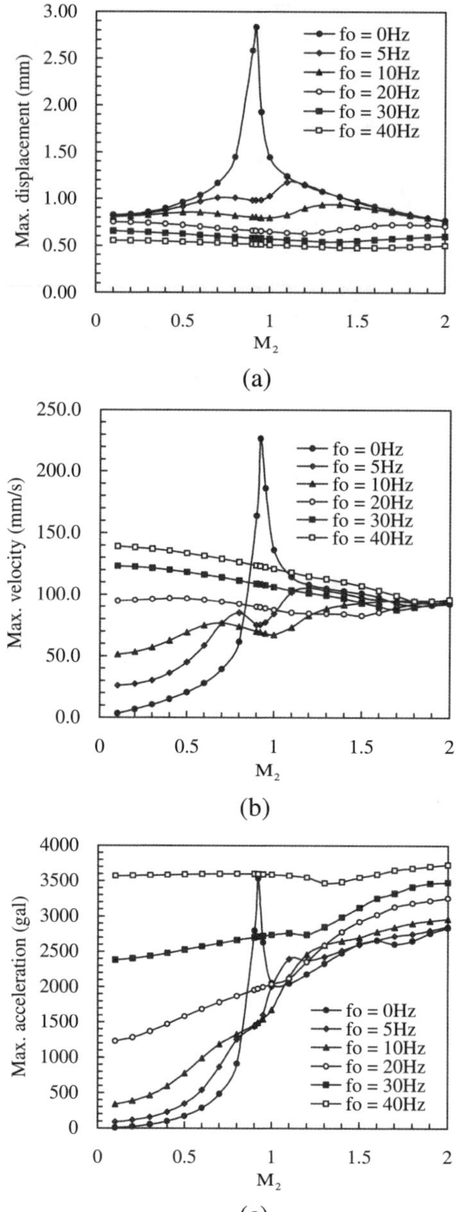

Fig. 2.21 Maximum responses induced by an elastically distributed load: (a) displacement; (b) velocity; (c) acceleration.

$$c_{cr} = \frac{\pm(f - f_0)}{f} c_R, \tag{2.43}$$

which can also be verified by the plot for the displacement $|\tilde{V}|$ versus speed c in Fig. 2.22 in frequency domain for f_0 = 10 Hz. For the case with f_0 = 0 Hz, the critical speed becomes equal to c_R, in consistence with what we have observed (see also Fig. 2.23). But for $f_0 \neq 0$ Hz, the critical speed depends not only on f_0, but also on the frequency $f = \omega / 2\pi$. However, as the time domain responses are computed as the superposition of the responses contributed by all the frequencies, the effect of critical speed can hardly be identified in the time domain for the case with $f_0 \neq 0$ Hz. In Fig. 2.24, it has been shown that the critical frequencies f_{cr} computed for given c and f_0 values are consistent with those computed from Eq. (2.42). Moreover, the values $1/f_{cr}$ computed from Fig. 2.24 are identical to the vibration period observed in Fig. 2.18.

The other observation from Fig. 2.21 is that for $f_0 \neq 0$ Hz, the influence of the self oscillation frequency f_0 on the maximum displacements is quite different from its influence on maximum velocities and accelerations. In Fig. 2.21(a), an increase of the frequency f_0 results in the decrease of the displacement, whereas in Figs. 2.21(b) and (c), the reverse is true for the velocity and acceleration. Moreover, for the cases with non-zero frequencies f_0, the higher the frequency of self oscillation f_0 of the moving load, the larger the magnitudes of the velocity and acceleration are. Nevertheless, it should be noted that all the results presented in Fig. 2.21 are based on the assumption that the amplitude of the wheel load remains constant, i.e., with T = 10 t, regardless of the variation of the self oscillation frequency f_0. For a moving static wheel load, this assumption is reasonable, but for a moving wheel load with non-zero self oscillation frequencies f_0, the amplitude can be far less than the wheel weight T. Thus, for the case with nonzero self oscillation frequencies f_0, the results presented herein serve merely as a qualitative illustration of the influence of the dynamic component of a vehicle load on ground vibrations.

In order to investigate the effect of response attenuation, the vertical displacement, velocity and acceleration along a line which is located 1 m beneath the x-axis for a range of 0 - 20 m were plotted for loads moving

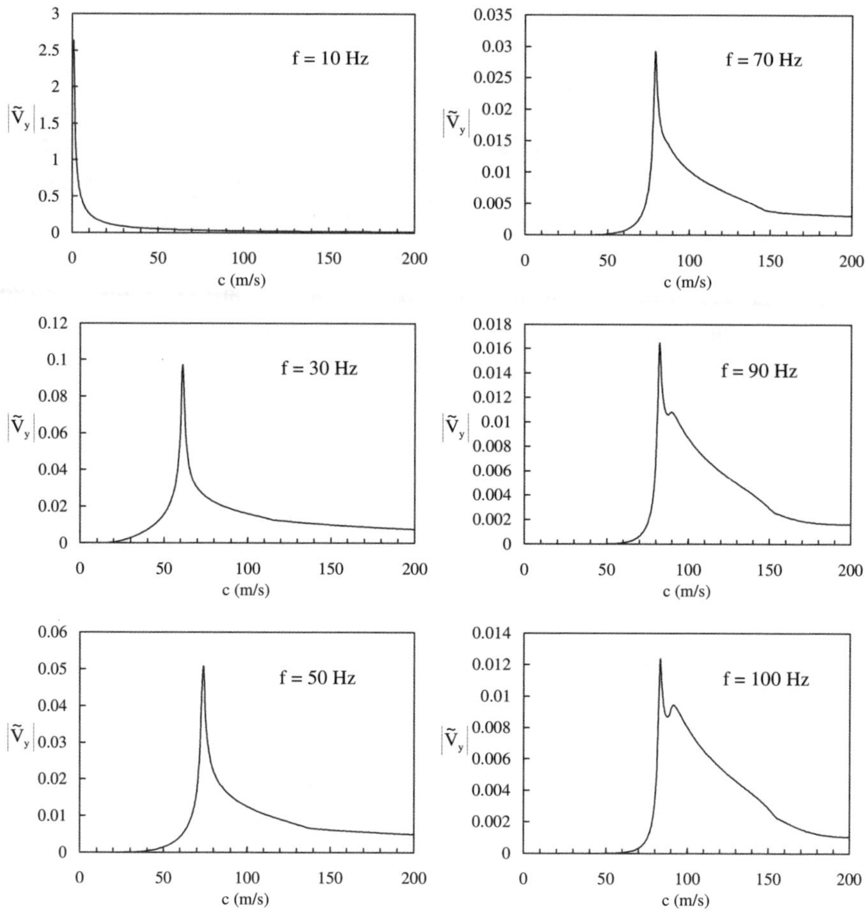

Fig. 2.22 Displacement in frequency domain for a moving point load with f_0 = 10 Hz and given f.

at the sub-critical speeds (c = 30 - 90 m/s) and trans-critical speeds (c = 110 - 170 m/s) with no self oscillation, i.e., with $f_0 = 0$, in Figs. 2.25 and 2.26, respectively. As can be seen, the attenuation trend is generally different for the two cases with sub- and trans-critical speeds. For the sub-critical speed case, the responses attenuate quite rapidly with respect to the distance, especially when the velocity and acceleration are concerned, whereas for the trans-critical speed case, the responses

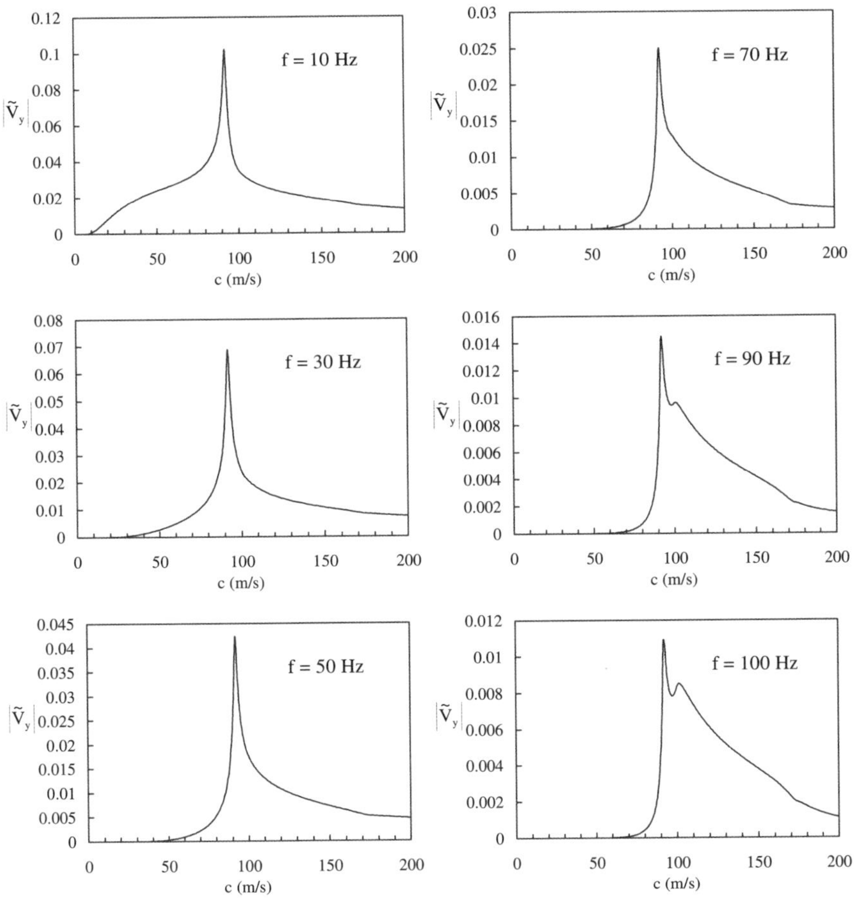

Fig. 2.23 Displacement in frequency domain for a moving point load with $f_0 = 0$ Hz.

attenuate much more slowly. Besides, for the sub-critical case, the response increases as the load speed increases, while for the trans-critical speed case, such a trend remains only true for acceleration, but is reversed for displacement and velocity.

To investigate the attenuation behavior of the responses for moving loads with self oscillation, the vertical responses computed along the line located at 1 m beneath the x-axis induced by a wheel load moving at speed $c = 70$ m/s, but with different self oscillation frequencies f_0, have been plotted in Fig. 2.27. As can be seen, the attenuation rate of the

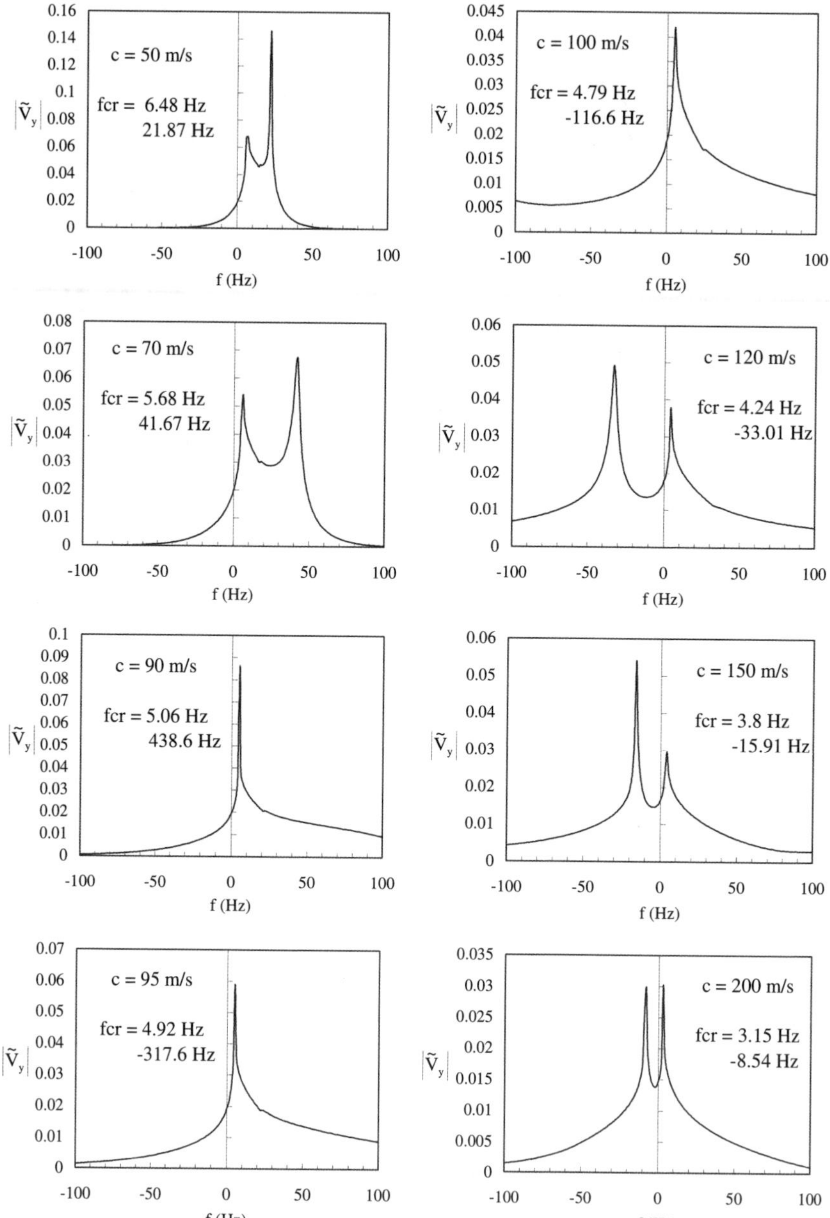

Fig. 2.24 Displacement in frequency domain for a moving point load with $f_0 = 10$ Hz and given c.

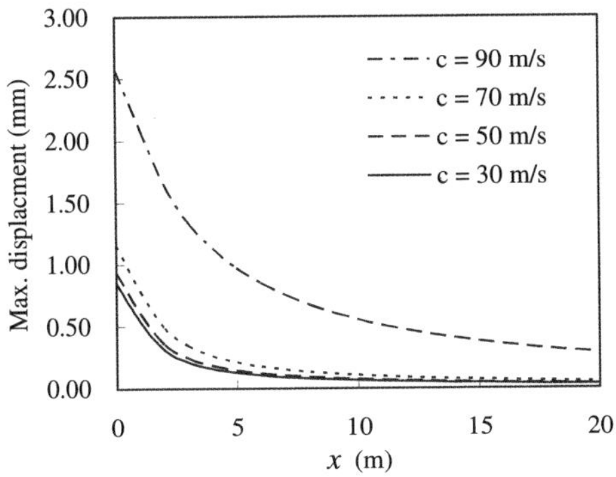

(a)

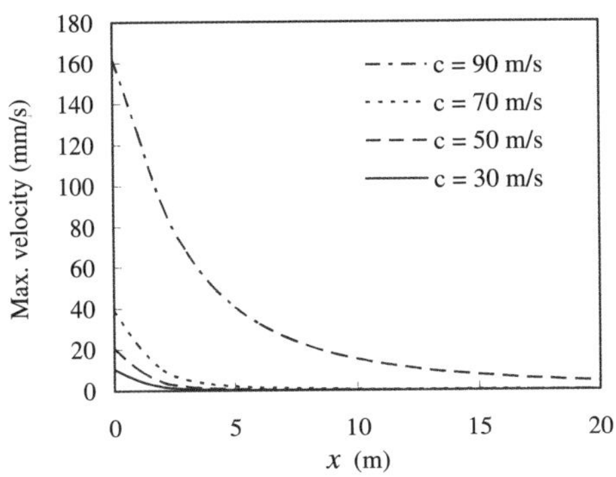

(b)

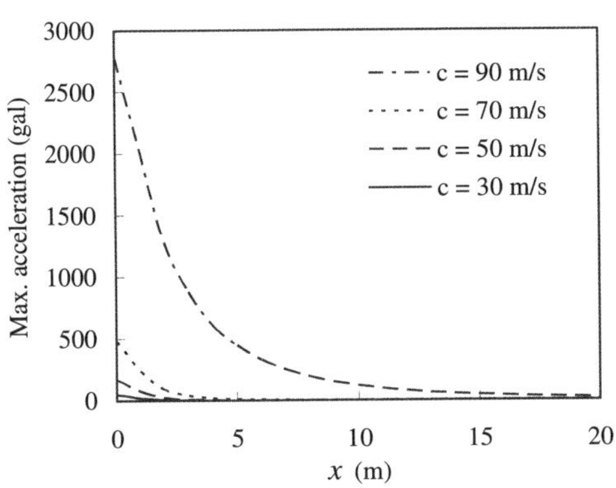

(c)

Fig. 2.25 Response attenuation for an elastically distributed moving load in sub-critical speed range: (a) displacement; (b) velocity; (c) acceleration.

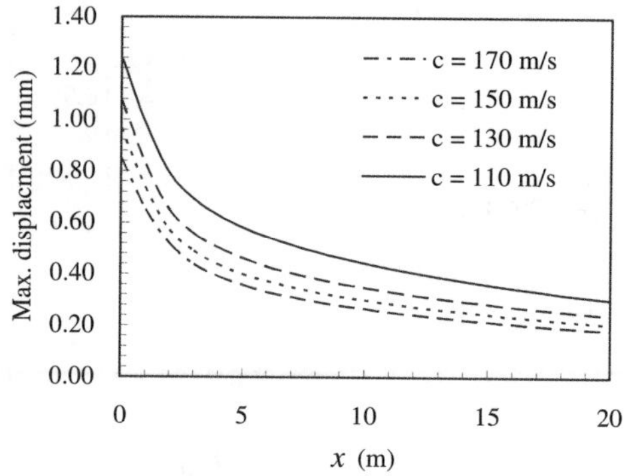

(a)

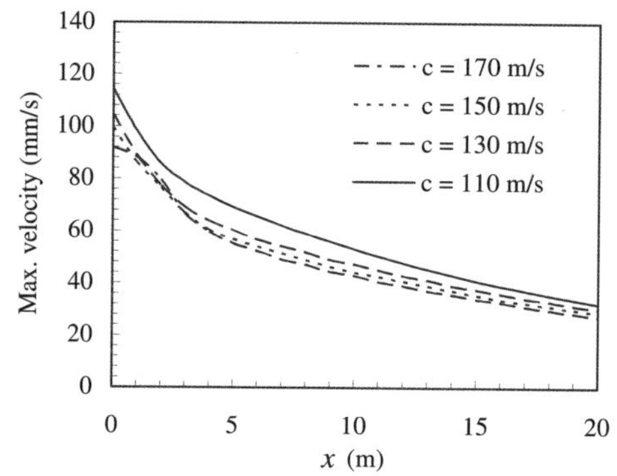

(b)

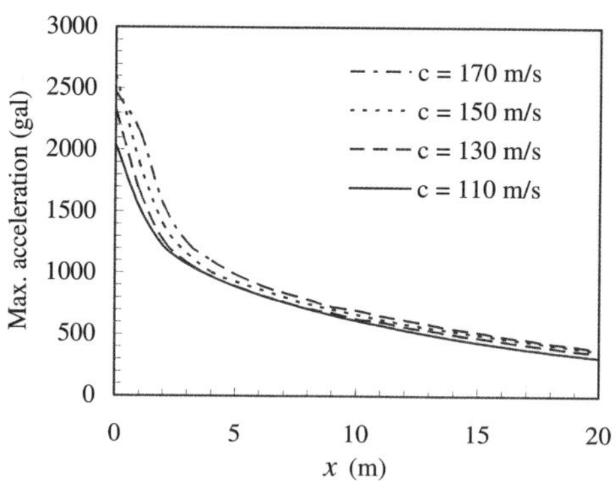

(c)

Fig. 2.26 Response attenuation for an elastically distributed moving load in trans-critical speed range: (a) displacement; (b) velocity; (c) acceleration.

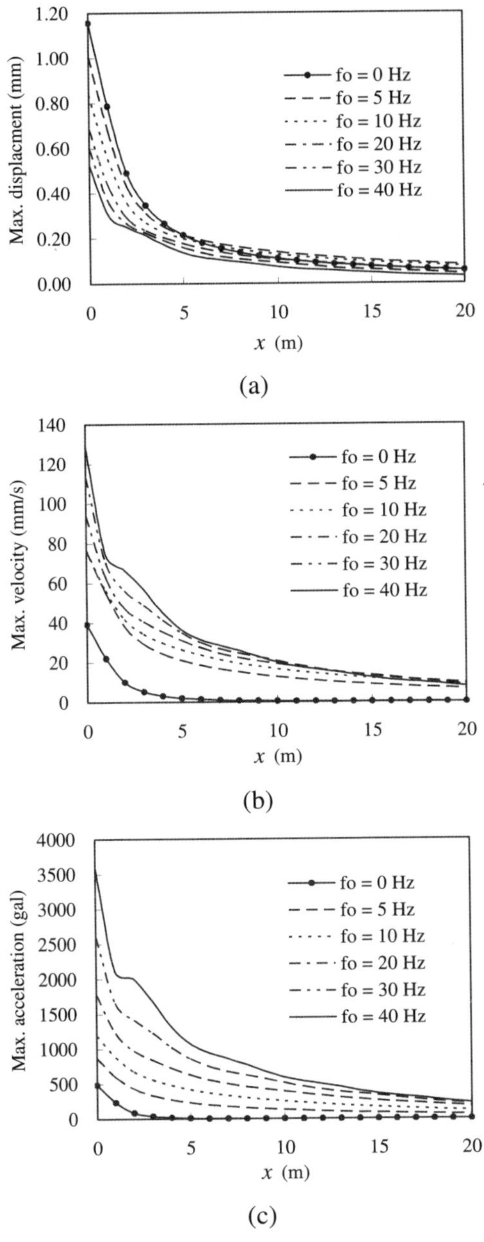

Fig. 2.27 Response attenuation for an elastically distributed moving load in trans-critical speed range: (a) displacement; (b) velocity; (c) acceleration.

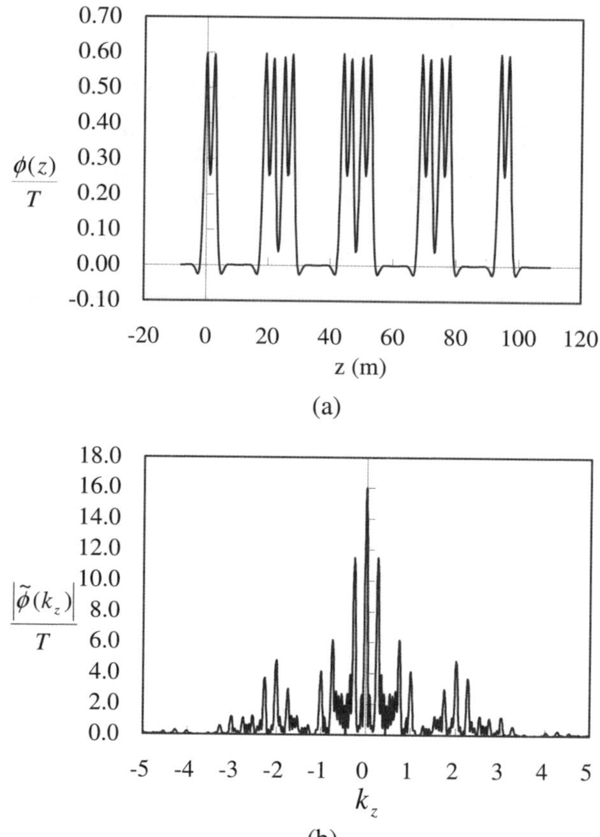

Fig. 2.28 A sequence of wheel loads: (a) load distribution function; (b) Fourier transform.

velocity and acceleration responses for the case with zero f_0 are larger than that for the case with non-zero f_0. In general, the attenuation rate on the displacement response is higher for the wheel loads with higher frequencies f_0 than that with lower frequencies f_0.

2.5.5 *A sequence of moving wheel loads*

As a final example, a moving load that takes into account the spatial distribution of the wheels of a real train will be considered. Based on the definition given in Eq. (2.36) and Fig. 2.3(b), the following dimensions are adopted: a = 2.56 m, b = 16.44 m and L = 25 m. The load distribution functions in the spatial and transformed domains for a carriage number of N = 4 were shown in Figs. 2.28(a) and (b). In Fig. 2.28(b), the load distribution function in the transformed (frequency) domain appears to be periodically distributed. Each peak in this figure relates to one passing frequency implied by one of the distances between wheels.

The effect associated with the number of carriages is investigated in Fig. 2.29, in which the maximum vertical responses with respect to different Mach numbers M_2 were plotted for the three cases: a single moving wheel load, four carriages, and ten carriages. From Fig. 2.29(a), it is observed that with the increase in the number of wheels or carriages, the displacement increases in all the speed range, but the difference between the cases with $N = 4$ and $N = 10$ is not so obvious. On the other hand, as can be seen from Figs. 2.29(b) and (c), basically no difference can be made for the velocity and acceleration responses as the carriage number N increases from 4 to 10, and the responses produced by a single wheel load appear to be even larger than the other two cases in the trans- and super-critical speed ranges. Based on the fact that little difference exists between the two cases with $N = 4$ and 10, the carriage number will be set to 4 in later chapters for simulation of the train loads to simplify the efforts of computation involved.

2.6 Concluding Remarks

In this chapter, a procedure for computing the responses of a visco-elastic half-space subjected to various moving vehicle loads on the surface has been presented. The moving load is assumed to consist of a static part, to simulate the gravitational effect, and a dynamic part, to simulate the self oscillation effect. Four different forms of the load distribution functions are considered. The influence of the load-moving speeds on the soil response is investigated for each of the four different types of moving loads over the sub-, trans-, and super-critical speed ranges.

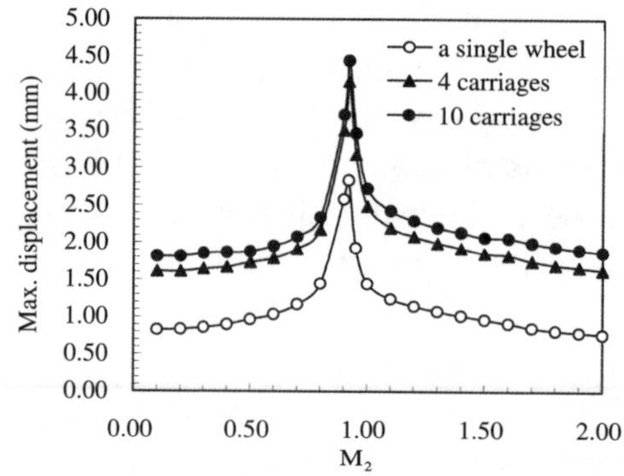

(a)

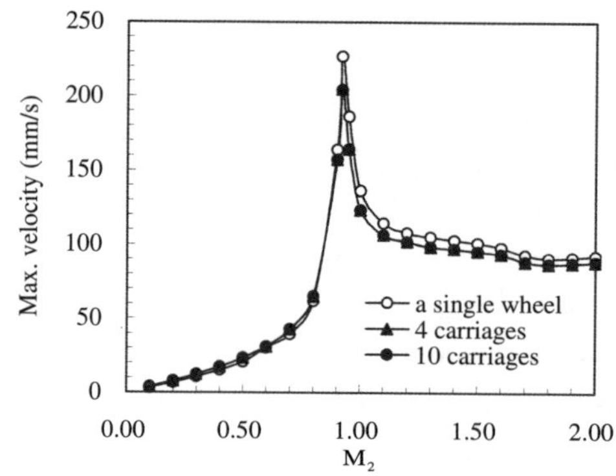

(b)

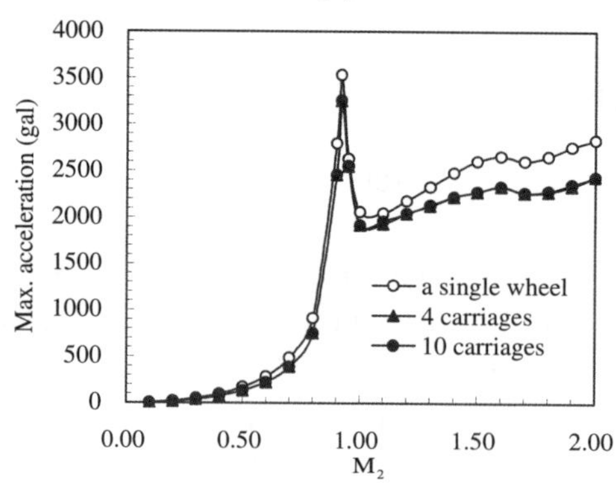

(c)

Fig. 2.29 Maximum vertical response caused by a sequence of wheel loads: (a) displacement; (b) velocity; (b) acceleration.

The following conclusions can be drawn from the numerical studies presented in this chapter:

(1) For the case with vertical moving loads, the critical speed is equal to the R-wave speed. For the cases with loadings applied along the load-moving direction, two obvious critical speeds can be observed, one is the R-wave speed and the other is the P-wave speed. And for the case with loadings applied along the direction transverse to the load-moving direction, the critical speed appears to be somewhat associated with the S-wave speed.
(2) Unlike the case with static moving loads, the critical speed for a moving load with a self oscillation component depends on the particular frequency considered. Because of this, no obvious resonance phenomenon can be observed in time domain.
(3) The dynamic component of a moving load can affect significantly the velocity and acceleration responses. As the self oscillation frequency increases, the amplitude of the velocity and acceleration increases as well.
(4) The attenuation rate of the soil responses for moving loads traveling in the trans- and super-critical speed ranges is smaller than that in the sub-critical speed range.
(5) The velocity and acceleration of the soils induced by a moving load that oscillates by itself with a higher frequency may attenuate faster than that with a lower frequency.
(6) An increase in the number of carriages constituting a train may result in increase of the displacement, but not of the velocity and acceleration of the soils.

Chapter 3

2D Finite/Infinite Element Method

Two drawbacks exist with prevoius infinite elements used for simulating the unbounded domains of semi-infinite problems. The first is the lack of an adequate measure for calculating the decay parameter. The second is the frequency-dependent property of the finite/infinite element mesh used for deriving the impedance matrices. Based on the properties of wave propagation, a scheme is proposed in this chapter for evaluating the decay parameter. In addition, it will be shown that by the method of dynamic condensation, the far-field impedance matrices for waves of lower frequencies can be obtained repetitively from the one for waves of the highest frequency considered, using an automatic mesh expansion scheme. Such an approach ensures that accuracy of the same order can be maintained for waves of all frequencies considered. Since the aforementioned two drawbacks have been removed, the proposed method is quite effective for dealing with the half-space problems encountered in practice, which may contain variations in geometry and materials.

3.1 Introduction

The use of infinite elements, in addition to finite elements, has been demonstrated to be a very effective tool for simulating interaction problems with unbounded domains. Following the conceptual works of Ungless (1973) and Bettess (1977), infinite elements have been widely applied to the solution of various wave propagation problems, such as the unbounded surface wave problems (Bettess and Ziekiewicz 1977; Astley 1983; Lau and Ji 1989), fluid-structure interaction problems (Saini *et al.*

1978; Park *et al.* 1991; Park *et al.* 1992), seepage problems (Zhao and Valliappan 1993; Honjo and Pokharel 1993), static geomechanic problems (Chow and Smith 1981; Beer and Meek 1981; Rajapakse and Karasudhi 1985), elastic wave propagation problems (Medina and Penzien 1982; Medina and Taylor 1983; Rajapakse and Karasudhi 1986; Zhang and Zhao 1987; Yang and Yun 1992; Karasudhi and Liu 1993; Yun *et al.* 1995), etc.

Various infinite elements have been proposed in the literature. Most of them are characterized by the fact that an exponentially decay term is multiplied to the shape functions associated with the direction extending to infinity to represent the amplitude attenuation effect of traveling waves. For the other directions with finite dimensions, basically the same shape functions as those of the parent finite elements are used.

The use of infinite elements for modeling the unbounded domains, and finite elements for modeling the near field of a soil-structure system has several advantages. First, the near field containing the railway track, structure, foundation, isolation devices, and underlying soils is always the focus of engineering concern. The use of finite elements offers a flexible means for modeling this region to meet the needs of structural designers. Second, the infinite elements used to represent the unbounded domain can be assembled in exactly the same way as that for the finite elements, while no additional degrees of freedom (DOFs) are introduced. Finally, the banded and symmetry properties of the system matrices are preserved when the infinite elements are used along with the finite elements, which is attractive the point of implementation of the computer codes.

However, existing infinite elements are not perfect for some reasons. One problem is that the *decay parameter* involved in the shape functions for simulating the amplitude attenuation effect of traveling waves with increasing distance was not clearly defined. The other problem is that the modeling of waves of different frequencies requires the use of elements and meshes of different sizes. Thus, it becomes improper to use a single finite/infinite mesh that is fixed in size to simulate the impedance of soils for waves of different frequencies. In this chapter, procedures will be presented to overcome the aforementioned two problems, based primarily on the works by Hung (1995) and Yang *et al.* (1996).

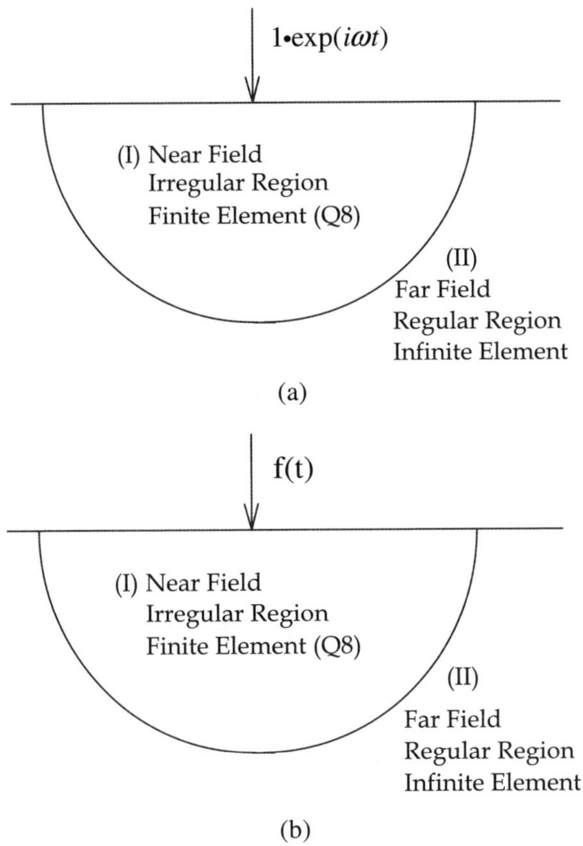

Fig. 3.1 Half-space model subjected to: (a) harmonic load; (b) time-varying load.

3.2 Formulation of the Problem

The fundamental problem considered is the one shown in Fig. 3.1, in which the *near field* with geometric and material irregularities will be modeled by finite elements, i.e., the conventional *quadratic 8-node* (*Q8*) *element*, and the *far field* extending to infinity will be modeled by the infinite element to be derived. In this study, we shall assume that the properties of the half-space are identical in the direction normal to the plane of the figure, and that the half-space is subjected to a harmonic line load on the surface. Consequently, the condition of *plane strain* can be adopted and the displacements of the half-space may be assumed to be harmonic as well.

Soils are highly nonlinear materials. However, for soil vibrations of rather small magnitudes, as those due to moving trains, but not those due to strong ground motions such as the earthquakes, we can neglect the nonlinear effects of the soils, and assume the soils to be isotropic and viscoelastic materials with hysteretic damping.

The following *principle of virtual displacements* applies to a finite element, an infinite element, or a solid body represented by finite and infinite elements. For an elastodynamic problem, by neglecting the body forces, the equation of motion for the solid body can be written as

$$\int_S t_i \delta u_i dA = \int_V \rho \ddot{u}_i \delta u_i dV + \int_v \tau_{ij} \delta \varepsilon_{ij} dV, \tag{3.1}$$

where an overdot denotes differentiation with respect to time, V and S denote the volume and surface area of the body of concern, δ denotes the variation of the quantity following, i.e., a virtual quantity, t_i denotes the surface tractions, u_i the displacements, ρ the mass density of the material, τ_{ij} the stresses, and ε_{ij} the strains. By assuming that there are n concentrated loads, the preceding equation can be rewritten as

$$\sum_{i=1}^{n} \{\delta u\}_i^T \{p\}_i = \int_V \{\delta u\}^T \rho \{\ddot{u}\} dV + \int_v \{\delta\varepsilon\}^T \{\tau\} dV, \tag{3.2}$$

where $\{u\}$ denotes the displacement field, $\{\delta u\}$ and $\{\delta\varepsilon\}$ denote the virtual displacements and virtual strains, respectively, assumed to be small in magnitudes, $\{p\}_i$ denotes the concentrated loads, and $\{u\}_i$ the corresponding displacements.

By the finite element strategy, the displacement field $\{u\}$ can be related to the nodal quantities as follows:

$$\begin{aligned} \{u\} &= [N]\{d\}, \\ \{\ddot{u}\} &= [N]\{\ddot{d}\}, \end{aligned} \tag{3.3}$$

where $[N]$ denotes the shape functions, and $[d]$ the displacement DOFs of the element. Substituting Eq. (3.3) into Eq. (3.2) yields:

$$\{\delta d\}^T \left[[M]\{\ddot{d}\} + [K]\{d\} - \{p\} \right] = 0, \tag{3.4}$$

where [*M*] and [*K*] denote the *mass* and *stiffness matrices* of the body of concern,

$$[M]=\int_V \rho[N]^T[N]dV, \tag{3.5a}$$

$$[K]=\int_V [B]^T[\tau]dV, \tag{3.5b}$$

where

$$[B]=[\partial][N]. \tag{3.6}$$

Here $[\partial]$ is the differential operator matrix related to the strains. By letting [*E*] denote the constitutive coefficient matrix, we can write the constitutive law as

$$\{\tau\}=[E]\{d\}. \tag{3.7}$$

It follows that the stiffness matrix can be written as

$$[K]=\int_V [B]^T[E][B]dV. \tag{3.8}$$

By taking into the arbitrary nature of the virtual displacements $\{\delta d\}$, we can rewrite Eq. (3.4) as follows:

$$[M]\{\ddot{d}\}+[K]\{d\}=\{p\}. \tag{3.9}$$

For a system under the harmonic loads, we can express the load $\{p\}$ on the right-hand side of the Eq. (3.9) as

$$\begin{aligned}\{p\}&=\{p_0\}e^{i\omega t},\\ \{d\}&=\{d\}e^{i\omega t},\end{aligned} \tag{3.10}$$

where $\{p_0\}$ denote the amplitude of the external loads and $\{d\}$ the amplitude of the nodal displacements. Substituting Eq. (3.10) into Eq. (3.9) yields the following *equation of motion*:

$$[S]\{d\}=\{p_0\}, \tag{3.11}$$

in which the *impedance* or *dynamic stiffness matrix* $[S]$ is

$$[S]=[K]-\omega^2[M]. \tag{3.12}$$

From Eq. (3.11), the displacement amplitudes can be solved:

$$\{d\} = [S]^{-1}\{p_0\} \tag{3.13}$$

for the problem with harmonic excitations.

The fundamental problem considered herein has been plotted in Fig. 3.1(b), in which $f(t)$ denotes the time history of the external load. The method of analysis for the half-space problem is carried out in the frequency domain. The so-called *frequency-domain analysis* hinges on application of the *Fourier transformation* to transforming the simultaneous partial differential equations of motion in time domain to the simultaneous algebraic equations of motion in frequency domain. From the latter the *frequency response function* can be solved for an applied unit load for each frequency in exactly in the same manner as the problem shown in Fig. 3.1(a). Realizing that the response functions are generally smooth in nature, we need not solve the problem for all the frequencies of concern, but only for some control points (frequencies) within the range of frequencies considered. Such results can then be interpolated to yield the transfer function for each frequency in the transformed space. By multiplying the response function by the spectral value of the external load after Fourier transformation, and then by applying the inverse Fourier transformation, we can obtain the response of the half-space in time domain corresponding to the external load $f(t)$.

3.3 Shape Functions and Matrices of Infinite Element

The kind of finite elements to be used in the near field is the traditional *quadratic 8-node* (*Q8*) *element*. For this element, the shape functions, method of integration, i.e., the *Gaussian quadrature*, and the element matrices are available in most finite element textbooks, see for instance Cook *et al.* (1988). No effort will be made herein to elaborate this part of derivation. In the following, we shall concentrate on derivation of the infinite element.

3.3.1 *Shape functions*

The kind of infinite elements to be used in the far field is defined in the frequency domain, which can be regarded as one kind of *isoparametric*

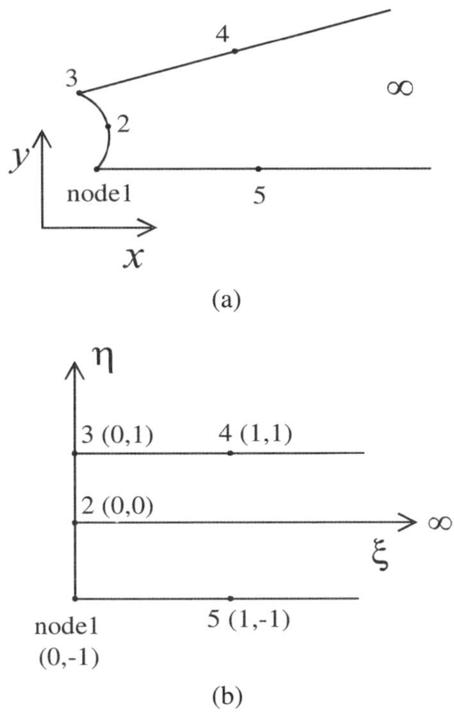

Fig. 3.2 Infinite element: (a) global coordinates; (b) local coordinates.

element or a variant of the Q8 element, as schematically shown in Fig. 3.2, in which part (a) denotes the infinite element in the *global coordinates* and part (b) in *local coordinates*.

The shape functions for the nodal coordinates of the infinite element are different from those for the nodal displacements. Let (x, y) denote the global coordinates and (ξ, η) the local or *natural coordinates* of the element. The coordinates (x, y) of a point within the element can be related to the nodal coordinates of the element as

$$x = \sum_{i=1}^{n} N'_i x_i, \quad y = \sum_{i=1}^{n} N'_i y_i, \tag{3.14}$$

where the shape functions N'_i are assumed to be linear in ξ and quadratic in η, i.e.,

$$N_1' = -\frac{(\xi-1)(\eta-1)\eta}{2}, \tag{3.15a}$$

$$N_2' = (\xi-1)(\eta-1)(\eta+1), \tag{3.15b}$$

$$N_3' = -\frac{(\xi-1)(\eta+1)\eta}{2}, \tag{3.15c}$$

$$N_4' = \frac{\xi(\eta+1)}{2}, \tag{3.15d}$$

$$N_5' = -\frac{\xi(\eta-1)}{2}. \tag{3.15e}$$

Similarly, the displacements (u, v) of a point within the element can be interpolated from the nodal quantities of the element as

$$u = \sum_{i=1}^{n} N_i u_i, \qquad v = \sum_{i=1}^{n} N_i v_i, \tag{3.16}$$

where the shape functions N_i are

$$\begin{aligned} N_1 &= \frac{\eta(\eta-1)}{2} P(\xi), \\ N_2 &= -(\eta-1)(\eta+1)P(\xi), \\ N_3 &= \frac{\eta(\eta+1)}{2} P(\xi). \end{aligned} \tag{3.17}$$

The function $P(\xi)$ in Eq. (3.17) is known as the *propagation function*,

$$P(\xi) = e^{-\alpha_L \xi} e^{-ik_L \xi}, \tag{3.18}$$

where α_L denotes the *amplitude decay factor* of displacement, which has the unit of $1/L$, and k_L is the *wave number*, both expressed in the local coordinates. In Eq. (3.18), the term $\exp(-\alpha_L \xi)$ is used to represent the amplitude attenuation due to wave dispersion, i.e., the so-called effect of *radiation damping*, and the term $\exp(-ik_L \xi)$ the phase decay due to wave propagation in the local coordinates. As can be seen from Eq. (3.17), the displacements of the infinite element are interpolated only along the direction of finite boundary, while in the direction with infinite boundary, the displacements are assumed to decay from the origin. Such

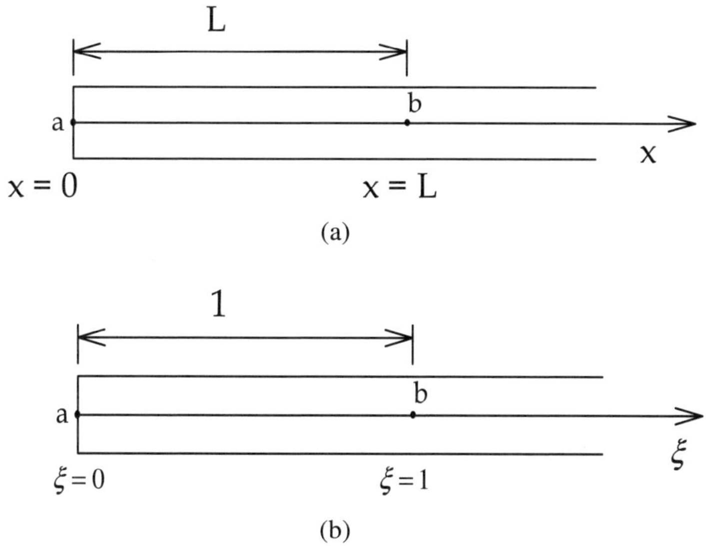

Fig. 3.3 One-dimensional mapping: (a) global coordinates; (b) local coordinates.

an approach has the advantage that no additional nodes are needed for the infinite side of the domain, of which the displacement is usually not of major concern.

In practice, the propagation function is available only in the global sense:

$$P(x) = \mathrm{e}^{-\alpha x}\mathrm{e}^{-ikx}, \tag{3.19}$$

where α denotes the amplitude decay factor of displacement and k the wave number in the global coordinates, i.e., $k = \omega / c$ with ω and c respectively denoting the frequency and velocity of the traveling waves. With reference to Fig. 3.3, let us denote the distance between points a and b as 1 and L in the local and global coordinates, respectively. The local coordinate ξ can be related to the global coordinate x as

$$\xi = \frac{x}{L}. \tag{3.20}$$

Substituting the preceding relation into Eq. (3.18) yields

$$P(x) = \mathrm{e}^{-\alpha_L x/L}\mathrm{e}^{-ik_L x/L}. \tag{3.21}$$

By comparing Eqs. (3.21) with (3.19), one can arrive at the following relations for the two parameters:

$$\begin{aligned} \alpha_L &= \alpha \cdot L, \\ k_L &= k \cdot L. \end{aligned} \tag{3.22}$$

3.3.2 *Element matrices*

For the infinite element shown in Fig. 3.2, the shape functions are those given in Section 3.3.1. By following the finite element procedure, the equation of motion can be derived for the infinite element as in Eq. (3.11), or

$$-\omega^2[M]\{\Delta\} + [K]\{\Delta\} = \{F\}, \tag{3.23}$$

where the nodal displacements $\{\Delta\}$ and nodal forces $\{F\}$ associated with the element shown in Fig. 3.2 are

$$\{\Delta\} = \begin{bmatrix} u_1 & v_1 & u_2 & v_2 & u_3 & v_3 \end{bmatrix}^T, \tag{3.24a}$$

$$\{F\} = \begin{bmatrix} F_{x1} & F_{y1} & F_{x2} & F_{y2} & F_{x3} & F_{y3} \end{bmatrix}^T. \tag{3.24b}$$

The mass matrix $[M]$ and stiffness matrix $[K]$ can be derived from Eqs. (3.5a) and (3.8) as follows:

$$[M]_{6\times 6} = \int_{-1}^{1}\int_{0}^{\infty} \rho [N]^T [N] t\,\mathrm{J}\, d\xi d\eta, \tag{3.25}$$

$$[K]_{6\times 6} = \int_{-1}^{1}\int_{0}^{\infty} [B]_{6\times 3}^T [E]_3^3 [B]_{3\times 6}\, tJd\xi d\eta, \tag{3.26}$$

where t is the thickness of the element and J is the determinant of the *Jacobian matrix* $[J]$:

$$J = \det[J]. \tag{3.27}$$

The Jacobian matrix is given as

$$[J]=\begin{bmatrix} N'_{1,\xi} & N'_{2,\xi} & N'_{3,\xi} & N'_{4,\xi} & N'_{5,\xi} \\ N'_{1,\eta} & N'_{2,\eta} & N'_{3,\eta} & N'_{4,\eta} & N'_{5,\eta} \end{bmatrix}\begin{bmatrix} x_1 & y_1 \\ x_2 & y_2 \\ x_3 & y_3 \\ x_4 & y_4 \\ x_5 & y_5 \end{bmatrix}. \tag{3.28}$$

The shape functions [N] involved in Eq. (3.25) and the constitutive coefficient matrix [E] in Eq. (3.26) can be given as

$$[N]=\begin{bmatrix} N_1 & 0 & N_2 & 0 & N_3 & 0 \\ 0 & N_1 & 0 & N_2 & 0 & N_3 \end{bmatrix}, \tag{3.29}$$

$$[E]=\frac{E}{(1+\upsilon)(1-2\upsilon)}\begin{bmatrix} 1-\upsilon & \upsilon & 0 \\ \upsilon & 1-\upsilon & 0 \\ 0 & 0 & \dfrac{1-2\upsilon}{2} \end{bmatrix}, \tag{3.30}$$

where υ is Poisson's ratio and E is Young's modulus of the material. In this study, the *hysteretic damping* suggested by Seed and Idriss (1970) will be adopted for the soil, by which the damping is assumed to depend on the strain, rather than on frequency of the soil. Moreover, the strains induced in the soil by the moving loads are much smaller than those by the earthquakes. Thus, it is proper to incorporate the material damping through replacement of Lamé's constants λ and G by $\lambda^* = \lambda(1+2i\beta)$ and $G^* = G(1+2i\beta)$, where β is the hysterestic damping ratio, as was done by previous researchers (Kausel and Roësset 1977; Chow and Smith 1981; Medina and Penzien 1982; Zhang and Zhao 1987; de Barros and Luco 1994; Grundmann *et al.* 1999). Thus for viscoelastic materials, the Young's modulus E in Eq. (3.30) is replaced by E^*,

$$E^* = E(1+2i\beta). \tag{3.31}$$

Define $[\Gamma]$ as the inverse of the Jacobian matrix $[J]$, i.e.,

$$[\Gamma]=[J]^{-1}. \tag{3.32}$$

We can write the matrix $[B]$ as follows:

$$[B]=\begin{bmatrix} \Gamma_{11} & \Gamma_{12} & 0 & 0 \\ 0 & 0 & \Gamma_{21} & \Gamma_{22} \\ \Gamma_{21} & \Gamma_{22} & \Gamma_{11} & \Gamma_{12} \end{bmatrix} \begin{bmatrix} N_{1,\xi} & 0 & N_{2,\xi} & 0 & N_{3,\xi} & 0 \\ N_{1,\eta} & 0 & N_{2,\eta} & 0 & N_{3,\eta} & 0 \\ 0 & N_{1,\xi} & 0 & N_{2,\xi} & 0 & N_{3,\xi} \\ 0 & N_{1,\eta} & 0 & N_{2,\eta} & 0 & N_{3,\eta} \end{bmatrix}. \tag{3.33}$$

At this point, all the component matrices required in computation of the mass matrix [*M*] in Eq. (3.25) and stiffnes matrix [*K*] in Eq. (3.26) have been made available.

3.3.3 *Damping property of materials*

When a material deforms due to stressing, part of the energy will be stored as strain energy, or referred to as *storage energy*, and the remaining will be dissipated through the internal friction of the material and referred to as *dissipated energy*. In order to model both the storage energy and dissipated energy, one can use the *complex shesar modulus* G^* to describe the constitutive law of the material, i.e.,

$$G^* = G(1+\mathrm{i}\eta), \tag{3.34}$$

where G is the shear modulus and η the loss factor. The traditional hysteretic damping ratio β equals half of the value of η. Thus,

$$G^* = G(1+2\mathrm{i}\beta). \tag{3.35}$$

Damping of this sort tends to increase as the strains of the material increase. According to Seed and Idriss (1970), the damping of soils depends much more on the material strains than on the frequency. For this reason, the hysteretic damping, rather than viscoelastic damping, will be adopted for soils in this study. This is exactly the type of damping adopted by most popular analysis programs such as SHAKE and FLUSH for soils.

3.3.4 *Method of numerical integration*

As can be seen from the element matrices presented in Section 3.3.2, due to presence of the term $p(\xi)=\exp[-(\alpha_L+ik_L)\xi]$, the shape functions involved in the mass matrix [*M*] and stiffness matrix [*K*] in Eqs. (3.25) and (3.26) are no longer polynomial functions of ξ and η. The upper and lower limits for integration along the ξ-direction have been changed from $(-1, 1)$ to $(0, \infty)$. The mass and stiffness matrices presented in Eqs. (3.25) and (3.26) represent one type of integrals of the following form:

$$\int_0^\infty F(\xi)\mathrm{e}^{-(\gamma+\mathrm{i}\lambda)\xi}d\xi, \tag{3.36}$$

which can no longer be computed using the traditional *Gaussian quadrature*.

To evaluate integrals with one infinity limit, Bettess and Zienkiewicz (1977) proposed a method similar to the Newton-Cotes integration method. The sampling points for this method can be arbitrarily selected, but are usually selected as points meeting the condition $\xi=\pi(2n+1)/(2\lambda)$, where n is an integer. By such an approach, the zero's in the real and imaginary parts of $\exp(-i\lambda\xi)$ can be circumvented. However, the value of λ depends on the wavelength considered, which implies that for an elastic body, we need to consider waves of different wavelengths. It will be highly time-consuming, if the sampling points have to be adjusted each time for different wavelengths.

In this study, we shall follow the suggestion by Chow and Smith (1981) and take the four sampling points with ξ = 2, 4, 6, 8. In other words, the same sampling points will be selected for waves of different wavelengths, but with different weights. The following is a summary of such a procedure.

Let us express the function $F(\xi)$ by the *Lagrange polynomials* as

$$F(\xi)=L_1F(2)+L_2F(4)+L_3F(6)+L_4F(8), \tag{3.37}$$

where

$$L_k=\sum_{\substack{j=1\\k\neq j}}^{4}\left(\frac{\xi_j-\xi}{\xi_j-\xi_k}\right), \tag{3.38}$$

with ξ_1, ξ_2, ξ_3, ξ_4 selected as 2, 4, 6, 8, respectively. Substituting

Eq. (3.37) into Eq. (3.36) yields

$$\int_0^\infty F(\xi)\mathrm{e}^{-(\gamma+\mathrm{i}\lambda)\xi}d\xi = F(2)W_1 + F(4)W_2 + F(6)W_3 + F(8)W_4, \tag{3.39}$$

where W_k are the weighted factors,

$$W_k = \int_0^\infty L_k \mathrm{e}^{-(\gamma+\mathrm{i}\lambda)\xi}d\xi. \tag{3.40}$$

Finally, by substituting Eq. (3.38) into Eq. (3.40), one obtains

$$\begin{aligned}
W_1 &= \tfrac{1}{24}\left(96\kappa - 52\kappa^2 + 18\kappa^3 - 3\kappa^4\right),\\
W_2 &= \tfrac{1}{8}\left(-48\kappa + 38\kappa^2 - 16\kappa^3 + 3\kappa^4\right),\\
W_3 &= \tfrac{1}{8}\left(32\kappa - 28\kappa^2 + 14\kappa^3 - 3\kappa^4\right),\\
W_4 &= \tfrac{1}{24}\left(-24\kappa + 22\kappa^2 - 12\kappa^3 + 3\kappa^4\right),
\end{aligned} \tag{3.41}$$

where

$$\kappa = \frac{1}{\gamma + i\lambda}. \tag{3.42}$$

The above procedure can be easily modified to include more sampling points. However, our numerical studies indicated that the use of more sampling points did not result in clearly improved overall results for the half-space problems, especially for the near field of concern. The reason is that the use of infinite elements to simulate the geometric attenuation or radiation damping of the half-space, as represented by the decay function in Eq. (3.36), is by itself an approximation. Thus, little improvement in accuracy can be gained for the overall results of the half-space problem by using more sampling points in computing integrals of the type in Eq. (3.36).

3.3.5 *Selection of amplitude decay factor* α

In the study by Zhang and Zhao (1987), the amplitude decay factor α, as given in Section 3.3.1, was regarded as a quantity without physical unit Such an assumption implies the dependence of computed results on the length unit used, which is not physically sound. In fact, the unit of α should be taken as $1/L$, with L representing the length of decay. This was exactly the case adopted by Bettess (1977), although no value was

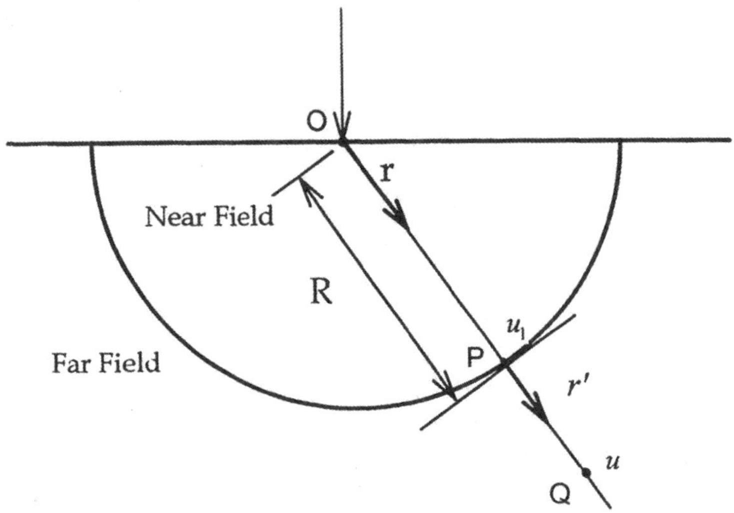

Fig. 3.4 Schematic diagram for determining the amplitude decay factor.

recommended by Bettess for *L*. In the following, we shall explain how to determine the optimal value for α.

According to the analytical solution given in Chapter 2, for an elastic half-space subjected to an *infinite line load*, the *body waves* traveling in the far field will decay at a rate proportional to $r^{-1/2}$, where r is the distance from the source to the point of concern. With reference to Fig. 3.4, let *O* denote the point of source of the half-space, *P* a point in the far field boundary with radius *R*, of which the displacement is u_1, and *Q* a third point along the direction of *OP*, but located in the far field, i.e., with radius $r > R$. The displacement *u* of point *Q* can be expressed as

$$u = u_1 \frac{r^{-1/2}}{R^{-1/2}}, \tag{3.43}$$

or, by expressing $r^{-1/2}$ in logarithmic form,

$$u = u_1 \cdot R^{1/2} e^{-(1/2)\log r}. \tag{3.44}$$

Let r' denote the distance from point *P* (the boundary of the far field) to the point *Q* of concern, as indicated in Fig. 3.4. We can replace r by

$R+r'$ and expand the logarithmic function in Eq. (3.44) by a series as follows:

$$u = u_1 \cdot \exp\left\{-\frac{1}{2}\left[\left(\frac{r'}{R}\right)+\frac{1}{2}\left(\frac{r'}{R}\right)^2+\ldots\right]\right\}. \tag{3.45}$$

Let $r'/R \ll 1$, i.e., let the point Q with distance r' be a point very near the boundary of the far field. For such a case, we need only retain the first term in Eq. (3.45), i.e.,

$$u = u_1 \cdot \exp\left(-\frac{r'}{2R}\right). \tag{3.46}$$

By comparing the preceding equation with the propagation function in Eq. (3.18), one finds that for the case of a *line load*, the amplitude decay factor α should be selected as

$$\alpha = \frac{1}{2R}. \tag{3.47}$$

Clearly, the factor α should not be given an arbitrary number. Instead, it should be proportional to the distance from the source to the boundary of the far field. Substituting Eq. (3.47) into Eq. (3.22) yields

$$\alpha_L = \frac{L}{2R}. \tag{3.48}$$

The above value for α has been derived for the regions where the body waves are dominant, i.e., for the regions well below the free surface of the half-space subjected to a surface line load. Since the *Rayleigh waves* do not decay on the free surface under the same loading condition, it is suggested that $\alpha = 0$ be used for regions near the free surface.

Similarly, for the case of a *point load*, the amplitude decay factor α should be selected as

$$\alpha = \frac{1}{R}, \tag{3.49}$$

for the regions well below the free surface, where the body waves are dominant. As for the free surface, where the *Rayleigh waves* are considered most important, the value of $1/(2R)$ should be adopted instead for α.

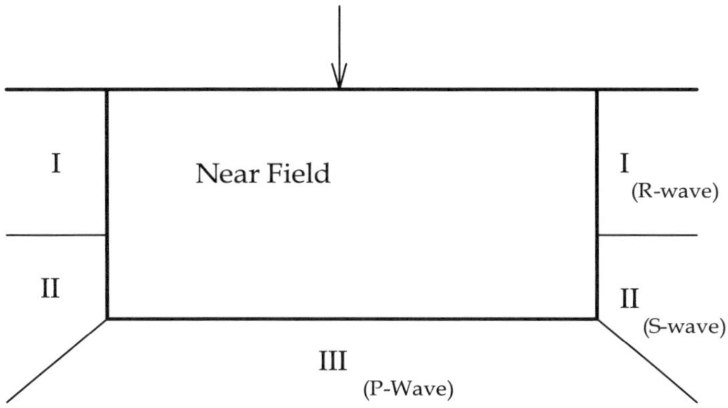

Fig. 3.5 Selection of wave numbers k.

3.3.6 *Selection of wave number k*

As was stated previously, the wave number k $(=\omega/c)$ depends on the wave speed c. It is realized that different waves have different speeds. For the surface water wave problems, for which the infinite elements have been developed (Bettess and Zienkiewics 1977), there is only a single frequency ω and speed c. The shape functions in Eq. (3.17), along with Eq. (3.18), with a single wave number k can be used for all the infinite elements.

However, for soil vibration problems, there exist three different types of waves, i.e., the Rayleigh (or surface) wave, longitudinal or dilatational wave (P-wave) and transverse or shear wave (S-wave). It is unlikely that the wave number k selected for each infinite element can duly represent the three different types of waves. In this regard, a reasonable approach is to consider only the dominant waves in each region of the domain (i.e., far field) extending to infinity. Based on the fact that Rayleigh waves are dominant near the free surface, and that the body waves are dominant in soils of greater depths (Gutowski and Dym 1976), it is suggested that for the case of an elastic half-space subjected to vertical *line loads*, for which the longitudinal waves will be dominant beneath the surface, the Rayleigh, shear and longitudinal waves numbers be used respectively for the three regions I, II and III indicated in Fig. 3.5. Besides, it is suggested

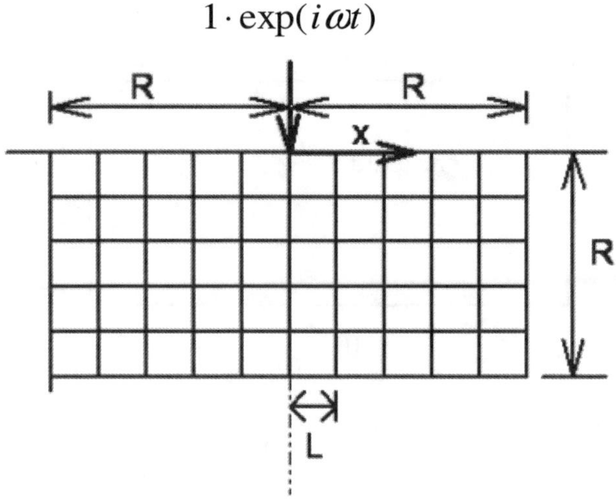

Fig. 3.6 Finite element mesh.

that the decay parameter α be set to 0 for region I and to $1/(2R)$ for regions II and III.

3.4 Mesh Range and Element Size

In order to obtain accurate dynamic results for an elastic half-space problem, it is essential that the finite element mesh used in the modeling be capable of simulating all the wave patterns involved. The purpose of this section is to conduct some convergence tests and to draw some guidelines from such tests for determining the element size and finite element mesh.

For the present purposes, let us consider the case of an elastic half-space subjected to a unit harmonic line load in Fig. 3.6. Let R denote the half-width of the finite element mesh used to model the half-space, and L the length of each element of which the mesh is composed. In theory, as the mesh size is large enough, the displacements computed for the half-space should converge to some values. In the numerical test, the half-width R is allowed to vary from $0.5\lambda_s$ to $5\lambda_s$, where λ_s is the length of the shear wave. Since the mesh with $R = 5\lambda_s$ covers a range that is wide

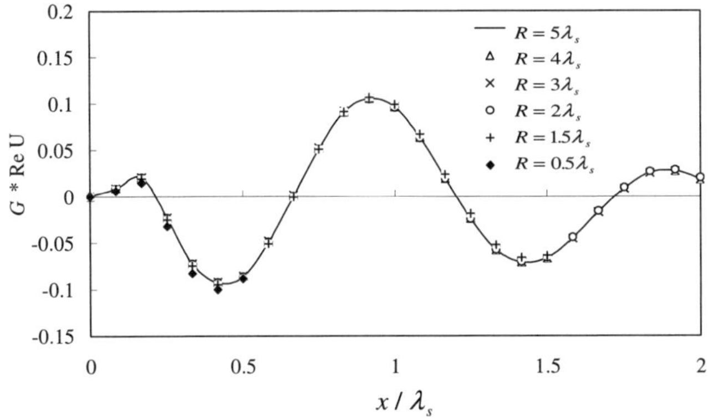

(a)

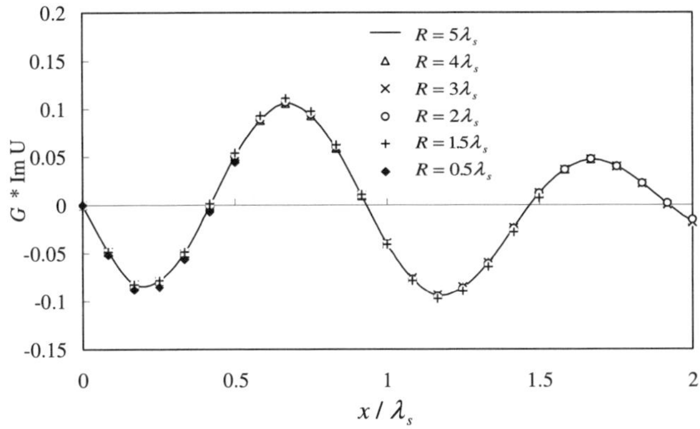

(b)

Fig. 3.7 Effect of mesh size *R* on horizontal displacement: (a) real part, (b) imaginary part.

enough, the solution obtained from this specific case can be used as the reference of comparison.

The horizontal and vertical displacements (multiplied by the shear modulus *G*) computed for the free surface have been plotted in Figs. 3.7 and 3.8, respectively. From these results, it is confirmed that the solution obtained with a half-width of $R = 1.0\lambda_s \sim 1.5\lambda_s$ can be considered as convergent, and that rather slight deviation has occurred with the solution obtained using a smaller width of $R = 0.5\lambda_s$.

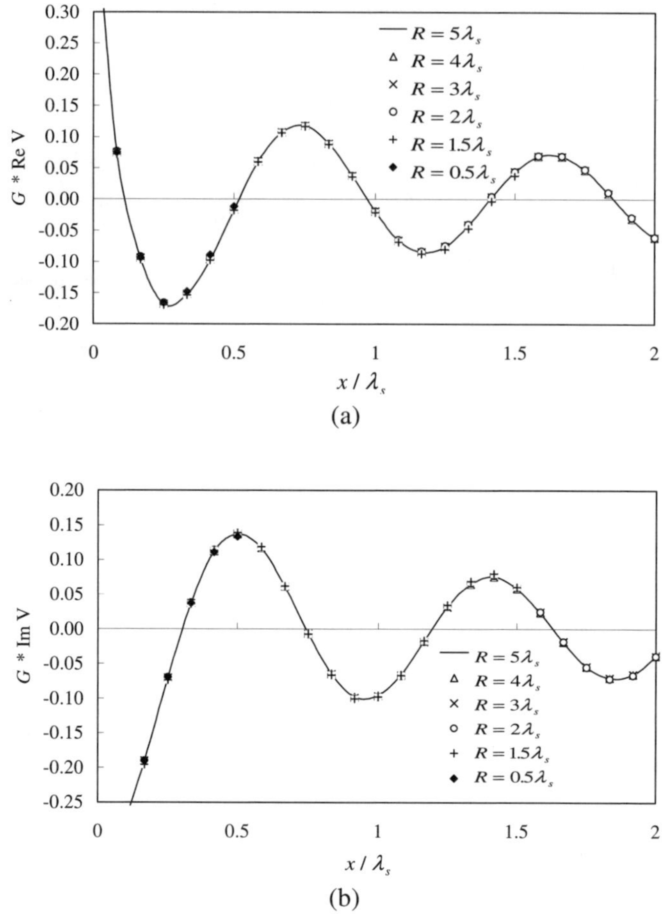

Fig. 3.8 Effect of mesh size R on vertical displacement: (a) real part, (b) imaginary part.

Figures 3.9 and 3.10 show the horizontal and vertical displacements of the free surface obtained using different element sizes L. Here, the solutions obtained with the smallest element size $L = \lambda_s / 20$ can be regarded as the basis of comparison. From these figures, it is concluded that for regions near the source, say, with a distance of less than $0.5\lambda_s$, an element size L of as small as $\lambda_s / 12$ should be used. However, if the response of the vicinity of the source of vibration is not of concern, then an element size of $L \leq \lambda_s / 6$ can generally be used.

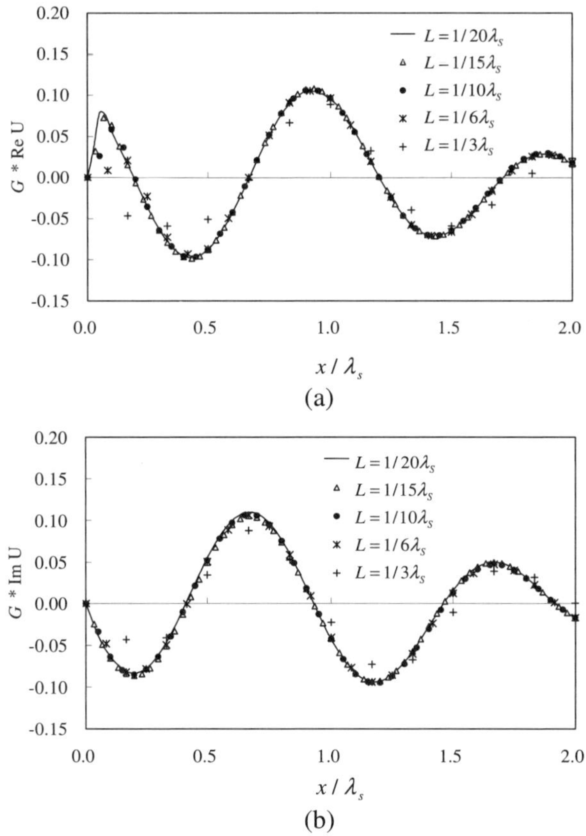

Fig. 3.9 Effect of element size L on horizontal displacement: (a) real part; (b) imaginary part.

From the above numerical studies, the following conclusions can be drawn:

(1) In general, the half-width or range R of the finite element mesh should be selected in the range of $\lambda_s \sim 1.5\lambda_s$ to guarantee the accuracy of solutions.

(2) In the region near the source, i.e., with a distance of less than $0.5\lambda_s$ from the source, the element size L selected should be less than $\lambda_s/12$, and in the remaining region of the half-space, the element size L selected should be less than $\lambda_s/6$. It has been demonstrated that quite accurate results can be obtained using the above element

sizes. It should be added that if we are not interested in the response of the half-space near the source, then the element size limit near the source can be relaxed and allowed to be no greater than $\lambda_s/6$.

3.5 Mesh Expansion by Dynmaic Condensation

As was stated in the preceding section, both the maximum element size L and minimum mesh size R required in the modeling depend on the wavelength λ_s, which in turn depends on the frequency ω, since they

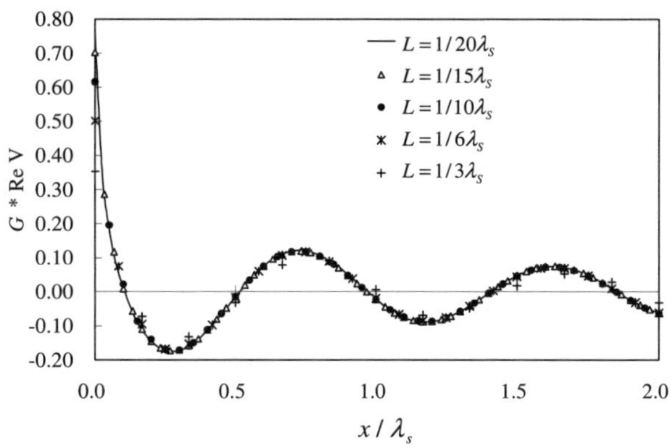

(a)

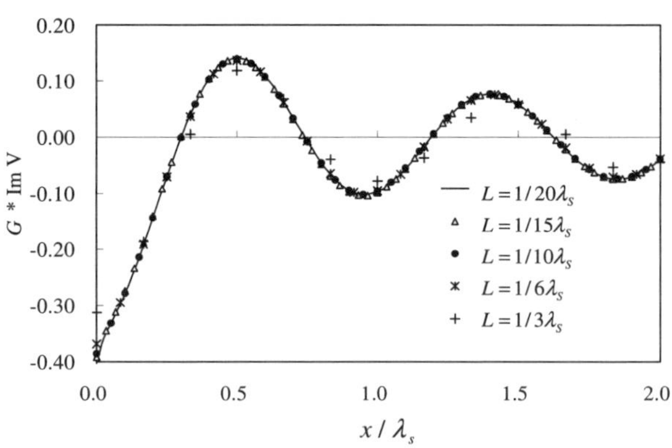

(b)

Fig. 3.10 Effect of element size L on vertical displacement: (a) real part; (b) imaginary part.

are related to each other by the relation $\lambda_s = 2\pi C_s / \omega$, where C_s is the velocity of the shear waves. Hence, for waves of lower frequencies (i.e., longer wavelengths), a finite element mesh of larger size R should be used. On the other hand, for waves of higher frequencies (i.e., smaller wavelengths), an element of smaller size L should be used, which also implies a smaller mesh. If we are only interested in the response of a system under a single frequency, the mesh created for that specific frequency is fixed.

However, for the analysis of half-space problems in the frequency domain, we need to consider the response of the system under each of the frequencies within the range of interest. If we use a mesh pattern for each of the frequencies, it would be highly impractical or uneconomical, because we may need to deal with a total of, say, dozens of mesh patterns. To this end, one alternative is to create a mesh pattern that can meet the diverse requirements for both the maximum element size L (as controlled by the highest frequency) and minimum mesh size R (as controlled by the lowest frequency). Unfortunately, this will usually lead to the use of a very large mesh with a huge number of small elements. As such, the effort required in preparing the mesh and the time spent in solution of the problem is generally tremendous. To resolve this problem, a simple scheme based on the *dynamic condensation* technique will be presented below for preparing the finite/infinite element mesh to meet the diverse demands of the high- and low-frequency waves.

For a two-dimensional problem, the *wave equation* is

$$\left(\frac{\partial^2}{\partial x^2} + \frac{\partial^2}{\partial y^2} \right) \Phi = \frac{\omega^2}{c^2} \frac{\partial^2 \Phi}{\partial t^2}, \tag{3.50}$$

where Φ denotes a scalar, ω the frequency and c the wave velocity. By letting

$$\xi = \frac{\omega x}{c}, \qquad \eta = \frac{\omega y}{c}, \tag{3.51}$$

the wave equation can be transformed to

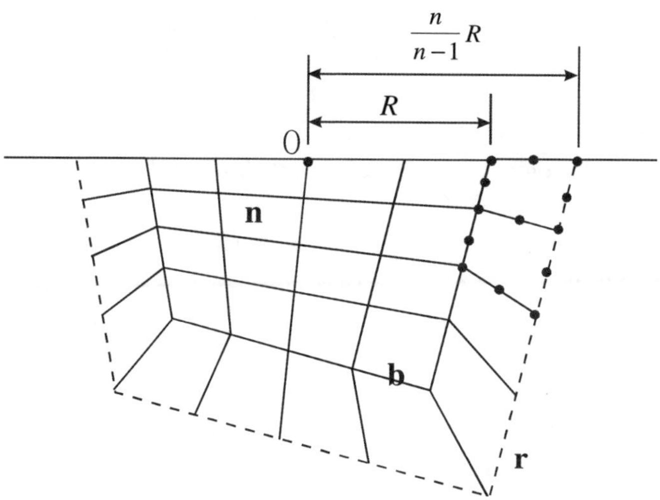

Fig. 3.11 Concept of dynamic condensation for mesh generation.

$$\left(\frac{\partial^2}{\partial \xi^2}+\frac{\partial^2}{\partial \eta^2}\right)\Phi=\frac{\partial^2 \Phi}{\partial t^2}. \tag{3.52}$$

Evidently, the solution to Eq. (3.52) is unique for a given set of ξ and η, or for specific values of $\omega x/c$ and $\omega y/c$. Since the wave velocity c is constant for a given material, it can be ascertained that the far-field impedance matrix [S], from which the solution is derived, is unique for given values of ωx and ωy. Such a property can also be observed from the analytical solutions for the surface displacements given in Eqs. (1.10) and (1.11) for a half-space subjected to a harmonic line load. Since the wave number k is equal to ω/c, the displacements u and v computed from the two equations are same for given value of ωx. In the following, we shall utilize this property to derive the impedance matrices for waves of different frequencies using the same finite/infinite element mesh.

As shown in Fig. 3.11, the symbols **n**, **r** and **b** respectively are used to denote the near field, far field, and the boundary between the near and far fields of an unbounded soil medium. Based on the aforementioned wave property, the far-field impedance computed for $\omega = n\Delta\omega$, where $\Delta\omega$ is a frequency increment and n an integer, at the boundary with a horizontal distance of R from the source (denoted as solid lines) should be equal to the far-field impedance for the neighboring frequency $\omega=(n-1)\Delta\omega$ at the boundary with a distance of $[n/(n-1)]R$ (denoted as dashed lines). In analysis, one may start by calculating the far-field impedance for the highest frequency $\omega=n\Delta\omega$ at the boundary with distance R, and set this impedance matrix as the one for $\omega=(n-1)\Delta\omega$ at the boundary with distance $[n/(n-1)]R$. Then, one can divide the region enclosed by the two boundaries into a number of quadratic quadrilateral (Q8) elements, with the distance between any two adjacent nodes at the outside boundary set equal to $n/(n-1)$ times that of its corresponding distance at the inside boundary. By condensing all the far-field DOFs to the inside boundary **b**, one can obtain the impedance matrix for the next highest frequency $\omega=(n-1)\Delta\omega$. The above procedure can be repeated to yield the far-field impedance matrices for all the remaining frequencies $\omega=(n-2)\Delta\omega$, $\omega=(n-3)\Delta\omega$, etc.

The following is the procedure for dynamic condensation. By letting $\omega=(n-1)\Delta\omega$ for each of the impedance matrices, the following can be written for the system shown in Fig. 3.11:

$$\begin{bmatrix} S_{\mathrm{nn}} & S_{\mathrm{nb}} & 0 \\ S_{\mathrm{bn}} & S_{\mathrm{bb}}+\bar{S}_{\mathrm{bb}} & \bar{S}_{\mathrm{br}} \\ 0 & \bar{S}_{\mathrm{rb}} & \bar{S}_{\mathrm{rr}}+S_{\mathrm{rr}}^{\mathrm{f}} \end{bmatrix} \begin{Bmatrix} v_{\mathrm{n}} \\ v_{\mathrm{b}} \\ v_{\mathrm{r}} \end{Bmatrix} = \begin{Bmatrix} P \\ 0 \\ 0 \end{Bmatrix}, \tag{3.53}$$

where P is the magnitude of external force, v_{n}, v_{b} and v_{r} are the displacement amplitudes associated with the DOFs of the near field, boundary, and far field, respectively, $\bar{S}_{\mathrm{bb}}$, $\bar{S}_{\mathrm{br}}$, $\bar{S}_{\mathrm{rb}}$ and $\bar{S}_{\mathrm{rr}}$ are the impedance matrices generated by the Q8 elements located between the two boundaries with distance R and $[n/(n-1)]R$, and $S_{\mathrm{rr}}^{\mathrm{f}}$ is the impedance matrix of the far field expressed in terms of the boundary nodes at position $[n/(n-1)]R$, which, as was stated above, is equal to the one for $\omega=n\Delta\omega$ at position R.

From the last line of Eq. (3.53), the following can be derived:

$$\{v_{\mathrm{r}}\} = -\left[\bar{S}_{\mathrm{rr}}+S_{\mathrm{rr}}^{\mathrm{f}}\right]^{-1}\left[\bar{S}_{\mathrm{rb}}\right]\{v_{\mathrm{b}}\}. \tag{3.54}$$

Substituting the preceding equation into the second line of Eq. (3.53) yields

$$\left[S_{\text{bn}}\right]\{v_{\text{n}}\}+\left\{\left[S_{\text{bb}}\right]+\left[\bar{S}_{\text{bb}}\right]-\left[\bar{S}_{\text{br}}\right]\left[\bar{S}_{\text{rr}}+S_{\text{rr}}^{\text{f}}\right]^{-1}\left[\bar{S}_{\text{rb}}\right]\right\}\{v_{\text{b}}\}=\{0\}. \tag{3.55}$$

Let the far-field impedance matrix condensed to the boundary **b** be denoted as

$$\left[S_{\text{bb}}^{\text{f}}\right]=\left[\bar{S}_{\text{bb}}\right]-\left[\bar{S}_{\text{br}}\right]\left[\bar{S}_{\text{rr}}+S_{\text{rr}}^{\text{f}}\right]^{-1}\left[\bar{S}_{\text{rb}}\right]. \tag{3.56}$$

Equation (3.53) can be rewritten,

$$\begin{bmatrix} S_{\text{nn}} & S_{\text{nb}} \\ S_{\text{bn}} & S_{\text{bb}}+S_{\text{bb}}^{\text{f}} \end{bmatrix}\begin{Bmatrix} v_{\text{n}} \\ v_{\text{b}} \end{Bmatrix}=\begin{Bmatrix} P \\ 0 \end{Bmatrix}. \tag{3.57}$$

Using the dynamic condensation procedure presented above, in the simulation of a soil-structure system, what one needs to do is to choose a finite/infinite element mesh that best simulate waves of the highest frequency, and then to establish from it the mesh for waves of the next lower frequency (i.e., with decrement $\Delta\omega$). The dynamic condensation is in effect a scheme for expanding the finite/infinite element mesh sizes, to meet the stricter needs of waves of lower frequencies. As the element DOFs located on the expanded boundary are condensed to the original boundary at the position *R*, the present procedure has the advantage that the total number of DOFs is kept constant throughout the process of solution.

3.6 Numerical Examples

This section has the objective of numerically evaluating the applicability of the numerical procedure developed in previous sections. To examine the accuracy of the proposed procedure in solving single-frequency problems, let us consider an elastic half-space subjected to a harmonic line load, also known as one of *Lamb's problems*. The Poisson's ratio is taken as $\nu=0.33$ and a finite element mesh of range $R=6\lambda_s$ and element size $L=\lambda_s/6$ is used.

For a damping ratio of $\beta=0$ and 0.05, the vertical displacements solved by the present procedure for the problem are compared with the analytical solutions (Ewing *et al.* 1957) in Figs. 3.12 and 3.13,

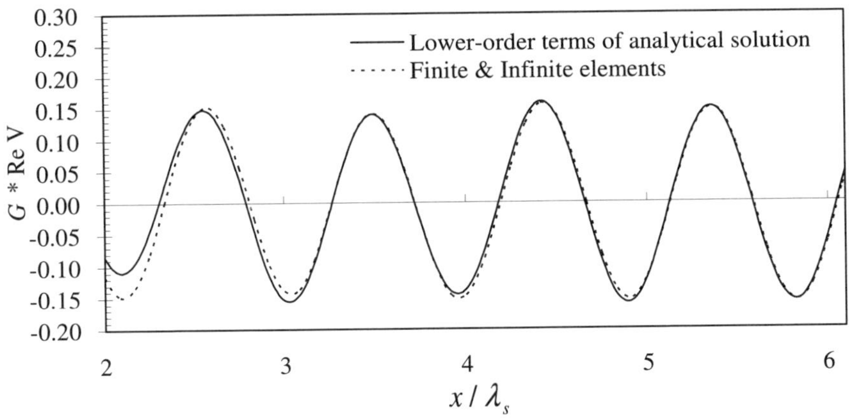

(a)

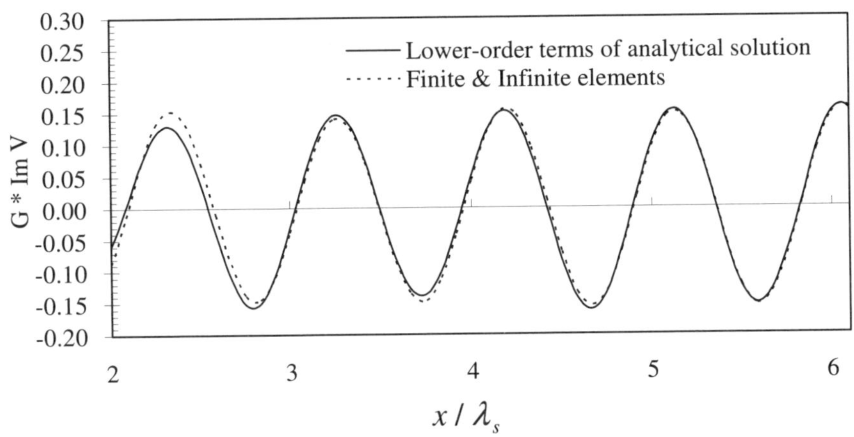

(b)

Fig. 3.12 Vertical displacement for Lamb's problem with $\nu = 0.33$ and $\beta = 0$: (a) real part; (b) imaginary part.

respectively. In these figures, the real and imaginary parts of the vertical displacements (multiplied by the shear modulus *G*) on the free surface of the half-space have been plotted against the non-dimensional horizontal axis x/λ_s for the range 2 to 6, with λ_s denoting the shear wavelength. Since only the first term (which is function of $x^{-3/2}$) has been retained for the analytical solution, the analytical solution is considered to be accurate only for larger *x*, but not in regions near the source. The fact that

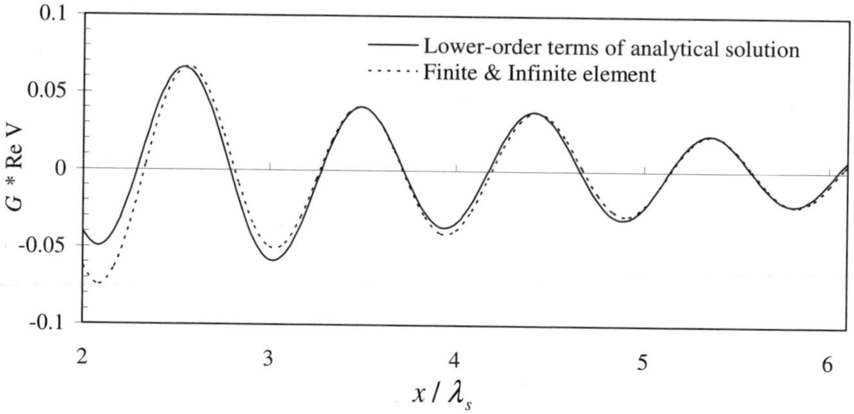

(a)

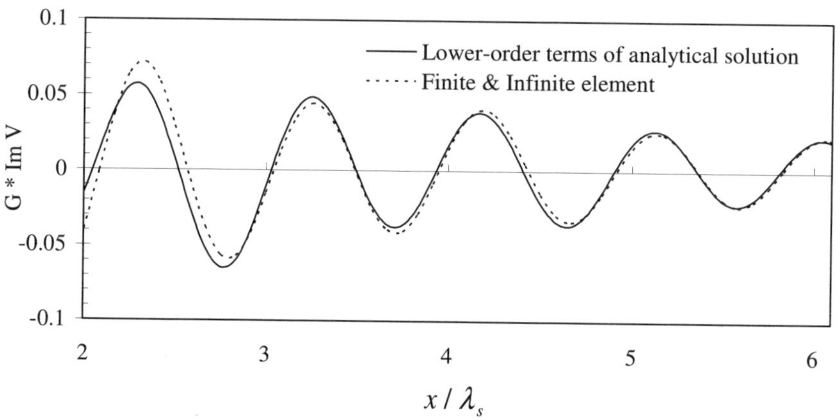

(b)

Fig. 3.13 Vertical displacement for Lamb's problem with $\nu = 0.33$ and $\beta = 0.05$: (a) real part; (b) imaginary part.

the numerical solutions agree very well with the analytical ones for increasing distance x clearly indicates the accuracy of the present procedure.

Next, let us investigate the *dynamic compliance*, which is the inverse of the dynamic stiffness or impedance, of a *massless rigid strip foundation* subjected to a vertical harmonic line load of varying frequencies. The finite element mesh used for waves of the highest frequency is selected as follows: $R = 1.1\lambda_s$, $L = \lambda_s / 15$ for $x < 0.4\lambda_s$ and

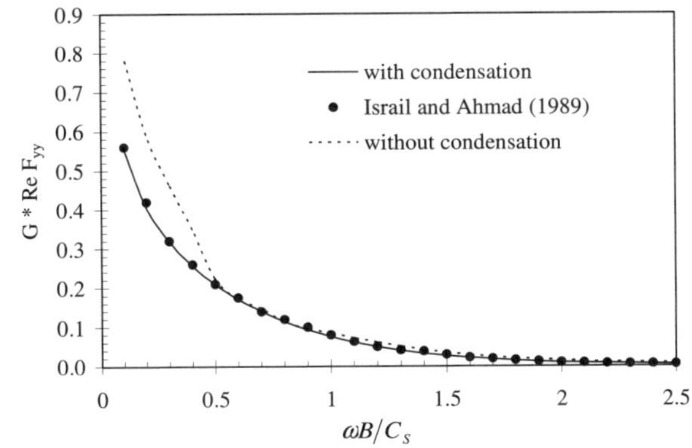

(a)

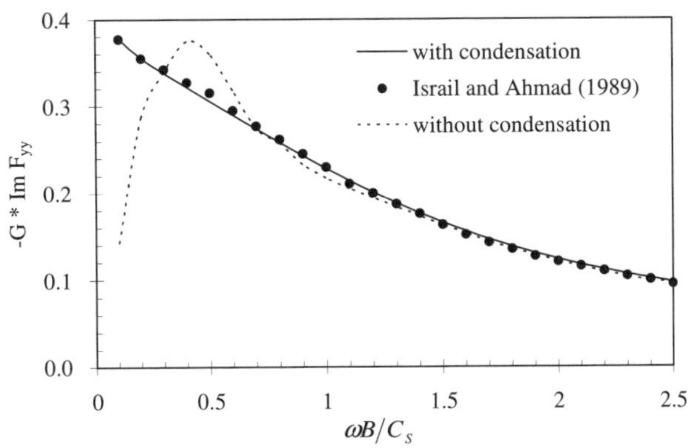

(b)

Fig. 3.14 Vertical dynamic compliance of massless strip foundation: (a) real part; (b) imaginary part.

$L = \lambda_s / 10$ for $x > 0.4\lambda_s$. The results obtained have been compared with those of Israil and Ahmad (1989) in Figs. 3.14(a) and (b), in which the horizontal axis denotes the non-dimensional frequency $\omega B / C_s$ (with B denoting the half-width of the foundation and C_s the velocity of the shear waves) and the vertical axis denotes the real and imaginary parts of the dynamic compliance multiplied by shear modulus G. The solid and dashed lines shown in these figures have been obtained by the present procedure respectively with and without dynamic condensation using a total of 169 elements of the Q8 type. From these figures, one observes that the solution obtained by the procedure with dynamic condensation are in excellent agreement with those of Israil and Ahmad (1989), while those with no condensation deviate significantly from those of Israil and Ahmad (1989) in the low-frequency range. From this example, the effectiveness of the present procedure can be readily appreciated.

3.7 Concluding Remarks

Infinite elements have frequently been used in the simulation of unbounded domains of semi-infinite problems. One characteristic of these elements is the use of a propagation function to represent the effect of amplitude decay of waves traveling to infinity. In this chapter, a procedure based on the property of wave propagation has been proposed for determining the decay parameter involved in the propagation function for waves of different types. Such a procedure is physically more rational than existing procedures. Guidelines for selecting the wave numbers for infinite elements located in different regions of the far field were also discussed. Moreover, the criteria for determining the element size and mesh range to meet the diverse needs of different wavelengths have been thoroughly evaluated.

Of most importance is the presentation of a dynamic condensation procedure for determining the far-field impedance matrices for waves of lower frequencies consecutively from the one selected for waves of the highest frequency. Such a procedure can also be regarded as an automatic mesh expansion scheme for waves of decreasing freqencies, based on the finite/infinite element mesh established for the highest frequency of concern. By such a procedure, the preparation of the finite/infinite element mesh appears to be virtually independent of the frequency content, at least from the point of view of program users. The reliability and accuracy of the proposed method has been demonstrated in the solution of some numerical examples.

Chapter 4

Characteristics of Foundation Vibrations

In this chapter, we shall focus on the vibration of the railway foundations simulated as strip foundations due to the moving trains. The train loads will be simulated as a combination of multi vertical harmonic loads. To begin, we shall study the case of a massless rigid strip foundation subjected to a vertical harmonic load. Such a case is of tutorial value, since it enables us to comprehend the influence of each soil parameter involved, while serving as a reference for verifying the validity of the numerical methods developed in Chapter 3. Later, we shall consider the more practical case of a non-massless elastic foundation subjected to the moving loads.

4.1 Introduction

Foundation is a medium used to support a structure such that the loads acting on the structure can be transmitted to the underlying soils. Thus, a proper understanding of the vibration characteristics of the foundation is essential to study of the general soil-structure interaction problems. Previous researches on the vibration of foundations have been focused on derivation of the dynamic stiffness and compliance of the foundation resting on a uniform half-space. For instance, the vibration of rectangular foundations was studied by Thomson and Kobori (1963). The vibration of infinitely long strip foundations was studied by Abascal and Domínguez (1986) and Israil and Ahmad (1989). Among the foundations that have been studied, *circular foundations* appear to be more frequently considered because of its axial symmetry, which makes it relatively easy to formulate the related theory. Typical previous works on circular

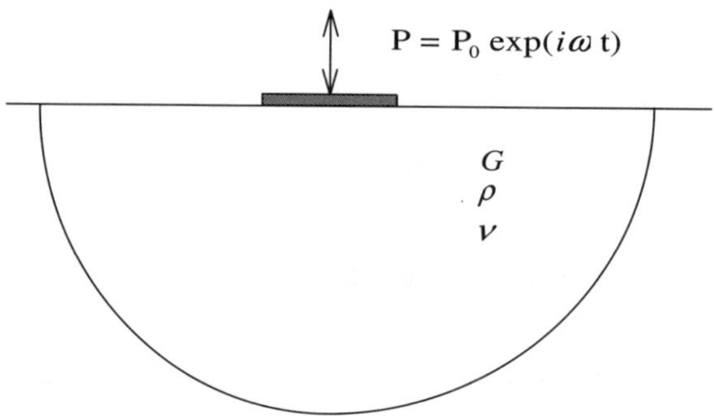

Fig. 4.1 Foundation resting on a half-space subjected to a harmonic load.

foundations include those of Luco and Westmann (1971) and Richart *et al.* (1970), among others.

In this chapter, we shall focus on the vibration of foundations or footings used as a part of railway tracks. They will be simulated as long strip foundations subjected to multi vertical harmonic loads as a representation of the train actions. To identify the key parameters of the underlying soils, we shall consider first a *massless rigid strip foundation* embedded in a uniform half-space subjected to a vertical harmonic load. Then, we shall proceed to deal with the more general case of a *non-massless elastic foundation.*

4.2 Dynamic Stiffness and Compliance of Foundation

Consider a strip foundation embedded in a uniform half-space subjected to a periodic load, as shown in Fig. 4.1:

$$P = P_0 \exp(i\omega t). \tag{4.1}$$

By letting $F(\omega)$ denote the *dynamic compliance* of the foundation, the vibration at the central bottom of the foundation can be written as

$$z(t) = P_0 \exp(i\omega t) F(\omega). \tag{4.2}$$

The dynamic compliance $F(\omega)$ is a complex function, which can be split into a real part $F_1(\omega)$ and an imaginary part $F_2(\omega)$ as

$$F(\omega) = F_1(\omega) + iF_2(\omega). \tag{4.3}$$

The real part $F_1(\omega)$ represents the displacement of the foundation that is in phase with the reaction, and the imaginary part $F_2(\omega)$ the displacement by a phase lag of $90°$.

Dividing the both sides of Eq. (4.2) by $F(\omega)$, one obtains

$$K(\omega)z(t) = P_0 \exp(i\omega t), \tag{4.4}$$

where the *dynamic stiffness matrix* $K(\omega)$ is

$$K(\omega) = \frac{1}{F(\omega)}. \tag{4.5}$$

The dynamic stiffness matrix $K(\omega)$ can also be split into a real part $K_1(\omega)$ and an imaginary part $K_2(\omega)$,

$$K_1(\omega) = \frac{F_1(\omega)}{F_1^2(\omega) + F_2^2(\omega)}, \tag{4.6a}$$

$$K_2(\omega) = -\frac{F_2(\omega)}{F_1^2(\omega) + F_2^2(\omega)}, \tag{4.6b}$$

where the real part $K_1(\omega)$ denotes the stiffness effect of the soil in resisting the external loads, and the imaginary part $K_2(\omega)$ the damping effect of the system, including those of the *geometry damping* (or *radiation damping*) and *material damping* (or *internal damping*).

For a typical linear dynamic system, i.e., for the massless spring-dashpot analog model shown in Fig. 4.2, the equation of motion can be written as follows:

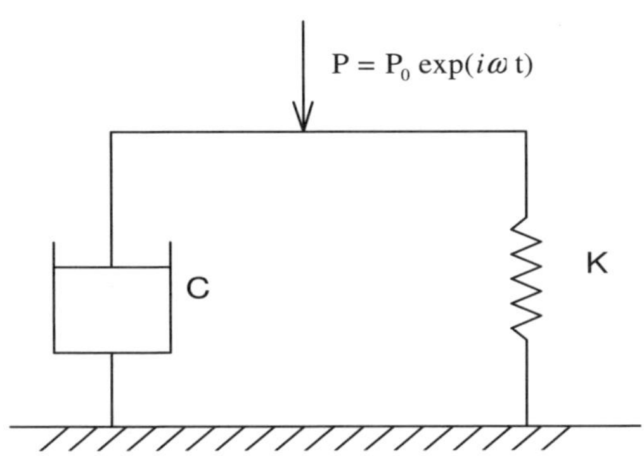

Fig. 4.2 Typical linear dynamic system.

$$C\dot{z}(t) + Kz(t) = P_0 \exp(i\omega t), \tag{4.7}$$

where K is the spring constant and C the damping coefficient. Substituting Eq. (4.2) into the preceding equation yields

$$(i\omega C + K)F(\omega) = 1. \tag{4.8}$$

Further, by substituting Eq. (4.3) for $F(\omega)$ into Eq. (4.8), one can separate the real and imaginary parts of the equation as

$$-\omega F_2(\omega)C + F_1(\omega)K = 1, \tag{4.9}$$

$$\omega F_1(\omega)C + F_2(\omega)K = 0. \tag{4.10}$$

The two parameters K and C can be solved from the preceding two simultaneous equations,

$$K = \frac{F_1(\omega)}{F_1^2(\omega) + F_2^2(\omega)} = K_1(\omega), \tag{4.11}$$

$$C = -\frac{F_2(\omega)}{\omega[F_1^2(\omega) + F_2^2(\omega)]} = \frac{K_2(\omega)}{\omega}. \tag{4.12}$$

Clearly, if the dynamic stiffness $K(\omega)$ is made available, then the soil can be modeled as the spring-dashpot unit shown in Fig. 4.2.

The above derivation has been made for the vertical vibration of the foundation. For the general case, there will be a total of six spring-dashpot units, with each corresponding to one of the six vibration modes, i.e., one vertical, two horizontal, two rocking, and one torsional mode. All the six modes can be described by relations similar to those presented above. For the vibration of soils induced by the moving trains, it is sufficient to consider only the vertical vibration mode, as is the case considered in this chapter.

4.3 Vibration of a Massless Rigid Strip Foundation

Based on the definitions for the dynamic stiffness and dynamic compliance presented in Section 4.2, a parametric study will be carried out for the effect of each of the soil parameters, i.e., Poisson's ratio, damping coefficient, rock depth, and soil shear modulus, on the vibration response (or the stiffness and damping properties) of the massless rigid

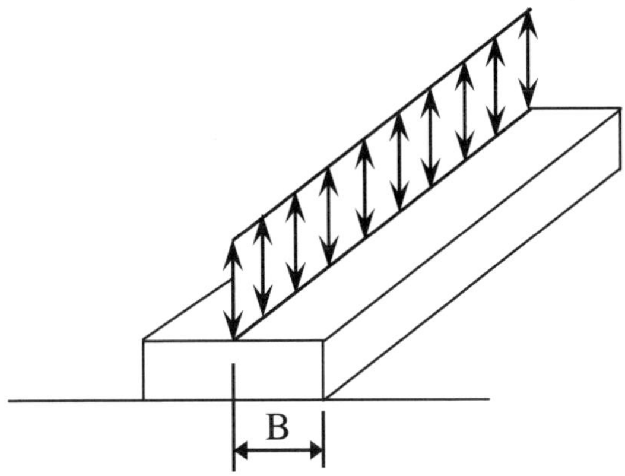

Fig. 4.3 An infinitely long strip foundation under vertical excitation.

strip foundation (Fig. 4.3) subjected to a vertical line load of various frequencies, using the method of analysis presented in Chapter 3.

Regarding the *dynamic compliance* of strip foundations, research has been previously conducted by Israil and Ahmad (1989). In Fig. 3.14 of Chapter 3, the dynamic compliance computed for a massless strip foundation by the 2D finite/infinite element approach has been shown to be in good agreement with that of Israil and Ahmad (1989). However, the frequency range considered by Israil and Ahmad (1989) is only up to $\omega = 2.5C_S / B$, where C_S is the shear wave speed of the soil. For the sake of completeness, a wider frequency range will be considered in the numerical study in this chapter, along with comparisons made for the various parameters involved.

By adopting the dynamic condensation procedure presented in Section 3.5, along with the guidelines for selecting the element size and mesh in Section 3.4, a finite/infinite element mesh with a free-surface width of only up to $4B$ will be used, where B denotes half of the foundation width. In particular, the relatively narrower mesh shown in Fig. 4.4(a) will be used for the uniform half-space, and the one with a wider range in Fig. 4.4(b) for the layered soils. It should be noted that the mesh adopted by Israil and Ahmad (1989) based on the boundary method has a free-surface width of $17B$, which is much wider than the ones used herein.

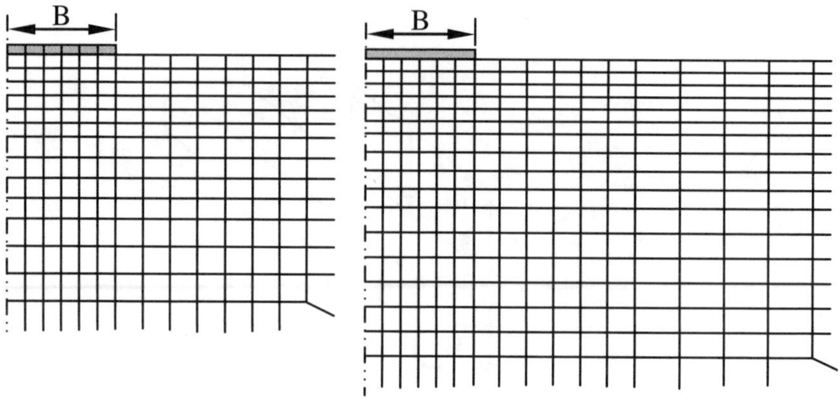

(a) (b)

Fig. 4.4 Finite/infinite element mesh: (a) uniform half-space; (b) layered soils.

The three soil models shown in Fig. 4.5 will be considered. Throughout the analyses in this chapter, the following properties will be adopted for the soil: Poisson's ratio ν = 0.33, and material damping coefficient $\beta = 0.05$, unless noted otherwise. For the case of nonuniform half-space, the depth of the top layered soil is taken as H = 2B, where B denotes half of the foundation width.

To make the computed results independent of the shear modulus of the soils, the shear modulus G of the top layered soil is used as the reference for normalization. In other words, the *dynamic stiffness* is divided by G and the *dynamic compliance* multiplied by G. Besides, the frequency is expressed as a nondimensional frequency parameter a_0,

$$a_0 = \frac{\omega B}{C_s}, \tag{4.13}$$

where C_s is the *shear wave velocity* of the top layered soil.

4.3.1 *Effect of bedrock depth (H/B)*

For the case of a soil layer resting on the bedrock, the incident body waves, i.e., compressional waves and shear waves, will be completely reflected when reaching the bedrock. Resonance will occur when the frequency of the excitational force is equal to the natural frequency of the

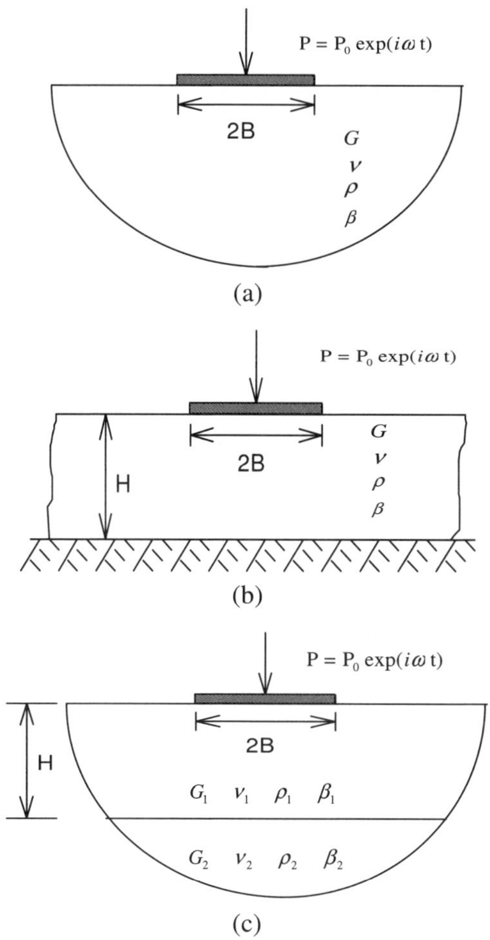

Fig. 4.5 Models of analysis: (a) uniform half-space; (b) soil layer resting on bedrock; (c) layered soils.

soil deposit. According to Wolf (1985), the *resonant frequency* for a vertical incident compressional wave traveling through a uniform single-layered soil is

$$f_c = (2n-1)\frac{C_p}{4H}, \tag{4.14}$$

where C_p is the *compressional wave velocity*, and H is the bedrock depth.

Meanwhile, the *resonant frequency* for a vertical incident shear wave is

$$f_s = (2n-1)\frac{C_s}{4H}, \tag{4.15}$$

where C_s is the *shear wave velocity*.

For the case of vertical excitations, the dominant waves generated are *compressional waves*. Thus, the resonant frequency should be slightly less than f_c, but greater than f_s. When expressed in the nondimensional form, the *resonant frequency* a_{0c} is

$$a_{0c} = (2n-1)\frac{\pi B C_p}{2HC_s}. \tag{4.16}$$

For a soil layer with a Possson's ratio of $\nu = 0.33$, the first (n = 1) nondimensional resonant frequencies for the compressional waves (P-waves) computed for the three bedrock depths of H/B = 1, 2, 3 are a_{0c} = 3.12, 1.56, 1.04, respectively, and the second frequencies are a_{0c} = 9.36, 4.68, 3.12, respectively.

Figure 4.6 show the variation of the *dynamic compliance* for the soil with various bedrock depths, in which part (a) represents the real part of the dynamic compliance, part (b) the imaginary part, and part (c) the displacement amplitude at the central point of the foundation bottom. As can be seen, the resonant frequencies correspond generally well to, but slightly lower than, the computed values a_{0c}. Moreover, the resonant frequencies decrease, while the resonant peaks increase, as the bedrock depth increases. For the case of a semi-infinite half-space, i.e., with $H/B = \infty$, no resonance occurs, as is expected.

Two forms of energy dissipation mechanism exist for the half-space. One form is through the *radiation damping*, caused by the transmission of energies by body and surface waves from the source to the boundaries in infinity. This is also regarded as the geometric attenuation effect of waves traveling to infinity. For fully elastic half-space, the geometric attenuation is the only mechanism for energy dissipation. The other form is through the *material damping*, which is caused by the cyclic oscillations of the soil particles located mainly beneath the foundation when subjected to external excitations. Such a form of energy dissipation

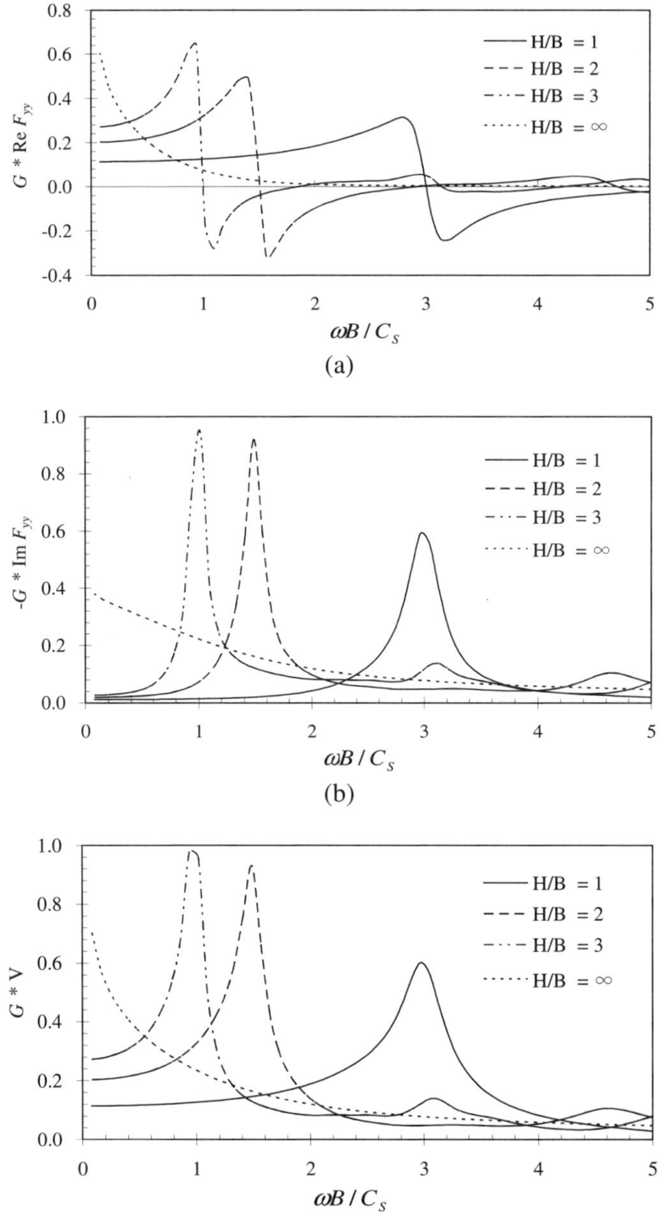

Fig. 4.6 Effect of bedrock depth on dynamic compliance: (a) real part; (b) imaginary part; (c) displacement amplitude at the central bottom of foundation.

is also known as *internal damping*, which will be prescribed by the *hysteretic damping ratio* β in this chapter.

For a soil deposit resting on the bedrock, the body waves transmitting through the deposit will be reflected when reaching the bedrock. No energy can be dissipated due to the geometric restraint for vertically traveling waves. For this problem, the vibration energy of the soil is dissipated either through material damping or through the radiation damping of surface waves traveling horizontally to the infinite boundaries on both sides of the deposit.

It should be noted that if the excitational frequency is smaller than the natural frequency of the soil deposit, no surface waves can occur (Dobry and Gazetas 1988). This has been referred to as the *cut-off effect of damping* for soil deposits resting on the bedrock. For this case, the only mechanism of energy dissipation is material damping. Furthermore, for the special case of an elastic soil deposit, i.e., with zero material damping, virtually no energy can be dissipated.

The above phenomenon can be explained using Fig. 4.7(b) for the imaginary part of the *dynamic stiffness* of the soil model. When the excitational frequency of the external force is smaller than the natural frequency of the soil, as indicated by the flat parts of the curves for the soil deposits, the effect of damping is very small, but not equal to zero, due to existence of material damping. For the special case of a semi-infinite half-space, i.e., with $H/B = \infty$, there exists no cut-off effect of damping, since no resonance can occur within the half-space. The other phenomenon observed from Fig. 4.7(b) is that when the frequency of the external force is greater than the natural frequency of the soil, the damping coefficient oscillates around the curve for the case with semi-infinite half-space, i.e., with $H/B = \infty$.

From Fig. 4.7(a) for the real part of the dynamic stiffness, one observes that for the semi-infinite half-space with $H/B = \infty$, the dynamic stiffness curve appears as a horizontal line, generally independent of the excitational frequencies. However, for the cases of soil deposits, the dynamic stiffness curves oscillate around the curve for the semi-infinite half-space with $H/B = \infty$. Of interest is the correlation of the oscillation behaviors observed in Figs. 4.7(a) and (b) respectively for the stiffness and damping properties of the system.

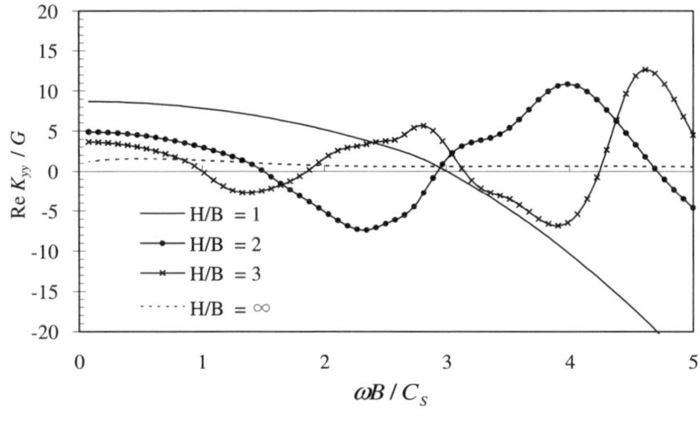

(a)

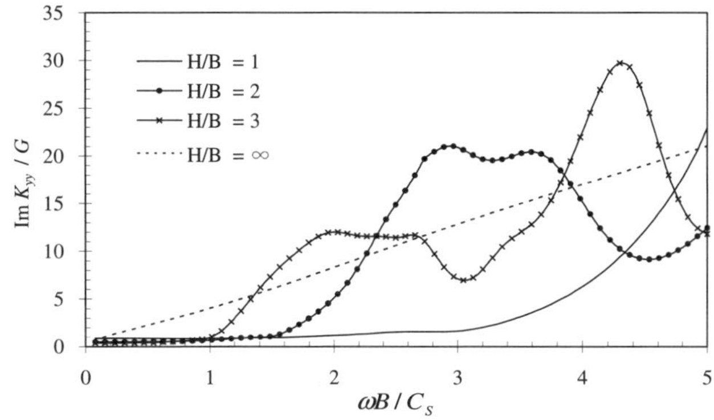

(b)

Fig. 4.7 Effect of bedrock depth on dynamic stiffness: (a) real part; (b) imaginary part.

4.3.2 *Effect of shear modulus ratio (G_1/G_2) of soil layers*

In this section, the depth of the bedrock is assumed to be $H/B = 2$. We shall evaluate the variation of the dynamic compliance and stiffness, as shown in Figs. 4.8 and 4.9, respectively, for the cases with the shear modulus ratio equal to G_1/G_2 = 1.0, 0.5, 0.25, 0.05, and 0.0. There are two extreme cases. One is the case with G_1/G_2 = 0.0, which represents a soil deposit resting on the bedrock, and the other is the case with G_1/G_2 = 1.0, which represents a uniform half-space. For all the other cases, the bottom layer is assumed to be harder than the top layer.

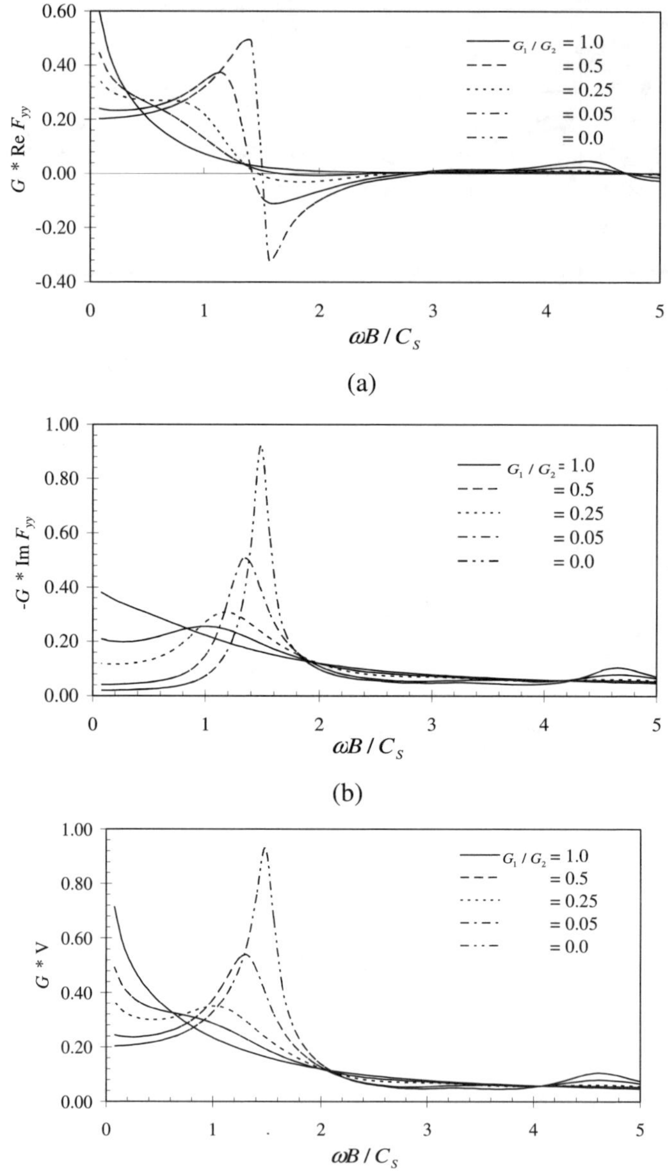

Fig. 4.8 Effect of shear modulus ratio on dynamic compliance: (a) real part; (b) imaginary part; (c) displacement amplitude at the central bottom of foundation.

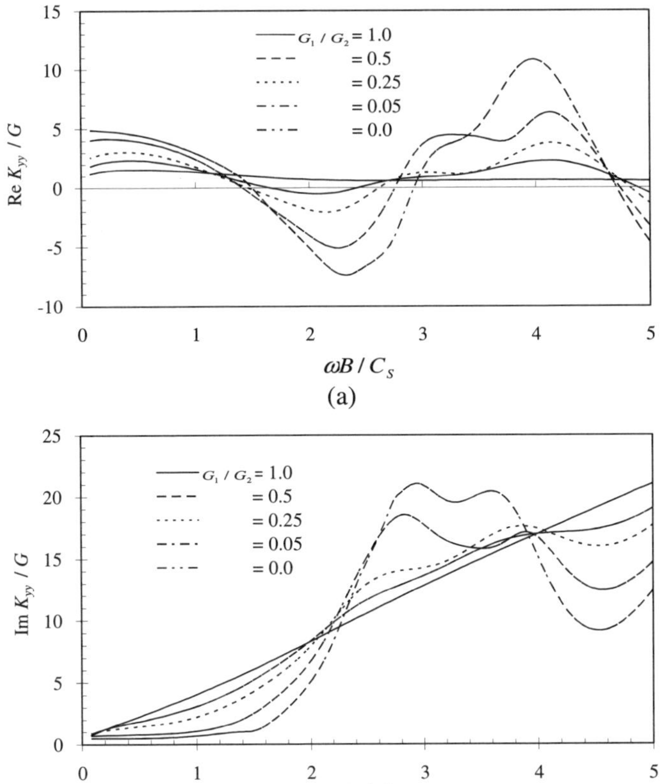

(b)

Fig. 4.9 Effect of shear mass ratio on dynamic stiffness: (a) real part; (b) imaginary part.

From Fig. 4.8, one observes that resonance occurs in each case, except for the uniform half-space. Furthermore, both the frequency and amplitude of the resonant peaks decrease as the shear modulus ratio G_1 / G_2 of the soil increases. This can be attributed to the fact that for two adjacent soil layers with different shear modulus ratios, the wave speed ratios in the two layers are different, and the portions energy carried by the reflected and refracted waves are also different. For the special case of a soil deposit with bedrock, i.e., $G_1 / G_2 = 0$, all the waves are reflected and therefore the peak reaches its maximum. An increase in the shear modulus ratio G_1 / G_2 implies a drop in the wave speed of the bottom layer and an increase in the equivalent depth of the bedrock, which implies a drop in the peak frequency.

Figure 4.9 shows the effect of shear modulus ratio on the dynamic stiffness of the soil. From Fig. 4.9(a), we observe that as the shear modulus ratio G_1/G_2 of the soil layers decreases, the real part of the dynamic stiffness oscillates around the curve for the uniform half-space, and becomes roughly equal to those of the latter at the first and second resonant frequencies. From Fig. 4.9(b), we observe that the *cut-off effect* of damping occurs when the excitational frequency is less than the first resonant frequency, as indicated by the flat portions of the curves on the left-hand side. However, as the shear modulus ratio G_1/G_2 increases, such a phenomenon becomes less obvious.

4.3.3 *Effect of Poisson's ratio*

Because vertical vibrations generate primarily compression waves, they will be significantly affected by Poisson's ratio ν of the soils. This can be appreciated from Eq. (2.6). It is easy to verify that as Poisson's ratio ν increases from 0.25 to 0.49, the compression wave speed increases by 4.12 times, and that when $\nu = 0.5$ (for incompressible half-space), the wave speed reaches infinity.

As shown in Fig. 4.10 for a uniform half-space, the increase in Poisson's ratio ν will result in a reduction of the dynamic compliance, due to the increase in restraint to the relatively larger lateral displacements for soils with larger Poisson's ratios.

Figure 4.11(a) shows the real part of the dynamic stiffness for a uniform half-space. For $\nu = 0.25$ and 0.33, basically no difference can be observed for the effect of Poisson's ratio on the dynamic stiffness of the half-space. Moreover, the dynamic stiffness remains approximately the same regardless of the variation in frequency. However, when Poisson's ratio ν equals 0.49, the dynamic stiffness of the half-space drops dramatically and even becomes negative, as the frequency increases. From Fig. 4.11(b) for the imaginary part of the dynamic stiffness, we observe that as Poisson's ratio ν increases, the damping of the half-space increases and is approximately proportional to the frequency.

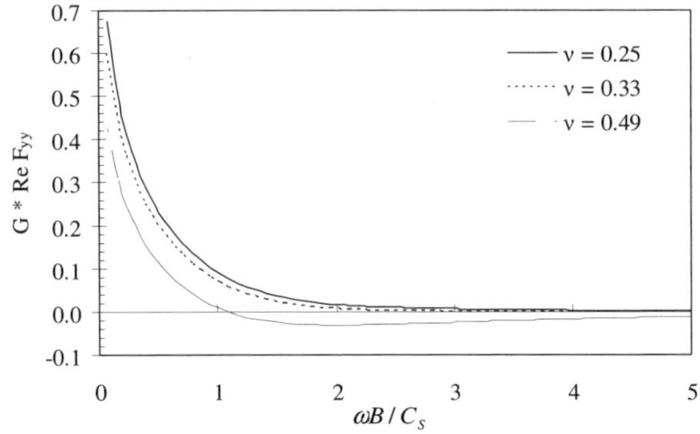

(a)

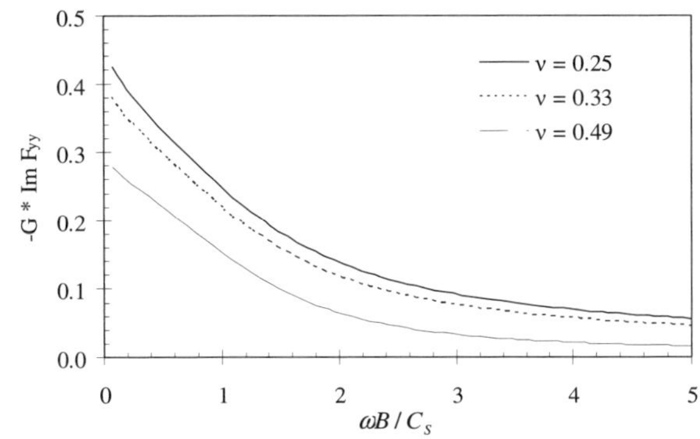

(b)

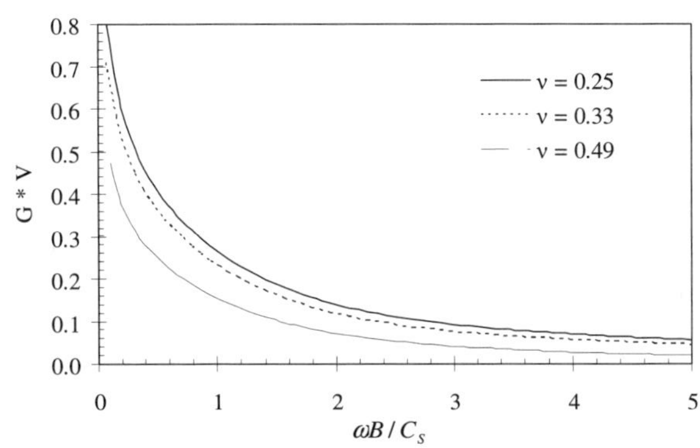

(c)

Fig. 4.10 Effect of Poisson's ratio on dynamic compliance (for uniform half-space): (a) real part; (b) imaginary part; (c) displacement amplitude.

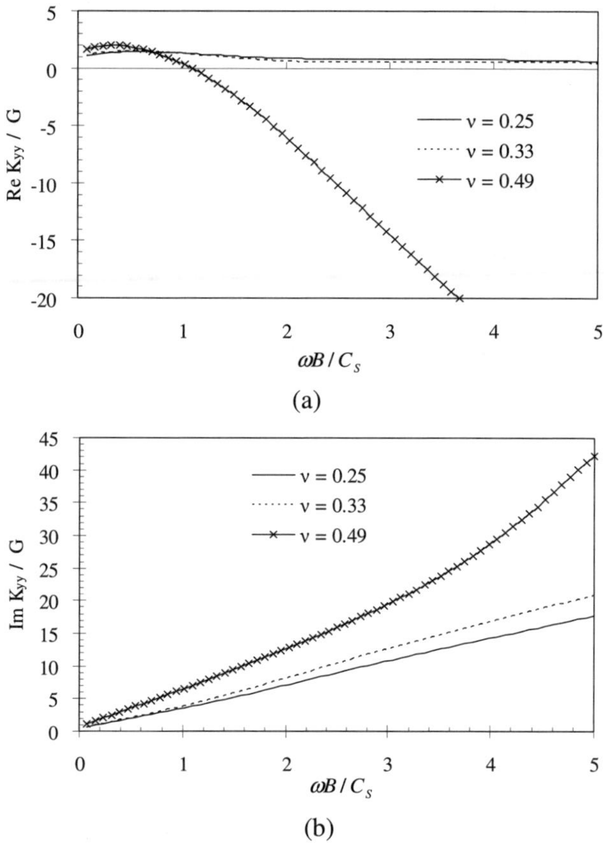

Fig. 4.11 Effect of Poisson's ratio on dynamic stiffness (for uniform half-space): (a) real part; (b) imaginary part.

Let us consider the special case of a soil deposit resting on a bedrock of depth $H/B = 2$. As was stated, an increase in Poisson's ratio ν causes an increase in the compressional wave speed. It follows that the resonant frequency of the soil deposit will be increased as well, as can be verified from Fig. 4.12. However, from the computed results, one also observes that the extent to which the resonant frequency is raised is not proportional to the extent of increase in the compressional wave speed. This indicates that the resonant frequency as given in Eq. (4.14) needs to be modified. In fact, according to Eq. (4.14), for the extreme case of

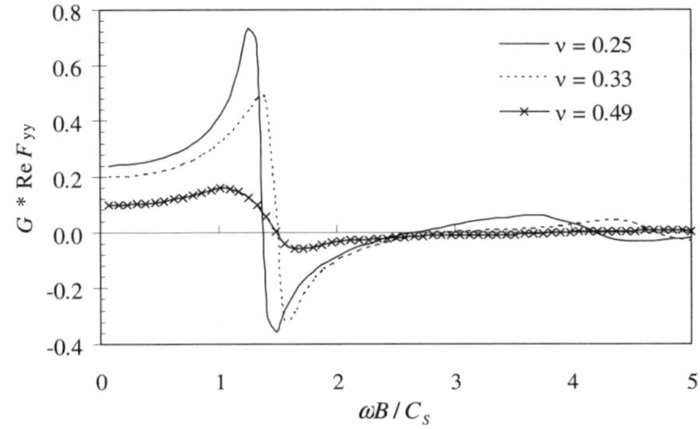

(a)

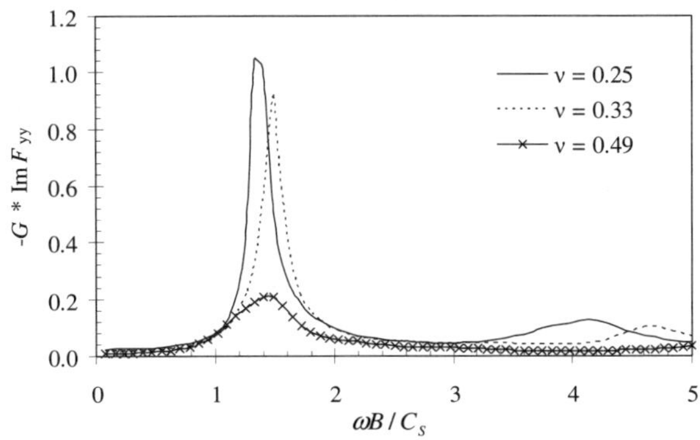

(b)

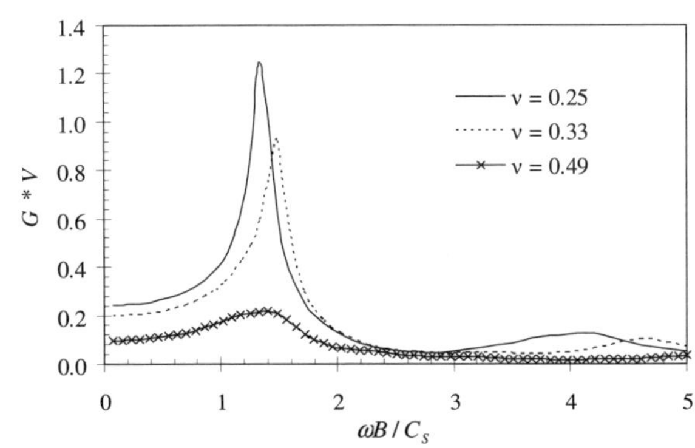

(c)

Fig. 4.12 Effect of Poisson's ratio on dynamic compliance (for soil deposit with bedrock at depth $H = 2B$): (a) real part; (b) imaginary part; (c) displacement amplitude.

$\nu = 0.5$, the frequency predicted is $f_c = \infty$, which can never occur in the physical world. This is another indication for the need to modify Eq. (4.14). Besides, similar to the case of a uniform half-space, an increase in Poisson's ratio will result in reduction of the dynamic compliance, as well as the peak amplitude.

4.3.4 *Effect of material damping ratio*

As can be seen from Fig. 4.13 for a uniform half-space, the influence of material damping ratio β on the *dynamic compliance* is restricted to the low-frequency range. In particular, the real part of the dynamic compliance, as shown in Fig. 4.13(a), decreases slightly with increasing material damping ratio β, while the imaginary part, as shown in Fig. 4.13(b), also increases slightly in response to the increase in material damping ratio β. The displacement amplitude also shows a trend of decrease with the increase in material damping ratio, as shown in Fig. 4.13(c), although within a very slight extent. From the results presented herein, it is concluded that for the uniform half-space, the effect of material damping ratio on the dynamic compliance can be generally neglected in practice.

From Fig. 4.14(a), we observe that for the uniform half-space, an increase in material damping ratio β will result in a drastic decrease of the *dynamic stiffness.* Such a fact has to be considered in engineering practice. On the other hand, from Fig. 4.14(b), it is observed that increasing the material damping ratio β can result in slightly better soil damping effect, though the level of improvement is generally negligible.

Let us now consider the case of a soil deposit resting on a bedrock with depth H/B = 2. The effects of material damping ratio β on the dynamic compliance and dynamic stiffness have been plotted in Figs. 4.15 and 4.16, respectively. From the results in Fig. 4.15, we observe that increasing the material damping ratio will result in a reduction of the dynamic compliance, especially for frequencies near the resonant frequency, as indicated in Fig. 4.15(a), as well as of the peak amplitude, as indicated in Fig. 4.15(c). It should be noted that the resonant frequency remains basically the same regardless of the variation in the material damping ratio β.

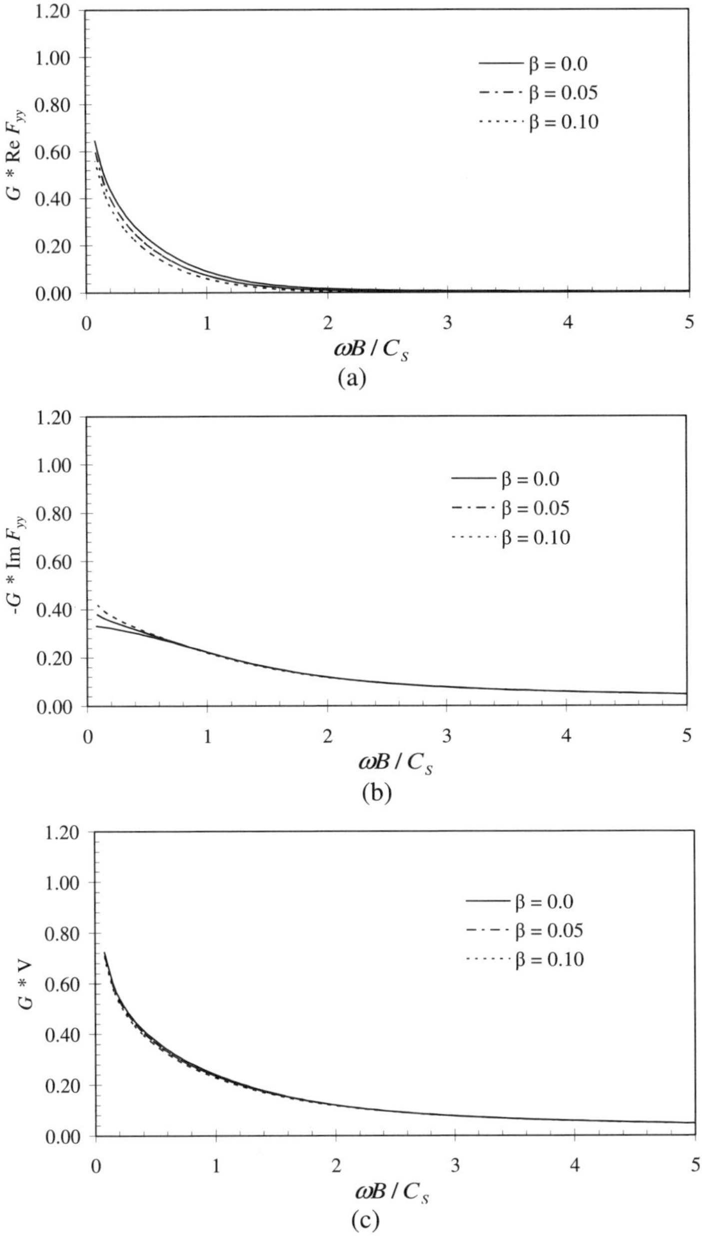

Fig. 4.13 Effect of material damping ratio on dynamic compliance (for uniform half-space): (a) real part; (b) imaginary part; (c) displacement amplitude.

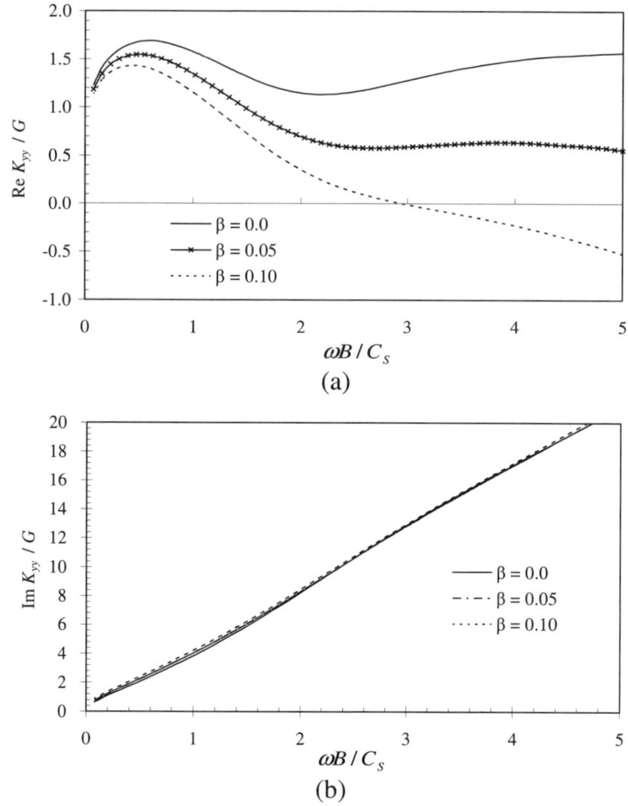

Fig. 4.14 Effect of material damping ratio on dynamic stiffness (for uniform half-space): (a) real part; (b) imaginary part.

As can be seen from Fig. 4.16(a), for the low frequency range, the material damping ratio has basically no influence on the dynamic stiffness. However, for the high frequency range, some oscillation in the dynamic stiffness curve may occur, especially when the damping ratio is very low. On the other hand, from Fig. 4.16(b), one observes that the *cut-off effect* exists clearly for the fully elastic case, i.e., for the case with zero damping ratio (β = 0.0). However, such an effect becomes not so obvious as the material damping ratio increases. This can be attributed to the fact that for the case $\beta \neq 0$, there always exists the effect of material damping, even in the absence of radiation damping (i.e., when the vibration frequency is less than the natural frequency).

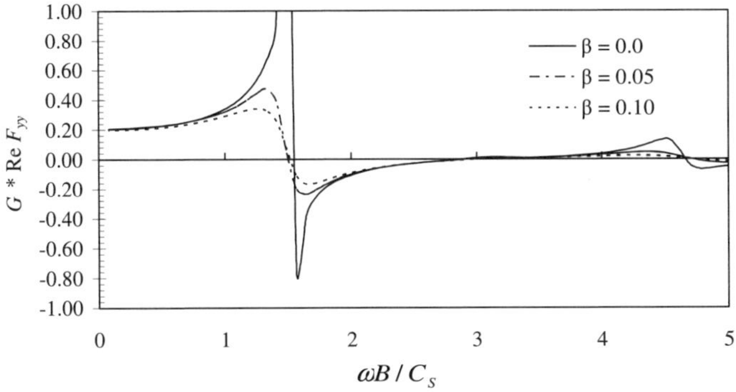

(a)

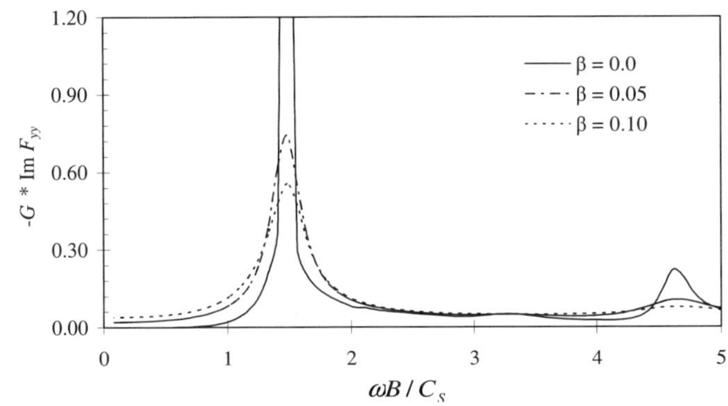

(b)

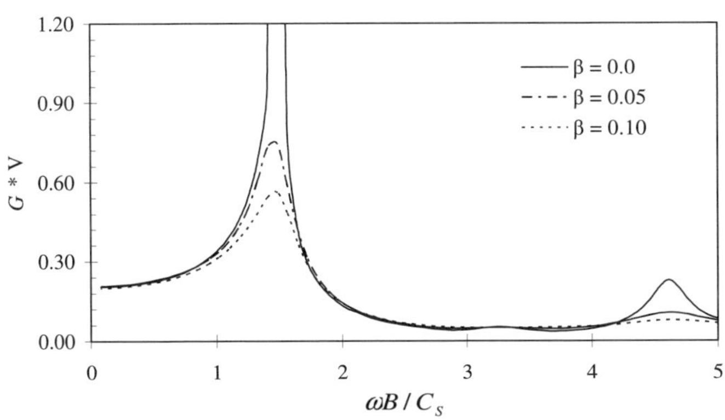

(c)

Fig. 4.15 Effect of material damping ratio on dynamic compliance (for soil deposit with bedrock at depth $H = 2B$): (a) real part; (b) imaginary part; (c) displacement amplitude.

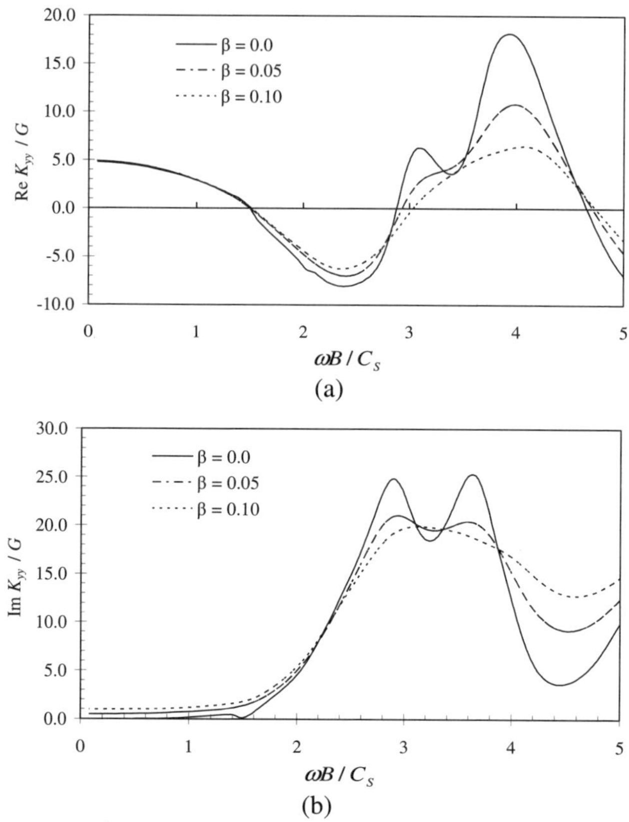

Fig. 4.16 Effect of material damping ratio on dynamic stiffness (for soil deposit with bedrock at depth $H = 2B$): (a) real part; (b) imaginary part.

4.4 Vibration of Rails and Ground under Harmonic Loads

In practice, rails are not fully rigid components. Thus, the assumption of rigid foundations adopted in the preceding section may not be fully realistic. In this section, analyses similar to those of the preceding section will be conducted at first, but with the assumption of *non-rigid strip foundations.* The material properties adopted for the foundation are the same as those for the rails in Laghrouche and Le Houedec (1994). The structural model is shown in Fig. 4.17 and the material properties of the foundation are listed in Table 4.1, in which E is Young's modulus.

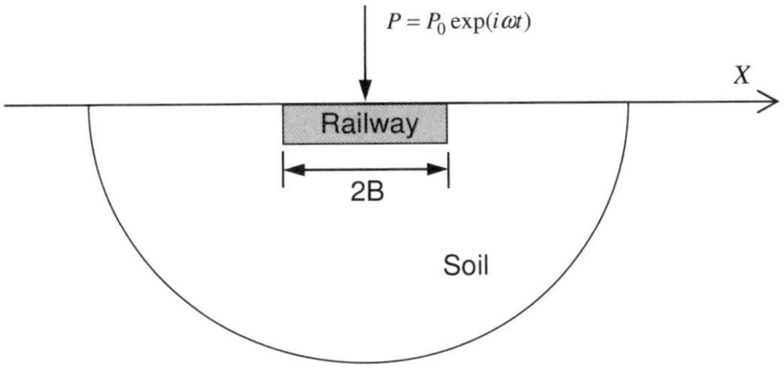

Fig. 4.17 Infinite railway subjected to a harmonic line load.

Table 4.1 Material properties of soil and railway.

	ρ (kg/m^3)	E (MPa)	ν	β (%)
Soil	1800	46	0.25	5
Railway	2400	13500	0.25	2

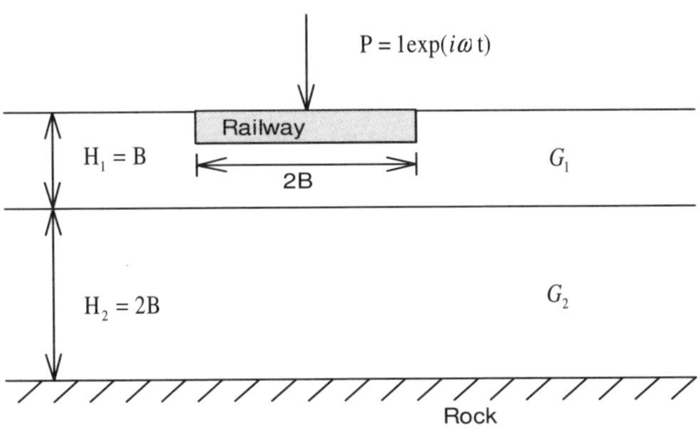

Fig. 4.18 Model of analysis.

However, all the results obtained for all the combinations of parametric values, along with the material data adopted for the foundation, show a trend very close to those of the preceding section, which therefore are omitted herein.

For the second example, we shall study the problem of a double layered soil resting on the bedrock, as shown in Fig. 4.18. The cases considered include two extreme ones, i.e., $G_1/G_2 = 1$ and $G_1/G_2 = 0$, indicating that the bedrock has a depth of 3*B* and *B*, respectively, of which the resonant frequencies are close to the values a_{0c} predicted by Eq. (4.16). As can be observed from Fig. 4.19, by increasing the shear modulus ratio G_1/G_2 of the soil layers from 0 to 1, the resonant frequency decreases significantly, while the peak amplitude increases slightly. Such a phenomenon can be attributed to the fact that the effective depth of the bedrock for the top soil layer approaches 3*B*, following the increase of the shear modulus ratio.

In the preceding section, we focused mainly on the vibration characteristics of the foundation used in supporting the railway track. As far as the environmental vibrations are concerned, one may also be interested in vibrations of the ground surface on the two sides of the railway, where residential buildings or high-precision factories may be located. For this reason, we shall investigate herein the variation of displacement amplitudes of the free surface on the two sides of the railway in the frequency domain. Two extreme cases are considered for the half-space. One is a uniform half-space, and the other is a soil layer resting on the bedrock, as shown in Fig. 4.20. The material properties adopted for the foundation are identical to those given in Table 4.1.

The tri-phase plot in Figure 4.21 shows the amplitude of vertical displacement of the free surface for the case of a uniform half-space. As can be seen, the displacement amplitude of the free surface decreases monotonically with the increase in frequency. For each specific frequency, the amplitude appears to be rather uniform on the foundation, and decreases gradually with the distance from the foundation.

Similarly, the tri-phase plot in Figure 4.22 shows the horizontal displacement amplitude of the free surface. Again, the amplitude decreases with the increase in frequency, but not as significant as that of the vertical displacement. Moreover, the magnitudes of the horizontal displacements are significantly smaller than those of the vertical displacements, say, by a difference of roughly 9 times. For this reason, only the vertical displacements of the ground surface or soils will be presented in latter chapters concerning the isolation of ground vibrations due to moving loads.

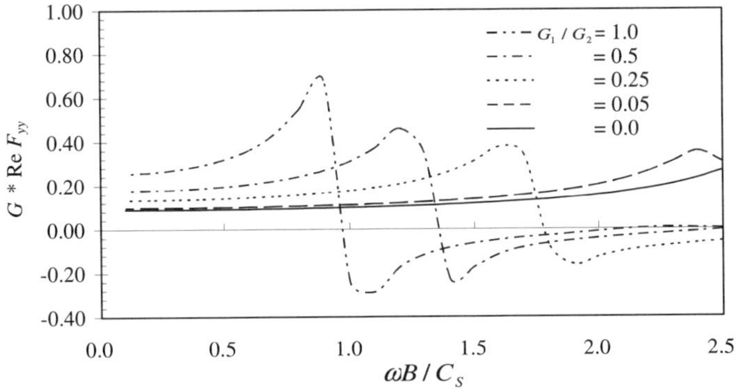

(a)

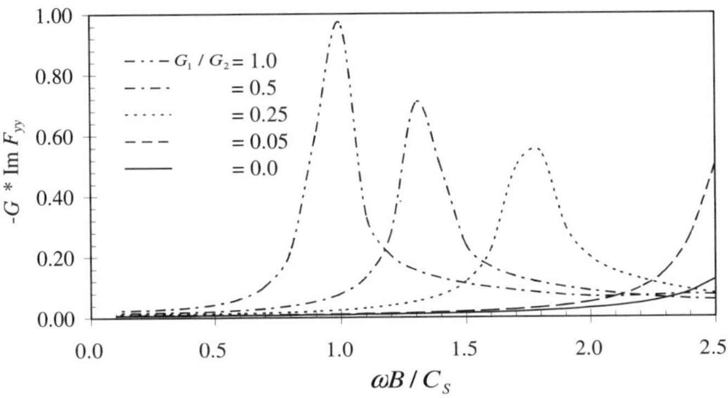

(b)

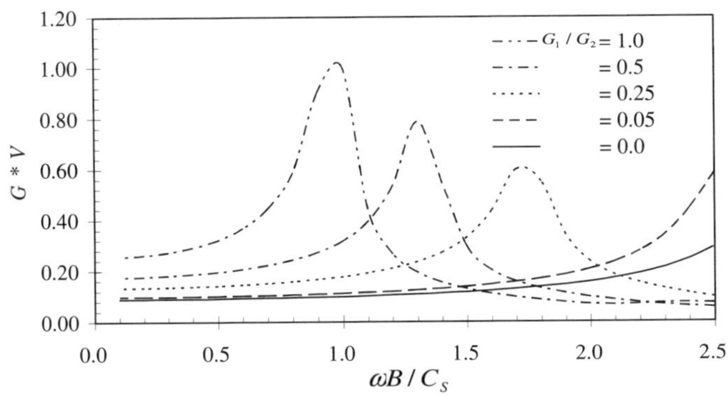

(c)

Fig. 4.19 Effect of shear modulus ratio of soil layers on displacement of railway: (a) real part; (b) imaginary part; (c) displacement amplitude.

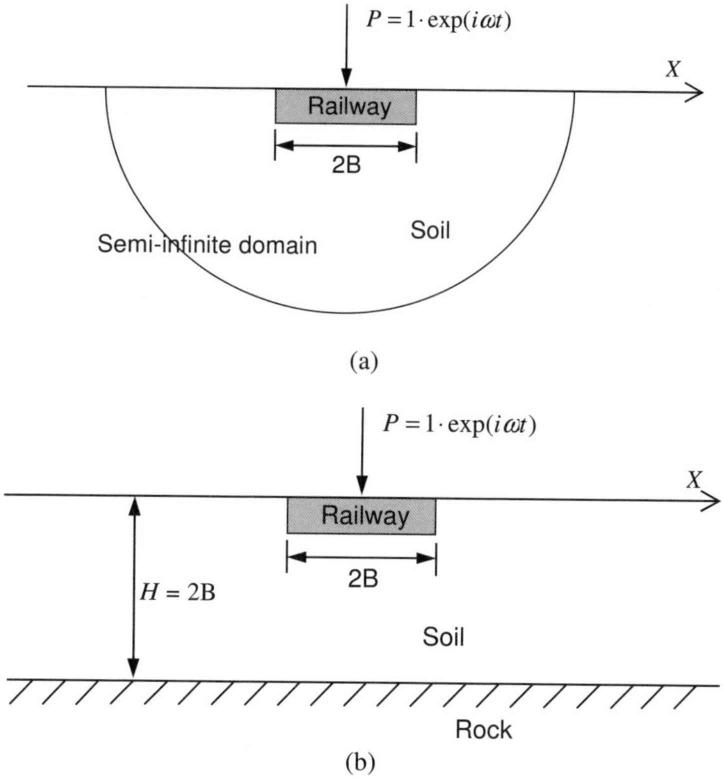

Fig. 4.20 Models of analysis: (a) uniform elastic half-space; (b) soil deposit resting on bedrock.

In contrast, Figs. 4.23 and 4.24 show the amplitudes of vertical and horizontal displacements, respectively, of the free surface for the case of a soil deposit with a bedrock of depth $H = 2B$. Clearly, when the non-dimensional frequency parameter equals 1.5, resonance occurs on both the vertical and horizontal displacements. For frequencies smaller than the resonant frequency, the amplitudes are generally small.

4.5 Application to Practical Problems

All the analyses in the preceding sections have been presented in terms of non-dimensional parameters. Although such results can be interpreted

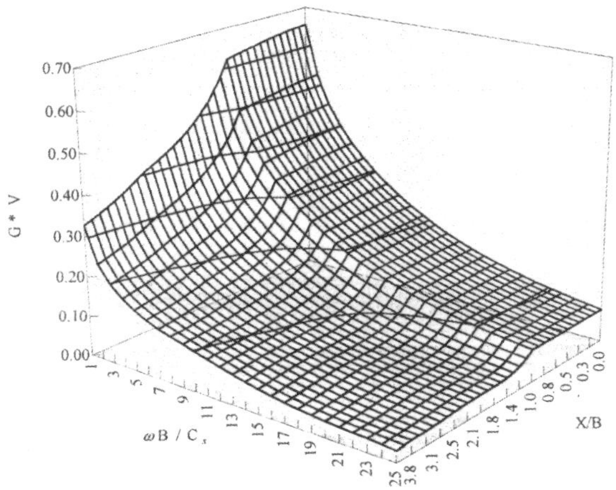

Fig. 4.21 Vertical displacement of free surface under different frequencies (uniform half-space).

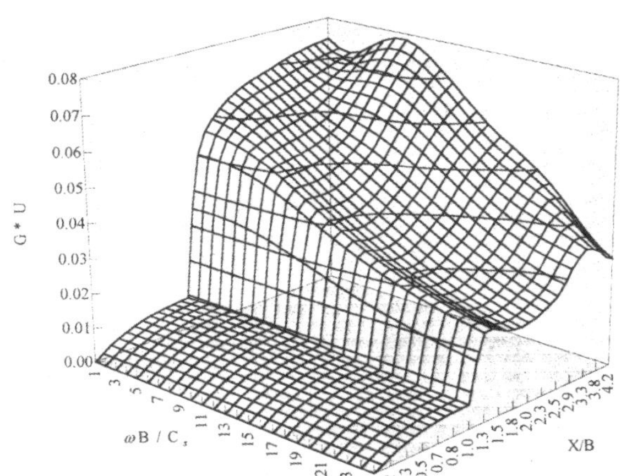

Fig. 4.22 Horizontal displacement of free surface under different frequencies (uniform half-space).

for a wide range of soil vibration problems, the real absolute values of the physical parameters are not directly made available, which makes it difficulty for engineers to have an intuition of the levels of vibrations for real physical problems.

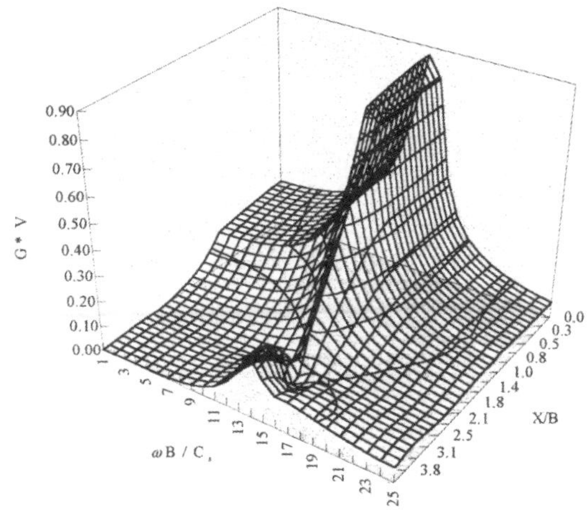

Fig. 4.23 Vertical displacement of free surface under different frequencies (with bedrock).

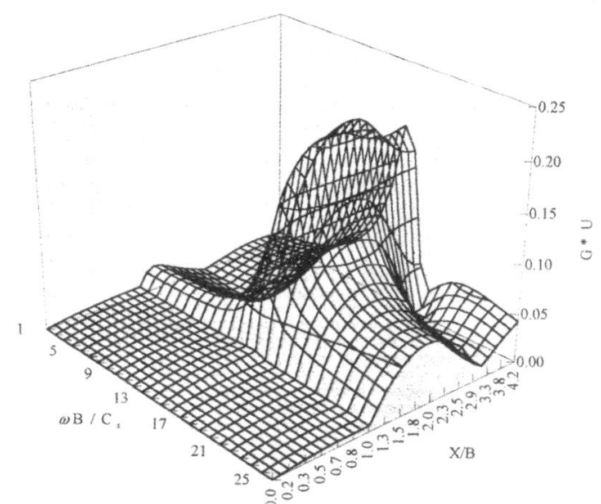

Fig. 4.24 Horizontal displacement of free surface under different frequencies (with bedrock).

In this section, we shall consider two real problems, i.e., problems with the value of each physical parameter explicitly given, and investigate the influence of elastic modulus of the flexible foundation on the displacement, velocity, and acceleration amplitudes of the free

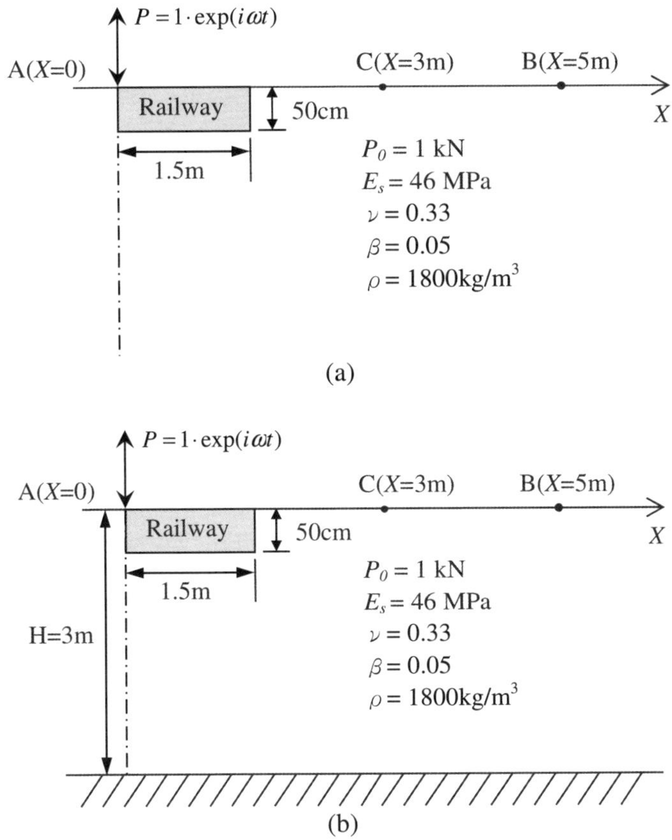

Fig. 4.25 More realistic models: (a) Problem 1; (b) Problem 2.

surface. The two problems considered are the uniform elastic half-space and the soil deposit resting on the bedrock, as shown in Figs. 4.25(a) and (b), respectively. Since the load acting on the railway is assumed to be a unit harmonic line force, the results obtained in the following can be regarded as the *transfer functions* for the displacement, velocity, and acceleration of the soil models in the frequency domain.

4.5.1 *Problem 1: Uniform half-space*

For the uniform half-space shown in Fig. 4.25(a), the vertical responses computed for points *A*, *B* and the horizontal responses computed for

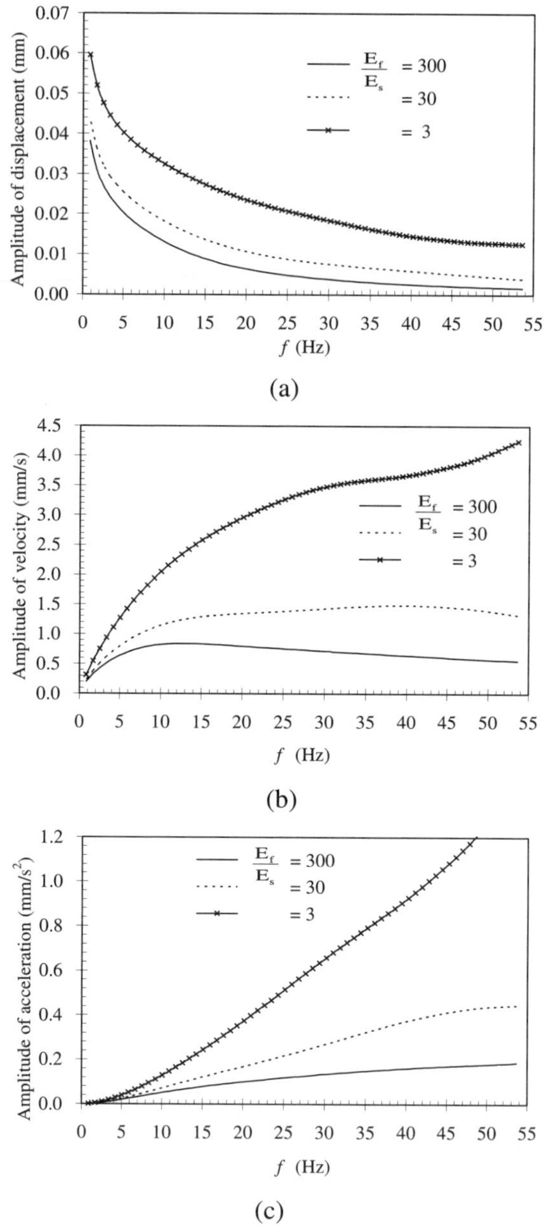

Fig. 4.26 Problem 1: Effect of elastic foundation on vertical responses of point *A*: (a) displacement; (b) velocity; (c) acceleration.

points *C*, *B* on the surface have been plotted in Figs. 4.26-4.27 and Figs. 4.28-4.29, respectively, in which E_s and E_f indicate Young's modulus of the soil and foundation, respectively.

It can be observed that the amplitudes of the soil responses increase as the foundation softens. Thus, the adoption of rigid foundations will underestimate the free surface responses. Besides, the variation of elastic modulus of the foundation has relatively smaller influence on the response amplitudes in the low-frequency range for the locations away from the foundation, such as point *B*. This can be attributed to the fact that for waves of low frequencies, the wavelengths are longer than the foundation width. It follows that the variation in the elastic modulus of the foundation can only affect the local soil response near the source, but not the response of the soil in the neighborhood. The other observation is that the displacement tends to be large in the low-frequency range, but the same is not true with the velocity and acceleration responses. Finally, the amplitude of both the vertical and horizontal accelerations of the free surface tends to be larger for higher frequencies.

For comparison, the horizontal responses of point B have been plotted in Fig. 4.29. By comparing this figure with Fig. 4.27, one observes that the horizontal displacement has a magnitude much smaller than the vertical one, say, by a difference of about 5 times. The velocity and acceleration of the horizontal response are also smaller than those of the vertical response. The other observation from Fig. 4.29 is that larger responses occur for foundations that are more flexible, similar to that for the vertical response.

4.5.2 *Problem 2: Soil deposit resting on bedrock*

The vertical responses computed of points A, B and the horizontal responses of C, B on the soil deposit with bedrock shown in Fig. 4.25(b) have been plotted in Figs. 4.30-4.31 and Figs. 4.32-4.33, respectively. As can be seen, resonance occurs on the responses due to existence of the bedrock, which is different from the case with uniform half-space. The peak vertical amplitudes occur at the frequencies 16 and 50 Hz, which are coincident with the resonant frequencies computed for the P-waves using the one-dimensional wave theory.

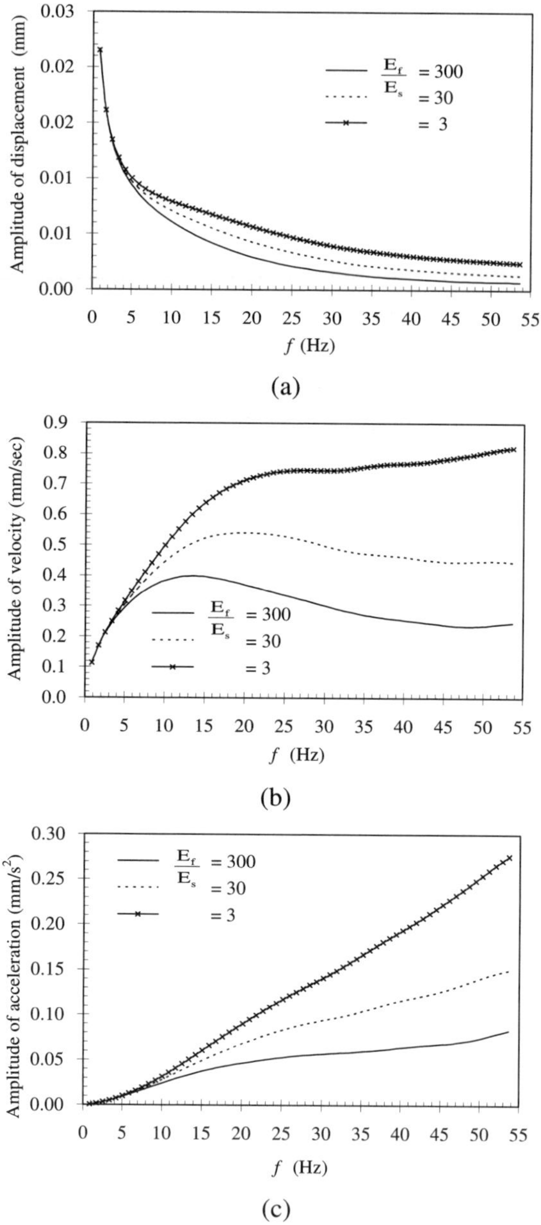

Fig. 4.27 Problem 1: Effect of elastic foundation on vertical responses of point *B*: (a) displacement; (b) velocity; (c) acceleration.

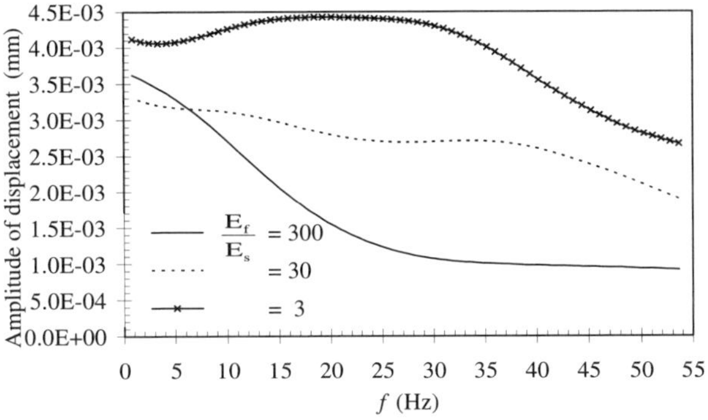

(a)

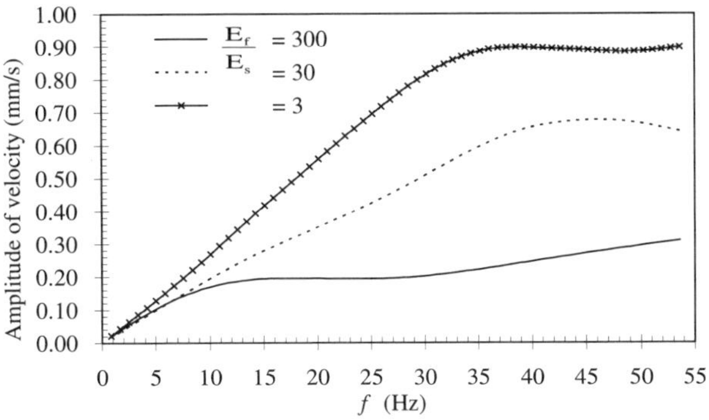

(b)

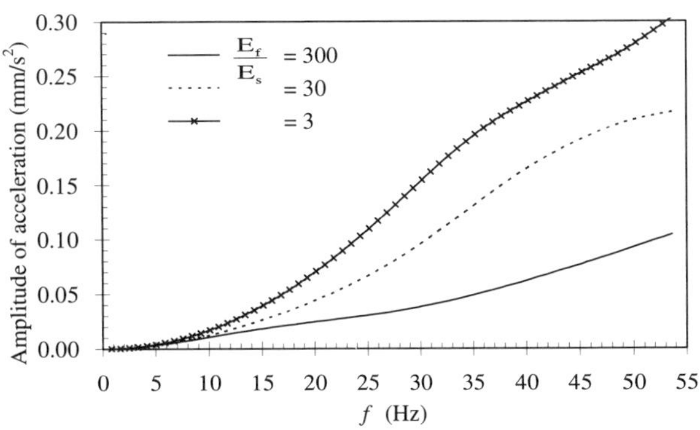

(c)

Fig. 4.28 Problem 1: Effect of elastic foundation on horizontal responses of point *C*: (a) displacement; (b) velocity; (c) acceleration.

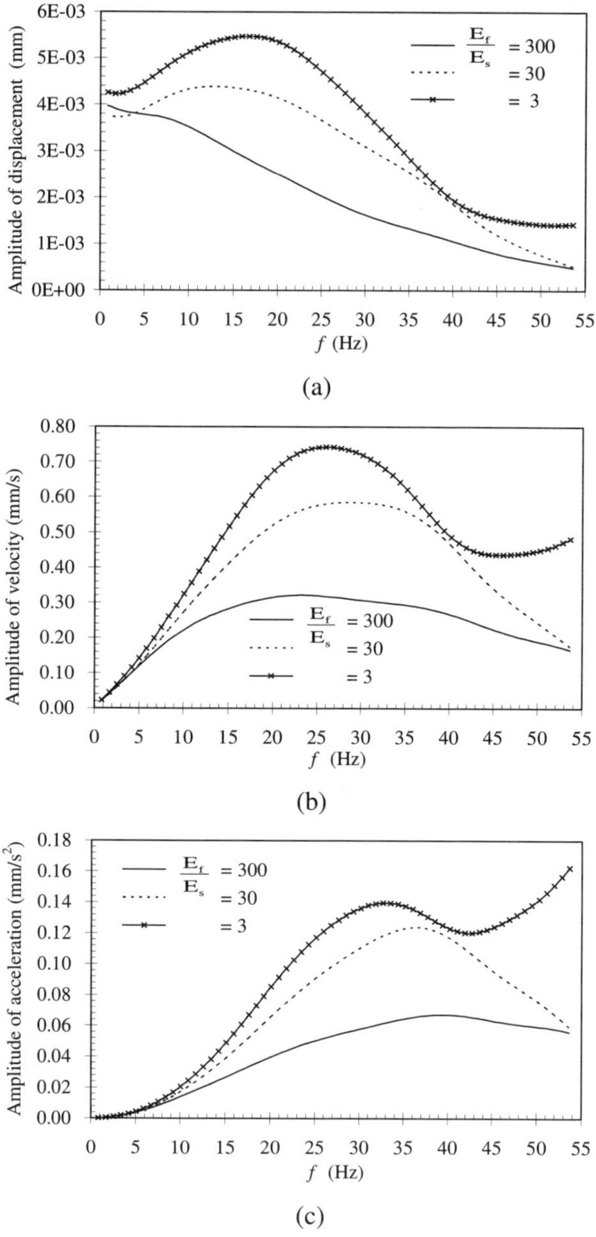

Fig. 4.29 Problem 1: Effect of elastic foundation on horizontal responses of point *B*: (a) displacement; (b) velocity; (c) acceleration.

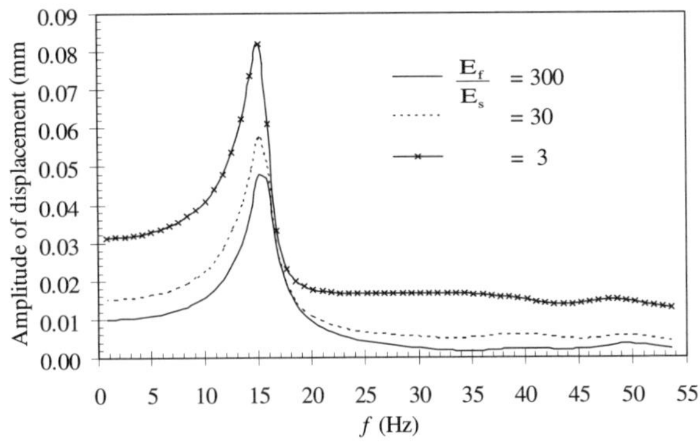

(a)

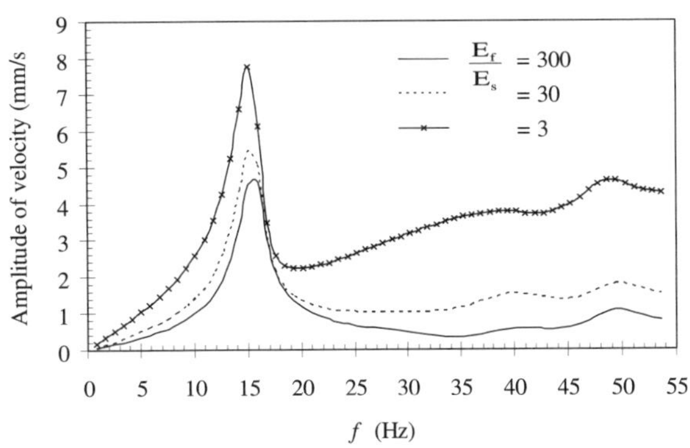

(b)

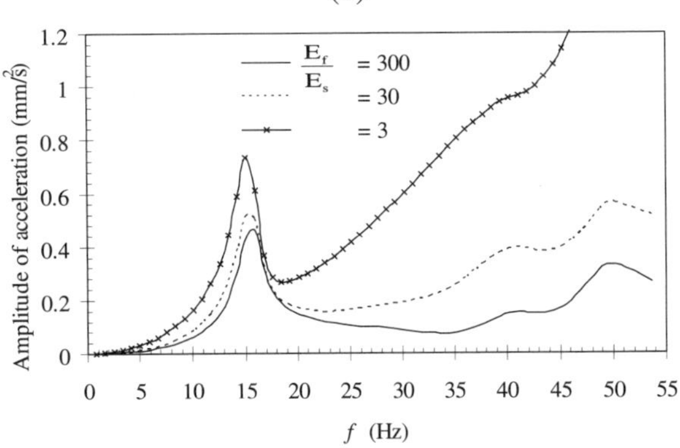

(c)

Fig. 4.30 Problem 2: Effect of elastic foundation on vertical responses of point *A*: (a) displacement; (b) velocity; (c) acceleration.

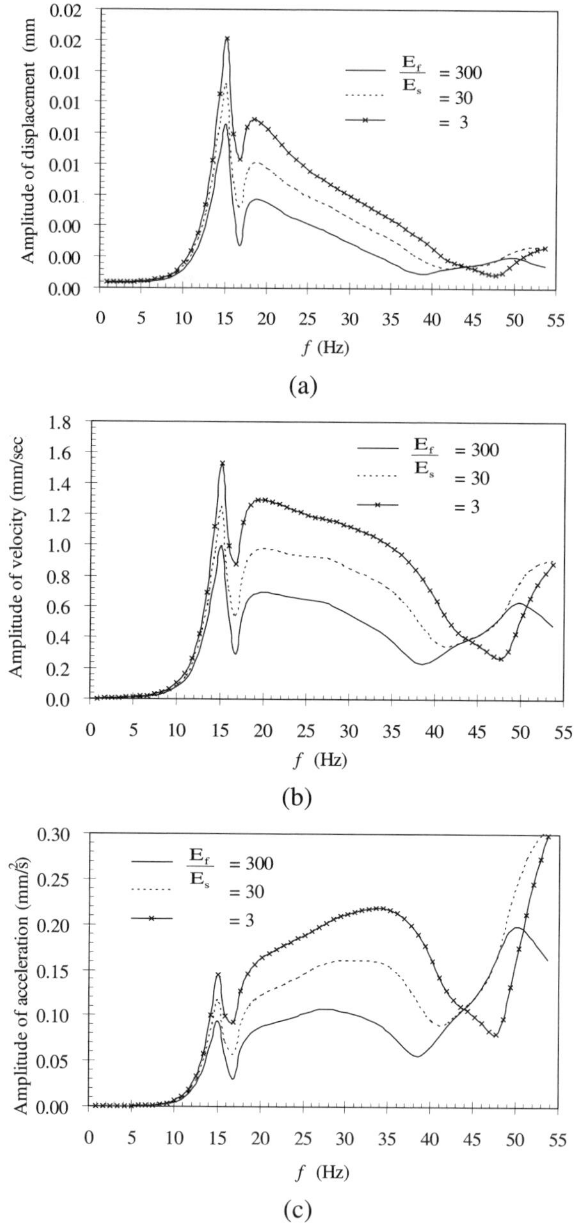

Fig. 4.31 Problem 2: Effect of elastic foundation on vertical responses of point *B*: (a) displacement; (b) velocity; (c) acceleration.

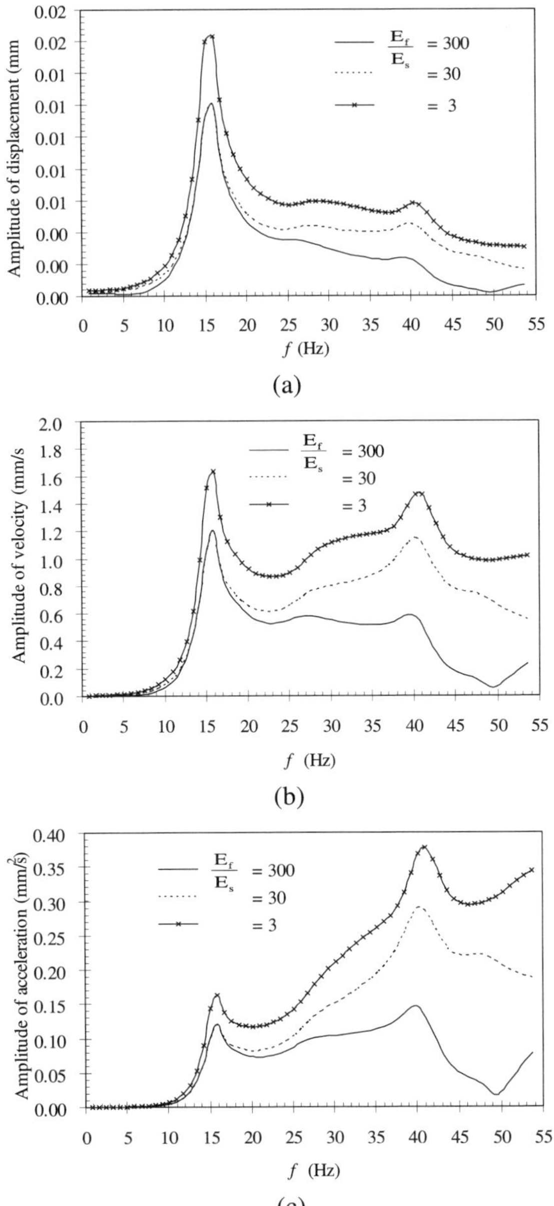

Fig. 4.32 Problem 2: Effect of elastic foundation on horizontal responses of point *C*: (a) displacement; (b) velocity; (c) acceleration.

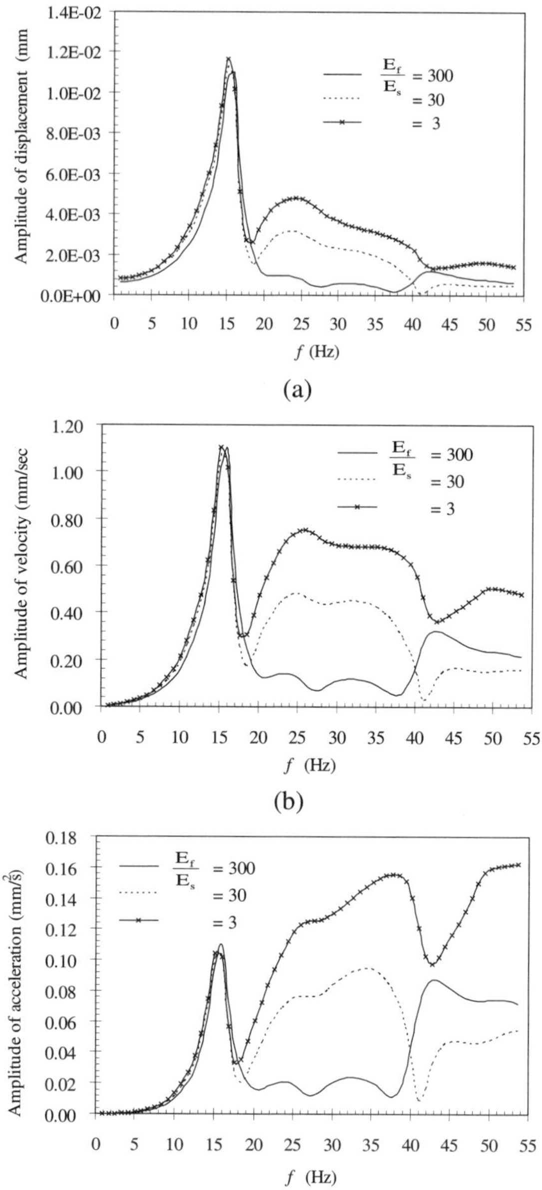

Fig. 4.33 Problem 2: Effect of elastic foundation on horizontal responses of point *B*: (a) displacement; (b) velocity; (c) acceleration.

From the above figures, one observes that the level of elasticity of the foundation can affect significantly the response amplitudes of the soil. The softer the foundation is, the higher the amplitude of the vertical response. However, the elastic modulus of the foundation has no influence on the resonant frequency, because the latter is determined primarily by the depth of the soil deposit.

The other observation from Figs. 4.31-4.33 is that the low-frequency displacement responses are fully suppressed, when compared with the case of uniform half-space in Figs. 4.27-4.29. This is largely due to the cut-off effect of the soil layer of finite depth, which tends to prevent the low-frequency vibrations from propagating outward.

In contrast, from the horizontal responses of points *C* and *B* plotted in Figs. 4.32 and 4.33, it is observed that the horizontal peak amplitudes occur at the frequencies 16, 25, and 40 Hz, among which the two frequencies 25 and 40 Hz correspond to the second and third resonant frequencies of the S-waves.

4.6 Concluding Remarks

The following conclusions can be drawn from the numerical studies performed in this chapter:

(1) When bedrock exists under a soil deposit, resonance will occur under certain frequencies. The resonant frequency decreases as the depth of the soil deposit increases, while the resonant peaks increase.
(2) When the frequency of the exciting force is less than the smallest peak frequency, no radiation damping can occur, as the waves are not allowed to travel outward due to the cut-off effect.
(3) When a soil layer is supported by harder soils underneath, rather than bedrock, resonance will occur as well. However, the resonant peak is lower and the resonant frequency smaller.
(4) Increasing Poisson's ratio of the soil can result in reduction of the response amplitude for a uniform half-space. When in resonance, the peak value will be reduced and the frequency increased.
(5) Increasing the material damping ratio of the soil can only influence the low-frequency response for a uniform half-space. It can also help

reducing the resonant amplitude, but not the resonant frequency of the soil deposit resting on bedrock.

(6) When the railway track is subjected to vertical harmonic forces, the vertical displacement of the free surface appears to be much larger than that of the horizontal displacement.

(7) In general, treating a foundation as a rigid foundation will result in an underestimate of the free surface response.

Chapter 5

Wave Barriers for Vibration Isolation of Foundations: Parametric Study

In places where residential buildings or factories are close to railway lines or highways, one important issue is how to isolate the foundations of buildings from ground-borne vibrations. In this chapter, we shall consider three types of wave barriers for isolating the foundation vibrations, i.e., the elastic foundation, open trench, and in-filled trench. The mathematical model adopted is the two-dimensional profile that contains the cross section of the railway, barrier and underlying soils, with the moving train loads simulated as a vertical harmonic line load. Concerning the effect of isolation, the following geometric and material parameters of each barrier are considered: the distance from the centerline of the railway, depth, width and thickness, damping ratio, shear modulus, mass density, Poisson's ratio, elastic modulus, etc. Also examined is the effectiveness of each barrier with respect to different excitation frequencies. Conclusions are drawn regarding the selection of optimal parametric values for the three barriers concerning isolation of train-induced ground vibrations.

5.1 Introduction

Ground-borne vibrations caused by traffic or machinery on buildings may be annoying to the residents or detrimental to the normal function of the high-precision equipment located inside. Thus, how to isolate the buildings alongside the railways or highways from ground-borne vibrations has been an issue of increasing concern to building designers. In general, the isolation of ground-borne vibrations for structures can be achieved through implementation of devices such as trenches, rubber foundations, sheet-pile walls and piles through the mechanism of interception, scattering, or diffraction of the incident waves.

With regard to the devices used for isolation of foundation vibrations, there are two general categories, i.e., *active isolation* and *passive isolation*. By active isolation, the device of isolation or barrier is installed around the source of excitation to directly cut off the energy emitted from the source. Since the isolation device is installed near the source, most of the body waves will be prevented from transmission outward. In contrast, by the concept of passive isolation, the isolation device is installed near the structure to be protected, so as to reduce the amount of vibrations transmitted from the soil to the structure of concern. Since the device is located at a distance from the source, the waves that can be isolated belong mainly to the Rayleigh or surface waves

With regard to isolation of ground-borne vibrations, numerous research works have been conducted in the past. Woods (1968) presented field survey results on the effectiveness of open trenches in reducing the amplitudes of vertical ground vibrations. Pao and Mow (1963) and Thau and Pao (1966) used analytical methods to investigate the refraction of waves around obstacles of spherical and parabolic shapes. Their solutions were confined to problems with simple geometries and idealized conditions, as closed-formed solutions cannot be easily obtained for complex, practical conditions.

Although the results obtained by experimental means appear to be most reliable and close to the real situations, a comprehensive field test may cost a lot. Moreover, for the half-space problems considered herein with wave barriers, it is generally difficult, if not impossible, to conduct small-scale model tests in the laboratories, and to extend from them useful results for practical applications. Based on the consideration of all related factors, numerical approaches turn out to be the most convenient means for investigating the wave propagation behaviors of a half-space with barriers or obstacles.

By the lumped mass method, Lysmer and Wass (1972) studied the effectiveness of a trench in reducing the horizontal shear wave motion induced by harmonic load acting on the rigid footing lying on a homogeneous soil layer. Segol *et al.* (1978) used finite elements along with special non-reflecting boundaries to investigate the isolation efficiency of open and bentonite-slurry-filled trenches in layered soils. Using a technique similar to that of Segol *et al.* (1978), Laghrouche and Le Houedec (1994) studied the effectiveness of an elastic mattress inserted under a railway track for reducing the traffic-induced ground vibrations.

All the aforementioned references involving numerical methods have found their applications for some problems. Nevertheless, due to the constraints inherent in the numerical approaches employed, they have inevitably assumed the existence of bedrock underlying the soil deposit in their models of analysis, while the quadratic complex-value problem has to be solved in certain cases, which requires a large amount of computation time.

In the past two decades, a great amount of studies on wave propagation problems have been performed by the boundary element method, including those listed in Beskos *et al.* (1986), Emad and Manolis (1985), Ahmad and Al-Hussaini (1991), and Al-Hussaini and Ahmad (1991), among others. One advantage with the boundary element method is that radiation damping associated with geometric attenuation of the far field can be automatically taken into account with no special treatment. However, this method is not suitable for modeling irregularities in geometry or material of the foundation and surrounding soils, which are often encountered in practical problems. To overcome such a drawback, the finite/infinite element method developed in Chapter 3 will be adopted herein to deal with the wave isolation problems for its versatility in accommodating various irregularities.

In the literature, most parametric studies on wave isolation problems were concerned with the passive isolation, in that the isolation devices are installed around the target structure to be protected. In contrast, relatively few works have been conducted along the lines of active isolation, for which the barrier is installed at a place closer to the source than to the structure to be protected. In this chapter, a parametric study will be conducted for the three wave barriers, i.e., the *elastic foundation*, *open trench*, and *in-filled trench*, for reducing the ground borne vibrations. The materials presented in this chapter were modified generally from those by Hung (1995) and Yang and Hung (1997).

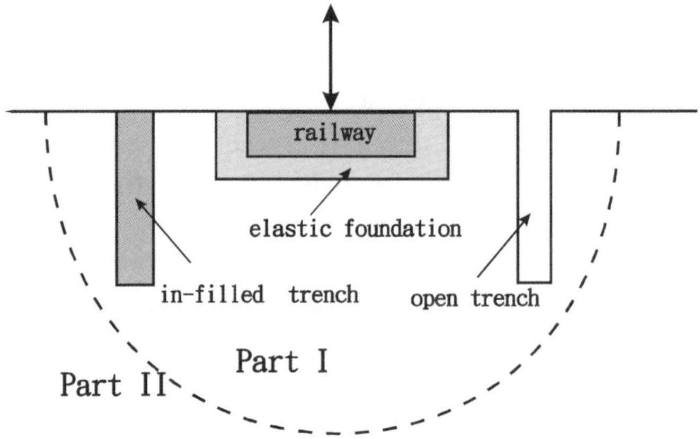

Fig. 5.1 Typical model of the problem considered.

5.2 Considerations in Parametric Studies

For brevity, all the three barriers, i.e., elastic foundation, open trench, and in-filled trench, are shown together in Fig. 5.1. However, in the following studies, we shall consider only one barrier at a time. By elastic foundation, we mean that some kind of soft materials are embedded directly under the railway track. By open trenches, we mean that two parallel empty trenches are constructed alongside the railway, arranged in a way symmetric to the centerline of the railway, such that only half of the problem needs to be considered in analysis. The in-filled trenches differ from the open trenches in that the former are not empty, but are filled with some materials, whether softer or stiffer than the neighboring soils.

As for the half-space shown in Fig. 5.1, the geometry is idealized by a near field (Part I) and a semi-infinite far field (Part II). The near field consists of the railway, wave barrier, and surrounding soils, which are generally irregular in terms of geometry and material. This region will be simulated by conventional finite elements, such as the 8-node quadratic elements. The far field contains soils with infinite boundaries, which may be homogeneous, or appear in the form of layered soils, with or without bedrock. This region can be easily dealt with using the infinite elements developed in Chapter 3.

Plane strain condition is assumed for the two-dimensional profile of the half-space. Since the traffic load acting at the centerline of the railway can always be expressed as a series of harmonic loads, only unit harmonic line loads will be considered for the train loads in this chapter. Besides, hysteretic damping is assumed for the soil, which is modeled as an isotropic viscoelastic medium.

The finite/infinite element approach described in Chapter 3 will be employed to investigate the influence of varying parameters upon the screening effect of the three types of wave barriers. By taking into account the property of symmetry, only half of the soil-structure system is considered in analysis, as shown in Fig. 5.2. The guidelines proposed in Chapter 3 for creating the finite element mesh and for selecting the wave numbers for the infinite elements are obeyed in the numerical study of this chapter. In particular, the meshes presented in Figs. 5.3(a) and (b), respectively, are adopted for the open and in-filled trenches, and for the elastic foundation.

In this chapter, we shall first investigate the effect of each material and geometry parameter of the three wave barriers, and then compare the advantages and disadvantages of each barrier. The screening effect of the wave barriers is evaluated using the *amplitude reduction ratio* A_r defined as (Beskos *et al.* 1986)

$$A_r = \frac{\text{displacement amplitude of ground surface with the barrier}}{\text{displacement amplitude of ground surface without the barrier}}. \tag{5.1}$$

If one is interested in the response of the soil over some range s beyond the barrier, the *average amplitude reduction ratio* $\overline{A}_r$ should be used instead:

$$\overline{A}_r = \frac{1}{s}\int A_r(x)dx. \tag{5.2}$$

Obviously, a smaller value of $\overline{A}_r$ implies that a better effect of isolation has been achieved by the barrier.

The material properties adopted for the parametric study, as listed in Table 5.1, have been obtained from Laghrouche and Le Houedec (1994), except those of the in-filled trenches. To avoid dependency of the analysis on the excitation frequency, all the geometric parameters used below will be normalized with respect to the Rayleigh wavelength λ_R.

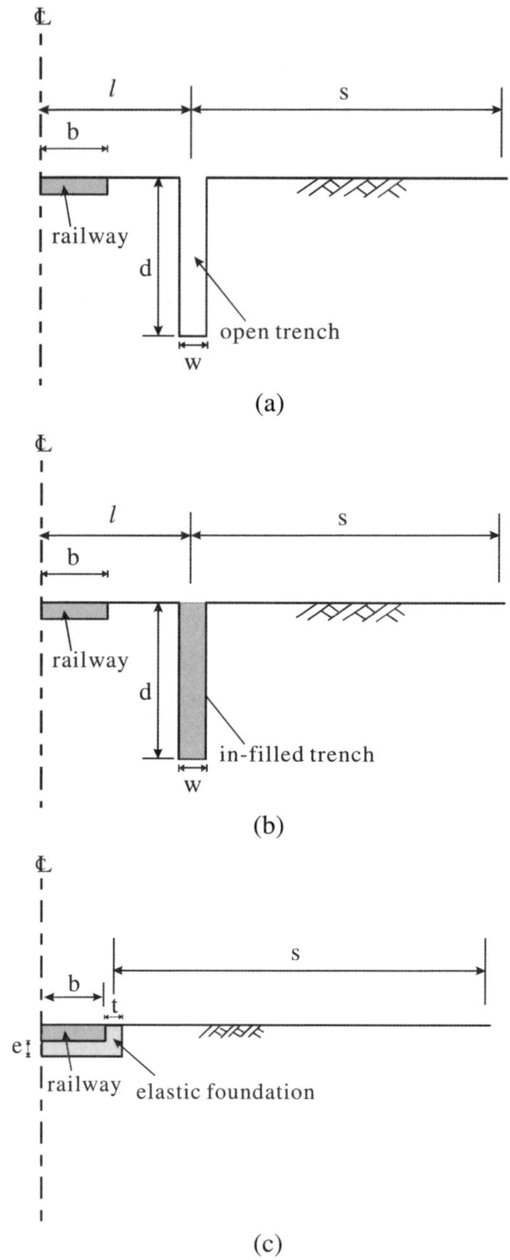

Fig. 5.2 Schematic representation of the problem: (a) open trench; (b) in-filled trench; (c) elastic foundation.

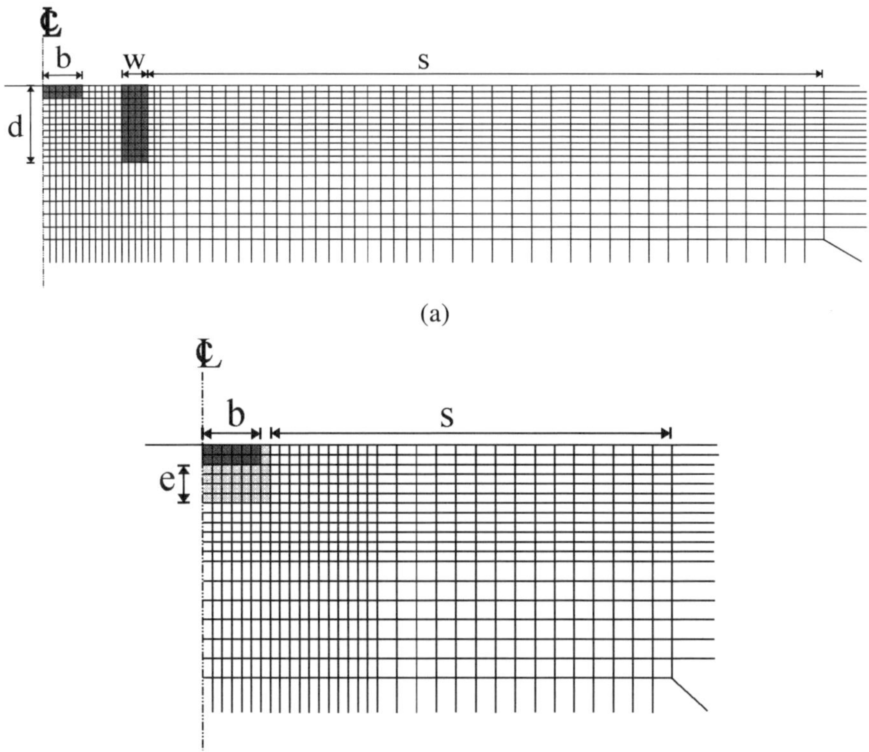

Fig. 5.3 Finite/infinite element mesh: (a) trench; (b) elastic foundation.

Table 5.1 Material properties.

Material	Young's modulus E (MPa)	Poisson's ratio ν	Density ρ (kg/m^3)	Damping ratio β
Soil	46	0.25	1800	0.05
Elastic foundation	1	0.25	150	0.10
In-filled trench	4600	0.25	2700	0.05
Railway	13500	0.25	2400	0.02

As can be seen, the elastic foundation has been intentionally made to be softer than the surrounding soil. In the parametric study, both the soil and railway properties will be kept constant. The railway is assume to have a half-width of $b = \lambda_R / 2$. Using the present data for the soil, the shear wave velocity is $C_s = (G/\rho)^{1/2} = (18.4 \times 10^6 / 1800)^{1/2}$ = 101.11 m/s, the

Rayleigh wave velocity is $C_R = C_s(0.87 + 1.12\nu)/(1+\nu) = 93.02$ m/s, and the Rayleigh wavelength is $\lambda_R = C_R / f = 93.02/31 = 3$ m.

The *impedance ratio* (IR) was often used by geotechnical engineers in distinguishing whether a barrier is softer or stiffer than the surrounding soil. The IR is defined as follows:

$$\mathrm{IR} = \frac{\rho_b V_b}{\rho_s V_s}, \tag{5.3}$$

where ρ_b and ρ_s denote the mass density of the barrier and surrounding soil, respectively, and V_b and V_s the wave velocities of the two media. For the case where the barrier is installed at a place away from the railway, both V_b and V_s should be interpreted as the Rayleigh wave velocities. On the other hand, if the barrier is installed near the railway, they should be interpreted as the body wave velocities instead.

5.3 Vibration Isolation by Elastic Foundation

For the elastic foundation shown in Fig. 5.2(c), the geometric parameters will be normalized with respect to the Rayleigh wavelength λ_R, e.g., mattress thickness $e = E \cdot \lambda_R$, vertical joint thickness $t = T \cdot \lambda_R$, inner width $b = B \cdot \lambda_R$, and range $s = S \cdot \lambda_R$, in order to make the results independent of the excitation frequencies. Unless otherwise noted, the dimensions of the elastic foundation are selected as $E = 1/2$, $T = 1/12$, $B = 1/2$, and $S = 3.5$.

From Figs. 5.4 and 5.5, the effect of the elastic foundation in isolating the vibrations can be readily appreciated. Figures 5.4(a)-(d) show the real and imaginary parts of the horizontal and vertical displacements of the ground surface for the cases with and without elastic foundation. As can be seen, both the real and imaginary parts of the displacements decrease dramatically due to existence of the elastic foundation for the region beyond the foundation. The same phenomenon can also be observed from the displacement amplitudes in Fig. 5.5 for the region outside the foundation. It should be noted that both the horizontal and vertical displacements are amplified at the place right above the elastic foundation. The vertical displacement is amplified even more seriously inside the railway track.

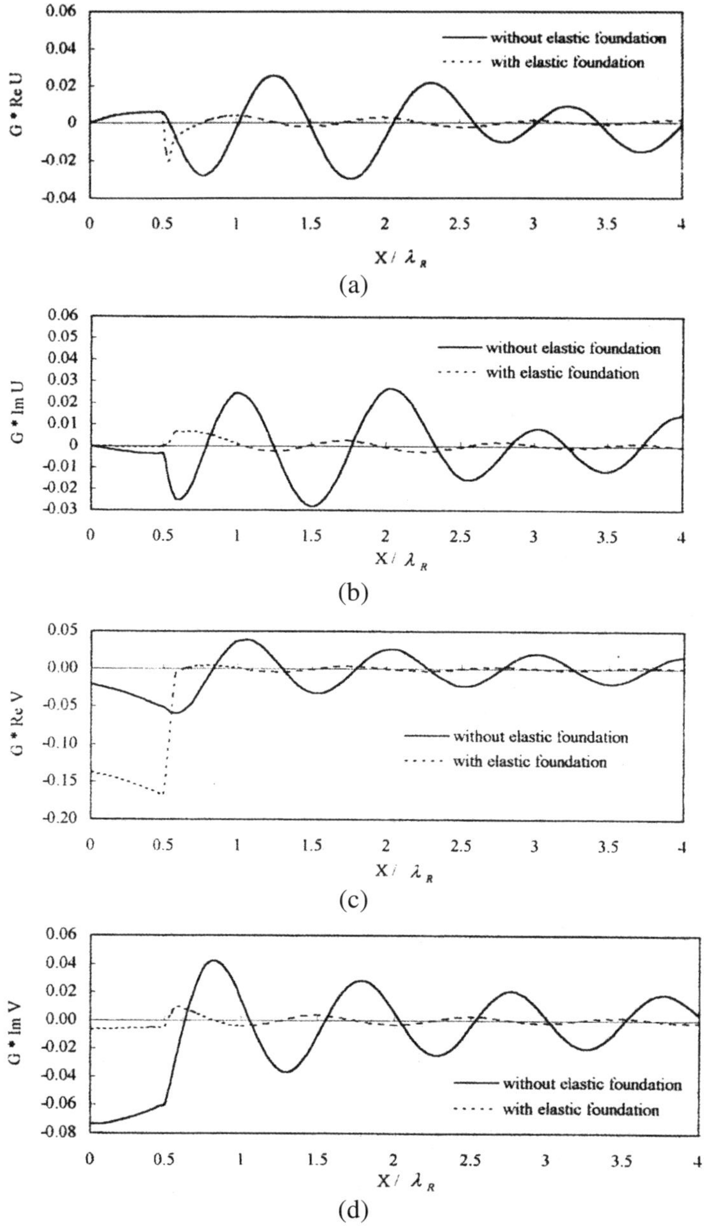

Fig. 5.4 Effect of elastic foundation (E = 1/12, T = 1/12): (a) real part of horizontal displacement; (b) imaginary part of horizontal displacement; (c) real part of vertical displacement; (d) imaginary part of vertical displacement.

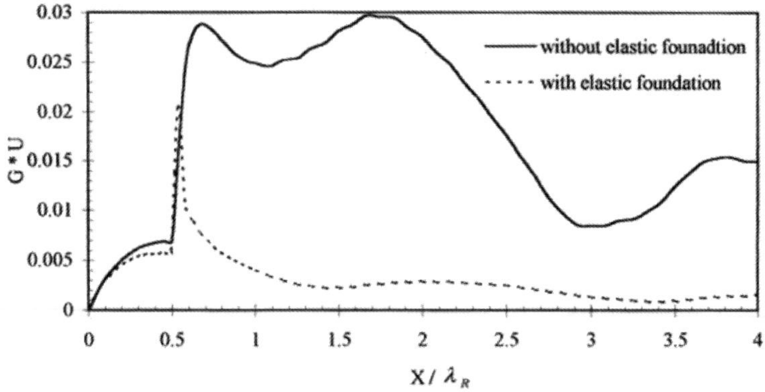

(a)

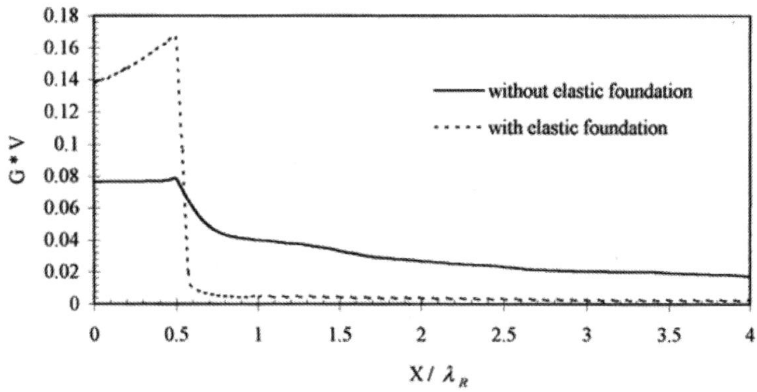

(b)

Fig. 5.5 Effect of elastic foundation (E = 1/12, T = 1/12): (a) horizontal displacement amplitude; (b) vertical displacement amplitude.

In the following subsections, we shall investigate the isolation effect of each parameter of the elastic foundation. Unless otherwise noted, all the data adopted for the parameters are those listed in Table 5.1.

5.3.1 *Young's modulus ratio* (E_s / E_e)

Let E_s and E_e denote the Young's modulus of the soil and elastic foundation, respectively. The ratio E_s / E_e represents the relative stiffness of the surrounding soil with respect to the foundation. A higher value of E_s / E_e implies that the elastic foundation is softer (than the soil),

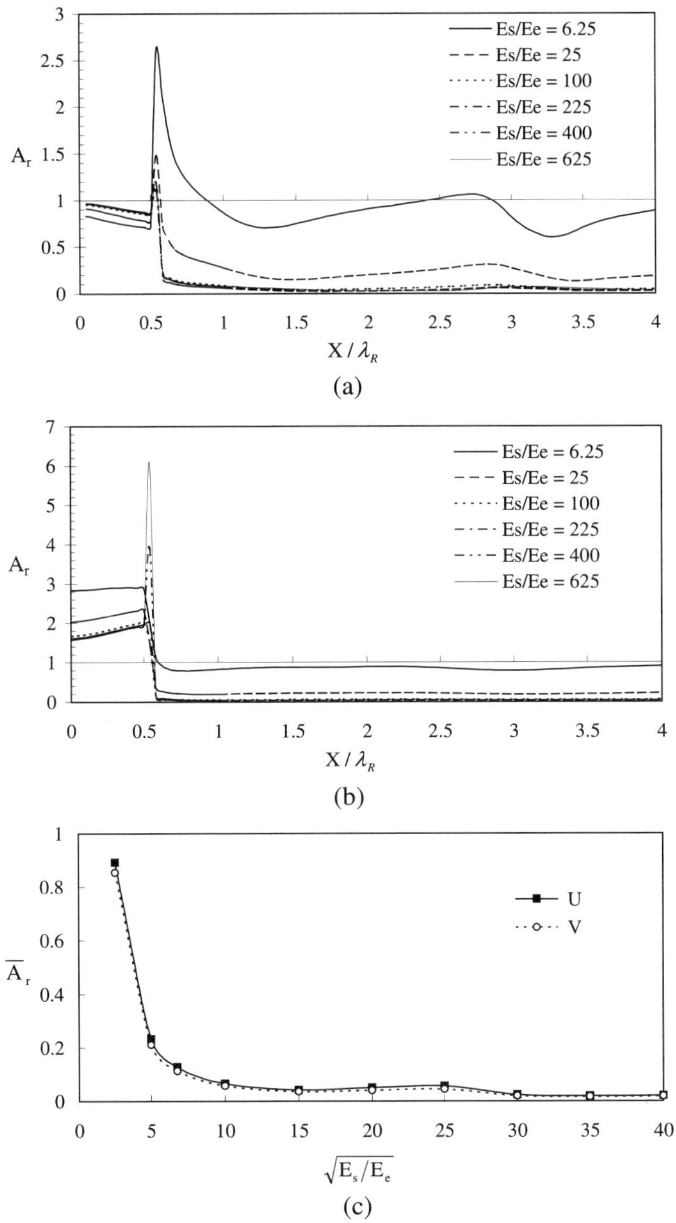

Fig. 5.6 Effect of Young's modulus of elastic foundation (IR < 1): (a) amplitude reduction ratio for horizontal displacement; (b) amplitude reduction ratio for vertical displacement; (c) averaged amplitude reduction ratio.

and has a lower wave velocity. Some of the results illustrating the effect of Young's modulus ratio were plotted in Fig. 5.6, in which parts (a) and (b) denote the amplitude reduction ratio A_r for the horizontal and vertical displacement, respectively, and part (c) represents the average amplitude reduction ratio $\overline{A}_r$ with respect to $\sqrt{E_s / E_e}$. As can be seen, the softer the elastic foundation is, the better the effect of isolation. However, the decreasing rate of $\overline{A}_r$ is not proportional to the increasing rate of E_s / E_e. When $\sqrt{E_s / E_e}$ is greater than 10, the decreasing rate of $\overline{A}_r$ becomes very slow. To be cost-effective, it is recommended that the value $\sqrt{E_s / E_e} = 10$ be used. On the other hand, from the amplitude reduction ratios $\overline{A}_r$ plotted in Fig. 5.6(c) for the horizontal displacement *U* and vertical displacement *V*, we observe that the effect of elastic foundations in isolating the train-induced vibrations is about the same for both displacements

The above analysis has been performed for the case where the elastic foundation is softer than the surrounding soil, i.e., with $E_e / E_s < 1$. As for the case where the elastic foundation is stiffer than the surrounding soil, i.e., with $E_e / E_s > 1$, the results of analysis have been plotted in Fig. 5.7. Clearly, the effect of isolation improves as the ratio $\sqrt{E_e / E_s}$ increases. However, the effect of isolation is not as good as that for the case with soft foundations, i.e., with $E_e / E_s < 1$.

The effect of isolation of elastic foundations has been plotted with respect to the impedance ratio IR (in logarithmic scale) in Fig. 5.8. As can be seen, for soft foundations, i.e., with IR < 1, by decreasing the impedance ratio, the average amplitude reduction ratio $\overline{A}_r$ decreases rapidly. On the other hand, for stiff foundations, i.e., with IR > 1, the stiffer the elastic foundation is, the better the effect of isolation by the elastic foundation. However, the effect of isolation for stiff foundations (with IR > 1) is not comparable to that for the soft foundation (with IR < 1). Another observation is that the effect of elastic foundations in isolating vibrations is about the same for both the horizontal and vertical displacements. In other words, for foundations that are soft compared with the surrounding soil, it is possible to achieve an average amplitude rate $\overline{A}_r$ smaller than 0.1. In contrast, the highest performance that can be achieved by stiff foundations is around $\overline{A}_r = 0.8$.

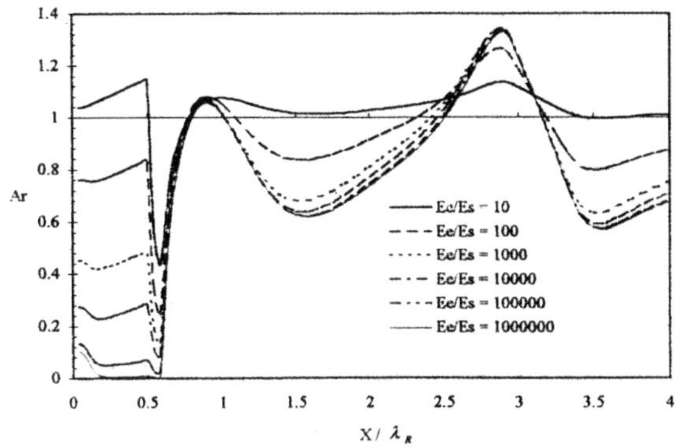

(a)

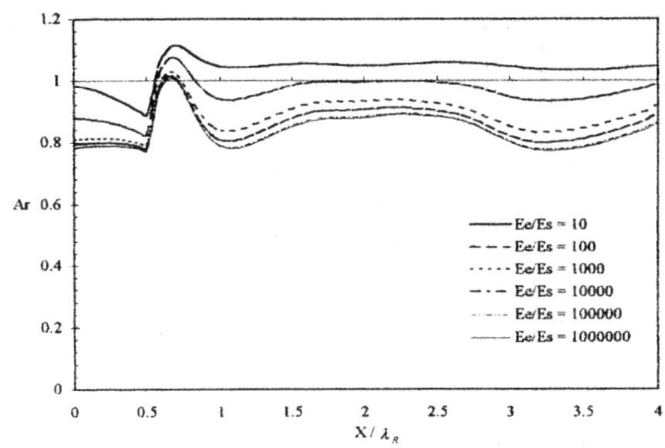

(b)

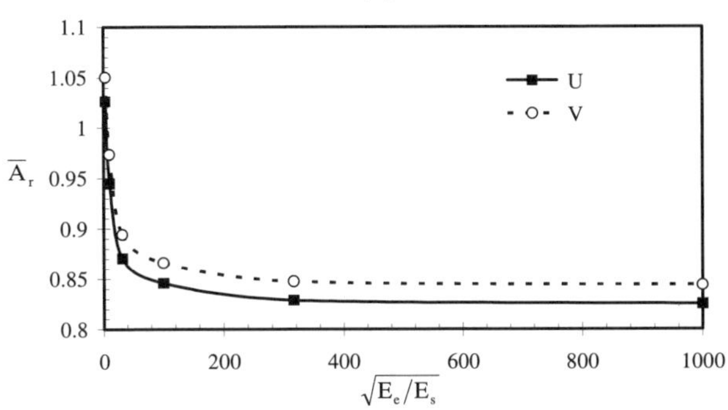

(c)

Fig. 5.7 Effect of Young's modulus of elastic foundation (IR > 1): (a) amplitude reduction ratio for horizontal displacement; (b) amplitude reduction ratio for vertical displacement; (c) averaged amplitude reduction ratio.

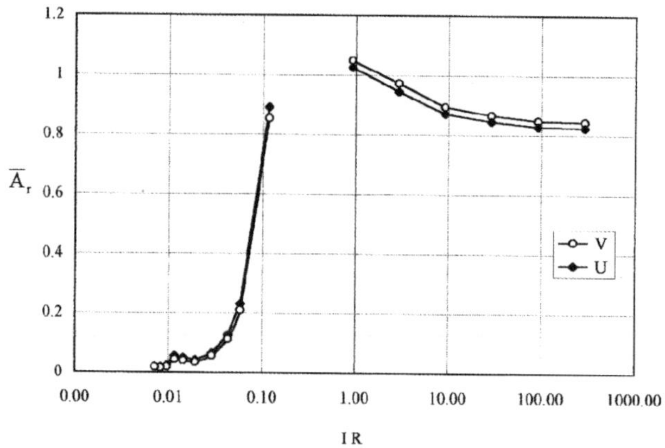

Fig. 5.8 Effect of impedance ratio of elastic foundation (by varying Young's modulus).

5.3.2 *Mass density ratio* (ρ_s / ρ_e)

Let ρ_s and ρ_e denote the mass density of the soil and elastic foundation, respectively. The average amplitude reduction ratio $\overline{A}_r$ has been plotted against the density ratio ρ_s / ρ_e in Fig. 5.9. In practice, the range of mass density ratio that can be utilized is quite narrow. For this reason, the mass density ratio ρ_s / ρ_e is allowed to vary from 6 to 16 in analysis. As indicated by the figure, slightly better effect of isolation can be achieved for larger density ratios ρ_s / ρ_e, though the range of improvement is rather small. Considering the fact that the mass density also affects the impedance ratio, we also plot the average amplitude reduction ratio $\overline{A}_r$ with respect to the impedance ratio (IR) in Fig. 5.10. From this figure, it can be appreciated that the effect of IR on the average amplitude reduction ratio is quite small. The reason is that a narrow range of mass density ratio ρ_s / ρ_e (i.e., from 6 to 16) allows the IR to vary only in a small range (i.e., from 0.035 to 0.060), which again implies a small range of variation for the average amplitude reduction ratio $\overline{A}_r$.

5.3.3 *Poisson's ratios* (ν_e, ν_s)

By fixing Poisson's ratio ν_s for the soil, the average amplitude reduction ratio $\overline{A}_r$ has been plotted against Poisson's ratio ν_e of the elastic

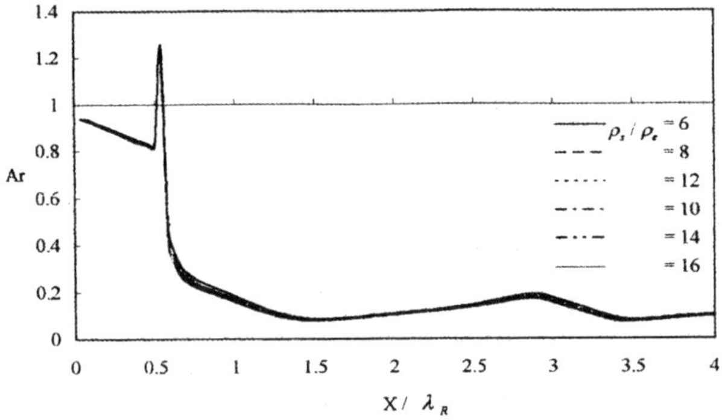

(a)

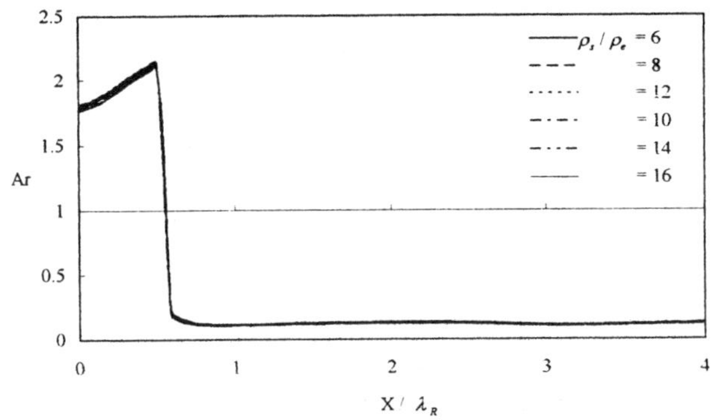

(b)

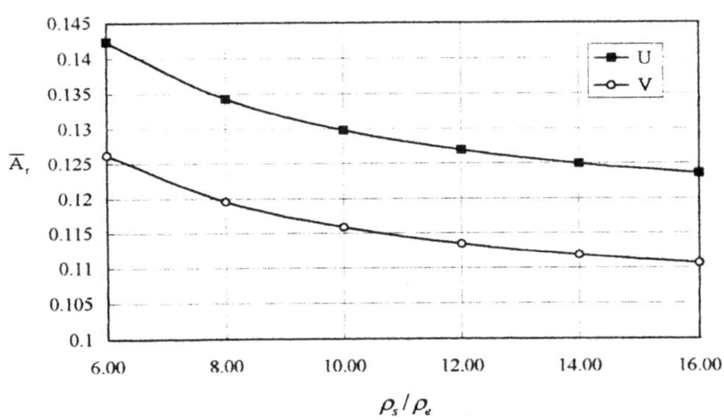

(c)

Fig. 5.9 Effect of mass density of elastic foundation: (a) amplitude reduction for horizontal displacement; (b) amplitude reduction for vertical displacement; (c) averaged amplitude reduction ratio.

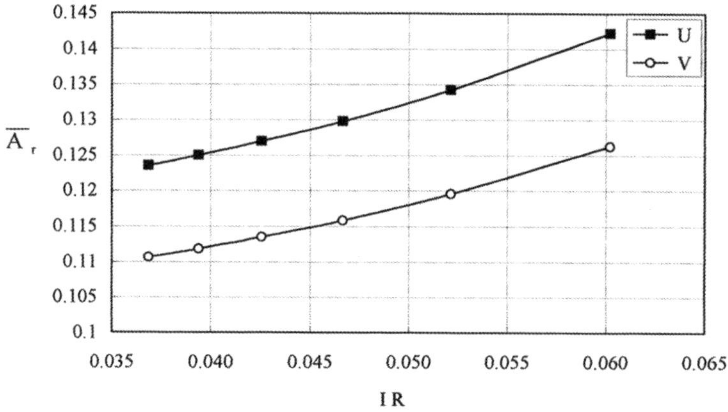

Fig. 5.10 Effect of impedance ratio of elastic foundation (by varying the mass density).

foundation in Fig. 5.11. As can be seen, for $\nu_e > 0.4$, the effect of isolation decreases significantly, due to the fact that large Poisson's ratios tend to increase the compression wave speed drastically, and that compression waves are crucial for half-space problems with vertical excitations. Such a phenomenon can be clearly observed from the results shown in Figs. 5.12(a) and (b) for the horizontal and vertical displacements, respectively, for some specific values of Poisson's ratios for the soil. Clearly, regardless of the Poisson's ratio for the soil, whenever $\nu_e > 0.4$, the effect of isolation tends to deteriorate drastically. Moreover, as Poisson's ratio of the soil increases, the effect of isolation decreases significantly. However, such a phenomenon does not exist for elastic foundations with $\nu_e < 0.4$. Hence, it is recommended that the Poisson's ratio always be kept below 0.4 for the elastic foundation in engineering practice.

5.3.4 *Material damping ratio* (β)

Figure 5.13 shows the results computed for the ground surface by allowing the damping ratio β of the elastic foundation to vary from 0 to 0.1. Clearly, for both the horizontal and vertical responses, the amplitude reduction ratio A_r or its average $\bar{A}_r$ remain nearly at constant values, regardless of the change in damping ratio. The implication herein is that

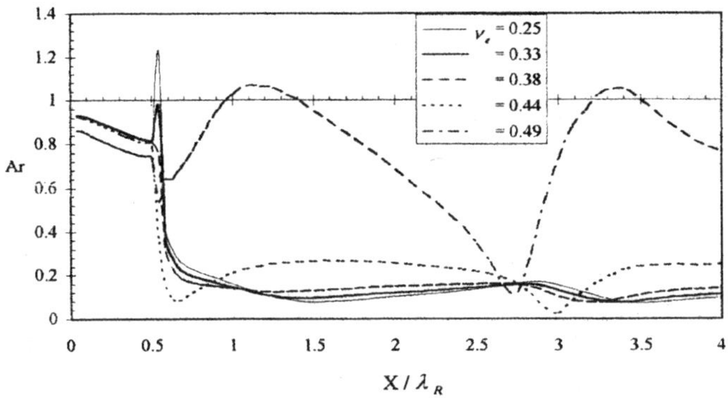

(a)

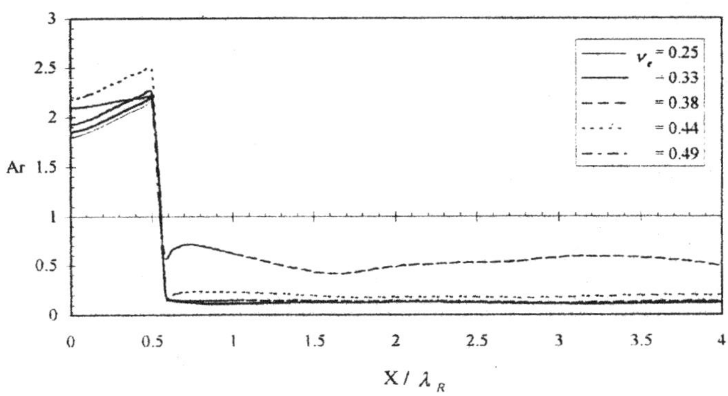

(b)

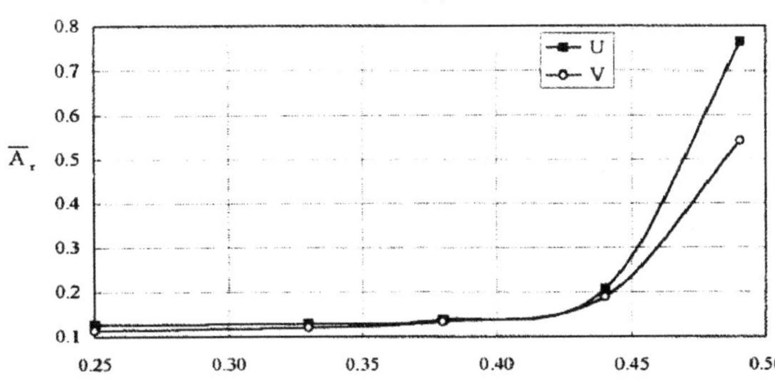

(c)

Fig. 5.11 Effect of Poisson's ratio of elastic foundation: (a) amplitude reduction for horizontal displacement; (b) amplitude reduction for vertical displacement; (c) averaged amplitude reduction ratio.

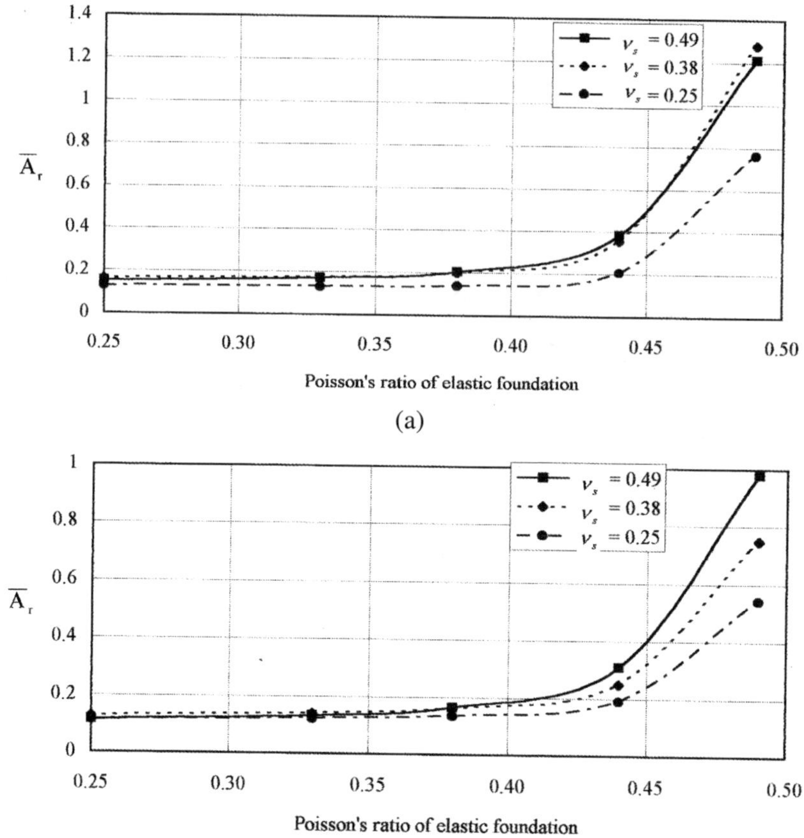

(b)

Fig. 5.12 Effect of Poisson's ratio of elastic foundation: (a) horizontal displacement; (b) vertical displacement.

as far as the vibration isolation is concerned, the effect of damping ratio can be ignored in practice.

5.3.5 *Normalized dimensions* (*T, E*)

As was stated in the beginning of this section, the geometric parameters of the elastic foundation are normalized with respect to the Rayleigh wavelength λ_R, i.e., mattress thickness $e = E \cdot \lambda_R$ and joint thickness $t = T \cdot \lambda_R$. By first fixing the mattress thickness *E*, the results computed

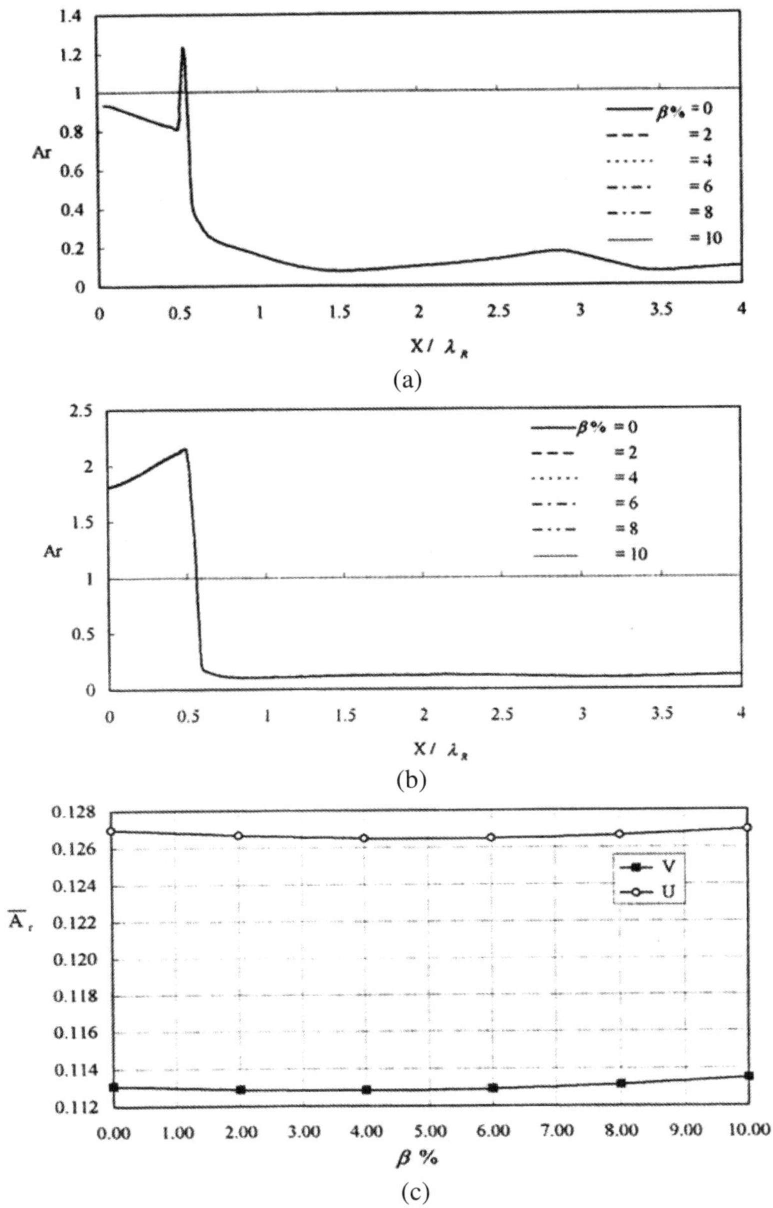

Fig. 5.13 Effect of damping ratio of elastic foundation: (a) amplitude reduction for horizontal displacement; (b) amplitude reduction for vertical displacement; (c) averaged amplitude reduction ratio.

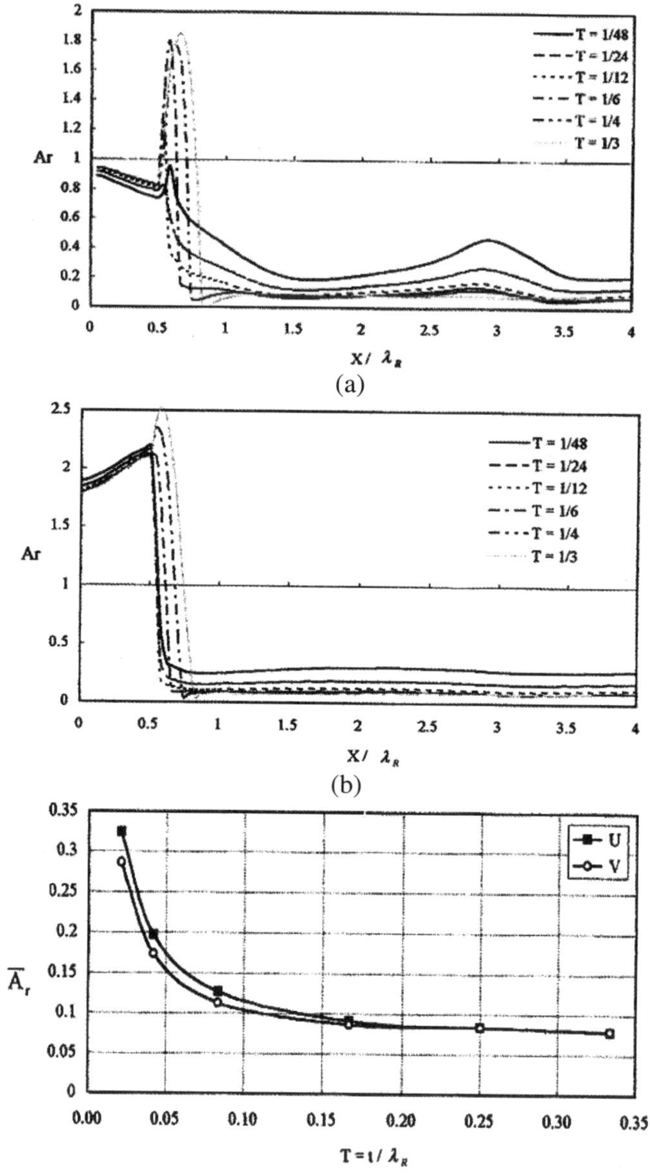

Fig. 5.14 Effect of joint thickness T of elastic foundation: (a) amplitude reduction for horizontal displacement; (b) amplitude reduction for vertical displacement; (c) averaged amplitude reduction ratio.

for various joint thicknesses T have been plotted in Fig. 5.14. It can be seen that both the horizontal and vertical displacements are amplified in the regions around the two sides of the railway (only one side is shown in the figure due to symmetry), and that for larger joint thickness T, there exist greater areas of amplification, while a better effect of isolation can be achieved in the average sense for the ground surface other than the regions near the railway, as indicated by Fig. 5.14(c).

On the other hand, we can also fix the value of the joint thickness T and allow the mattress thickness E of the elastic foundation to vary over a certain range. The results have been presented in Fig. 5.15. Clearly, the use of a larger thickness E for the foundation will result in reduction of the vertical response of the railway. Thus, if the vibration of the railway is of concern, an elastic foundation of large thickness should be used. Also, from Fig. 5.15(c), we observe that the use of a larger thickness E for the foundation need not result in a better effect of isolation. In fact, there exists an optimal value for the foundation thickness E for which the effect of isolation appears to be maximal.

Such a phenomenon can be observed from the responses plotted in Fig. 5.16 with respect to the mattress thickness E, given the joint thickness T. Clearly, for each value of joint thickness T, there exists an optimal E value, which is especially true as the joint thickness T increases. For instance, for $T = 1/3$, the optimal mattress thickness that should be selected is $E = 1/12$.

5.3.6 *Bedrock depth H*

Let the depth H of the bedrock vary from $0.5\lambda_R$ to $2.0\lambda_R$, where λ_R is the Rayleigh wavelength of the soil layer. Also, let the non-dimensional bedrock depth be denoted as $h = H \cdot \lambda_R$. The results of analysis have been plotted in Fig. 5.17, in which the case with $H = \infty$ denotes a uniform half-space, exactly the same as the case considered in all previous sub-sections. Although the bedrock depth has some influence on the isolation efficiency of the elastic foundation, the level of influence is generally small. It follows that the results presented previously for the elastic foundation in a uniform half-space can be generally applied to the cases with bedrock without causing much errors.

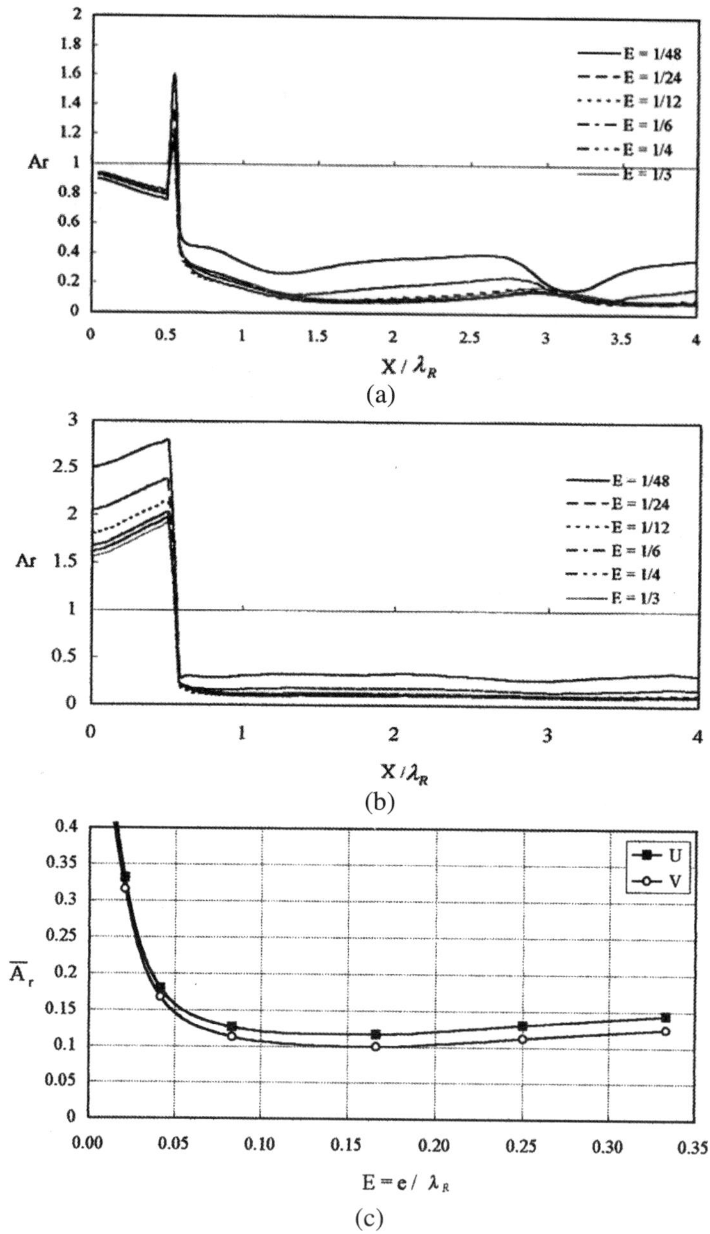

Fig. 5.15 Effect of mattress thickness E of elastic foundation: (a) amplitude reduction for horizontal displacement; (b) amplitude reduction for vertical displacement; (c) averaged amplitude reduction ratio.

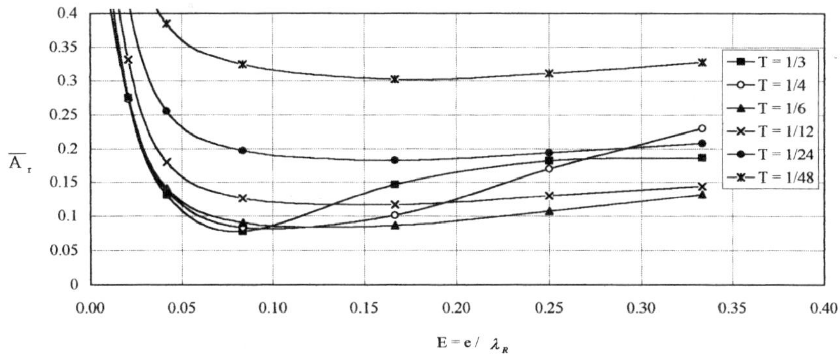

(a)

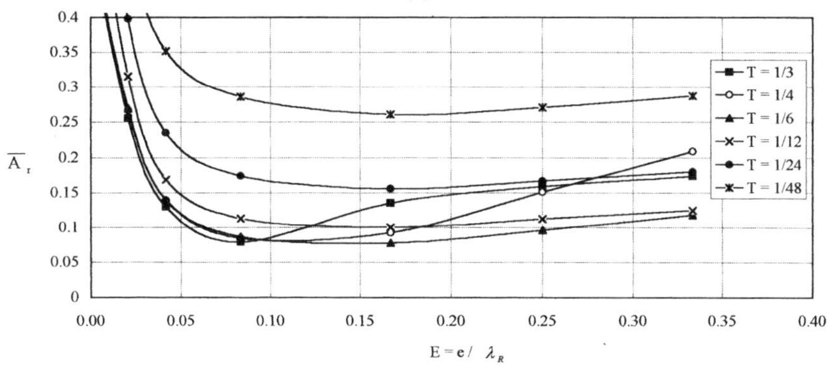

(b)

Fig. 5.16 Effect of dimensions of elastic foundation: (a) horizontal displacement; (b) vertical displacement.

5.4 Vibration Isolation by Open Trenches

With reference to the open trenches shown in Fig. 5.2(a), the following parameters are adopted: distance from the centerline of the railway $l = L \cdot \lambda_R$, depth $d = D \cdot \lambda_R$, width $w = W \cdot \lambda_R$, half width of railway $b = B \cdot \lambda_R$, and range of measurement $s = S \cdot \lambda_R$, where λ_R is the Rayleigh wavelength, and *L*, *D*, and *W* are non-dimensional parameters. Unless otherwise specified, the following data are adopted for the open trenches: distance $L = 1$, depth $D = 1$, width $W = 1/3$, and range of measurement $S = 9$. The material properties adopted for the open trenches and railway have been listed in Table 5.1.

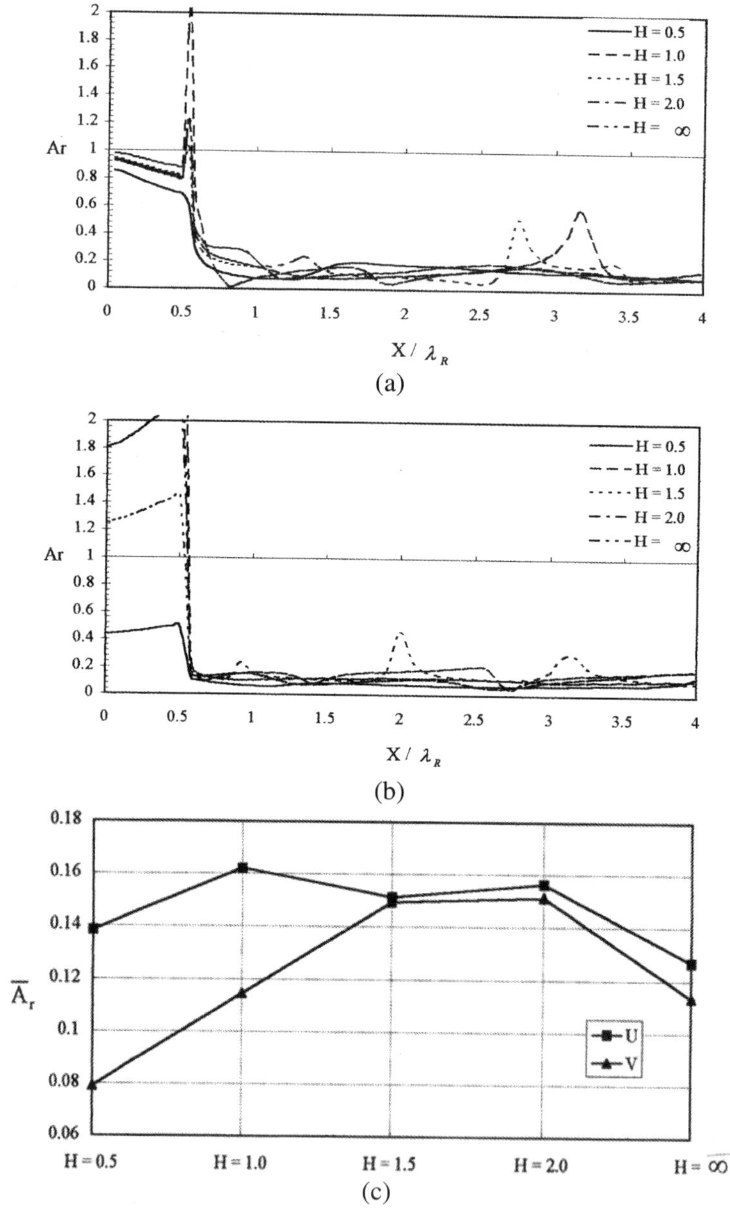

Fig. 5.17 Effect of bedrock depth *H* under elastic foundation: (a) amplitude reduction for horizontal displacement; (b) amplitude reduction for vertical displacement; (c) averaged amplitude reduction ratio.

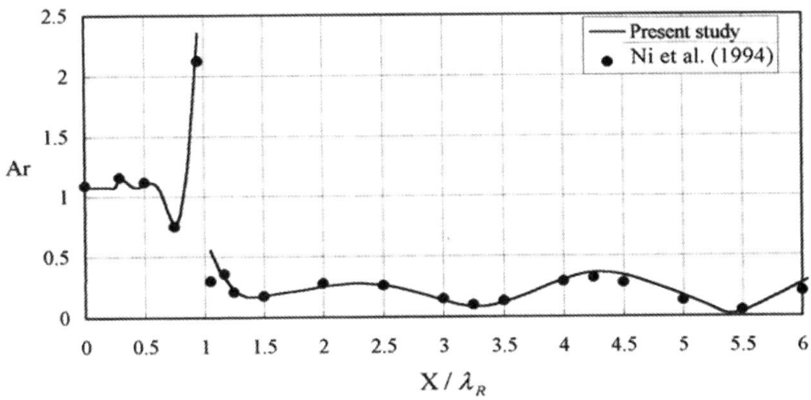

Fig. 5.18 Test case 1 for open trenches ($W = 0.1$, $D = 1$, $L = 1$).

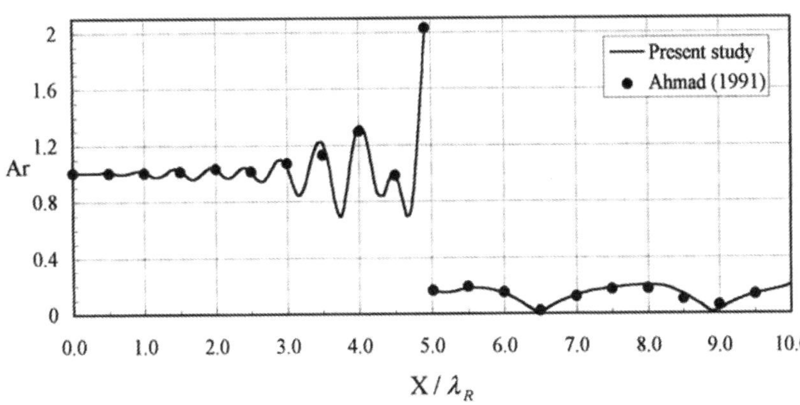

Fig. 5.19 Test case 2 for open trenches ($W = 0.1$, $D = 1$, $L = 5$).

To verify the correctness of the procedure, the results obtained by the present approach for the first test case with $D=1$, $W=0.1$, $L=1$, $B=0.25$, $\nu=0.25$, and $\beta=0.06$ have been compared with the those of Ni *et al.* (1994) in Fig. 5.18. As can be seen, good agreement has been achieved between the two results. In the second test, the following data are adopted for the open trenches: $D=1$, $W=0.1$, $L=5$, and $B=0.25$. The results obtained have been compared with those of Ahmad and Al-Hussaini (1991) in Fig. 5.19. Again, the present results coincide very well with those by Ahmad and Al-Hussaini (1991) using the boundary element method.

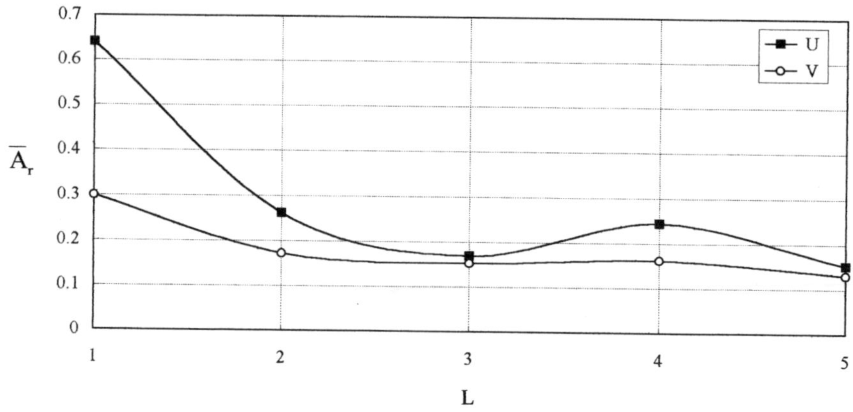

Fig. 5.20 Effect of distance of open trenches from source.

From Fig. 5.19, it is interesting to note that the response amplitude of the surface located inside and adjacent to the two open trenches has been drastically amplified. For the surface located between the railway and the open trench on each side, the amplitude oscillates up and down around the level for the same half-space without open trenches. The distance between any two adjacent peaks or valleys in this region equals exactly half of the Rayleigh wavelength. From such an analysis, one observes that waves are reflected by the inner face of each trench. The oscillatory response of the surface displacement is caused by superposition of the incident and reflected waves in this region. In fact, the wave propagation behavior near the inner face of the open trench on each side should be regarded as a total reflection, but no refraction, since the open trenches contain no material in itself for transmission of waves.

In the following, we shall investigate the influence of each parameter of the open trenches in mitigating the outward-transmitting waves from the source via the half-space. The data adopted in analysis are those listed in Table 5.1, unless noted otherwise.

5.4.1 *Distance L between the railway and open trench*

The results obtained by varying the distance *L* between the railway centerline and the open trench on each side have been plotted in Fig. 5.20, where U and V respectively represent the horizontal and vertical responses. As can be seen, the results are about the same for all values of L, except for the case with L equal to one, where the isolation effect appears to be the worst. The reason for this is that as the trenches are located close to the source (e.g., when $L = 1$), the body waves play a role more important than the surface waves. Since the body waves decay more slowly in the downward direction than the surface waves, a great portion of the body waves can pass beneath the trench, making the screening effect of the open trenches not so efficient. On the other hand, for larger distance (e.g., for $L \geq 2$), the effect of isolation becomes rather efficient, mainly due to the rather smaller influence of body waves in this regions and the rapid decrease of surface waves in the direction downward, which implies that only a smaller amount of waves can pass beneath the trenches (compared with the case with $L = 1$). Besides, the results also demonstrate that the open trenches tend to isolate the vertical vibration more effectively than the horizontal vibration.

5.4.2 *Depth D and width W of open trench*

By changing the normalized width W of the trenches, the results computed for the horizontal and vertical response of the ground surface have been plotted with respect to the depth D in Figs. 5.21(a) and (b), respectively. It can be seen that the isolation effect is not good for a depth D smaller than 0.5. The same results were redrawn in Fig. 5.22 with respect to the width W for different values of depth D. From this figure, it can be observed that for shallower trenches, i.e., with $D = 0.25$, the effect of isolation deteriorates as the width W increases. For this case, a wider trench means a larger area of surface, which allows a higher percentage of body waves to be converted to the surface waves, thereby aggravating the effect of isolation. On the other hand, for deeper trenches, e.g., when $D \geqq 0.5$, the influence of trench width W becomes quite limited.

One lesson from the above analysis is that for shallow trenches, much more Rayleigh waves can transmit via the bottom of the trench. Thus, the influence of trench width W cannot be ignored, if a trench filled with water, whether flowing or not, is to be used as a wave barrier. However,

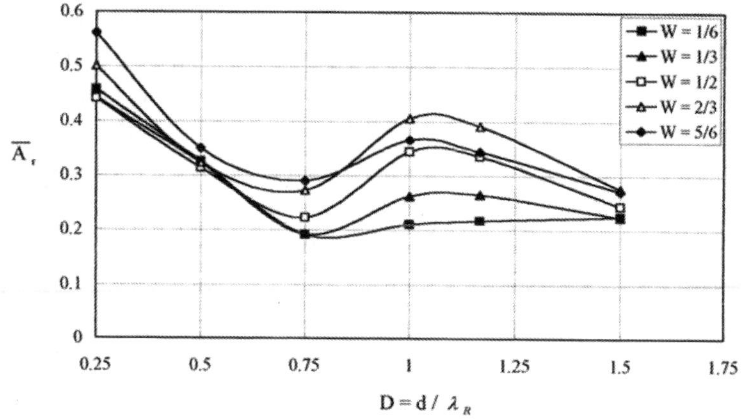

(a)

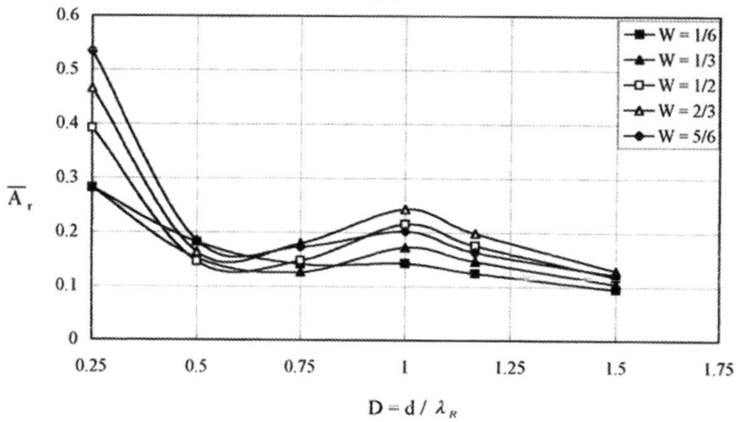

(b)

Fig. 5.21 Effect of depth D of open trenches for given width W (L = 2): (a) amplitude reduction for horizontal displacement; (b) amplitude reduction for vertical displacement.

for deeper trenches, most of the Rayleigh waves are reflected by the trench. As a result, the influence of trench width W is quite limited.

5.5 Vibration Isolation by In-Filled Trenches

The geometrical parameters of the in-filled trenches have been shown in Fig. 5.2(b), with the relevant data listed in Table 5.1. They are all normalized in a way identical to that for the open trenches, to make the results independent of the excitation frequencies.

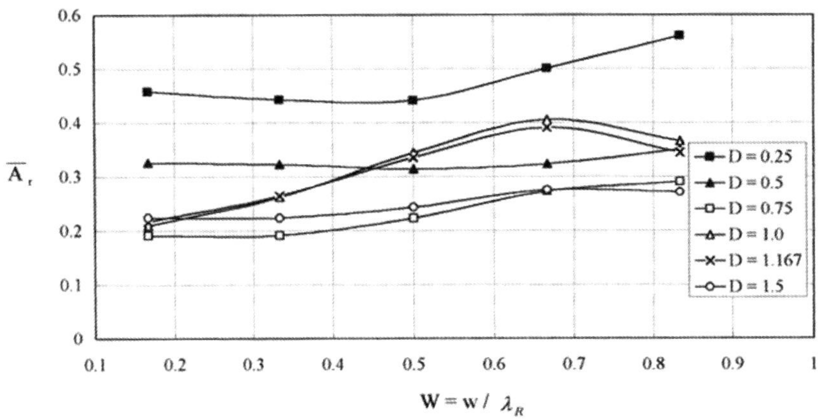

(a)

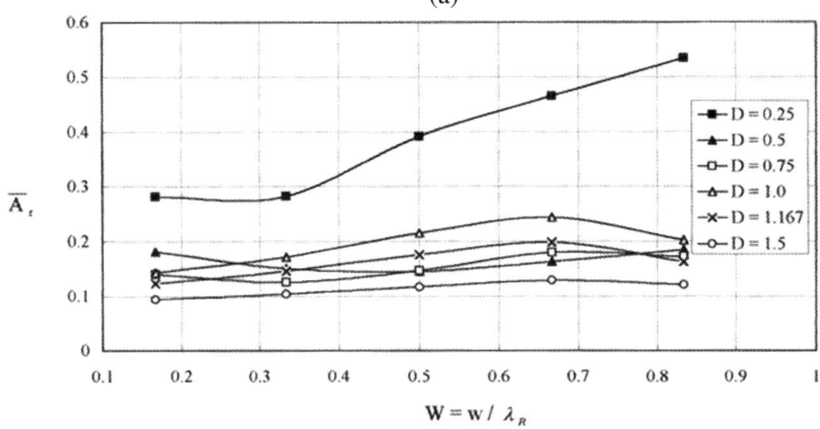

(b)

Fig. 5.22 Effect of width W of open trenches for given depth D (L = 2): (a) amplitude reduction for horizontal displacement; (b) amplitude reduction for vertical displacement.

The general effect of isolation of in-filled trenches can be appreciated from the results presented in Figs. 5.23(a)-(d) for the real and imaginary parts of the horizontal and vertical displacements. As can be seen, for both the real and imaginary parts, the surface responses are greatly reduced for the region shielded by the in-filled trenches, coupled by a forward-shifting phase. This can be attributed to the fact that the in-filled trenches are physically stiffer than the soil, which allows the waves to propagate at a faster speed within the trench bodies themselves, thereby resulting in elongation of the wavelength.

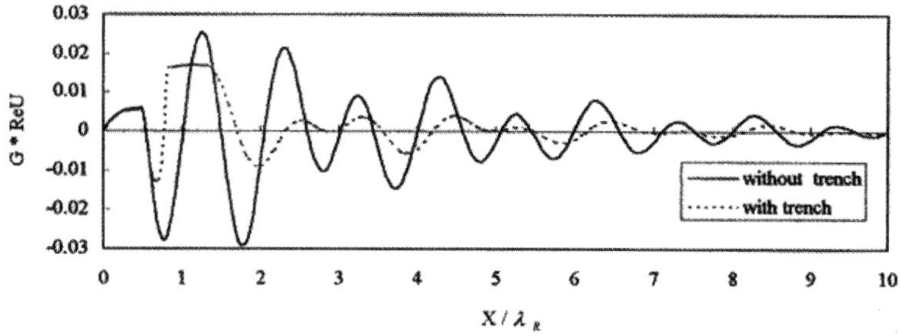

(a)

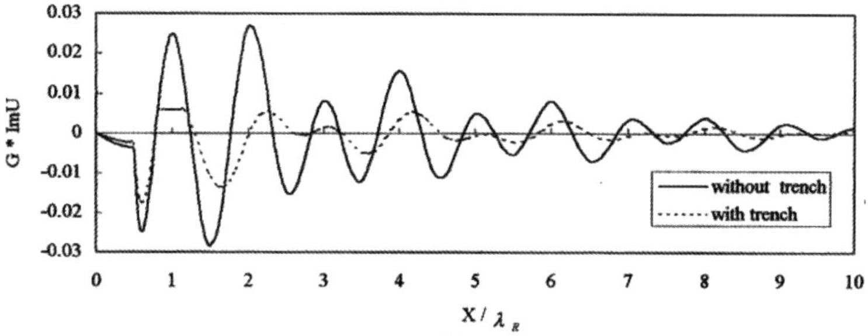

(b)

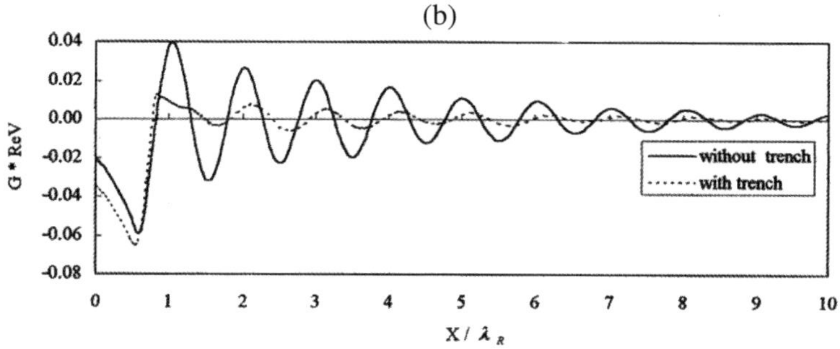

(c)

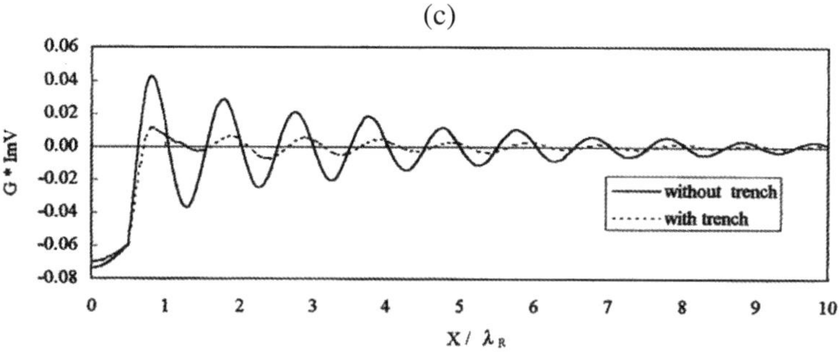

(d)

Fig. 5.23 Effect of in-filled trenches (L = 1, D = 1, W = 1/3): (a) real part of horizontal displacement; (b) imaginary part of horizontal displacement; (c) real part of vertical displacement; (d) imaginary part of vertical displacement.

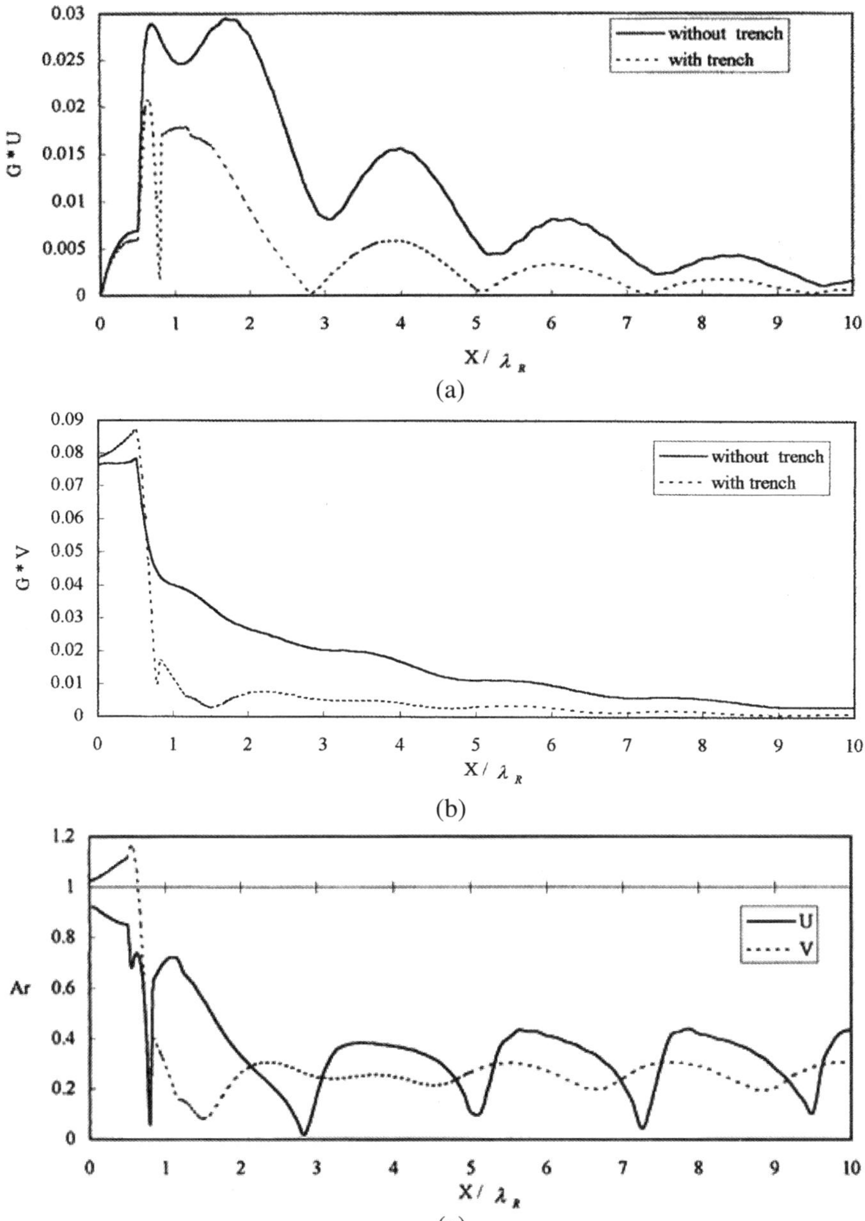

Fig. 5.24 Effect of in-filled trenches (L = 1, D = 1, W = 1/3): (a) horizontal displacement amplitude; (b) vertical displacement amplitude; (c) amplitude reduction ratio.

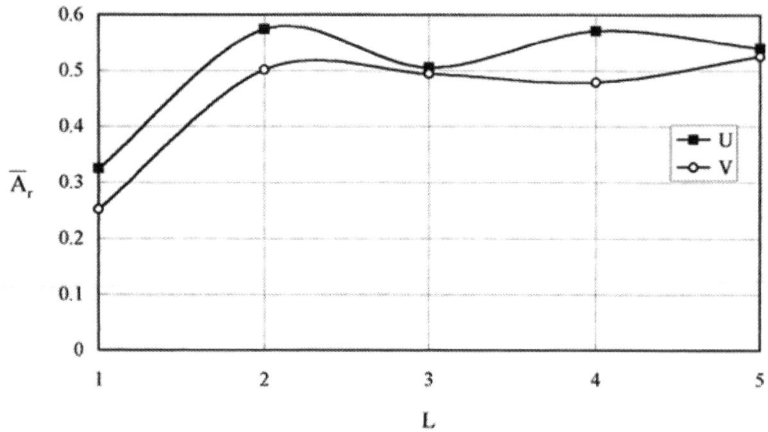

Fig. 5.25 Effect of distance of in-filled trench from source.

On the other hand, from the response amplitudes plotted in Fig. 5.24, one observes that for the soils located right above the trenches and railway, the amplification effect is not as obvious as that for the case with elastic foundation. Thus, if the response amplitude of the railway track is of concern, the in-filled trench appears to be a better choice, compared with the elastic foundation.

In the following, we shall investigate the influence of each parameter of the in-filled trenches on the ground response concerning the isolation of vibrations. All the data adopted for the materials and geometry are those given in Table 5.1.

5.5.1 *Distance L between the railway and in-filled trench*

As shown in Fig. 5.2(b), we shall use l to denote the distance from the in-filled trenches to the centerline of railway, which can be normalized as $L = l / \lambda_R$, with λ_R denoting the Rayleigh wavelength. By allowing L to vary from 1 to 5, one can obtain the results shown in Fig. 5.25. Similar to those for open trenches, for the case with distance L greater than 2, the influence of distance L appears to be rather small. When the distance L reduces to 1, the effect of isolation is greatly improved. The reason for this is similar to, but more complicated than that for open trenches, due to the fact that incident waves may not only be reflected, but also

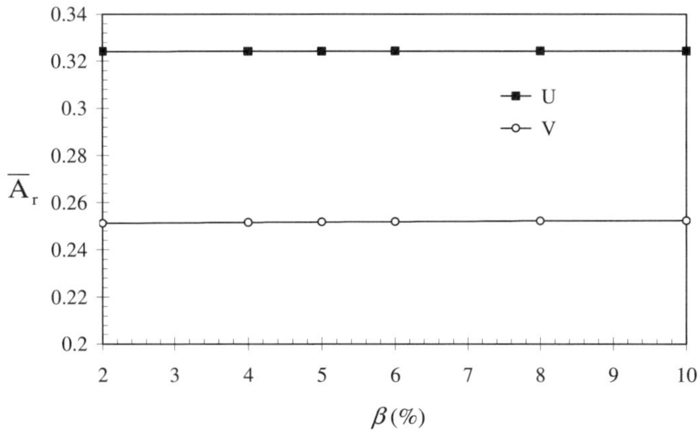

Fig. 5.26 Effect of damping ratio of in-filled trenches.

refracted when hitting the inner face of the in-filled trenches. The occurrence of optimal isolation at distance $L = 1$ is due to the combined effect of cancellation for the incident, reflected and refracted waves.

5.5.2 *Material damping ratio* β

By allowing the damping ratio β of the in-filled material to vary from 0.02 to 0.10, the results computed for vibration reduction have been presented in Fig. 5.26. For the horizontal vibration, the average amplitude reduction ratio $\overline{A}_r$ is around 0.324, and for the vertical vibration, it is around 0.252, with very little range of variation. Thus, it is concluded that the effect of material damping of the in-filled material in vibration reduction can be ignored in practice.

5.5.3 *Shear modulus ratio* (G_{sb}/G_{ss})

Let G_{sb} and G_{ss} respectively denote the shear modulus of the trench material and neighboring soil. From Fig. 5.27, one observes that following the increase in $\sqrt{G_{sb}/G_{ss}}$, better effect of isolation can be achieved, although there exists a limit for further improvement. For the horizontal vibration, the limit value for $\overline{A}_r$ is 0.32, which can hardly be improved for materials with $\sqrt{G_{sb}/G_{ss}} > 6$, and for vertical vibration,

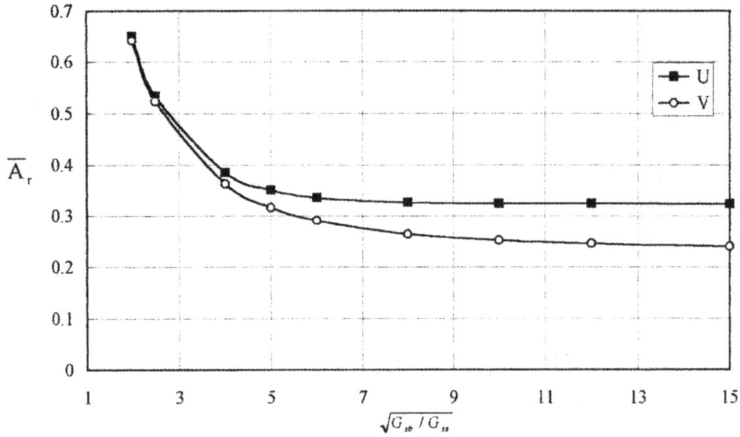

Fig. 5.27 Effect of shear modulus ratio of in-filled trenches (IR > 1).

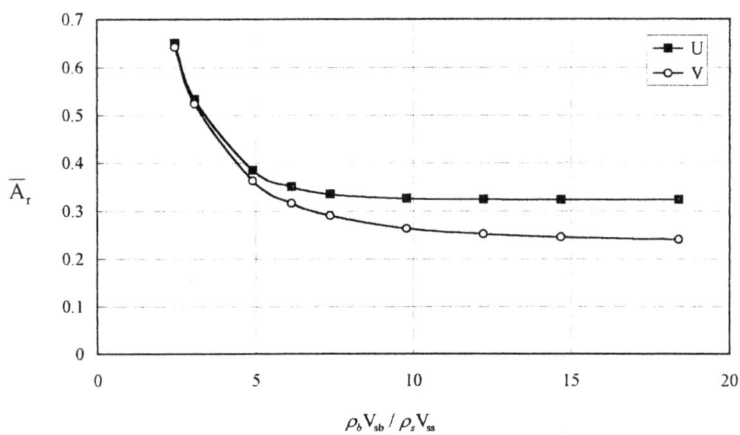

Fig. 5.28 Effect of impedance ratio (IR > 1) of in-filled trenches (by varying the shear modulus).

the limit value for $\overline{A}_r$ is 0.24, which is reached when $\sqrt{G_{sb}/G_{ss}} > 8$ for the material. For the case considered herein, the optimal shear modulus ratio for vibration reduction by in-filled trenches is $G_{sb}/G_{ss} = 64$.

In Fig. 5.28, the average amplitude reduction ratio $\overline{A}_r$ has been plotted against the impedance ratio (IR), as defined in Eq. (5.3), through variation of the shear modulus. The trend revealed herein is similar to that shown in Fig. 5.27.

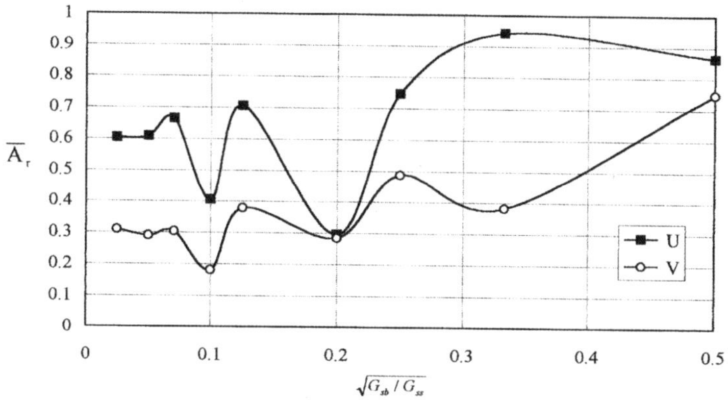

Fig. 5.29 Effect of shear modulus ratio of in-filled trenches (IR < 1).

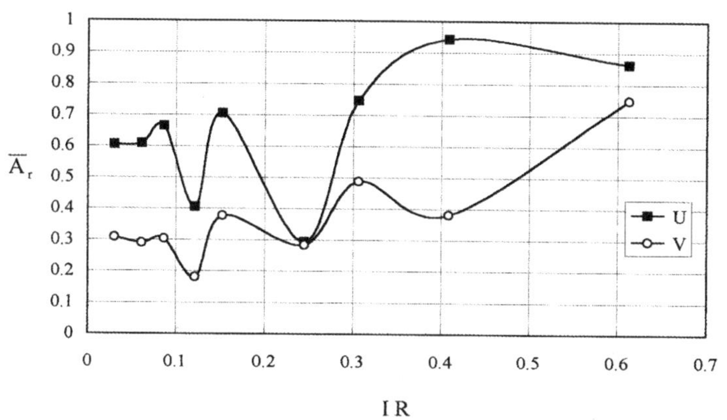

Fig. 5.30 Effect of impedance ratio (IR < 1) of in-filled trenches (by varying the shear modulus).

The above analysis has been conducted for in-filled trenches that are stiffer than the surrounding soil, i.e., with IR > 1. In the following, we shall consider the case where the in-filled trenches are softer than the surrounding soil, i.e., with IR < 1. From the results plotted in Fig. 5.29, it can be observed that for softer trenches, the average amplitude reduction ratio $\overline{A}_r$ oscillates in a drastic manner with respect to the increase in shear modulus parameter $\sqrt{G_{sb}/G_{ss}}$. Further, when $\sqrt{G_{sb}/G_{ss}}$ of the in-filled material approaches zero, the value of $\overline{A}_r$ approaches that for the open trenches (see Fig. 5.20). This is evidently true, since the open

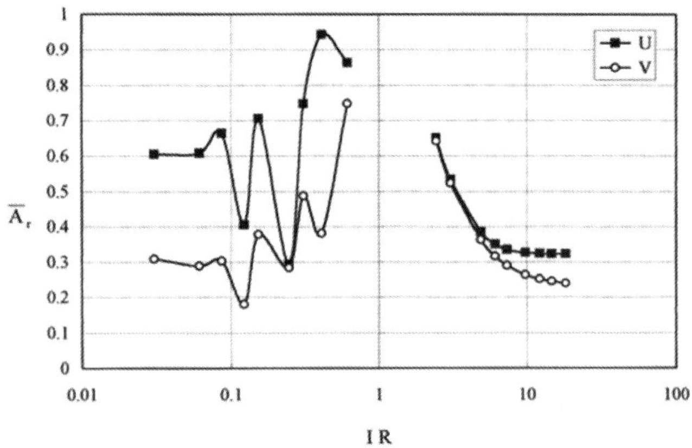

Fig. 5.31 Effect of impedance ratio of in-filled trenches (by varying the shear modulus).

trenches are merely a special case of the in-filled trenches with IR = 0. In Fig. 5.30, the average amplitude reduction ratio $\overline{A}_r$ of the ground surface has been plotted with respect to the impedance ratio by varying the shear modulus.

Combining the results in Figs. 5.28 and 5.30, we obtain the figure shown in Fig. 5.31, in which the abscissa for IR has been given in the logarithmic scale. Although good effect of isolation can be achieved for trenches with some specific IR values less than 1, the result is considered unreliable, due to appearance of the oscillatory behavior in $\overline{A}_r$. For this reason, softer trenches are not recommended for general engineering applications.

5.5.4 *Mass density ratio* (ρ_b / ρ_s)

Let ρ_b and ρ_s denote the mass density of the barrier and surrounding soil, respectively. As can be seen from the results given in Fig. 5.32, the average amplitude reduction ratio $\overline{A}_r$ is inversely proportional to the density ratio ρ_b / ρ_s, indicating that better effect of isolation can be achieved for larger ratio of ρ_b / ρ_s. The results in this figure were also redrawn in terms of the impedance ratio (IR) in Fig. 5.33. Clearly, as the IR value increases, the effect of isolation is improved. By comparing Fig. 5.33 with Fig. 5.28, one observes that the mass density is a parameter

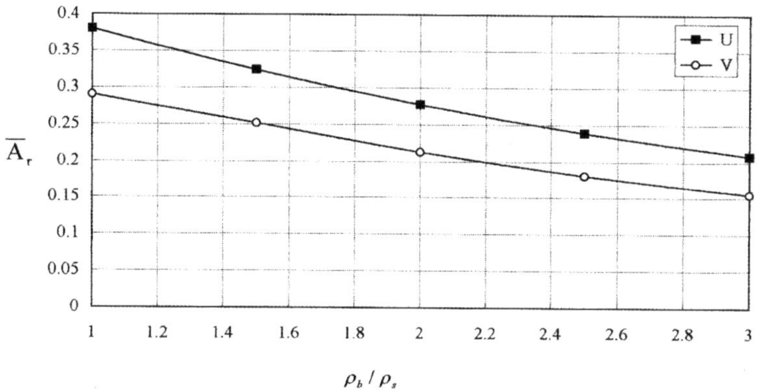

Fig. 5.32 Effect of mass density of in-filled trenches.

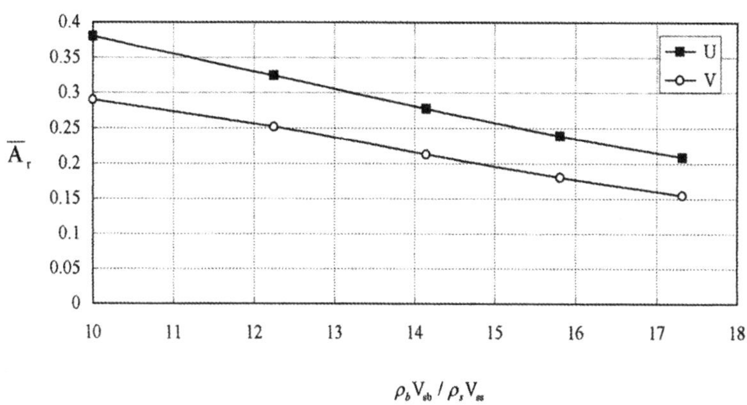

Fig. 5.33 Effect of impedance ratio of in-filled trenches (by varying the mass density).

more influential than the shear modulus in the range $10 < \mathrm{IR} < 17$. However, it is also realized that the range of mass density for selection of in-filled materials is quite small in practice. Thus, the difference between the maximum and minimum values of $\overline{A}_r$ shown in Fig. 5.33 is not as large as that shown in Fig. 5.28.

5.5.5 *Poisson's ratios* (ν_b, ν_s)

The Poisson's ratios of the trenches (ν_b) and soil (ν_s) are adopted as the variables. The results obtained for the horizontal and vertical responses

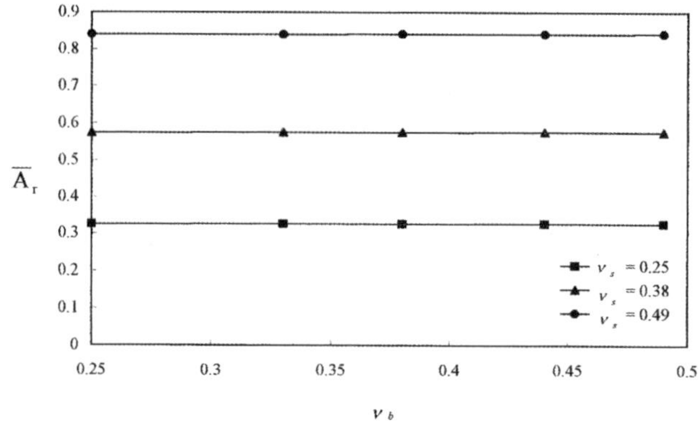

(a)

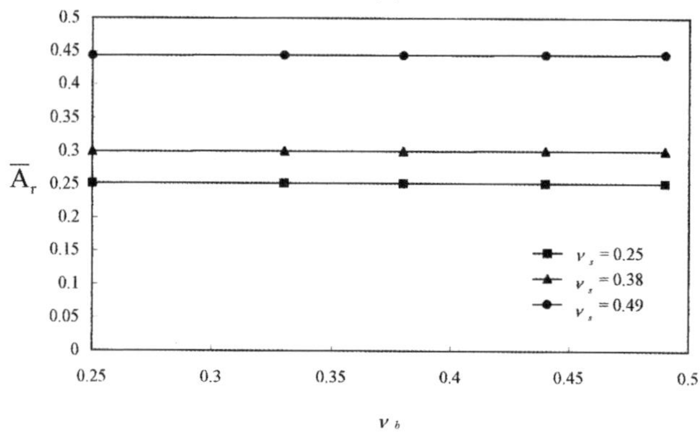

(b)

Fig. 5.34 Effect of Poisson's ratio of in-filled trenches on average amplitude reduction ratio: (a) horizontal; (b) vertical.

have been drawn in Figs. 5.34(a) and (b), respectively with respect to Poisson's ratio ν_b of the trenches. Clearly, the Poisson's ratio ν_b of the in-filled trench has basically no effect on the surface response. However, an increase in Poisson's ratio ν_s of the soil causes the isolation effect to decline significantly. The reason for this is that increasing Poisson's ratio of the soil will lead to the rise of the velocity and wavelength of the compressional waves, as indicated by the wavelength equation: $\lambda_p = 2\pi C_p / \omega$. It therefore requires deeper trenches to achieve the same degree of isolation. Such a phenomenon is more pronounced for the horizontal than vertical response, due to the relatively large contribution of the compressional waves propagating in the horizontal direction.

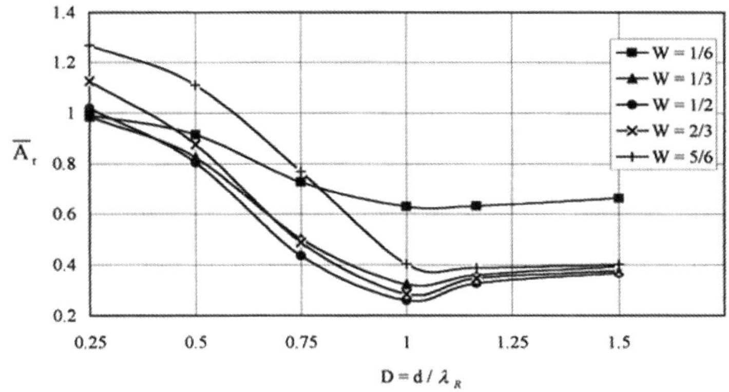

(a)

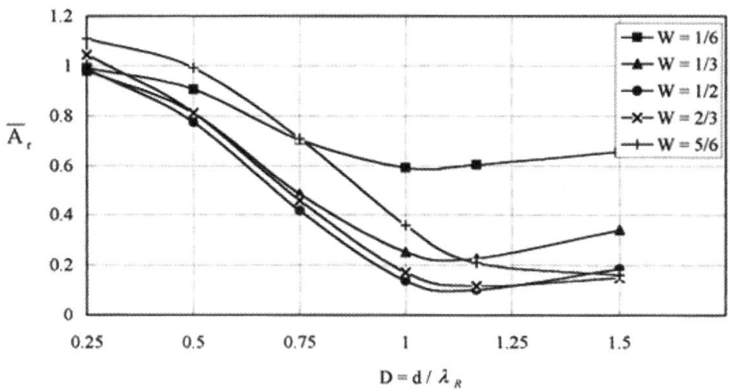

(b)

Fig. 5.35 Effect of depth *D* of in-filled trenches for given width *W* on average amplitude reduction ratio: (a) horizontal; (b) vertical.

5.5.6 *Depth D and width W of in-filled trench*

For trenches that are stiffer than the surrounding soil, i.e., with IR > 1, the average amplitude reduction ratio $\overline{A}_r$ has been plotted with respect to the trench depth *D* in Figs. 5.35(a) and (b) for the horizontal and vertical displacements, respectively, for different trench widths *W*. Clearly, better isolation can be achieved by deeper trenches. However, only marginal improvement can be made for trenches with depths *D* greater than 1, due to the fact that for trenches with certain depths, the bottom part of the trenches can hardly affect the response of the soil on the free surface.

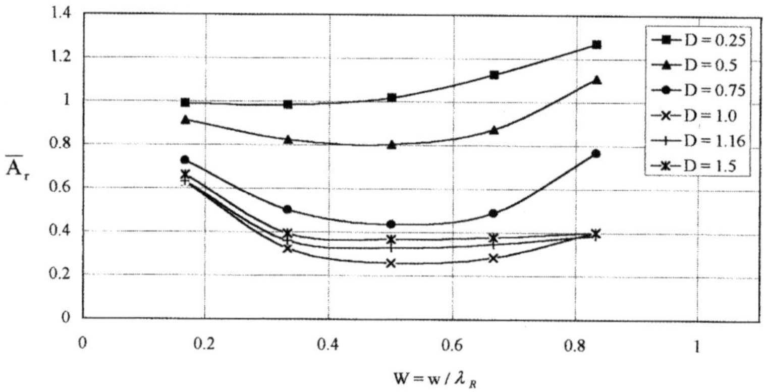

(a)

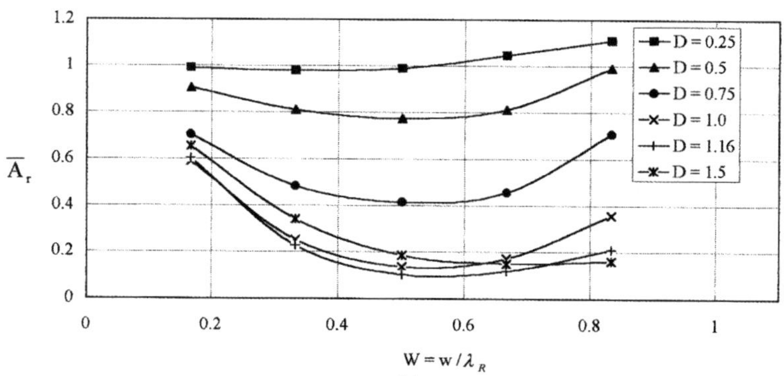

(b)

Fig. 5.36 Effect of width *W* of in-filled trenches for given depth *D* on average amplitude reduction ratio: (a) horizontal; (b) vertical.

In Fig. 5.36, the surface responses have been plotted with respect to the width *W* of the trenches, given the trench depth *D*. As can be seen, for different trench depths, the effect of trench width on isolation is generally different. Namely, for trenches that are not deep enough, the effect of isolation decreases as the trench width increases, while for trenches that are deep enough, better isolation effect can be achieved by increasing the trench width. In general, the depth of the trench is a parameter more crucial than the width. For trenches with a depth *D* less than 0.5, basically little effect of isolation can be achieved, regardless of the width used.

5.6 Effect of Frequencies of Traffic Loads

In the preceding parametric studies, all the geometric parameters have been normalized with respect to the Rayleigh wavelength, so that the results obtained are independent of the excitation frequency of the driving loads and can be applied directly to design of barriers for each specific frequency. However, as the traffic loads cover a wide range of frequencies, it becomes necessary to consider the influence of various frequencies on the effectiveness of isolation.

In this section, the screening effect of the elastic foundation, open trenches, and in-filled trenches with real dimensions will be studied for traffic loads of various frequencies. Assume the soil underlying the railway to be a uniform elastic half-space, with the shear and dilatational wave velocities as 101.11 and 175.19 m/s, respectively. The railway has a half-width of b = 1.5 m and depth of d = 50 cm. As shown in Fig. 5.2(c), the depth and thickness of the elastic foundation are chosen as e = 0.5 m and t = 0.5 m. For the open trenches, the depth and width are d = 3.0 m and w = 0.5 m, and the distance from each of the trenches to the railway center is l = 3.0 m. And for the in-filled trenches, the following data are adopted: d = 3.0 m, w = 1.0 m, and l = 3.0 m; other material parameters not specified are the same as those listed in Table 5.1.

The average amplitude reduction ratios computed for the three barriers have been compared in the frequency domain in Fig. 5.37. As can be seen, for waves of higher frequencies, the three barriers are

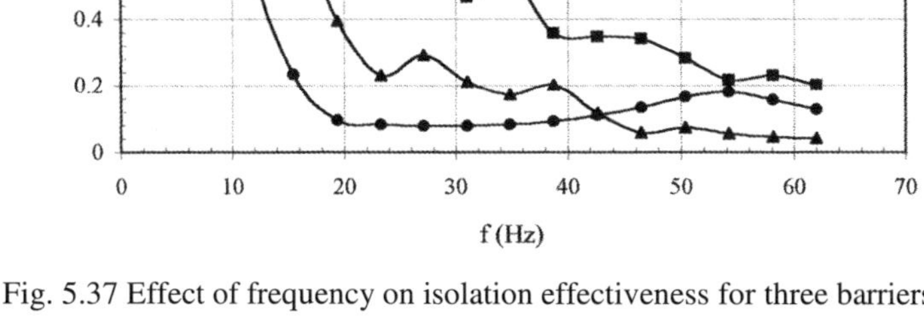

Fig. 5.37 Effect of frequency on isolation effectiveness for three barriers.

generally effective in reducing the vibration amplitudes; however, for waves of lower frequencies, the response can be adversely magnified using either the open trenches or elastic foundation.

5.7 Concluding Remarks

A systematic study of the parameters affecting the effectiveness of the three wave barriers, i.e., the elastic foundation, open trenches, and in-filled trenches, in isolating traffic-induced vibrations has been performed. The conclusions acquired from the parametric study in the preceding sections can be summarized as follows:

(1) The effectiveness of each barrier in isolating the ground-borne vibrations depends largely on its dimensions relative to the wavelength of the underlying soil, especially to those of the body waves when active isolation is concerned. In general, it is impractical to use the barriers presented herein to isolate vibrations with long wavelengths.
(2) For elastic foundations, the mattress should have a thickness of $E > 0.05$ so that acceptable isolation effect can be achieved.
(3) For open trenches, the normalized trench depth is the dominant factor; the normalized width is generally not important, except for shallow trenches.
(4) For in-filled trenches, the normalized depth and width should meet the condition of $D > 1$ and $W > 0.3$ in order to achieve satisfactory effect of isolation.
(5) For practical purposes, the influence of density and damping ratio of the three barriers, as well as the location of trenches, on the isolation of train-induced vibrations can in generally be ignored.
(6) Poisson's ratio is a critical factor for the active isolation problems, since it can affect considerably the compressional wavelength.
(7) In-filled trenches that are stiffer than the surrounding soil perform generally better than softer ones. For elastic foundations, the reverse is true.
(8) All the three wave barriers are suitable for isolating vibrations associated with waves of higher frequencies.

Chapter 6

Vibration Reduction of Buildings Located Alongside Railways

In this chapter, the finite/infinite element approach presented in Chapter 3 is applied to analysis of practical soil-structural systems. Central to the present study is the adoption of the mesh expansion scheme presented in Section 3.5 for repetitively generating the impedance matrix for the next lower frequency from the current frequency considered. The structure considered is assumed to be of the box type and the soil is composed of two layers. By varying the shear modulus of the bottom layer, both a uniform half-space and a soil deposit with bedrock can be simulated. A parametric study is conducted to investigate the effect of *open* and *in-filled trenches* in reducing the structural vibration caused by a passing train simulated as a harmonic line load. The key parameters that dominate the performance of wave barriers in reducing the structural vibrations are identified. The results presented herein serve as a useful guideline for the design of open and in-filled trenches with regard to wave reduction.

6.1 Introduction

As the high-technology community enters the so called "age of the nanometer," traffic-induced vibrations in buildings or high-precision factories have become an issue of great concern. Most highly developed cities or metropolises in the world have encountered the problem that transportation constructions inevitably come across or close to vibration-sensitive residential or industrial buildings. Although the vibrations induced by adjacent traffic may not result in collapse of structures in the way as earthquakes do, they have been known to cause malfunction of delicate instruments located inside the buildings. In vibration-sensitive environment, it is important that the major vibration sources be identified, analyzed and isolated, as part of the facility design process.

The ground-borne vibration due to the railway traffic has been a subject of increasing research in recent years due to the construction of highspeed railways and subways worldwide. By and large, previous research carried out along these lines can be classified into four categories.

The first category relates to design of special track systems as a first aid to mitigate the transmission of vibrations directly generated by the passing trains. Elaborate vehicle-track interaction models have been used in this regard. For instance, the *floating slab track* (Grootenhuis 1977; Wilson *et al.* 1983; Balendra *et al.* 1989), which is composed of concrete slab track and resilient supporting elements, has been proved to be a very effective means for isolating the vibration at frequencies above the resonance frequency of the floating slab system.

The second category is to investigate the influence of train speed on ground vibrations and to investigate the propagation of vehicle-induced vibrations via the soils to areas alongside the railways. Both theoretical methods with the assumption of a uniform elastic half-space and experimental investigations have been employed in this regard (Gutowski and Dym 1976; Dawn and Stanworth 1979; Alabi 1992; Krylov and Ferguson 1994; Krylov 1995; Heckl *et al.* 1996; Madshus *et al.* 1996).

The third category is focused on the design of buildings with certain features to mitigate the traffic-induced vibrations. For instance, theoretical solutions were obtained by Takahashi (1985, 1986a,b) using the plate elements to simulate a box-type structure subjected to a harmonic line load on the surface of a viscoelastic half-space under different boundary conditions. An optimal choice of the structure thickness and material was proposed based on the parametric study considered.

The last category is concerned with the installation of various *wave barriers*, such as open and in-filled trenches (Woods 1968; Lysmer and Waas 1972; Segol *et al.* 1978; Emad and Manolis 1985; Beskos *et al.*

1986; Al-Hussaini and Ahmad 1991, 1996; Ahmad *et al.* 1996; Yang and Hung 1997), buried concrete plates (Schmid *et al.* 1991; Antes and von Estorff 1994), or aligned piles (Boroomand and Kaynia 1991), between the railways and the buildings to be protected. These barriers are not only helpful for isolating the vibration caused by passing trains, but also for reducing ground-transmitted waves generated by other vibration sources, such as machines, vehicles, blasting, etc. In this chapter, only the fourth category will be dealt with.

Concerning the isolation of structures from ground-borne vibrations, a large volume of research has been conducted in the past using either analytical or experimental methods. In early studies employing the analytical approaches, restrictions were often imposed on the geometry and material properties of the problem considered, as closed-form solutions cannot be readily obtained for other complex conditions. On the other hand, although the results obtained by the experimental methods are considered to be most reliable and close to real situations, an exhausted field test may cost a lot.

Starting from the mid 1970s, various numerical methods emerged as effective tools for solving the wave propagation problems. By the lumped mass method, Lysmer and Waas (1972) studied the effectiveness of a trench in reducing the horizontal shear wave motion induced by a harmonic load acting on the rigid footing lying on the horizontal soil layer. Segol *et al.* (1978) used finite elements along with special non-reflecting boundary to investigate the isolation efficiency of open and bentonite-slurry-filled trenches in layered soils. The axisymmetric infinite elements that are capable of dealing with multiple wave components have been employed by Yang and Yun (1992) and Yun and Kim (1995) to deal with soils of unbounded domains. Using the finite/infinite element scheme, the isolation effects of open trenches, in-filled trenches and elastic foundations were investigated by Yang and Hung (1997) in their parametric study.

In the last decade, a great amount of studies on wave propagation problems were performed by the boundary element method, including Beskos *et al.* (1986), Al-Hussaini and Ahmad (1991, 1996), Ahmad *et al.* (1996), among others. One advantage of the boundary element method is that the radiation damping resulting from geometric attenuation of waves traveling to infinity can be accurately taken into account. However, it is not suitable for simulating the realistic situations where irregularities may exist either in the geometry or materials of the structures and underlying soils.

As far as vibrations of the buildings, foundations, and surrounding soils are concerned, a finite element representation remains the most convenient choice, considering its versatility in treating various irregularities. In the study by Yang *et al.* (1996) or in Chapter 3, it has been demonstrated that radiation damping of the far field can be accurately simulated by the Q8-compatible infinite element derived. In this chapter, the same finite/infinite element method will be adopted. To take advantage of both types of elements, the structure and soils in the near field will be modeled by the finite elements, and the radiation property of the far field by the infinite elements. Such an approach is attractive in that both the finite and infinite elements can be assembled using the conventional finite element procedures, with no additional degrees of freedom required for the far field.

In the literature, most studies concerning wave isolation were focused on reducing the surface vibration. In comparison, relatively few works have been reported on the reduction of building response using open and in-filled trenches. The purpose of this chapter is to conduct a parametric study on the open or in-filled trenches for reducing the traffic-induced vibrations on buildings. The soil-structure model considered herein is two-dimensional. The building is simulated as a box structure, and the moving train as a unit harmonic line load. Layered soils with or without bedrock are considered. A parametric study is first conducted for both the open and in-filled trenches in reducing the building vibration at each specific frequency. Then, the screening effect of the trenches over a wide range of frequencies is examined.

Previously, rather few studies have been carried out to evaluate the screening effect of trenches over a wide range of frequencies, due to the diverse requirements existing for the extent and spacing of the finite element grid under different frequencies. However, such a problem can be easily overcome using the dynamic condensation technique presented in Section 3.5. By this technique, the far-field impedance matrices for the entire range of frequencies can be repetitively derived from the mesh

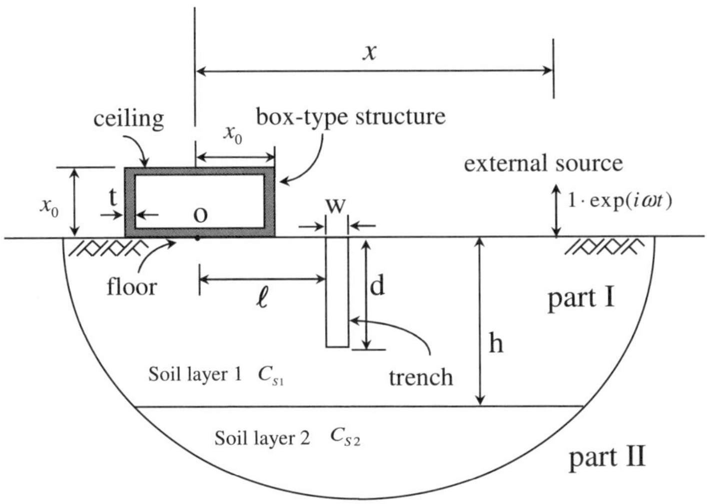

Fig. 6.1 Typical soil-structure model.

established for the highest frequency of the problem considered. The materials presented in this chapter have been rewritten primarily from the paper by Hung *et al.* (2001).

6.2 Problem Formulation and Basic Assumptions

The soil-structure system considered is shown in Fig. 6.1, which consists of a near field (part I) and a semi-infinite far field (part II). Typically, Part I contains the region and components that may be irregular in geometry or material, including the building, railway (i.e., source of vibration), wave barriers, and underlying soils, all of which will be represented by the 8-node quadratic (Q8) element as mentioned before. Part II covers the soils in the far field extending to infinity, which will be modeled by the Q8-compatible infinite element derived in Chapter 3.

To simulate the action of moving train loads, a line load is applied at the center of the railway. Such an approximation is acceptable provided that the point of interest from the railway track is approximately less than $1/\pi$ times the length of train (Gutowski and Dym 1976). Moreover, since a general traffic load can always be transformed into the frequency domain and expressed in a series of harmonic functions, only harmonic line loads are considered in this chapter. It follows that the displacements generated within the soil-structure system may be assumed to oscillate harmonically with time as well. Based on the assumption that the length of the building in the direction parallel to the railway track is relatively long compared with the width and height of the building on the profile perpendicular to the railway track, plain strain condition is assumed to apply. The soil is assumed to be an isotropic viscoelastic medium (with hysteresis damping), with or without bedrock.

According to Chapter 3, for the case of a uniform half-space subjected to a line load on the free surface, the displacement amplitude decay factor α appearing in Eq. (3.19) should be selected as $\alpha = 1/(2R)$ for modeling the regions where the body waves are dominant, where R denotes the distance between the source of vibration and the far field boundary. Since the Rayleigh waves do not decay on the free surface under the same loading condition, it is suggested that $\alpha = 0$ be used for regions near the free surface.

6.3 Scheme for Generating Finite/Infinite Element Mesh

As was stated, the near field including the building and trench will be modeled by the Q8 plane elements, and the far field by the Q8-compatible infinite elements. For a soil vibration problem, the maximum element size and minimum mesh size required depend on the wavelength or the frequency ω of the (dominant) traveling waves considered. Hence, for waves of lower frequencies, a finite element mesh of larger extent R should be used. On the contrary, for waves of higher frequencies, an element of smaller size L should be used. Based on the convergence tests conducted in Section 3.4, the following are the requirements for the finite element mesh that should be used to ensure the accuracy of the computed results: element size $L \le \lambda_s / 6$ and mesh extent $R \ge 0.5\lambda_s$, where λ_s denotes the shear wavelength. See Fig. 3.6 for definitions of the element size L and mesh extent R.

As was pointed out in Chapter 3, it is rather difficult, or at least computationally inefficient, to create a finite/infinite element mesh that

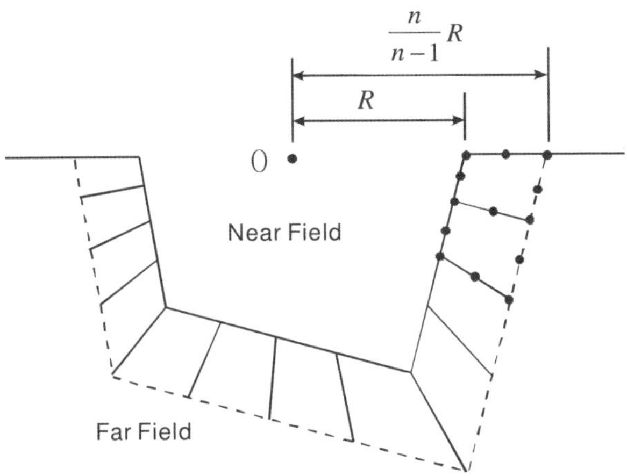

Fig. 6.2 Schematic diagram for condensation to the inner boundary.

can meet the diverse needs of waves of both low and high frequencies for the soil-structure system considered. The following is a summary of the procedure presented in Section 3.5 based on the dynamic condensation technique for generating the far field impedance matrices for the full range of frequencies considered using exactly the finite/infinite element mesh established for the highest frequency of concern.

Consider the two far fields as indicated by the one with a solid boundary and the other one with a dashed boundary in Fig. 6.2. Let the two far fields be similar with respect to point O, in the sense that along each radial line originating from point O, the ratio of the distance between point O and the point on the dashed line to the distance between point O and the corresponding point on the solid line remains equal to $n/(n-1)$, where n is an integer, assuming the material properties to be identical along each radial direction. Let $\Delta\omega$ denote a constant frequency increment. Based on the property observed for the non-dimensional wave equation presented in Eq. (3.52), it can be ascertained that for the two-dimensional problem, the far field impedance $[S]$ computed for the frequency $\omega = (n-1)\Delta\omega$ at the outer boundary (dashed line) with an extent of $[n/(n-1)]R$ should be equal to the far field impedance $[S]$ for the frequency $\omega = n\Delta\omega$ at the inner boundary (solid line) with distance R.

In analysis, one may start by calculating the far field impedance [S] for the highest frequency $\omega = n\Delta\omega$ at the inner boundary with distance R, i.e., by assembling the structural impedance matrices of the infinite elements over the inner boundary, and set the far-field impedance [S] for $\omega = (n-1)\Delta\omega$ at the outer boundary with distance $[n/(n-1)]R$ equal to the impedance [S] for $\omega = n\Delta\omega$. Then, one can divide the region enclosed by these two boundaries into a number of Q8 elements, with the distance between any two adjacent nodes on the outer boundary set equal to $n/(n-1)$ times that of its corresponding distance on the inner boundary.

By condensing all the far-field degrees of freedom, including those of the newly inserted Q8 elements and those on the outer boundary, to the nodes on the inner boundary, one can obtain the impedance matrix [S] for the next highest frequency $\omega = (n-1)\Delta\omega$ at the inner boundary. The above procedure can be repeated to yield the far-field impedance matrices [S] for all the remaining frequencies, i.e., $\omega = (n-2)\Delta\omega$, $\omega = (n-3)\Delta\omega$, ..., etc. It should be noted that although the location of the outer boundary moves as the value $n/(n-1)$ changes, such a condensation process can be easily carried out by internal computer codes. Consequently, only the finite element mesh for the near field need be established prior to analysis, while the rest can be easily dealt with automatically by the computer program.

6.4 Parametric Studies for Open Trenches

For the purpose of wave reduction, an open trench will be constructed between the vibration source and the building to be protected, as was schematically shown in Fig. 6.1. In this section, a uniform half-space, i.e., with no bedrock, is assumed for the soils underlying the building. The screening effect of various parameters of open trenches on waves of a specific frequency will be investigated. Based on the criteria stated earlier for mesh generation, a finite/infinite element mesh that meets all the demands was created, as shown in Fig. 6.3. The screening effect of the wave barrier can be evaluated using the *amplitude reduction ratio* A_r defined as

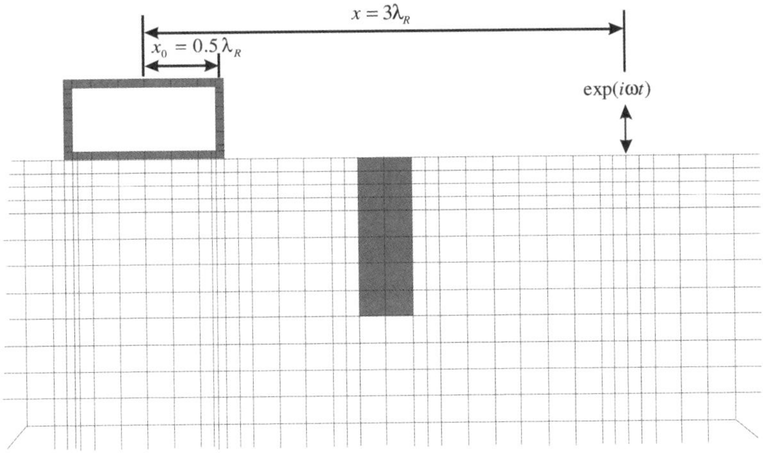

Fig. 6.3 Finite/infinite element mesh (with no bedrock).

$$A_r = \frac{d_a}{d_b}, \tag{6.1}$$

where d_a denotes the average displacement amplitude over the floor (or ceiling) surface of the structure with the wave barrier, and d_b the corresponding average displacement amplitude for the corresponding case with no barrier. Obviously, a smaller value of A_r implies that a better effect of isolation has been achieved by the barrier.

In what follows, the effect of each parameter of the trench will be studied. To make the analysis results independent of the excitation frequency, all the geometric parameters used and defined in Fig. 6.1 will be normalized with respect to the Rayleigh wavelength λ_R of the soil,

$$\begin{aligned} w &= W \cdot \lambda_R, \\ d &= D \cdot \lambda_R, \\ l &= L \cdot \lambda_R, \\ x_0 &= X_0 \cdot \lambda_R, \\ t &= T \cdot \lambda_R, \\ x &= X \cdot \lambda_R, \\ h &= H \cdot \lambda_R, \end{aligned} \tag{6.2}$$

where the parameters W, D, L, X_0, T, X and H are dimensionless. Unless otherwise noted, $D = 1$ and $L = 3/2$ will be adopted exclusively for both the open and in-filled trenches, and $W = 1/3$ and $W = 1/2$ for the open and in-filled trench, respectively. In the parametric study, the soil and building properties, as well as the location of the external load, will be kept constant, i.e., $X_0 = 1/2$, $T = 1/20$ and $X = 3$ will be adopted throughout the analysis and the soil extending to infinity ($H \to \infty$) is assumed to be homogeneous and viscoelastic.

All the material properties assumed for the standard case have been listed in Table 6.1. Using the present data, the shear wave velocity C_s computed for the soil is 160 m/s, and the Rayleigh wave velocity is 150 m/s.

Table 6.1 Material properties of soil-structure model.

	Shear modulus G (MPa)	Young's modulus E (MPa)	Poisson's ratio ν	Density ρ (kg/m^3)	Damping ratio β (%)
Soil	43.52	-	0.33	1700	5
In-filled trench	1840	-	0.25	2700	5
Structure	-	21000	0.2	2300	2

6.4.1 *Normalized distance L from the structure*

The results computed for the ceiling and floor of the structure by varying the distance L between the structure and the trench have been plotted in Fig. 6.4. As can be seen, the difference between the ceiling (i.e., building) and floor (i.e., ground) responses with regard to the effect of isolation is not really noticeable. For both the ceiling and floor responses, it can be observed that the isolation efficiency of the trench deteriorates when it is located either close to the source (i.e., with $L = 2.5$) or to the structure (i.e., with $L = 0.75$). One possible reason for this is that at places near the external source, the body waves play a role more important than the surface waves. Since the body waves decay slowly downward from the surface, a great portion of these waves can pass through below the trench

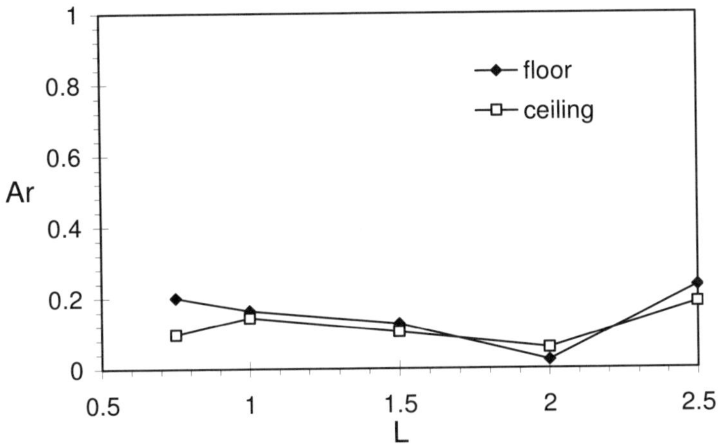

Fig. 6.4 Effect of normalized distance from the structure to open trench.

screening. On the other hand, for the case when the open trench is located near the building, the unstable nature of the trench walls will cause the adjacent building to vibrate without restraint and therefore reduce the effect of isolation.

6.4.2 *Normalized depth D and width W of trench*

By changing the depth D of the open trench, the results computed for the vertical response of the floor of the structure, which can also be interpreted as the ground response, have been plotted against the normalized width W in Fig. 6.5. As can be seen, for shallow trenches, say, with $D = 1/3$ and 2/3, the greater the trench width W, the worse the effect of isolation is. In contrast, for deep trenches, say, with $D = 1$, 4/3, 3/2, greater width may result in better screening effect, although the phenomenon is only marginal. One reason for this is that as the depth D is small, the use of wider trenches, which means wider free surface, allows the body waves to be transformed into the surface waves, which suffer little geometric attenuation and can travel quite far. As a consequence, the influence of width becomes adversely pronounced. Besides, the figure indicates that for the depth D greater than 1, the influence of the trench depth becomes rather minor. However, for small

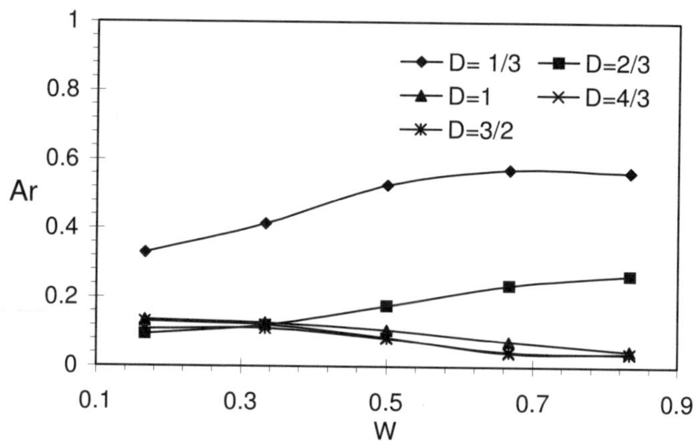

Fig. 6.5 Effect of normalized depth and width of the open trench on floor response.

values of D, say, for $D \leq 2/3$, an increase in the trench depth can improve significantly the effect of isolation. It should be added that a trend similar to the one given in Fig. 6.5 exists for the ceiling response, which is not shown here for brevity.

6.5 Parametric Studies for In-Filled Trenches

The screening effects of various parameters of the in-filled trench shown in Fig. 6.1 on waves of a specific frequency will be investigated in this section. Again, a uniform half-space is considered. The finite/infinite element mesh used is similar to the one in Section 6.4, i.e., the one given in Fig. 6.3.

6.5.1 *Normalized distance L from the structure*

The amplitude reduction ratios A_r for both the ceiling and floor of the building have been plotted against the distance L between the building and in-filled trench in Fig. 6.6. From this figure, a trend opposite to that of the open trench is observed, that is, the screening effect appears to be greater when the in-filled trench is located either close to the source (with $L = 2.5$) or to the building (with $L = 0.75$). It is hard to explain why the in-filled trench shows better isolation effect when located closer to

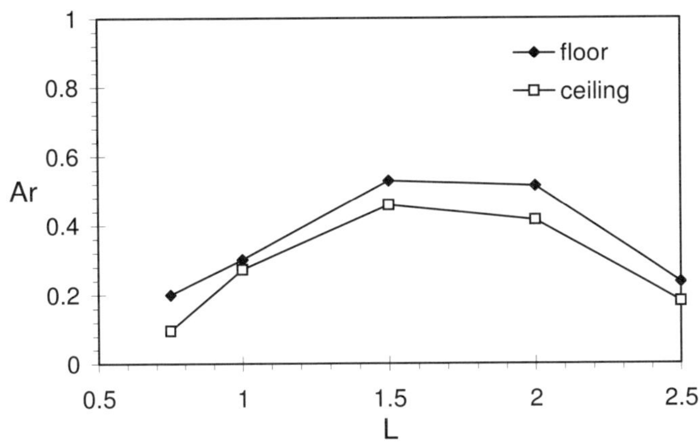

Fig. 6.6 Effect of normalized distance from the structure to in-filled trench.

the source, i.e., with L = 2.5, due to the complex nature of the wave propagation phenomenon, which may involve reflection, refraction, diffraction, mode conversion of waves and soil-structure interaction. Nevertheless, it is natural to see that the in-filled trench performs better when located closer to the building, as it is stiffer than the surrounding soils, which tends to constrain the adjoining building from being affected by incident waves. The other observation from Fig. 6.6 is that a better effect of isolation can be generally achieved for the ceiling than for the floor using the in-filled trench. This is an indication of the effectiveness of the in-filled trench in reducing the vibrations transmitted upward inside the building.

6.5.2 *Normalized depth D and width W of trench*

To investigate the effect of the trench dimensions in reducing the floor response of the structure, the influence of the trench width W is investigated for different values of trench depth D. As can be seen from Fig. 6.7, the deeper the trench, the better the isolation effect is. This figure also reveals that for shallow trenches, say, with $D \leq 2/3$, an increase in the trench width does not necessarily lead to better isolation. The result for the ceiling response is similar to the one given in Fig. 6.7 for the floor response, which is not shown here.

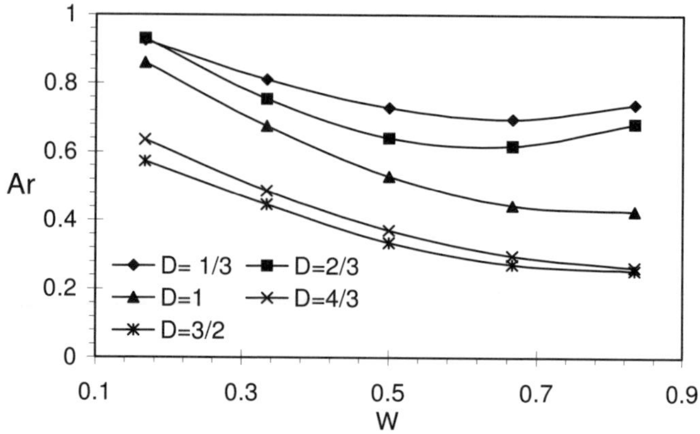

Fig. 6.7 Effect of normalized depth and width of the in-filled trench on floor response.

6.5.3 *Impedance ratio of in-filled trench*

The *impedance ratio* (IR), as defined in Eq. (5.3), is a parameter widely used by geotechnical engineers to distinguish whether a wave barrier is soft or hard with respect to the surrounding soil, also given below:

$$\mathrm{IR} = \frac{\rho_b V_b}{\rho_s V_s}, \tag{6.3}$$

where ρ_b and ρ_s denote the mass density of the barrier and the soil, respectively, and V_b and V_s the corresponding shear wave velocities. Since the shear wave velocity V can be related to the shear modulus and mass density, i.e., $V = \sqrt{G/\rho}$, the preceding equation can be given in the following equivalent form:

$$\mathrm{IR} = \sqrt{\frac{\rho_b G_b}{\rho_s G_s}}, \tag{6.4}$$

where G_b and G_s denote the shear modulus of the in-filled trench and the underlying soil, respectively.

In this section, only the shear modulus G_b of the in-filled trench is allowed to vary, while the shear modulus G_s of the soil and the mass densities for both the trench and soil are kept constant. The amplitude reduction ratio A_r for both the ceiling and floor responses has been

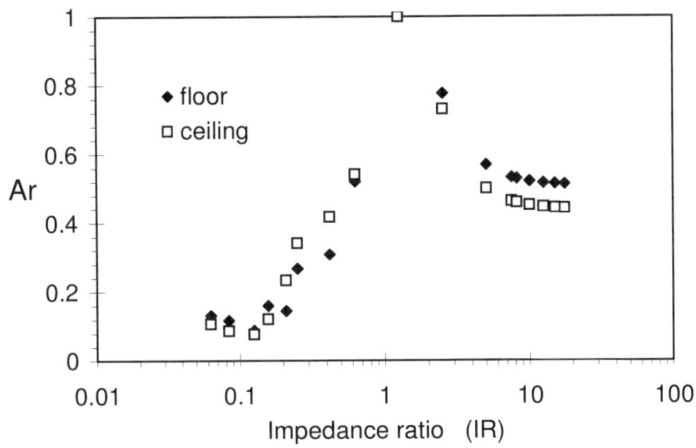

Fig. 6.8 Effect of impedance ratio for the in-filled trench.

plotted against the impedance ratio IR in Fig. 6.8. As can be seen from the right-hand part of the figure, for trenches with IR > 1, i.e., for trenches stiffer than the soil, the effect of isolation can be improved by increasing the impedance ratio IR. However, the amplitude reduction ration A_r tends to approach an asymptotic value of 0.5 and 0.45 for the floor and the ceiling, respectively, as the barrier gets stiffer and stiffer. In general, the value of IR = 7 can be regarded as an optimal choice for hard barriers.

On the other hand, from the left-hand part of Fig. 6.8, it can be observed that for barriers with IR < 1, i.e., softer than the soil, the amplitude reduction ratio A_r declines dramatically as the impedance ratio IR becomes smaller and reaches a minimum of 0.05 for IR = 0.11. A scrutiny of this figure reveals that as the impedance ratio IR reduces to zero, the amplitude reduction ratio A_r approaches the value given in Fig. 6.5 for the open trench with the same dimensions of W = 1/2 and D = 1. Such a phenomenon can be easily conceived since the open trench is nothing but a special case of the in-filled trench with zero shear modulus, i.e., with IR = 0. From the point of construction, an in-filled trench with IR < 1 can be achieved using properly designed soil-bentonite mix as the fill material.

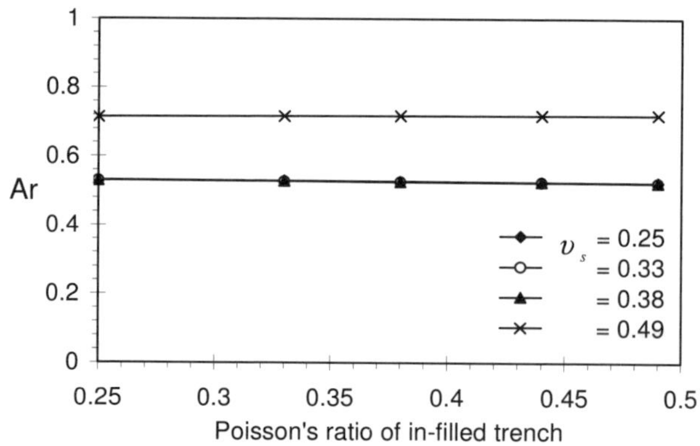

Fig. 6.9 Effect of Poisson's ratio for the in-filled trench.

6.5.4 *Poisson's ratios* (ν_b, ν_s)

The amplitude reduction ratio A_r for the floor response over different Poisson ratios of the trench (ν_b) and the underling soil (ν_s) has been plotted in Fig. 6.9, in which the results for ν_s = 0.25, 0.33, 0.38 are almost coincident. Clearly, for soils with Poisson's ratio ν_s in the range of 0.25 to 0.38, no difference can be observed for the isolation response.

However, as Poisson's ratio ν_s of the soil equals 0.49, the efficiency of isolation declines drastically. Such a phenomenon can be attributed to the relation between the Poisson's ratio of the soil and the compressional wave velocity C_p and therefore the wavelength. The compressional wave velocity C_p can be related to the shear wave velocity C_s as:

$$C_P = \sqrt{\frac{2(1-\nu_s)}{(1-2\nu_s)}} C_S . \tag{6.5}$$

As Poisson's ratio ν_s increases from 0.25 to 0.38, the compressional wavelength increases by 1.31 times, while as Poisson's ratio ν_s increases from 0.25 to 0.49, the compressional wavelength increases by more than 4 times. Hence, for the case with ν_s = 0.49, deeper trenches are required to achieve the same degree of isolation. Besides, the figure indicates that changing ν_b causes basically no influence on the isolation response. The above observations remain valid for the isolation of ceiling response, which is not shown here. As a side note, the effects of the damping ratio and mass density of the soil on the isolation of building response are not presented herein, since they are generally small and not suitable for practical applications.

6.6 Effect of Frequencies and Soil Conditions

In the preceding parametric studies, all the results obtained are independent of the frequency of the applied load, because all the geometric parameters considered have been normalized with respect to the Rayleigh wavelength of the soil. However, as the traffic loads cover a wide range of frequencies and the building size x_0 remains fixed in practice, the parametric studies conducted in the preceding section under the assumption that the half-width of the building equals half of the Rayleigh wavelength of the soil, i.e., $x_0 = \lambda_R / 2$, appear to be insufficient or unrealistic for the purpose of engineering applications.

In this section, the screening effect of the open and in-filled trenches will be investigated in a more realistic manner considering a range of frequencies for the real soil conditions. The model adopted herein is exactly the one given in Fig. 6.1. For the present purposes, however, all the geometric parameters will be normalized respect to the half-width x_0 of the building, rather than the Rayleigh wavelength λ_R of the soil.

In particular, the thickness of each side of the building is assumed to be $t = x_0 / 10$, the vibration source is located at a distance of $x = 4.5x_0$ from the center of the building, and the depth of soil layer 1 is $h = 2x_0$. For the open trench, the depth and width are selected as $d = 1.5x_0$, $w = 0.25x_0$, and the distance from the trench to the center of the building is $l = 1.25x_0$. As for the in-filled trench, the following data are used: $d = 1.5x_0$, $w = 0.5x_0$, and $l = 1.25x_0$. All the material properties for soil layer 1, as well as those for the structure and the trench, are the same as those listed in Table 6.1.

Two different conditions will be considered for the soil. One is a uniform half-space soil with no bedrock, which can be achieved by setting the shear wave velocity of layer 2 equal to layer 1, i.e., $C_{s2} = C_{s1} = 160$ m/s. The other is a soil layer supported by bedrock, which can be achieved by assigning a rather high value to the shear wave

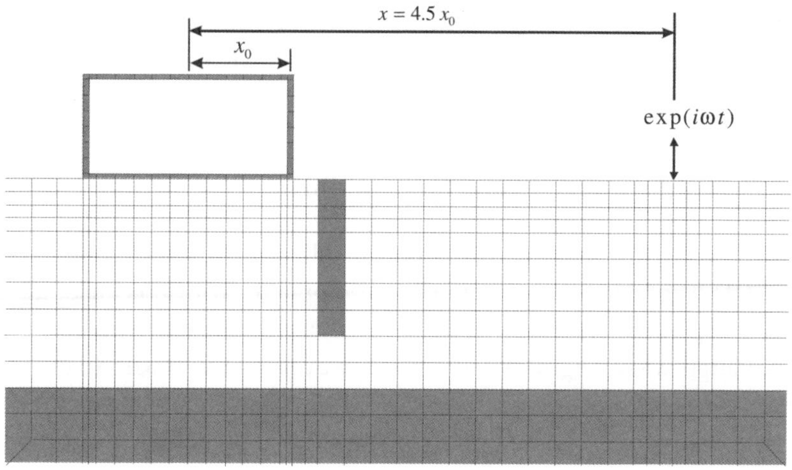

Fig. 6.10 Finite/infinite element mesh (with bedrock).

velocity of layer 2, say, using $C_{s2} = 1200$ m/s. Through a careful examination of the present data for the soil and structure, as well as the requirements for the mesh range and element size in Section 3.4, a finite/infinite element mesh was created for the soil-structure system as shown in Fig. 6.10.

The non-dimensional frequency factor $\omega x_0 / C_s$ is used as a frequency parameter, where ω denotes the excitation frequency and C_s the shear wave velocity of soil layer 1. The average response of both the ceiling and floor of the structure will be analyzed. Let V denote the vertical displacement and G the shear modulus of soil layer 1. In addition to the normalized vertical response $V \bullet G$, a log-scale *vibration acceleration level* (VAL) with unit dB will also be adopted in expressing the computed results,

$$\text{Relative VAL}[\text{dB}] = 20\log_{10}\frac{\textit{calculated acceleration}}{\textit{reference acceleration}}, \tag{6.6}$$

where the reference acceleration is obtained from the response of a reference analysis in which no trench is present. Obviously, the relative VAL serves as a good indicator for expressing the effectiveness of a trench used in reducing the building vibrations.

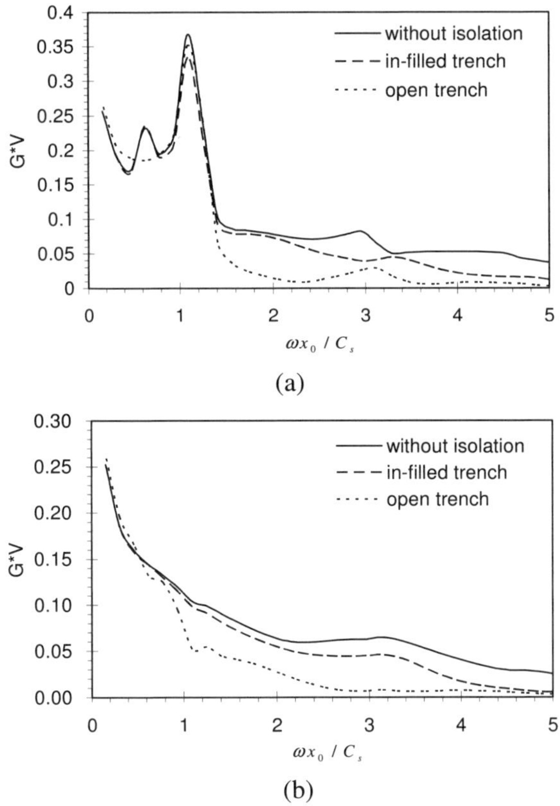

Fig. 6.11 Effect of frequency for soil with no bedrock on vertical displacements of: (a) ceiling; (b) floor.

6.6.1 *Soil with no bedrock*

For the case of a uniform half-space with no bedrock, the vertical responses computed of the ceiling and the floor of the building under different frequencies have been plotted in Figs. 6.11(a) and (b), respectively. As can be seen, the resonant responses occur at $\omega x_0 / C_s =$ 0.6 and 1.1 on the ceiling, but no similar resonant response can be observed on the floor. The lack of consistence in the resonant response of the ceiling and floor can be attributed to the fact that the floor is in direct contact with the soil and thus is reflective of the property of the soil underlying, but the ceiling is not in direct contact with the soil.

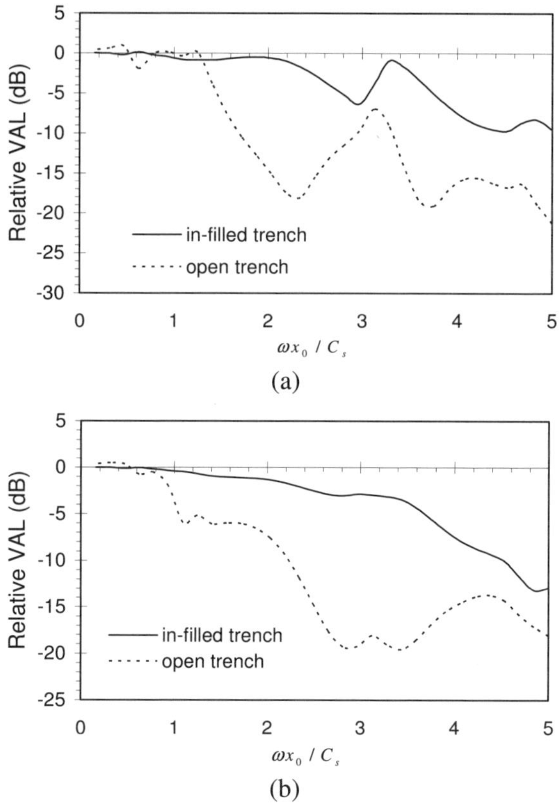

Fig. 6.12 Effect of frequency for soil with no bedrock on VAL response of: (a) ceiling; (b) floor.

Consequently, the resonance frequencies occurring on the ceiling should be interpreted as the natural frequencies of the building.

Corresponding to Figs. 6.11(a) and (b), the effectiveness of the trenches in reducing the ceiling and floor responses has been plotted in terms of the relative VAL in Figs. 6.12(a) and (b), respectively. As can be seen, even though the trenches are quite effective for reducing vibrations of the high-frequency components, basically no isolation effect can be achieved for vibrations of low frequencies, say, in the range with $\omega x_0 / C_s < 1.3$ for the ceiling and with $\omega x_0 / C_s < 1.0$ for the floor. Such a result can be interpreted using the relation: $\lambda = 2\pi C / \omega$, where λ is the wavelength and C the wave velocity. As the soil properties remain constant for the problem considered, so does the wave velocity C. It follows that a lower frequency ω implies a longer wavelength λ. Hence, to achieve the same level of isolation, deeper trenches must be used for waves of lower frequencies, in order to accommodate the relatively long wavelengths. Moreover, as was revealed by Fig. 6.11, the responses of the building for the case with no bedrock are dominated by the low-frequency components. Thus, if one is interested in mitigating the low-frequency response of the problem, the trenches should not be regarded as an effective tool.

Besides, the results also demonstrate that the open trench tends to isolate the vibration more effectively than the in-filled trench, and that no clear trend exists for the efficiency of screening with relation to the increase in frequency, especially for the open trench with $\omega x_0 / C_s > 2.5$. This implies that as the wavelength relative to trench depth is smaller than a certain level, an increase of the trench dimension does not necessarily lead to better isolation, which is consistent with the observations made in the preceding section.

6.6.2 *Soil with bedrock*

To simulate the effect of underlying bedrock, the shear wave velocity of soil layer 2 is assumed to be eight times that of layer 1. The absolute values of the vertical displacement of the ceiling and floor of the building versus the non-dimensional frequency parameter have been plotted in Figs. 6.13(a) and (b), respectively. By comparing the results for the present case with those for the case with no bedrock in Fig. 6.11, one observes that pretty low responses occur at lower frequencies for the present case with bedrock. This can be attributed to the fact that no vibration eigenmodes can be induced below the *cut-off frequency* of the soil stratum. According to Wolf (1985), the cutoff frequency for the vertical injected compressional wave is $C_p/(4h)$ (which is equivalent to $\omega x_0 / C_s = (\pi/4)(C_p/C_s) = 1.56$) and for the shear wave is $C_s/(4h)$ (which is equivalent to $\omega x_0 / C_s = \pi/4 = 0.79$).

Another observation is that, unlike the case with no bedrock, the floor may set in *resonance* with the soil stratum, with rather large peak response induced. Here, the resonance frequency of the soil stratum is

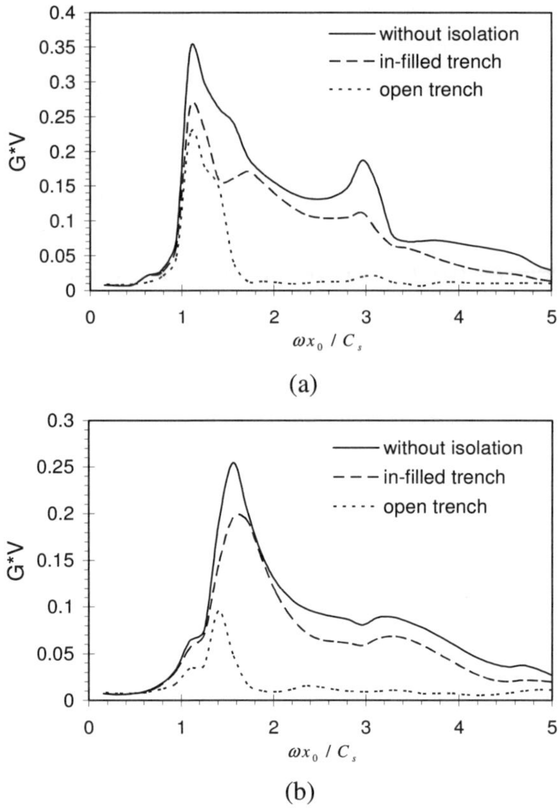

Fig. 6.13 Effect of frequency for soil with bedrock on vertical response of: (a) ceiling; (b) floor.

$(2n-1)C_p/(4h)=(2n-1)\cdot 1.56$ = 1.56, 4.68, ..., for the compressional waves, and $(2n-1)C_s/(4h)=(2n-1)\cdot 0.79$ = 0.79, 2.36, ..., for the shear waves. The resonant frequency of the ceiling should be regarded as that associated with the natural frequency of the building.

Corresponding to Figs. 6.13(a) and (b), the effectiveness of the trenches in reducing the ceiling and floor response has been plotted in terms of the relative VAL in Figs. 6.14(a) and (b), respectively. A comparison of these figures with those for the case with no bedrock in Fig. 6.12 indicates that similar trend exists between these two cases, implying that the existence of bedrock causes basically no effect on the efficiency of isolation of the trenches. However, as was revealed by

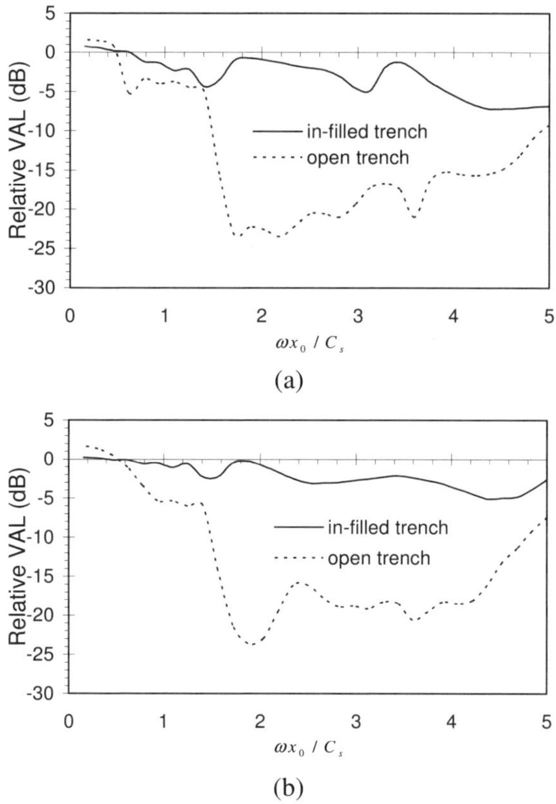

Fig. 6.14 Effect of frequency for soil with bedrock on VAL response of: (a) ceiling; (b) floor.

Fig. 6.13, the building responses are already quite small in the low frequency range due to presence of bedrock. Thus, the potential drawback of trenches in reducing the low-frequency components becomes tolerable for the case with bedrock

6.7 Concluding Remarks

The following conclusions can be drawn from the numerical results presented in this chapter:

(1) In order to achieve a good effect of isolation, the open or in-filled trenches should have a depth of the same order as that of the

Rayleigh wave length of the soil. As a result, the isolation of ground-borne vibrations by trenches is effective only for moderate to high frequency vibrations.

(2) For the cases studied, the open trench performs better than the in-filled trench in reducing the ground-borne vibrations. However, the open trench is inferior to the in-filled trench because of its relatively higher difficulty in construction and higher cost in maintenance.

(3) Soils with large Poisson's ratios can reduce the effect of isolation of in-filled trenches, due to the fact that large Poisson's ratios will result in considerably longer compressional wavelength.

(4) The stiffer (or softer) an in-filled trench with respect to the surrounding soil, the better the effect of isolation is.

(5) Although the ceiling (building) response differs significantly from the floor (ground) response due to involvement of the building frequencies, the efficiency of trenches in isolating the ceiling and floor responses appears to be similar.

(6) For soils with bedrock, the response of the building at frequencies lower than the cutoff frequencies becomes rather small, as compared with the case without bedrock. However, the response may become more pronounced at frequencies equal to or higher than the cutoff frequency because of the resonant effect of the soil stratum. Both the cutoff effect and the resonance effect should be considered if an artificial bedrock is to be installed under the structure or the source to mitigate the vibrations induced by trains.

Chapter 7

2.5D Finite/Infinite Element Method

In this chapter, a 2.5D finite/infinite element procedure is presented for dealing with the three-dimensional ground vibrations induced by the moving loads using only a two-dimensional profile. Besides the two degrees of freedom (DOFs) per node conventionally used for plane strain problems, an extra DOF is introduced to account for the out-of-plane wave transmission. As usual, the profile of the half-space is divided into a near field and a semi-infinite far field. The near field containing loads, railway tracks, buildings, and soils is simulated by finite elements, while the far field covering the soils with unbounded domain by infinite elements. The effect of radiation damping is considered by including an amplitude decay function in the shape functions associated with the infinite axis of the infinite element. The accuracy of the present method is verified through comparison of the results obtained for some typical problems with the analytical solutions.

7.1 Introduction

The vibration of a half-space subjected to moving loads is a problem frequently encountered in practice. The moving loads considered may be generated by vehicles traveling over the ground surface or through an underground tunnel. The problem of train-induced vibrations on the ground has received increasing attention from researchers recently due to the popularity of high-speed railways and subways in different parts of the world.

Most of the early studies on this subject were conducted by analytical or semi-analytical methods. Eason (1965) investigated the three-dimensional steady-state problem for a uniform half-space subjected to forces moving at sub-critical speeds. Gakenheimer and Miklowitz (1969) derived the transient displacements in the interior of an elastic half-space for a normal point load that is suddenly applied and then moves with constant speed on the free surface, considering the sub-, trans- and super-critical speed cases. The steady-state response of the same problem was presented by Frýba (1972) in an integral form. Using a method similar to Eason's, Alabi (1992) studied the response of an elastic half-space to an oblique point load moving on the free surface. A parametric study was carried out to numerically investigate the effects of load speed, distance and ground depth for the sub-critical speed case. By considering the effect of layered soils, de Barros and Luco (1994) proposed a procedure for obtaining the steady-state displacements and stresses within a visco-elastic half-space generated by a buried or surface point load moving with arbitrary constant speeds. Recently, Grundmann *et al.* (1999) studied the response of a layered half-space to a single periodic moving load, as well as a simplified train load. All the aforementioned works have brought some insight into the problem of ground vibrations induced by the moving loads, yet restrictions were inevitably imposed on the geometry and materials of the problem considered, as close-formed solutions cannot be readily obtained for complex conditions using the analytical approaches.

In contrast, the finite element method is known to be capable of dealing with the irregularities or variations in geometry and materials, including the embedded structures and the natural layering of soil deposits. However, due to the use of element meshes that are of finite sizes, the finite element method was criticized for its handicap in simulating the effect of radiation damping caused by the loss of energy for waves traveling to infinity.

In this chapter, a numerical procedure will be presented for dealing with the ground vibrations induced by moving loads. As most of the structural analysis programs commercially available have been written in finite element codes, it is desirable that such a procedure be established within the framework of finite element methods, so that compatibility can be maintained with existing analysis programs. To overcome the drawback of the finite element method in simulating the radiation damping, the finite/infinite element approach presented in Chapter 3 will be adopted with modification to account for the load-moving effect. As usual, the near field of the soil-structure system will be modeled by finite elements, and the far field by infinite elements.

Previously, a number of methods have been employed for modeling the far field with infinite domain, including the boundary element method (Beskos 1987, 1997; Estorff and Kausel 1989), consistent boundary (Lysmer and Wass 1972; Kausel *et al.* 1975), various transmitting boundaries (Lysmer and Kuhlemeyer 1969; Kausel and Roësset 1977), consistent infinitesimal finite-element cell method (Wolf and Song 1996), and infinite elements (Ungless 1973; Bettess 1977; Bettess and Zienkiewicz 1977; Beer and Meek 1981; Chow and Smith 1981; Medina and Penzien 1982; Rajapakse and Karasudhi 1986; Zhang and Zhao 1987; Lau and Ji 1989; Park *et al.* 1991; Zhao and Valliappan 1993; Yun and Kim 1995; Yang *et al.* 1996). Among these methods, the infinite elements can be formulated in a way similar to the finite elements and thus easily incorporated in existing finite element programs. For this reason, the infinite element approach appears to be most attractive to engineers working in the FEM-dominated environment, even though such an approach may not be as accurate as the boundary element method.

For an elastic half-space subjected to a moving load, a plane strain model that is *two-dimensional* (2D) cannot simulate the Mach radiation effect in the load-moving direction, whereas a *three-dimensional* (3D) model is generally costly to apply, since it requires a tremendously large amount of computation time. By assuming the material and geometric properties to be uniform along the load-moving direction, only a profile of the half-space normal to that direction need to be considered. However, if the load-moving effect in the third dimension is to be included, then the problem is more complicated then two-dimensional and simpler than three-dimensional.

In studying the response of an underground structure to traveling seismic waves, Hwang and Lysmer (1981) considered the relation between the displacements of two neighboring nodes along the traveling direction of waves, and used a condensation procedure to reduce the 8-node three-dimensional solid element to a 4-node plane element, but with three DOFs per node. The traveling effect of waves can be faithfully retained because of the existence of the third DOF. Hanazato *et al.* (1991) later applied the same concept to the analysis of traffic-induced vibrations. In their study, the far field was modeled by thin-layered elements. However, due to the restriction inherent with thin-layered elements, only systems with horizontal soil layers or bedrock underneath were analyzed.

The formulation to be presented in this chapter for the finite/infinite elements can be regarded as a 2.5D approach (Hung 2000; Yang and Hung 2001). This approach differs from the previous 2D approach in that a third DOF along the load-moving direction is added to each node of the originally 2D finite and infinite elements, resulting in the so-called 2.5D elements. By such a procedure, the ground vibrations induced by the moving loads, which are 3D in nature, can be simulated using a model that is basically 2D in geometry. Again, the variations of material and geometric properties on the profile of the half-space can be taken into account with no difficulty.

7.2 Formulation of the Problem and Basic Assumptions

Consider a series of vehicles moving with speed c along one line, i.e., the z-axis, on the ground surface or through an underground tunnel (Fig. 7.1). The moving vehicles will be represented by an external load of the general form $\psi(x, y)\phi(z)$, which can be determined as the interaction forces existing between the wheels and rails in space domain at a fixed time. The effect of wheel intervals along the z-axis can be incorporated through the *load distribution function* $\phi(z)$. Using the coordinates shown in Fig. 7.1, the external load can be given as

$$f(x, y, z, t) = \psi(x, y)\phi(z - ct)q(t), \tag{7.1}$$

where $q(t)$ represents the dynamic component of the loading induced by the mechanical system of the vehicles and/or by rail surface unevenness. By the Fourier transformation, $q(t)$ can be decomposed into a series of harmonic functions. For the sake of simplicity, only a single harmonic term, i.e., $\exp(i\omega_0 t)$, is considered in the following derivation. Here ω_0 is regarded as the *self oscillation frequency* of the moving load. It should

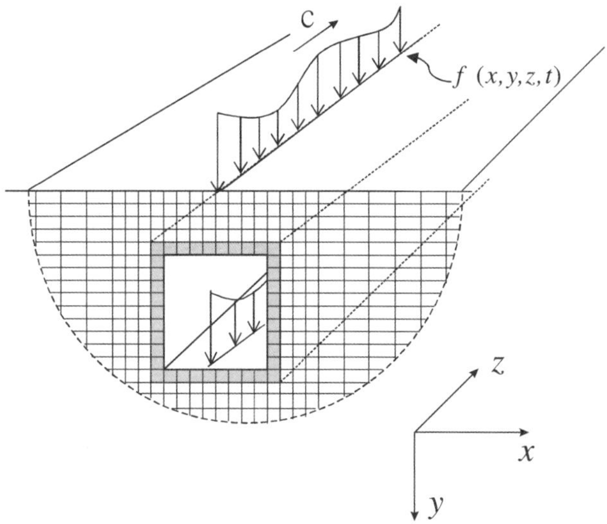

Fig. 7.1 Typical structure of analysis.

be noted that when $\omega_0 = 0$ or when $q(t) = 1$, the external load amounts to a quasi-static loading with a fixed pattern, which is the case typical for vehicles with no self oscillation moving over rails with smooth surface.

By performing the Fourier transformation to Eq. (7.1), one can express the external load in frequency domain as

$$\tilde{f}(x, y, z, \omega) = \frac{1}{c}\psi(x, y)\exp(-ikz)\tilde{\phi}(-k), \tag{7.2}$$

in which

$$k = \frac{\omega - \omega_0}{c}, \tag{7.3}$$

and $\tilde{\phi}(k)$ is the Fourier transform of the load distribution $\phi(z)$, defined as

$$\tilde{\phi}(k) = \frac{1}{2\pi}\int_{-\infty}^{\infty} \phi(z)\exp(-ikz)\,dz. \tag{7.4}$$

On the other hand, by the inverse Fourier transformation, the external load in time domain can be recovered as

$$f(x, y, z, t) = \int_{-\infty}^{\infty} \frac{1}{c}\tilde{\phi}(-k)\psi(x, y)\exp(-ikz)\exp(i\omega t)\,d\omega. \tag{7.5}$$

The preceding equation shows that the external load can be expressed as the integration of a series of harmonic functions.

For a linear system, the total steady-state response of the half-space can be obtained by superposing the responses generated by all harmonic functions of the external load. Let $H(i\omega)$ denote the complex response function for each harmonic function of the external load, $\psi(x,y)\exp(-ikz)\exp(i\omega t)$. The total response of the half-space in time domain is

$$d(x,y,z,t)=\int_{-\infty}^{\infty}\frac{1}{c}\tilde{\phi}(-k)H(i\omega)\exp(i\omega t)\,d\omega. \qquad (7.6)$$

The way to obtain the response function $H(i\omega)$ will be given in the next section by the finite/infinite element approach in frequency domain. In this chapter, the total response of the system in Eq. (7.6) will be obtained by the fast Fourier transformation, while the Fourier transform of the load distribution function $\phi(z)$ in Eq. (7.4) by an analytical approach.

7.3 Procedure of Derivation for Finite/Infinite Elements

Assume the material and geometric properties of the system in Fig. 7.1 to be identical along the load-moving or z-direction. In response to the external load $\psi(x,y)\exp(-ikz)\exp(i\omega t)$, the displacements u, v and w along the three axes of the half-space can be expressed as

$$\begin{aligned} u(x,y,z,t)&=\hat{u}(x,y)\exp(-ikz)\exp(i\omega t),\\ v(x,y,z,t)&=\hat{v}(x,y)\exp(-ikz)\exp(i\omega t),\\ w(x,y,z,t)&=\hat{w}(x,y)\exp(-ikz)\exp(i\omega t), \end{aligned} \qquad (7.7)$$

where $\hat{u}$, $\hat{v}$ and $\hat{w}$ denote the displacements of the profile along the three axes, which are independent of the load-moving direction z. In this chapter, the displacements $\hat{u}$, $\hat{v}$ and $\hat{w}$ will be evaluated using the finite/infinite element approach to be presented in the following section, by which the variations in the geometry and materials of the soil profile can be easily taken into account.

As revealed by Eq. (7.7), the influence of the variable z has been separated from the two variables x and y by the term $\exp(-ikz)$. It follows that the three-dimensional response of the half-space can be computed using merely the elements discretized over the *xy* plane of the profile, on the condition that the half-space is uniform along the load-moving direction. This is certainly one advantage offered by the present approach, which enables us to compute the three-dimensional response of the half-space considering the load-moving effect, by using a simple two-dimensional profile. It is for this reason that such an approach is referred to as *2.5D approach* (Hung 2000, Yang and Hung 2001).

The displacements $\hat{u}$, $\hat{v}$ and $\hat{w}$ of each element of the profile can be interpolated as follows:

$$\hat{u} = \sum_{i=1}^{n} N_i u_i, \quad \hat{v} = \sum_{i=1}^{n} N_i v_i, \quad \hat{w} = \sum_{i=1}^{n} N_i w_i, \tag{7.8}$$

where N_i is the displacement shape function, *n* the number of nodes for each element, e.g., $n = 8$ for a quadratic 8-node (Q8) plane element. The coordinates *x* and *y* within the element can be expressed as

$$x = \sum_{i=1}^{n} M_i x_i, \qquad y = \sum_{i=1}^{n} M_i y_i, \tag{7.9}$$

where M_i is the shape function for the coordinates, which represent the mapping of the element from the global coordinates *xy* to the local coordinates $\xi\eta$.

Substituting the displacement field in Eq. (7.7), along with Eq. (7.8), into the equation of virtual work for a half-space discretized into a number of elements, following the procedure presented in Section 3.2, the equation of motion in frequency domain can be written as

$$([K] - \omega^2 [M])\{D\} = \{F\}, \tag{7.10}$$

in which $\{F\}$ denotes the vector of external loads and $\{D\}$ the vector of nodal displacements, and [*K*] and [*M*] are the global stiffness and mass matrices, respectively, both obtained by assembly of the corresponding element matrices [*k*] and [*m*]:

$$\begin{aligned} [m] &= \iint \rho [N]^T [N] tJ d\xi d\eta, \\ [k] &= \iint [B]^T [E] [\bar{B}] tJ d\xi d\eta, \end{aligned} \tag{7.11}$$

where ρ is mass density, *J* the determinant of the *Jacobian matrix* [*J*] defined as

$$[J]=\begin{bmatrix}\sum M_{i,\xi}x_i & \sum M_{i,\xi}y_i \\ \sum M_{i,\eta}x_i & \sum M_{i,\eta}y_i\end{bmatrix}. \tag{7.12}$$

The Jacobian J can be regarded as the scale factor that yields the area $dxdy$ from $d\xi d\eta$. $[B]$ is the matrix that relates the stresses to the strains, that is,

$$[B]_{6\times 3n}=[\Gamma]_{6\times 9}\times\begin{bmatrix} N_{1,\xi} & 0 & 0 & \dots & 0 \\ N_{1,\eta} & 0 & 0 & \dots & 0 \\ -ikN_1 & 0 & 0 & \dots & 0 \\ 0 & N_{1,\xi} & 0 & & 0 \\ 0 & N_{1,\eta} & 0 & & 0 \\ 0 & -ikN_1 & 0 & & 0 \\ 0 & 0 & N_{1,\xi} & & N_{n,\xi} \\ 0 & 0 & N_{1,\eta} & & N_{n,\eta} \\ 0 & 0 & -ikN_1 & \dots & -ikN_n \end{bmatrix}_{9\times 3n}. \tag{7.13}$$

Here, the terms $-ikN_1$, $-ikN_2$, etc., have originated from the strains (involving $\partial u/\partial z$, $\partial v/\partial z$, $\partial w/\partial z$) through differentiation of the displacements u, v, w that contain the exponential term $\exp(-ikz)$. The matrix $[\Gamma]$ is the inverse of $[J]$, $[\bar{B}]$ is the conjugate of matrix $[B]$, and $[E]$ is a material property matrix,

$$[E]=\frac{E}{(1+\upsilon)(1-2\upsilon)}\begin{bmatrix} 1-\upsilon & \upsilon & \upsilon & 0 & 0 & 0 \\ \upsilon & 1-\upsilon & \upsilon & 0 & 0 & 0 \\ \upsilon & \upsilon & 1-\upsilon & 0 & 0 & 0 \\ 0 & 0 & 0 & \frac{1-2\upsilon}{2} & 0 & 0 \\ 0 & 0 & 0 & 0 & \frac{1-2\upsilon}{2} & 0 \\ 0 & 0 & 0 & 0 & 0 & \frac{1-2\upsilon}{2} \end{bmatrix}, \tag{7.14}$$

where υ is *Poisson's ratio*. For viscoelastic materials, *Young's modulus* E can be replaced by $E^*=E(1+2i\beta)$, where β is the *hysteretic damping ratio* of the material.

The shape functions given in Eqs. (7.8) and (7.9) may be selected to be identical to those of the conventional plane finite/infinite elements. Take the Q8 element as an example, by substituting the shape functions of the conventional Q8 element into the above equations, a Q8-based finite element for modeling the case with moving loads can be established. In this case, the size of the element matrices given in Eq. (7.11) becomes 24×24, instead of 16×16 as for the conventional one. The mass matrices remain real and symmetric, but the stiffness matrices turn out to be complex and asymmetric.

It should be noted that the above procedure from Eqs. (7.8) to (7.14) holds true for both the infinite and finite elements, since they differ only in the domain of integration and shape function. For an infinite element, the integration limit along the ξ-direction in Eq. (7.11) is from 0 to ∞, instead of from –1 to 1 as for the conventional finite element. In addition, the mass matrix for the infinite element is complex, instead of real, as for the finite element.

Through assembly of the stiffness and mass matrices for all the finite and infinite elements, the equation of motion in Eq. (7.10) can be established. In particular, the load vector $\{F\}$ represents the nodal forces corresponding to the moving load function $\psi(x, y)$ in Eqs. (7.2) and (7.5) with a magnitude of unity, i.e., $\psi(x, y) = 1 \cdot \delta(x - x_0)\delta(y - y_0)$, to represent the location of the unit moving load on the xy plane. The displacements $\{D\}$ solved from Eq. (7.10) should be interpreted as the *frequency response function* $H(i\omega)$ for the displacement. Accordingly, the displacement response in time domain can be computed from Eq. (7.6) using the fast Fourier transformation. If the velocity and acceleration responses are desired, then the function $H(i\omega)$ in Eq. (7.6) should be replaced by $i\omega\{D\}$ and $(i\omega)^2\{D\}$, respectively.

7.4 Wave Numbers for the Case with Moving Loads

For a finite element, the displacement field is often approximated by simple polynomials because of their relative ease in computation and because the errors induced from rough shape functions can always be minimized through mesh refinement. However, for an

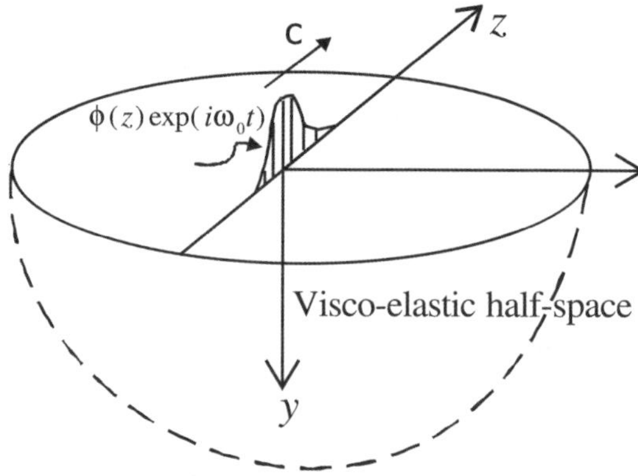

Fig. 7.2 A line load moving on the free surface of a half-space.

infinite element, whose major role is to compensate for the inadequacy of finite elements in dealing with radiation damping, it is meaningless to refine the mesh along the direction leading to infinity. Therefore a much more accurate shape function should be adopted for that direction. Historically, this has been achieved by adopting the analytical solution derived for a special case of the problem considered. In this chapter, the analytical responses as given in Eq. (2.23) for a uniform visco-elastic half-space subjected to a moving harmonic point load will be further studied and considered in determining the shape functions of the infinite element of interest.

Consider a vertical harmonic point load P_y (i.e., directed along the y-axis) with frequency ω_0 moving on the surface of an elastic half-space that covers the domain with $y > 0$, as shown in Fig. 7.2. The external load P_y can be written as

$$P_y(x, y, z, t) = \delta(x)\delta(y)\phi(z - ct)\exp(i\omega_0 t). \tag{7.15}$$

By letting $\hat{P}_x = 0$ and $\hat{P}_z = 0$, one can obtain from Eqs. (2.23), (2.27) and (2.38) the final expression for the displacements of the half-space as follows:

$$u_y = \int_{-\infty}^{\infty} \frac{1}{c} \tilde{\phi}(-k)\exp(-ikz)\tilde{u}_y(\omega)\exp(i\omega t)\,d\omega, \tag{7.16a}$$

$$v_y = \int_{-\infty}^{\infty} \frac{1}{c} \tilde{\phi}(-k) \exp(-ikz) \tilde{v}_y(\omega) \exp(i\omega t)\, d\omega, \tag{7.16b}$$

$$w_y = \int_{-\infty}^{\infty} \frac{1}{c} \tilde{\phi}(-k) \exp(-ikz) \tilde{w}_y(\omega) \exp(i\omega t)\, d\omega, \tag{7.16c}$$

in which $k = (\omega - \omega_0)/c$; u, v, w are the displacement components in time and space domains along the direction of x, y, z, respectively; the subscript y denotes the coordinate axis along which the load is applied; and $\tilde{u}(\omega)$, $\tilde{v}(\omega)$ and $\tilde{w}(\omega)$ are given as follows:

$$\begin{aligned}
\tilde{u}_y(\omega) &= \frac{1}{2\pi\mu} \int_{-\infty}^{\infty} \frac{ik_x}{2Q(k_x)} \left[(k^2 + k_x^2 - \tfrac{1}{2} k_S^2) e^{-m_1(k_x)y} - m_1 m_2 e^{-m_2(k_x)y} \right] \\
&\quad \exp(ik_x x)\, dk_x, \\
\tilde{v}_y(\omega) &= \frac{1}{2\pi\mu} \int_{-\infty}^{\infty} \frac{m_1}{2Q(k_x)} \left[(k^2 + k_x^2 - \tfrac{1}{2} k_S^2) e^{-m_1(k_x)y} - (k^2 + k_x^2) e^{-m_2(k_x)y} \right] \\
&\quad \exp(ik_x x)\, dk_x, \\
\tilde{w}_y(\omega) &= \frac{1}{2\pi\mu} \int_{-\infty}^{\infty} \frac{ik}{2Q(k_x)} \left[(k^2 + k_x^2 - \tfrac{1}{2} k_S^2) e^{-m_1(k_x)y} - m_1 m_2 e^{-m_2(k_x)y} \right] \\
&\quad \exp(ik_x x)\, dk_x,
\end{aligned} \tag{7.17}$$

where μ is shear modulus and the *wave numbers* for the compressional and shear wave velocity respectively are defined as

$$k_P = \frac{\omega}{c_P}, \qquad k_S = \frac{\omega}{c_S}, \tag{7.18}$$

with c_P and c_S denoting the compressional and shear wave velocity of the uniform elastic body, respectively, and

$$\begin{aligned}
Q(k_x) &= \left(k^2 + k_x^2 - \tfrac{1}{2} k_S^2\right)^2 - m_1 m_2 \left(k^2 + k_x^2\right), \\
m_1(k_x) &= \left(k^2 + k_x^2 - k_P^2\right)^{1/2}, \\
m_2(k_x) &= \left(k^2 + k_x^2 - k_S^2\right)^{1/2}.
\end{aligned} \tag{7.19}$$

It should be noted that if the material damping of the half-space is taken into account, then both c_P and c_S, and of course k_P and k_S, become complex numbers. By comparing Eqs. (7.16) with (7.6), one observes from Eq. (7.16) that $\tilde{u}(\omega)$, $\tilde{v}(\omega)$ and $\tilde{w}(\omega)$ should be interpreted as the analytical solution for the *frequency response function* $H(i\omega)$.

By a similar procedure, the analytical expressions of the frequency response transfer function for a load directed along the *x*-axis, but moving along the *z*-direction, can be obtained as

$$\tilde{u}_x(\omega)=\int_{-\infty}^{\infty}\frac{-m_2}{4\pi\mu Q}\left\{k_x^2 e^{-m_1 y}-\left[k^2+k_x^2-\tfrac{1}{2}k_S^2+\frac{k^2}{m_2^2}(k^2+k_x^2-\tfrac{1}{2}k_S^2-2m_1m_2)\right]e^{-m_2 y}\right\}$$
$$\exp(ik_x x)\,dk_x,$$

$$\tilde{v}_x(\omega)=\int_{-\infty}^{\infty}\frac{ik_x}{4\pi\mu Q}\left[m_1m_2e^{-m_1 y}-(k^2+k_x^2-\tfrac{1}{2}k_S^2)e^{-m_2 y}\right]\exp(ik_x x)\,dk_x,$$

$$\tilde{w}_x(\omega)=\int_{-\infty}^{\infty}\frac{-kk_x}{4\pi\mu Qm_2}\left[m_2^2e^{-m_1 y}+(k^2+k_x^2-\tfrac{1}{2}k_S^2-2m_1m_2)e^{-m_2 y}\right]$$
$$\exp(ik_x x)\,dk_x,$$

$$(7.20)$$

and those for the moving load directed aong the *z*-axis as

$$\tilde{u}_z(\omega)=\int_{-\infty}^{\infty}\frac{-kk_x}{4\pi\mu Qm_2}\left[m_2^2e^{-m_1 y}+(k^2+k_x^2-\tfrac{1}{2}k_S^2-2m_1m_2)e^{-m_2 y}\right]$$
$$\exp(ik_x x)\,dk_x,$$

$$\tilde{v}_z(\omega)=\int_{-\infty}^{\infty}\frac{ik}{4\pi\mu Q}\left[m_1m_2e^{-m_1 y}-(k^2+k_x^2-\tfrac{1}{2}k_S^2)e^{-m_2 y}\right]\exp(ik_x x)\,dk_x,$$

$$\tilde{w}_z(\omega)=\int_{-\infty}^{\infty}\frac{-m_2}{4\pi\mu Q}\left\{k^2e^{-m_1 y}-\left[k^2+k_x^2-\tfrac{1}{2}k_S^2+\frac{k_x^2}{m_2^2}(k^2+k_x^2-\tfrac{1}{2}k_S^2-2m_1m_2)\right]e^{-m_2 y}\right\}$$
$$\exp(ik_x x)\,dk_x.$$

$$(7.21)$$

For the case when the load velocity *c* approaches infinity, the variable *k* is zero, as indicated by the relation $k=(\omega-\omega_0)/c$. In this case, the present three-dimensional problem with moving loads reduces to the plane strain problem with a harmonic line load. For instance, by letting k equal to zero, the displacements $\tilde{w}_y(\omega)$ and $\tilde{w}_x(\omega)$ in Eqs. (7.17) and (7.20) vanish, and the other displacements ($\tilde{u}_y$, $\tilde{v}_y$) and ($\tilde{u}_x$, $\tilde{v}_x$) reduce exactly to the ones induced by a harmonic line load acting on the surface of a uniform elastic half-space along the y- and x-axis, respectively (Graff 1973).

Similarly, if the variable k in Eq. (7.21) is set to zero, as for infinite load velocity, the displacements $\tilde{u}_z(\omega)$ and $\tilde{v}_z(\omega)$ vanish, and $\tilde{w}_z(\omega)$ reduces exactly to the one induced for the out-of-plane displacement (with respect to the profile of the plane strain problem) by a harmonic line load directed along the z-axis acting on the surface of the uniform elastic half-space. This has the indication that *a harmonic line load for the plane strain problem can be interpreted as a special case of a point load moving at a speed approaching infinity in the three-dimensional sense.*

For a plane strain problem with a non-moving line load, an infinite element has been presented in Chapter 3. It is possible to extend this infinite element to include the effect of moving loads in the third direction by taking into account the influence of the variable k. One inportant characteristic of the infinite element derived in Chapter 3 is that the wave numbers of the Rayleigh waves ($k_R = \omega / c_R$), shear waves and compressional waves are all included in the wave propagation functions to account for the different traveling speeds of these waves in the far field. Therefore, the first step herein is to find the new wave numbers that are suitable for the case with moving loads.

By setting the function $Q(k_x)$ in Eq. (7.19a) equal to zero, which will result in the famous *Rayleigh equation* if the variable k is dropped, the new *wave number* for the Rayleigh waves can be found as $\sqrt{{k_R}^2 - k^2}$. Likewise, by setting $m_1(k_x)$ and $m_2(k_x)$ in Eqs. (7.19b) and (7.19c) equal to zero, the new wave numbers for the compressional and shear waves can be obtained as $\sqrt{{k_P}^2 - k^2}$ and $\sqrt{{k_S}^2 - k^2}$, respectively. By denoting the new wave numbers as k_i'

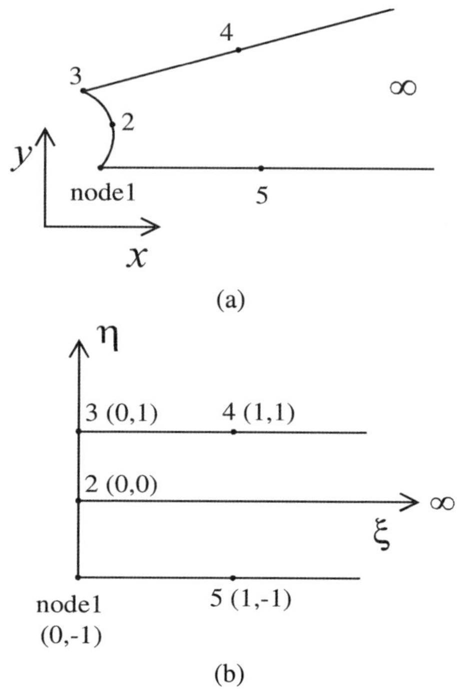

Fig. 7.3 Infinite element: (a) global coordinates; (b) local coordinates.

$$k_i' = \sqrt{\left(\frac{\omega}{c_i}\right)^2 - \left(\frac{\omega - \omega_0}{c}\right)^2}, \tag{7.22}$$

where the subscript i represent R, P or S waves and c is the load-moving velocity, a new infinite element for modeling the waves generated by a load moving on the surface of the half-space can be established.

7.5 Shape Functions of Infinite Element

The mapping shape functions M_i adopted for the coordinates of the infinite element shown in Fig. 7.3 are identical to those given in Eq. (3.15) for the 2D infinite element, i.e.,

$$M_1 = -\frac{(\xi - 1)(\eta - 1)\eta}{2}, \tag{7.23a}$$

$$M_2 = (\xi - 1)(\eta - 1)(\eta + 1), \tag{7.23b}$$

$$M_3 = -\frac{(\xi - 1)(\eta + 1)\eta}{2}, \tag{7.23c}$$

$$M_4 = \frac{\xi(\eta + 1)}{2}, \tag{7.23d}$$

$$M_5 = -\frac{\xi(\eta - 1)}{2}. \tag{7.23e}$$

As stated previously, the wave numbers in Eqs. (3.17) and (3.18) for the displacement shape functions of the 2D infinite element should be replaced by the one k_i' derived in Eq. (7.22) for the present moving load case, namely,

$$\begin{aligned} N_1 &= \frac{\eta(\eta-1)}{2} e^{-\alpha\xi} e^{ik'\xi}, \\ N_2 &= -(\eta-1)(\eta+1) e^{-\alpha\xi} e^{ik'\xi}, \\ N_3 &= \frac{\eta(\eta+1)}{2} e^{-\alpha\xi} e^{ik'\xi}, \end{aligned} \tag{7.24}$$

where α denotes the displacement amplitude decay factor, as a representation of the geometric attenuation, and the sign of the exponent of the term $e^{ik'\xi}$ was adjusted to account for the complex property of the wave number k_i', compared with Eq. (3.18).

By substituting the shape functions in Eqs. (7.23) and (7.24) into Eq. (7.11), the stiffness matrix $[k]$ and mass matrix $[m]$ for the present infinite element can be obtained. As was noted in Chapter 3, to evaluate integrals for the direction extending to infinity, i.e., the ξ-direction, the special integration scheme devised by Bettess and Zienkiewicz (1977) was adopted.

7.6 Wave Propagation Properties for Different Vehicle Speeds

For the case with zero load frequency, i.e., with $\omega_0 = 0$, one observes from Eq. (7.24), along with Eq. (7.22), that for the *sub-critical speed* case, i.e., when the load-moving speed c is less than the Rayleigh wave velocity c_R, all the wave numbers k_R', k_P' and k_S' turn out to be imaginary, meaning that no waves will propagate outward. Similarly, for the *trans-critical speed* case ($c_S < c < c_p$), only the shear waves and Rayleigh waves can travel outward. Such a phenomenon was previously noted by Dieterman and Metrikine (1996), which provides some reference for selection of the wave numbers k_i' for the infinite elements considered. The above phenomenon can also be observed from the frequency reponses given in Eq. (7.17), as will be explained below.

For the case with frequency f = 32 Hz, ω_0 = 0 Hz, and $\beta = 0.01$, the real-part displacements in frequency domain calculated from Eq. (7.17) for various vehicle speeds c along the x-axis at y = 1 m (to represent near-surface response) and along the y-axis (to represent deep-soil response) were plotted in Figs. 7.4 and 7.5, respectively. In these figures, $\tilde{V} = 2\pi\mu\tilde{v}(i\omega)/c$ and the P-, S-, and R-wave speeds of the soil considered are taken as 173.2, 100 and 92 m/s, respectively. The integration in Eq. (7.17) was carried out using appropriate subroutines provided by IMSL.

As can be seen from Fig. 7.4 for the displacements near the surface, distinct wave shapes can be observed for speed $c > c_R$ and no waves are generated for speed $c < c_R$. By carefully examining the wavelength shown in this figure, one finds that the wavelength is exactly equal to the value calculated by $\lambda_R' = 2\pi / k_R'$, i.e., the wavelength corresponding to the Rayleigh waves with wave number k_R' given in Eq. (7.22). In contrast, for the displacements deeply underneath the surface as given in Fig. 7.5, waves can be generated only for $c > c_S$. For this case, when $c_P > c > c_S$, the wavelength shown in Fig. 7.5 is roughly equal to $\lambda_S' = 2\pi / k_S'$, i.e., the wavelength of the S-waves. When $c > c_P$, the waves appear as a combination of the S- and P-waves.

Likewise, for the case with non-zero load frequencies $\omega_0 = 2\pi f_0$, Figs. 7.6 and 7.7 show the real-part displacements in frequency domain along the x-axis at y = 1 m and along the y-axis, respectively, for f = 32 Hz and vehicle speed c = 70 m/s. Again, for the displacements near the surface (Fig. 7.6), k_R' is a crucial variable. Namely, for $k_R' > 0$, a distinct wave shape with wavelength $\lambda_R' = 2\pi / k_R'$ can be seen, but for $k_R' < 0$ (f_0 = 64 Hz), no waves can be generated. On the other hand, for

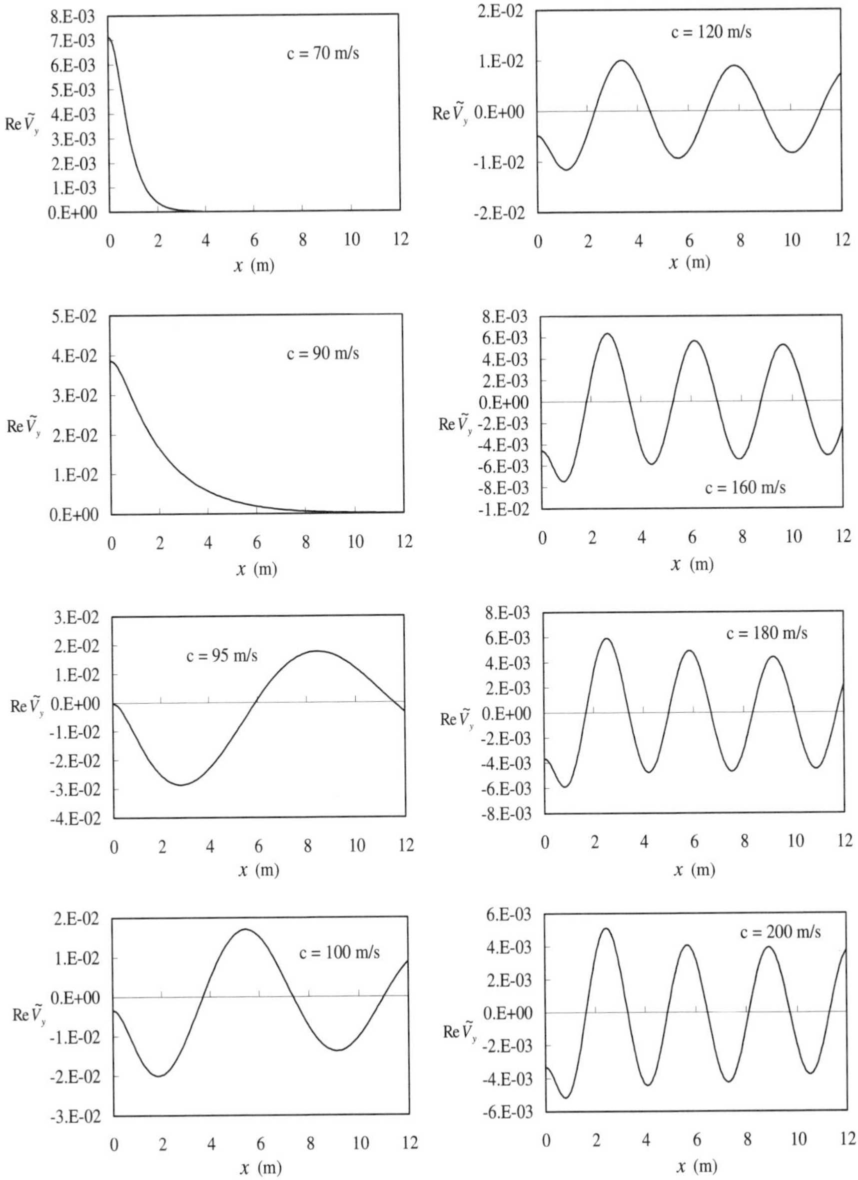

Fig. 7.4 Real-part frequency-domain displacements $\tilde{V}_y$ for the case with $f = 32$ Hz, $f_0 = 0$ Hz, $\beta = 0.01$ along the x-axis at $y = 1$ m.

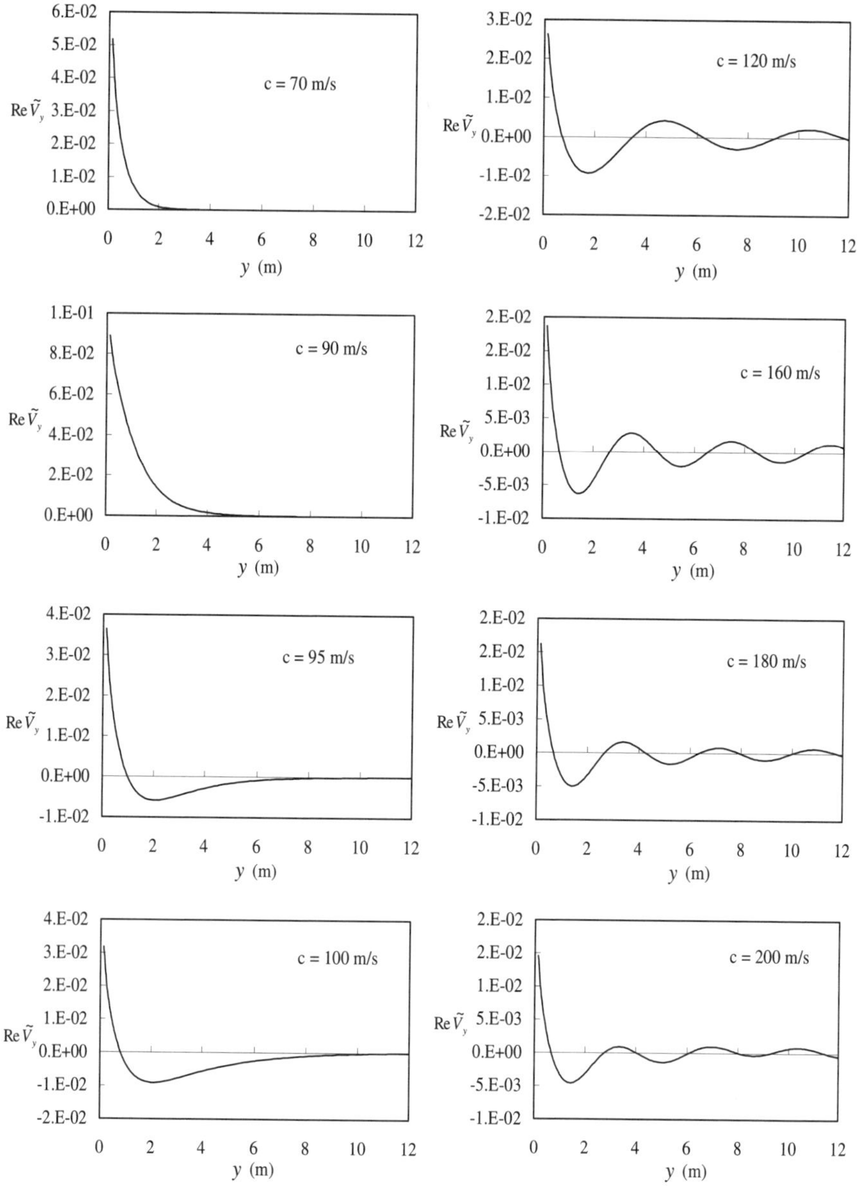

Fig. 7.5 Real-part frequency-domain displacements $\tilde{V}_y$ for the case with $f = 32$ Hz, $f_0 = 0$ Hz, $\beta = 0.01$ along the *y*-axis.

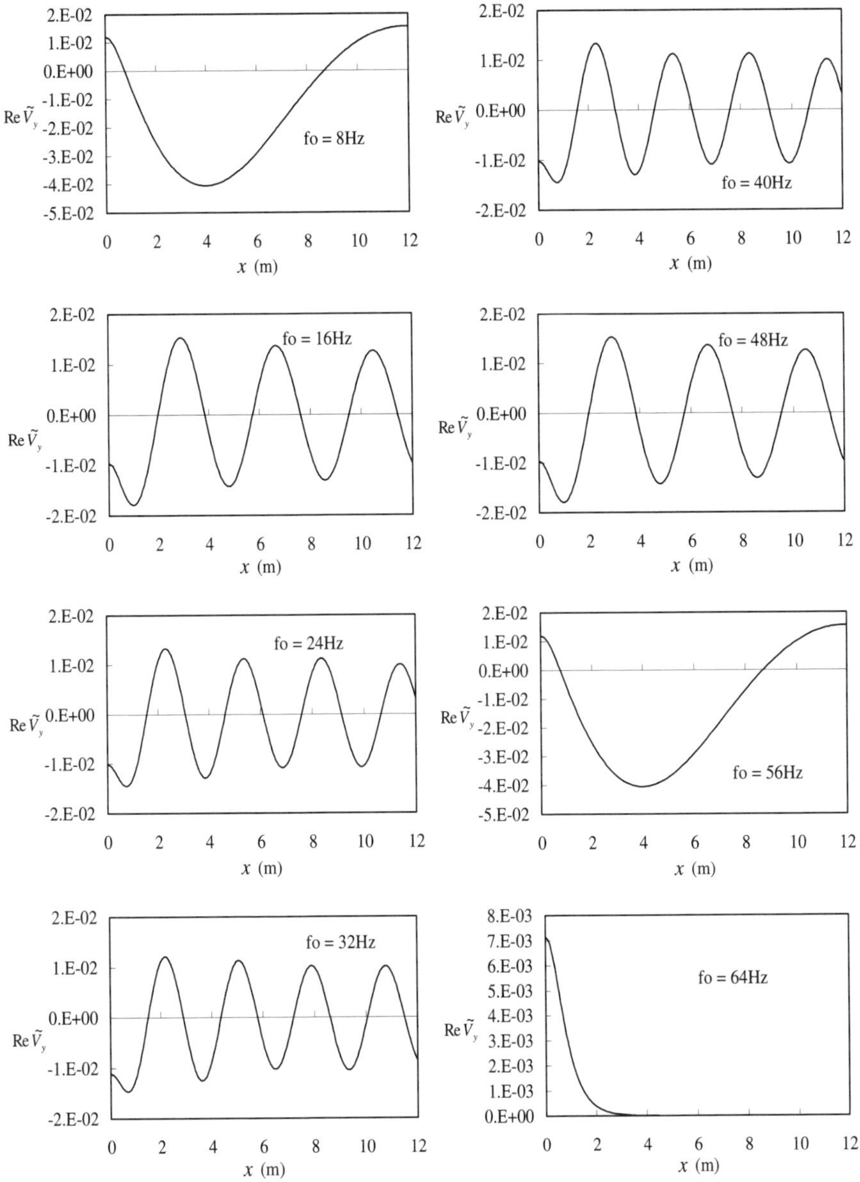

Fig. 7.6 Real-part frequency-domain displacements $\tilde{V}_y$ for the case with $f = 32$ Hz, $c = 70$ m/s, $\beta = 0.01$ along the x-axis at $y = 1$ m.

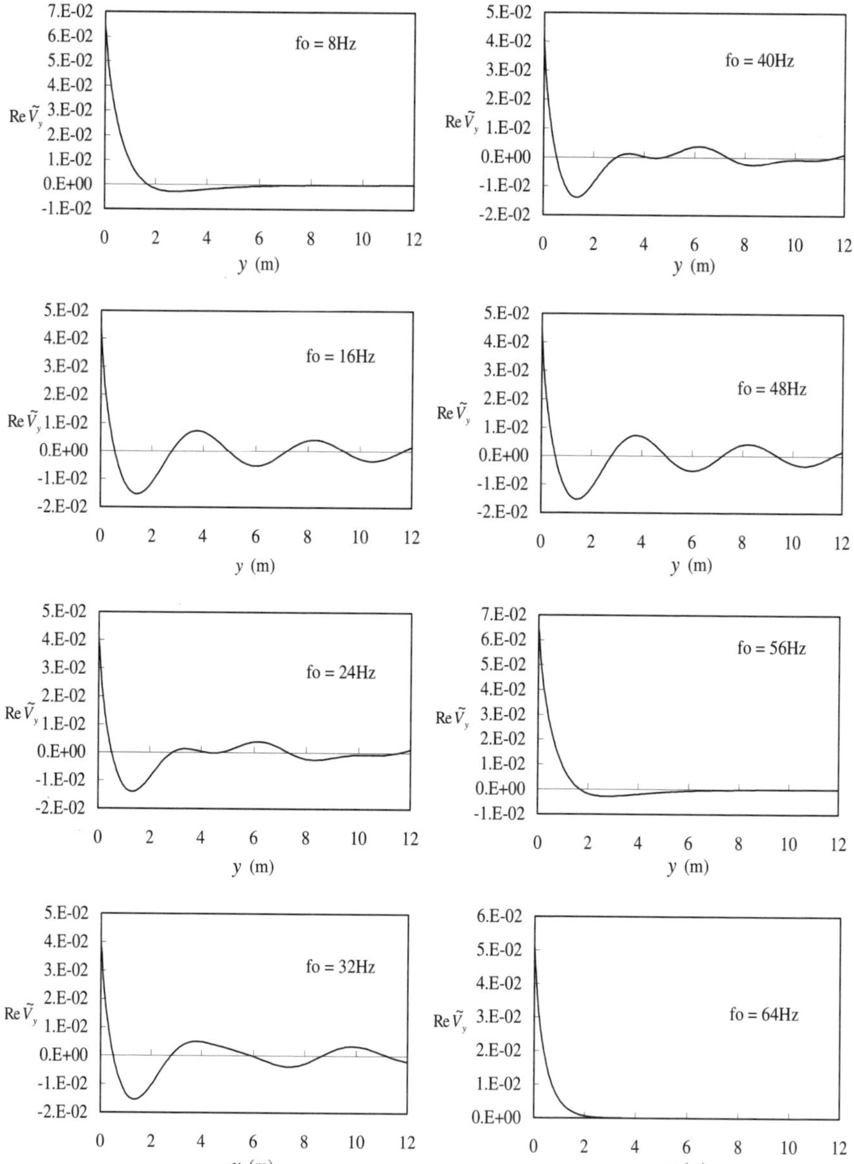

Fig. 7.7 Real-part frequency-domain displacements $\tilde{V}_y$ for the case with $f = 32$ Hz, $c = 70$ m/s, $\beta = 0.01$ along the y-axis.

the displacements along the y-axis, i.e., deep into the soil (Fig. 7.7), both k_S' and k_P' are crucial variables: for $k_S' < 0$ (with f_0 = 8, 56, 64 Hz), no waves can be generated; for $k_S' > 0$ but $k_P' < 0$ (f_0 = 16, 48 Hz), a distinct wave with wavelength $\lambda_S' = 2\pi / k_S'$ can be observed; while for $k_S' > 0$ and $k_P' > 0$ (with $f_0 = 24$, 32, 40 Hz), both S- and P-waves can be generated. All of the aforementioned observations indicate that the Rayleigh waves are dominant near the free surface, and that the body waves (S- and P-waves) are dominant at greater depths.

7.7 Selection of Element Size and Mesh Range

Concerning selection of the mesh and element sizes for the 2D analysis, the details have been given in Section 3.4. For accuracy of solution, the mesh size is determined by the wavelength of the shear waves of the highest frequency ω considered and the mesh range should be large enough to allow waves to propagate. In general, the element size should meet the requirement $L \leq \lambda_S / 6$ and the mesh range should meet the requirement $R \geq 0.5\lambda_S$, where the shear wave length λ_S is $\lambda_S = 2\pi / k_S$, with $k_S = \omega / c_S$.

As for the 2.5D approach considered in this chapter, the same requirements can be followed, except that the wavelength λ_S should be replaced by λ_S' to consider the effect of the moving loads, that is, $L \leq \lambda_S' / 6$ and $R \geq 0.5\lambda_S'$, where $\lambda_S' = 2\pi / k_S'$ and the *wave number* k_S' for the shear wave is

$$k_S' = \sqrt{\left(\frac{\omega}{c_S}\right)^2 - \left(\frac{\omega - \omega_0}{c}\right)^2}. \tag{7.25}$$

Owing to involvement of the three parameters ω, ω_0 and c in Eq. (7.25), the selection of the mesh range for the 2.5D approach is much more complicated than that of the 2D approach, as will be discussed below.

Consider first the case when the load frequency is zero, i.e., $\omega_0 = 0$. For this case, the *wave number* k_S' in Eq. (7.25) reduces to

$$k_S' = \omega\sqrt{\left(\frac{1}{c_S}\right)^2 - \left(\frac{1}{c}\right)^2}. \tag{7.26}$$

Thus, we can use the wavelength λ_S' ($=2\pi/k_S'$) of the highest frequency ω considered to determine the minimum mesh range required, and then adopt the mesh expansion scheme, based on the dynamic condensation technique, presented in Section 3.4 to generate the finite/infinite element mesh for the next highest frequency considered. Such a procedure should be repeated for all the frequencies considered.

It should be noted that whenever the vehicle speed c is less than the Rayleigh wave speed c_R (which also means $c < c_S$), k_S' becomes an imaginary number and no waves can propagate outward, as was mentioned previously in relation to Fig. 7.4. For this case, the influence of the mesh range is quite small. However, for the case when the vehicle speed c is slightly larger than the Rayleigh wave speed c_R (as indicated by the case with $c = 95$ m/s in Fig. 7.4), the wavelength λ_S' becomes quite large for small frequencies ω. For this case, the mesh expansion scheme should be adopted to generate the mesh for smaller frequencies, in order to meet the condition of $R \geq 0.5\lambda_S'$.

Next, let us consider the case when the load frequency is not equal to zero, i.e., $\omega_0 \neq 0$. Owing to the involvement of the load frequency ω_0, a constant, in Eq. (7.25), the dynamic condensation technique presented in Section 3.4 for expanding the element mesh becomes invalid. In analysis, we use the maximum value computed for k_S' from Eq. (7.25) to determine the minimum wavelength and correspondingly the maximum element size L. Also, the minimum value of k_S' computed from Eq. (7.25) is used to determine the maximum wavelength and therefore the minimum mesh range R.

It should be noted that for the case with non-zero load frequencies ω_0, the response in frequency domain is not symmetrical any more, i.e., $H(\mathrm{i}\omega) \neq H(-i\omega)$. Thus, the frequency domain analysis has to be conducted from $-\omega$ to $+\omega$ (as can be observed from the transfer function in Figs. 8.6-8.9 to be presented later), before the results can be converted to the time domain by the inverse Fourier transformation.

Again, when the following condition is met, both k_R' and k_S' become imaginary numbers, implying that no waves can propagate outward:

$$\left(\frac{\omega}{c_R}\right)^2 - \left(\frac{\omega - \omega_0}{c}\right)^2 < 0. \tag{7.27}$$

In other words, when either of the following conditions is met for the case with $c < c_R$,

$$\omega > \frac{\omega_0}{1 - \frac{c}{c_R}} \quad \text{or} \quad \omega < \frac{\omega_0}{1 + \frac{c}{c_R}}, \tag{7.28}$$

And for the case with $c > c_R$, when the following condition is met:

$$\omega < \frac{\omega_0}{1 + \frac{c}{c_R}}, \tag{7.29}$$

no waves will propagate outward, which can also be observed from the transfer function in Figs. 8.6-8.9 to be presented later. Such a property will be utilized in selection of the frequency range and in determining the wave number k_S', element size *L*, and mesh range *R* to make the finite/infinite element mesh more efficient.

As can be seen from the above discussions, two parameters are crucial to selection of element size and mesh range, i.e., the element size $L \le \lambda_S / 6$ and mesh range $R \ge 0.5\lambda_S$. In fact, the restriction on element size *L* is caused mainly by the shape functions used in the finite element representation, while the mesh range *R* is related to the shape functions used by the infinite element. By comparing the numerical results obtained by the present 2.5D approach with the analytical ones, it is confirmed that the shape functions used by the infinite element are generally consistent with the theoretical ones. Thus, the requirement for the mesh range $R \ge 0.5\lambda_S$ is not as strict or as important as that for the element size $L \le \lambda_S / 6$. If the range *R* selected is slightly less than $0.5\lambda_S$, the solution obtained is still quite accurate and acceptable. For problems of wave propagation, where a large finite element mesh is always required, quite accurate solutions can be obtained with no recourse to the mess expansion scheme mentioned above.

7.8 Selection of Wave Number k_i'

There are three different wave numbers, as given in Eq. (7.22), but for an infinite element with only a single node along the direction leading to infinity, it is unlikely to consider simultaneously all the three different types of waves. In this regard, a rational approach is to consider only the type of waves that is dominant in each region of the far field of concern. Based on the fact that Rayleigh waves are dominant near the free surface, and that the body waves are dominant at greater depth, it is suggested that the wave numbers k_R' associated with Rayleigh waves be used for the regions near the free surface, while for the other regions, the wave numbers k_P' and k_S' associated with body waves be used.

In the literature, some researchers have developed infinite elements that are capable of dealing with multiple traveling waves, by adding two additional nodes along the direction leading to infinity. It is true that such a procedure can be adopted herein for simulating the effect of multiple waves. However, after some extensive tests, it was found that such a 'revised' infinite element did not add any accuracy to the solution, compared with the 'original' infinite element, but only increased the complexity of calculation. For this reason, only the 'original' infinite element, i.e., the one with a single node lying along the direction leading to infinity, will be adopted in the present analysis.

7.9 Selection of Amplitude Decay Factor α of Displacement

Let R denote the radial distance from the source of excitation to the far field boundary. In Section 3.3.5, it is suggested that for a half-space subjected to a point load on the free surface, $\alpha = 1/R$ be used for the body waves, and $\alpha = 1/2R$ for the Rayleigh waves; whereas for the case of a line load, $\alpha = 1/2R$ and $\alpha = 0$ be used instead. As was mentioned previously, the effect of moving point load at speed c is between those of a point load and of a line load. When the load speed c equals zero, the moving point load in frequency domain reduces to a harmonic point load. On the other hand, when the load speed c approaches infinity, it will behave as a harmonic line load. Consequently, the amplitude decay factor α of displacement should be chosen between these two cases depending on the value of $k = (\omega - \omega_0)/c$. In this chapter, the following amplitude decay factors are adopted:

$$\alpha_R = \frac{1}{2R}\frac{k^2}{k^2 + k_R^2}, \tag{7.30a}$$

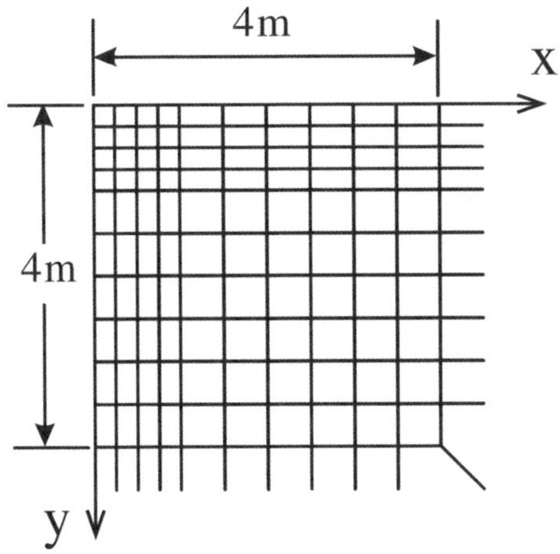

Fig. 7.8 Finite and infinite element mesh.

$$\alpha_P = \frac{1}{2R} + \frac{1}{2R}\frac{k^2}{k^2 + k_S^2}, \tag{7.30b}$$

$$\alpha_S = \frac{1}{2R} + \frac{1}{2R}\frac{k^2}{k^2 + k_P^2}, \tag{7.30c}$$

for the Rayleigh waves, compressional waves and shear waves, respectively.

7.10 Verification of the Present Approach

In this section, we shall evaluate the applicability of the numerical procedure developed in previous sections. To examine the accuracy and efficiency of the procedure, this section will be further separated into five subsections to verify the various capacities of the procedure. The basic problem considered herein is a uniform visco-elastic half-space subjected to a harmonic load with pattern $\phi(z)$ moving at constant speed c (Fig. 7.2), of which the compressional, shear, and Rayleigh wave speeds are taken as 173.2, 100, and 92 m/s, respectively. As shown in Fig. 7.8, only a half of this system is modeled by the finite/infinite element mesh due to symmetry of the problem. The far field is modeled by infinite elements, and the near field by 8-node finite elements.

It should be noted that the mesh plotted in Fig. 7.8 has been created based on the requirements for the maximum element size L and minimum mesh extent R described above for the highest frequency of the soil considered, namely, $L \le \lambda_S' / 6$ and $R \ge 0.5\lambda_S'$, where $\lambda_S' = 2\pi / k_S'$, for the problem considered. In terms of accuracy and computation, such a mesh is likely to be most economical, which will be adopted throughout all the numerical studies in this section, regardless of the load moving speed. The displacements in frequency domain along the three axes will be described in a normalized sense by $\tilde{U} = 2\pi\mu\tilde{u}(i\omega)/c$, $\tilde{V} = 2\pi\mu\tilde{v}(i\omega)/c$ and $\tilde{W} = 2\pi\mu\tilde{w}(i\omega)/c$; and those in time-domain by $U = 2\pi\mu u(t)$, $V = 2\pi\mu v(t)$ and $W = 2\pi\mu w(t)$.

7.10.1 *Responses in frequency domain for moving loads at sub-, trans- and super-critical speeds*

The finite and infinite element equations presented in this chapter are all given in frequency domain. Thus, to verify the accuracy of these elements, a comparison of the displacement response in frequency domain should be conducted at first. The damping ratio of the half-space is taken as $\beta = 0.05$. The vertical displacements in frequency domain ($f = \omega / 2\pi = 32$ Hz) for the case with the load moving at velocity c = 90 m/s (sub-critical), 120 m/s (trans-critical), and 200 m/s (super-critical) have been compared with the analytical solutions of Eq. (7.17) in Figs. 7.9-7.11, respectively, where the inverse transform with respect to k_x for the analytical solutions was computed using appropriate quadrature routines available in IMSL. In each figure, part (a) shows the displacements along the x-axis at y = 1 m, and part (b) the displacements along the y-axis, with the solid dots and solid lines representing the real part of the displacements, and the harrow circles and dashed lines the imaginary part.

As can been seen, for the three ranges of sub-, trans- or super-critical speeds, the present finite/infinite element scheme can simulate the waves traveling behavior accurately. In particular, distinct wave shapes can be observed from parts (a) for the displacements along the x-axis near the free surface, indicating that the Rayleigh waves, which are dominant near the free surface, can be very well simulated. The Rayleigh wavelength is

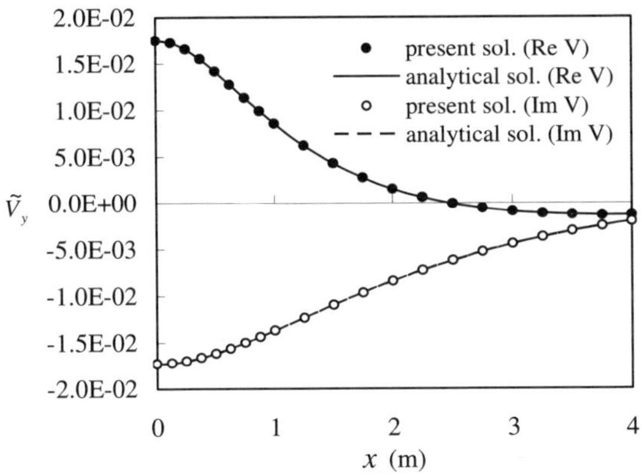

(a)

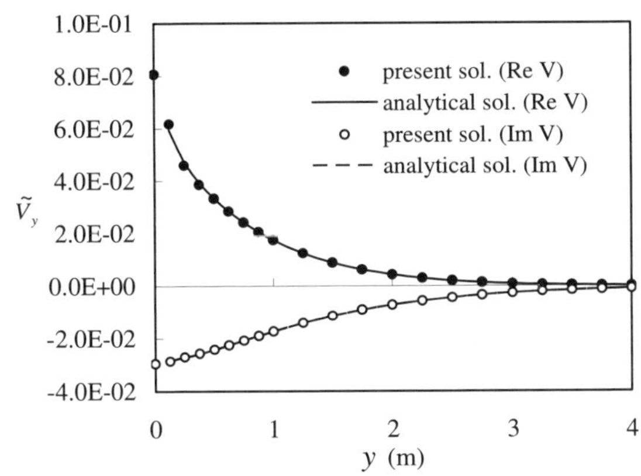

(b)

Fig. 7.9 Frequency-domain displacements $\tilde{V}_y$ for the case with f = 32 Hz, c = 90 m/s, $\beta = 0.05$: (a) along the x-axis at y = 1 m; (b) along the y-axis.

$\lambda_R' = 2\pi / k_R'$. For the sub-critical speed case with c = 90 m/s [Fig. 7.9(a)], λ_R' is imaginary, which means that the Rayleigh waves decay exponentially along the free surface. For the trans- and super-critical speed cases with c = 120 and 200 m/s [Figs. 7.10(a), 7.11(a)], the computed Rayleigh wavelengths are 4.48 and 3.24 m, respectively, which appear to be in exact agreement with the ones observed from the figures.

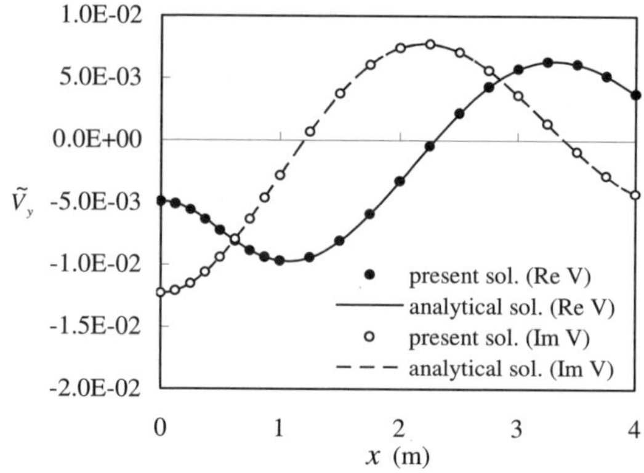

(a)

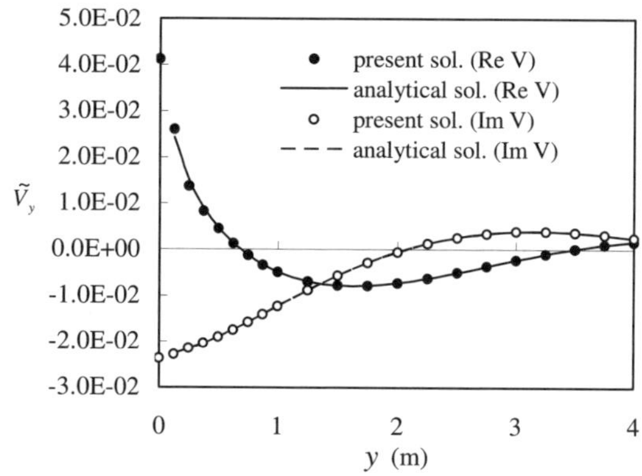

(b)

Fig. 7.10 Frequency-domain displacements $\tilde{V}_y$ for the case with f = 32 Hz, c = 120 m/s, $\beta = 0.05$: (a) along the x-axis at y = 1 m; (b) along the y-axis.

7.10.2 *Responses in frequency domain for moving loads with self oscillation*

By letting $\omega_0 \neq 0$, the self oscillation frequency of the moving load will be taken into account. A damping ratio of $\beta = 0.05$ is adopted for the soil. For the case with a vertical harmonic load moving at speed c = 120 m/s, self oscillation frequency of $f_0 = \omega_0 / 2\pi$ = 16 Hz, and system frequency of f = 32 Hz, the displacements in frequency domain along the

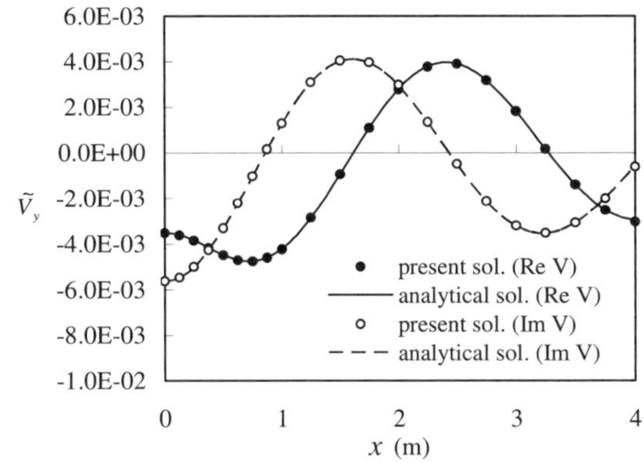

(a)

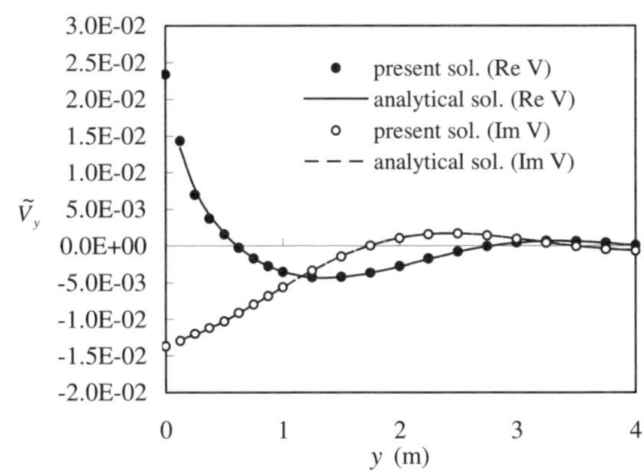

(b)

Fig. 7.11 Frequency-domain displacements $\tilde{V}_y$ for the case with f = 32 Hz, c = 200 m/s, $\beta = 0.05$: (a) along the x-axis at y = 1 m; (b) along the y-axis.

three axes have been computed and compared with the analytical ones in Figs. 7.12-7.14. As can be seen, the present solutions agree excellently with the analytical ones.

7.10.3 *Effectiveness and accuracy of condensation procedure*

The effectiveness and accuracy of the condensation procedure for generating the far-field impedance matrices will be demonstrated herein.

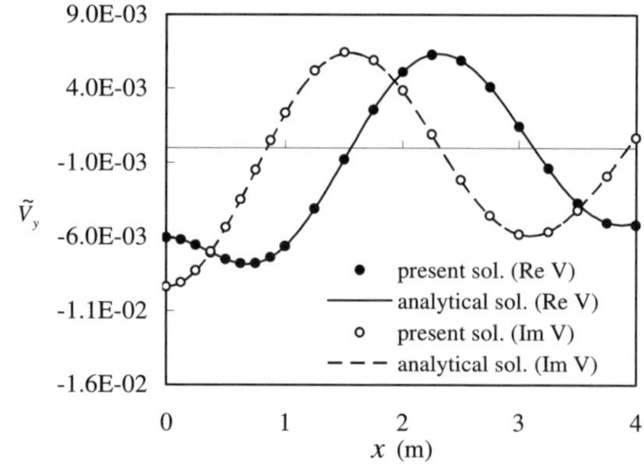

(a)

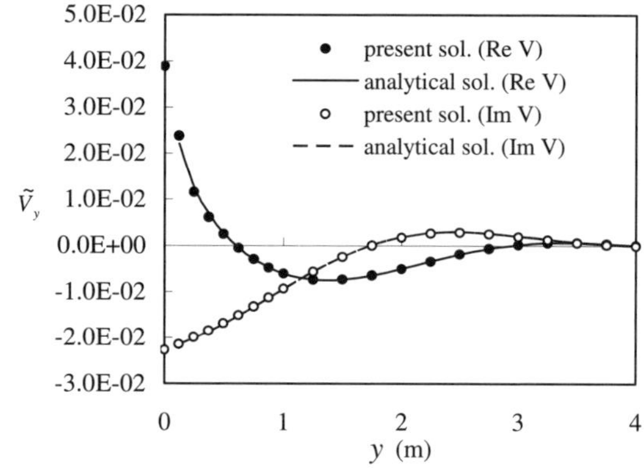

(b)

Fig. 7.12 Frequency-domain displacements $\tilde{V}_y$ for the case with f = 32 Hz, f_0 = 16 Hz, c = 120 m/s, $\beta = 0.05$: (a) along the x-axis at y = 1 m; (b) along the y-axis.

Assume the damping ratio of the soil system to be 0.05. For the case with a vertical non-pulsating load (f_0 = 0 Hz) moving at speed c = 120 m/s and for a small soil frequency of $f = 4$ Hz, the solutions obtained by the condensation procedure starting from the highest frequency f = 32 Hz of the soil have been plotted in Figs. 7.15-7.17, along with the analytical solutions obtained from Eq. (7.17). As can be seen, even though the wave length is very long (λ_R' = 45 m) for the present case, a small element mesh with a range of R = 4 m can still be used to produce very

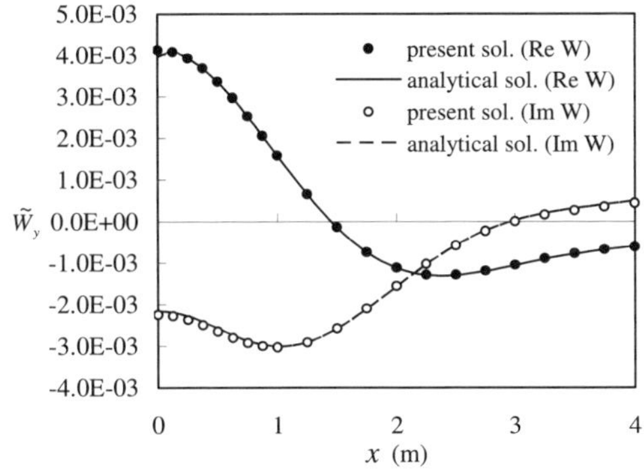

(a)

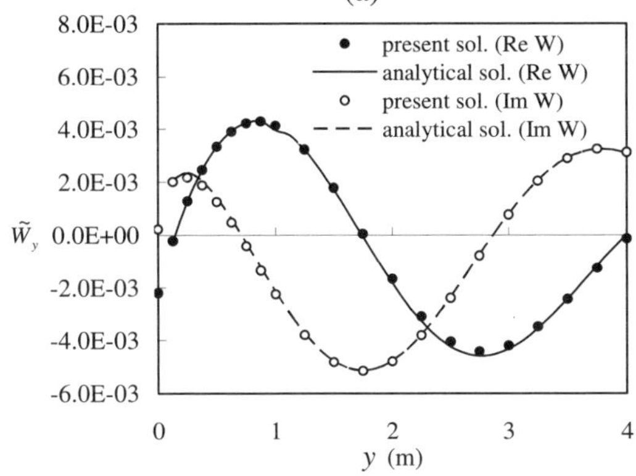

(b)

Fig. 7.13 Frequency-domain displacements $\tilde{W}_y$ for the case with f = 32 Hz, f_0 = 16 Hz, c = 120 m/s, $\beta = 0.05$: (a) along the x-axis at y = 1 m; (b) along the y-axis.

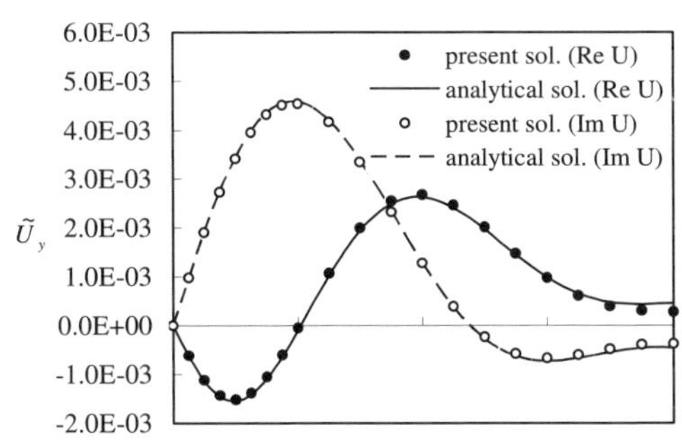

Fig. 7.14 Frequency-domain displacements $\tilde{U}_y$ for the case with f = 32 Hz, f_0 = 16 Hz, c = 120 m/s, $\beta = 0.05$: (a) along the x-axis at y = 1 m; (b) along the y-axis.

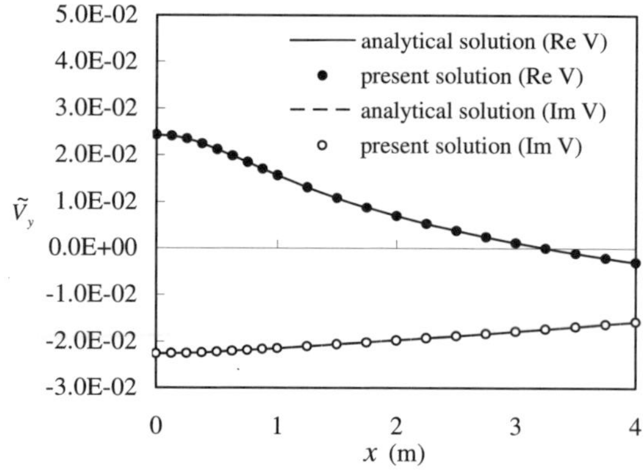

(a)

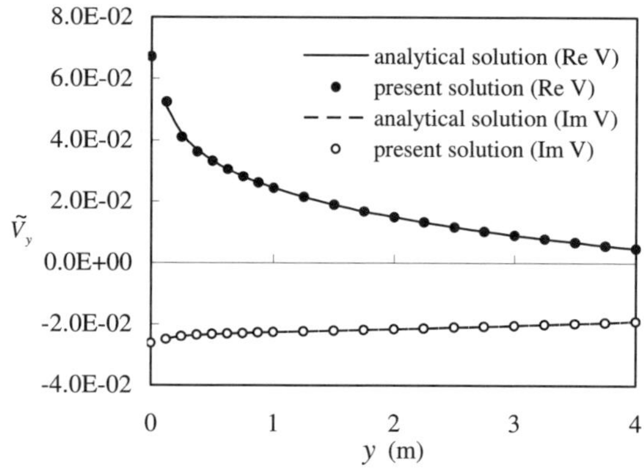

(b)

Fig. 7.15 Frequency-domain displacements $\tilde{V}_y$ for the case with f = 4 Hz, f_0 = 0 Hz, c = 120 m/s, $\beta = 0.05$: (a) along the x-axis at y = 1 m; (b) along the y-axis.

accurate solutions. This can be attributed mainly to adoption of the condensation procedure, by which the finite/infinite element mesh is enlarged each time when the expansion procedure is repeated consecutively for the next smaller frequency.

7.10.4 *Responses in time domain for sub-critical speed case*

For the present purposes, $\phi(z) = \delta(z)$, $\omega_0 = 0$ and $\beta = 0$ are assumed, and a frequency range of 0 - 200 Hz is considered for the inverse fast

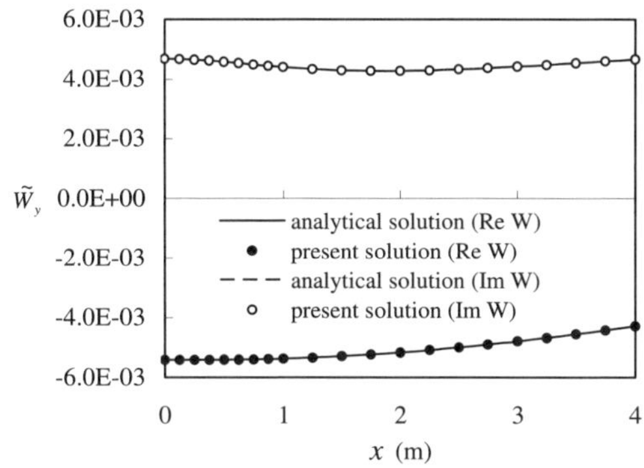

(a)

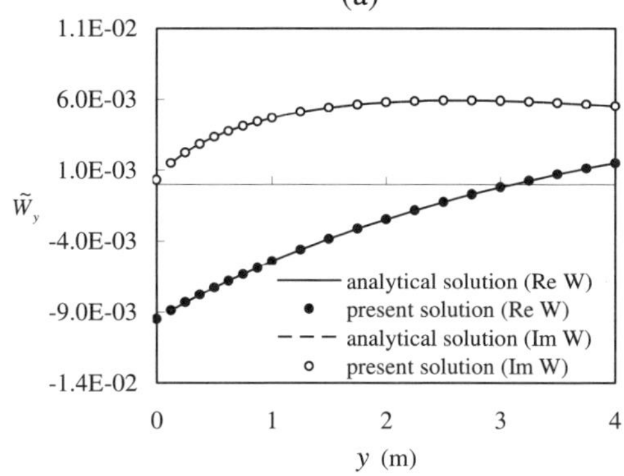

(b)

Fig. 7.16 Frequency-domain displacements $\tilde{W}_y$ for the case with f = 4 Hz, f_0 = 0 Hz, c = 120 m/s, $\beta = 0.05$: (a) along the x-axis at y = 1 m; (b) along the y-axis.

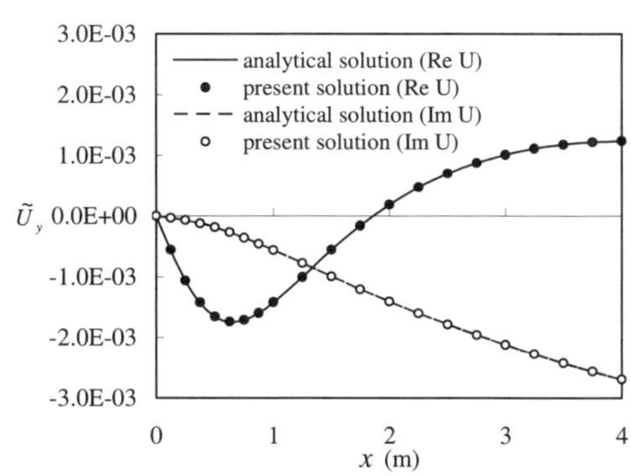

Fig. 7.17 Frequency-domain displacements $\tilde{U}_y$ for the case with f = 4 Hz, f_0 = 0 Hz, c = 120 m/s, $\beta = 0.05$: (a) along the x-axis at y = 1 m; (b) along the y-axis.

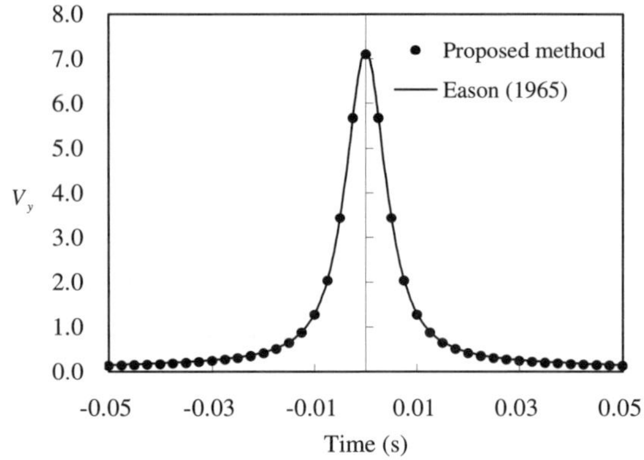

(a)

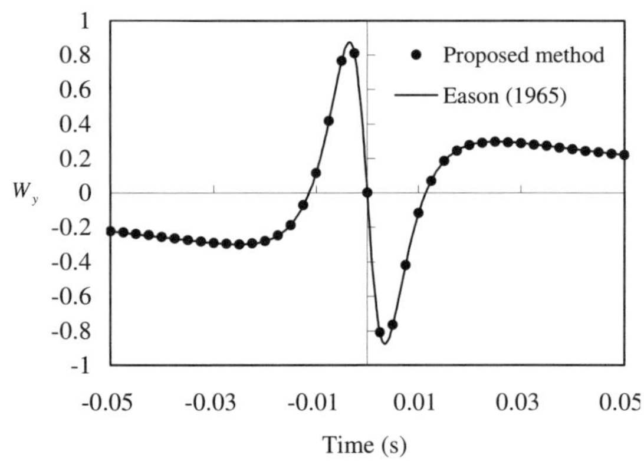

(b)

Fig. 7.18 Time-domain displacements for the case with c = 90 m/s, $\beta = 0$: (a) vertical displacement V_y; (b) longitudinal displacement W_y.

Fourier transform. To verify the accuracy of the entire solution procedure, including the element formulation and computation of the inverse fast Fourier transform, the time-domain displacements obtained for an interior point right under the origin, i.e., at (0, 1, 0 m), of a uniform elastic half-space subjected to a point load moving at a sub-critical speed (c = 90 m/s) on the free surface are compared with those obtained by Eason (1965) in Figs. 7.18-7.20. Here, the subscript of the displacements denotes the direction along which the point load is applied. Thus, the three figures sequentially represent the displacements induced by the

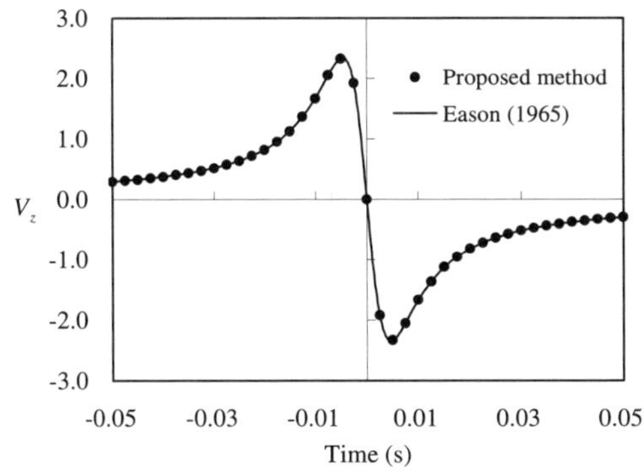

(a)

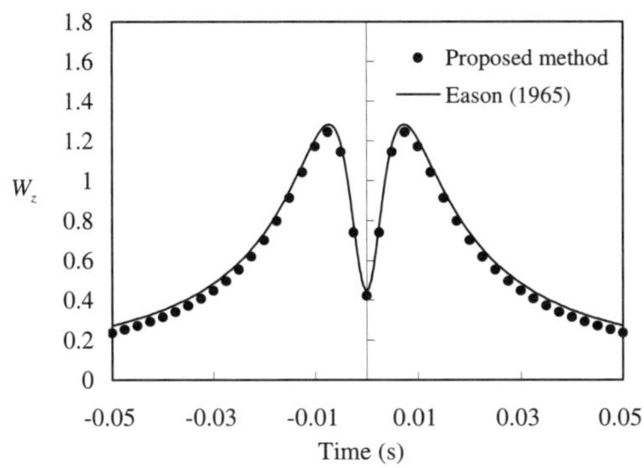

(b)

Fig. 7.19 Time-domain displacements for the case with c = 90 m/s, $\beta = 0$: (a) vertical displacement V_z; (b) longitudinal displacement W_z.

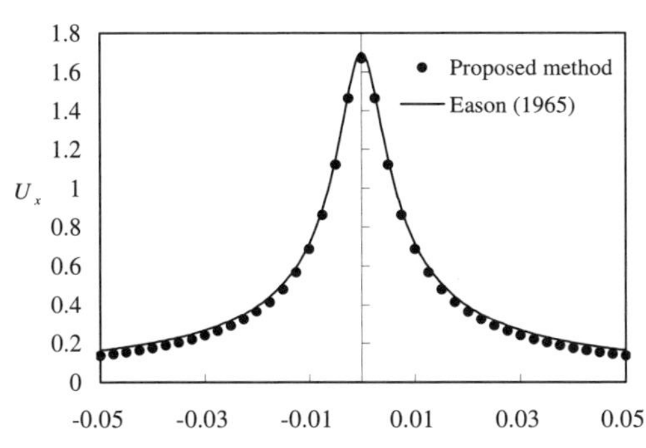

Fig. 7.20 Time-domain displacement U_x for the case with $c = 90$ m/s, $\beta = 0$.

moving load directed along the y-, z- and x-axis. The time $t = 0$ indicates the instant at which the point load passes through the profile $z = 0$. Evidently, the present results agree excellently with those by Eason (1965). For the sub-critical speed case, the responses are either symmetric or anti-symmetric with respect to time $t = 0$.

7.10.5 *Responses in time domain for trans-critical speed case*

As a final test, a hysteretic damping ratio of $\beta = 0.05$ is used for the soil system with the range of frequency considered from 0 to 200 Hz. The displacements obtained for an interior point of the uniform visco-elastic half-space, i.e., at (0, 1, 0 m), subjected to a point load moving at a trans-critical speed (c = 120 m/s) on the surface are compared with the analytical ones in Figs. 7.21-23. The analytical solutions shown in Figs. 7.21-7.23 for the moving load applied along the each axis were obtained from Eq. (2.23) using numerical integration. The inverse transform with respect to ω for the analytical solutions were performed by the fast Fourier transform with a cut-off frequency of $f = 400$ Hz.

As can be seen from these figures, the displacements obtained by the present 2.5D finite/infinite element procedure agree excellently with the analytical ones. One feature revealed by these figures is that for the trans-critical speed case, the surface response tends to arrive at a time later than that of the sub-critical speed case, as a result of the *Mach radiation effect*. Moreover, the response curves are no longer symmetric or anti-symmetric with respect to $t = 0$.

7.11 Case Study

One advantage of the finite element-based algorithm is the possibility to obtain all the surface responses at the same time. In this section, some simple cases will be studied by the finite/infinite element method presented in this chapter, with all the surface responses given in three-dimensional plots. The uniform visco-elastic half-space is characterized by the S-wave speed $c_S = 100$ m/s, P-wave speed c_P = 173.2 m/s, R-wave speed $c_R = 92$ m/s, and a damping ratio of $\beta = 0.02$.

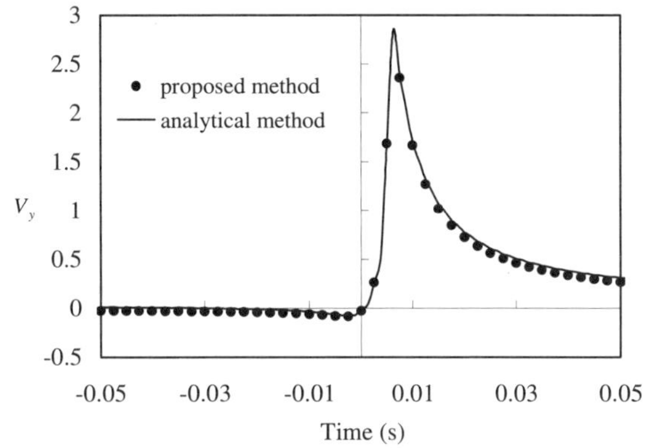

(a)

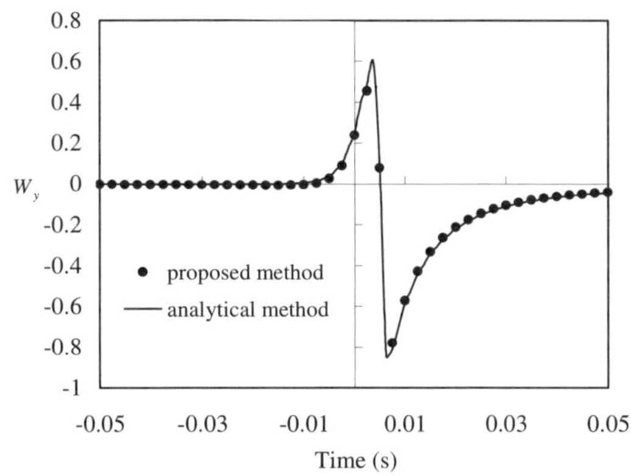

(b)

Fig. 7.21 Time-domain displacements for the case with c = 120 m/s, $\beta = 0.05$: (a) vertical displacement V_y; (b) longitudinal displacement W_y.

Figures 7.24(a)-(c) show the spatial distribution of the ground displacement field generated by a wheel load moving at speed c = 70, 100, and 120 m/s, respectively. For the present case, the wheel load is taken to be T = 10 t and the characteristic length taken as α = 0.8 m based on the definition given in Eqs. (2.30) and (2.31). As can be seen, for the load moving on the free surface of the uniform half-space with the sub-critical speed $c < c_R$ (c = 70 m/s), radiated wave field do not exist, only localized quasi-static field around the source can be observed. On the other hand, for the cases with speeds larger than the critical

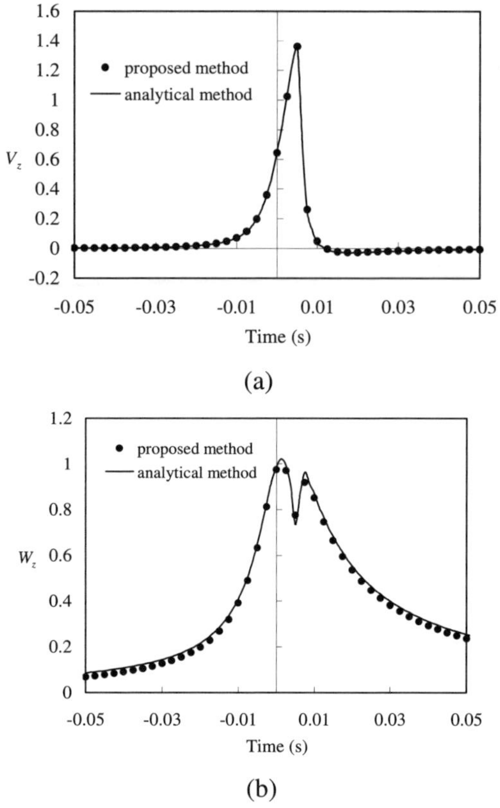

Fig. 7.22 Time-domain displacements for the case with c = 120 m/s, $\beta = 0.05$: (a) vertical displacement V_z; (b) longitudinal displacement W_z.

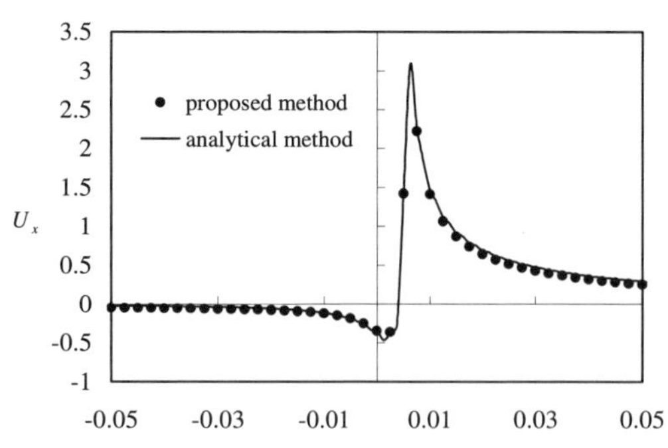

Fig. 7.23 Time-domain displacement U_x for the case with c = 120 m/s, $\beta = 0.05$.

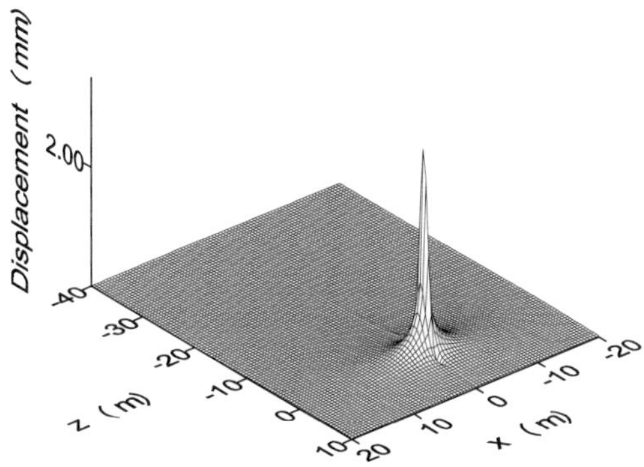

(a)

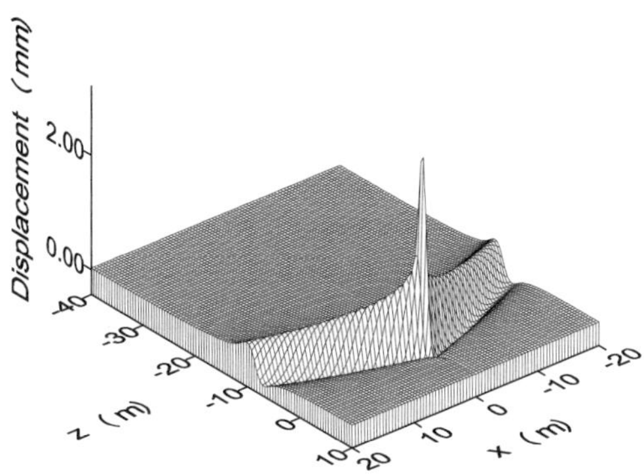

(b)

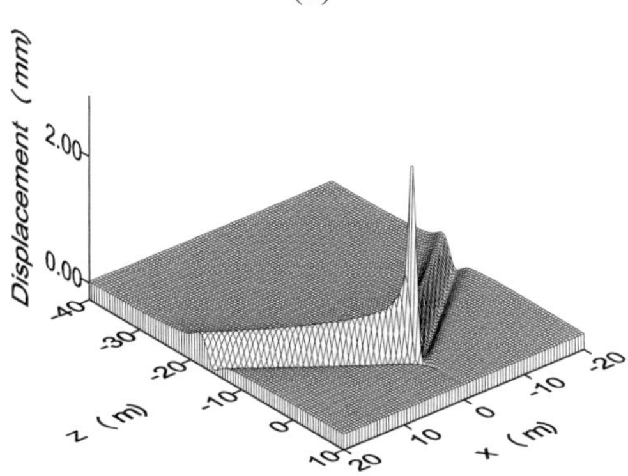

(c)

Fig. 7.24 Displacements on the surface (y = 0 m) caused by a moving wheel load with speed: (a) c = 70 m/s; (b) c = 100 m/s; (c) c = 120 m/s.

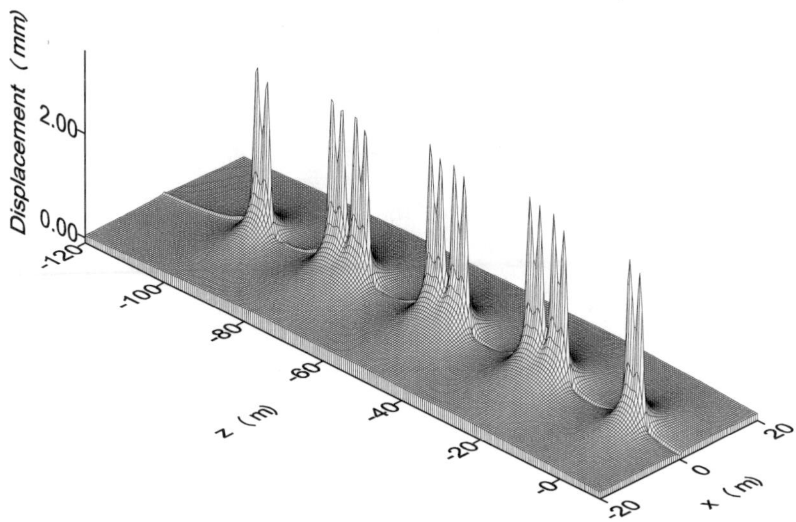

(a)

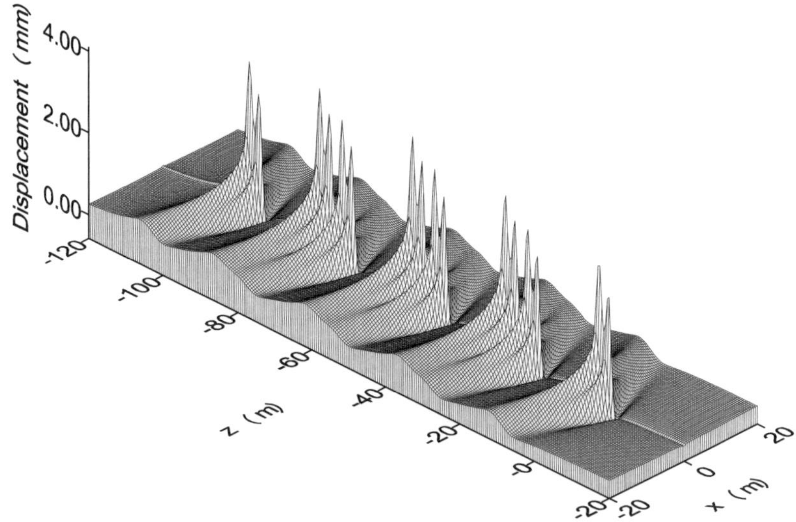

(b)

Fig. 7.25 Displacements on the surface ($y = 0$ m) caused by a sequence of moving wheel loads with speed: (a) $c = 70$ m/s; (b) $c = 100$ m/s.

speed c_R, as the ones given in parts (b) and (c) with speed c = 100 and 120 m/s, respectively, one observes a clear *Mach cone* in the displacement field. The relation between the angle of the cone and the speed is $\theta = 2\sin^{-1}(c_R / c)$. As the speed grows, the angle of cone behind the source becomes smaller.

Figure 7.25 shows the displacements on the ground surface caused by a moving train with a carriage number of N = 4. Based on the definition for the moving loads given in Eq. (2.33) and Fig. 2.3(b), the following dimensions are adopted: a = 2.56 m, b = 16.44 m and L = 25 m. Similar trend as Fig. 7.24 can be observed for Fig. 7.25.

For a moving wheel load with self oscillation, Figs. 7.26-28 show the real-part displacement field of the ground surface caused by the wheel load with an oscillation frequency of $f_0 = 10$ Hz and moving at speed c = 50, 70, and 100 m/s, respectively, in which parts (a) and (b) represent the spatial graphs and contour lines, respectively. Clearly, waves of different wavelengths exist ahead of and behind the wheel load. The wavelength of waves in front of the wheel load is generally smaller than that behind the wheel load, as a result of the *Doppler effect* mentioned in Chapter 2.

7.12 Concluding Remarks

In this Chapter, a 2.5D finite/infinite element approach has been presented for computing the steady state responses of a uniform visco-elastic half-space subjected to the moving loads. The elements derived are basically the same as the plane strain elements presented in Chapter 3, except that an additional DOF is introduced at each node to account for the effect of wave propagation along the direction normal to the profile considered.

Owing to adoption of the assumption that the half-space is uniform in terms of material and geometric properties, the 2.5D finite/infinite element procedure presented herein is particularly suitable for application to analysis of ground vibrations caused by trains moving along the subways, for which the tunnel structure is generally long compared with the cross section of the tunnel, and the soils surrounding

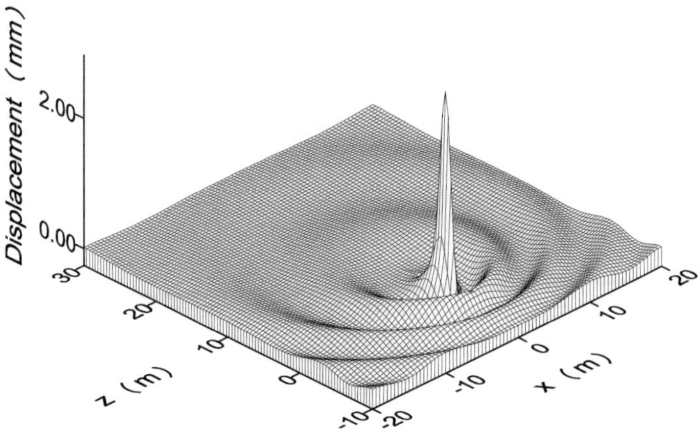

(a)

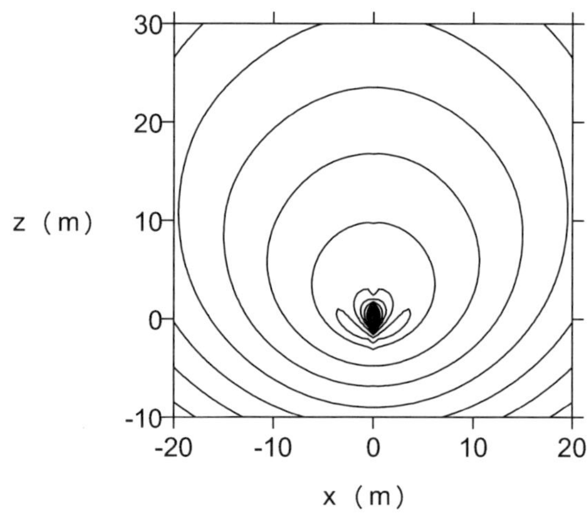

(b)

Fig. 7.26 Real-part displacements on the ground surface caused by a moving wheel load with f_0 = 10 Hz and c = 50 m/s: (a) spatial graph; (b) contour lines.

the tunnel structure may be assumed to be uniform along the tunnel direction.

The reliability and accuracy of the elements and procedure presented have been demonstrated in the study of several numerical examples. The following are the conclusions drawn from the numerical studies:

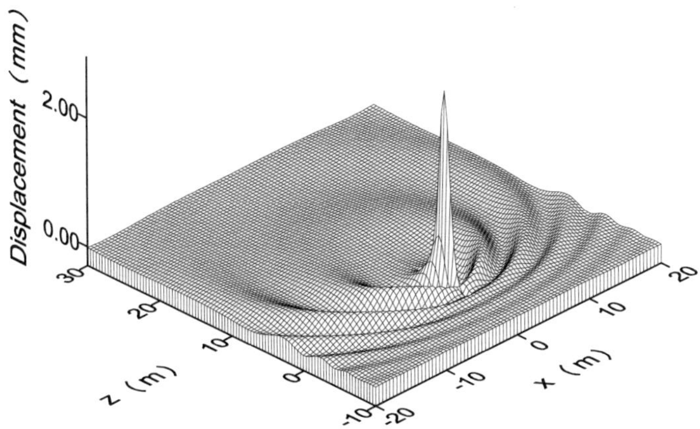

(a)

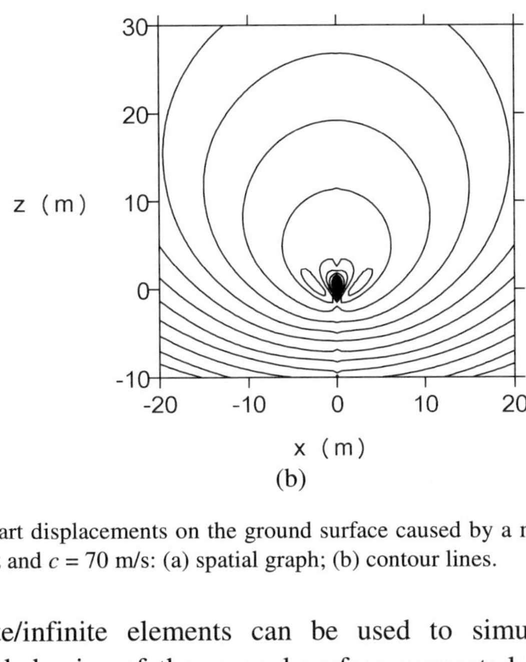

(b)

Fig. 7.27 Real-part displacements on the ground surface caused by a moving wheel load with f_0 = 10 Hz and c = 70 m/s: (a) spatial graph; (b) contour lines.

(1) The finite/infinite elements can be used to simulate the wave traveling behavior of the ground surface generated by the moving loads at all range of speeds with rather good accuracy, including the sub-, trans- and super-critical speed ranges.

(2) For moving loads with no self oscillation frequency (i.e., with $\omega_0 = 0$), with the aid of the condensation procedure presented in Section 3.5,

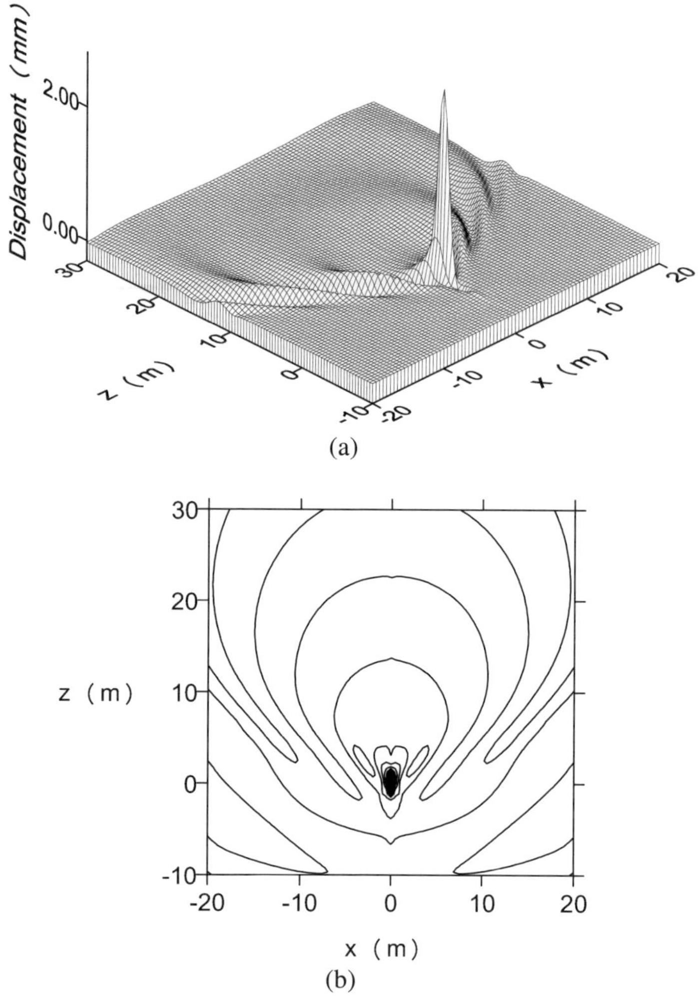

Fig. 7.28 Real-part displacements on the ground surface caused by a moving wheel load with f_0 = 10 Hz and c = 100 m/s: (a) spatial graph; (b) contour lines.

a finite/infinite element mesh that is good for the highest frequency considered can be used to yield very accurate solutions for all the lower frequencies.

(3) For moving loads with nonzero self oscillation frequencies (i.e., $\omega_0 \neq 0$), when either of the following conditions is met, $\omega > [\omega_0/(1-c/c_R)]$ and $\omega > [\omega_0/(1+c/c_R)]$, no waves will propagate

outward. Such a property can be utilized in selection of the frequency range and in determining the wave number k_s', element size L, and mesh range R to make the finite/infinite element mesh more efficient.

(4) The requirement for the mesh range, i.e., $R \geq 0.5\lambda_s$ is not as strict or as essential as that for the element size, i.e., $L \leq \lambda_s / 6$. If the range R selected is slightly less than the limit of $0.5\lambda_s$, the solution obtained can still be quite accurate and acceptable.

Chapter 8

Ground Vibration Due to Moving Loads: Parametric Study

The 2.5D finite/infinite element approach presented in the preceding chapter will be adopted to study the vibration attenuation characteristics of the ground due to moving loads. As was stated previously, by assuming the half-space to be uniform along the train-moving direction, only a 2D profile perpendicular to the train-moving direction needs to be considered in analysis. An extensive parametric study is carried out in this chapter. The parameters considered include the shear wave velocity, damping ratio, Poisson's ratio, layer thickness of the soils, and the train speed. A moving train is idealized as a sequence of moving wheel loads that may vibrate by themselves with some specific frequencies. The results to be presented include the vibration attenuation of the maximum responses in time domain, the transfer function of the soil layers in frequency domain, and the displacement field on the ground surface in space domain.

8.1 Introduction

The theory of wave propagation for the half-space presented in Chapter 2 is applicable only to problems of which the homogeneous soil layer has a thickness that is large compared with the wavelength of the traveling waves. With the 2.5D finite/infinite element procedure presented in the preceding chapter, we are able to analyze problems of a great variety. Besides the case of a uniform half-space subjected to moving trains, we shall consider as well in this chapter the cases of a single or multi soil layers that may or may not be superimposed on a bedrock, so as to obtain results of wider applicability to practical problems. Moreover, we shall focus our attention on the vibration attenuation characteristics of the surface points, in addition to that underneath the moving loads, as was concerned in Chapter 2. In order to identify the key parameters governing the ground response induced by the moving trains, a comprehensive parametric study will be performed.

In the literature, a great amount of research has been carried out on the dynamic response of soils with different geometric or material properties under a point or line load, which include Israil and Ahmad (1989), Ewing *et al.* (1957), Wolf (1985, 1988), and so on. In these studies, the important properties of layered soils and various effects of soil materials on the dynamic response have been investigated. However, the external loads considered by most previous researchers are all acting at the same location, i.e., with the effect of moving loads excluded. Although the fundamental effects of soils obtained for the point or line load cases remain valid qualitatively for the cases with moving loads, different frequency contents need to be considered for the moving loads with varying speeds, while the degree to which the soil properties may affect can be quite different. On the other hand, although some previous researches have been conducted on the problem of moving loads, as were reviewed in Chapter 1, most of them were concentrated on the effect of moving speed. Relatively few have been extended to include the effect of soil properties on ground vibrations.

The purpose of this chapter is to perform an extensive parametric study to investigate the effect of soil properties on the surface vibration characteristics due to high-speed moving trains. The 2.5D finite/infinite element procedure of analysis presented by Yang and Hung (2001) or Chapter 7 is adopted, which allows us to consider the variations of soil properties in an easy manner. The parameters to be considered include the shear wave velocity, damping ratio, Poisson's ratio, layer thickness of soils, and the train speed. The passage of a train over the soil profile is idealized as a sequence of moving wheel loads that may vibrate by themselves with some specific frequencies. The results to be presented include the vibration attenuation of the maximum responses in time domain, the transfer function of the soil layers in frequency domain, and the displacement field on the ground surface in space domain.

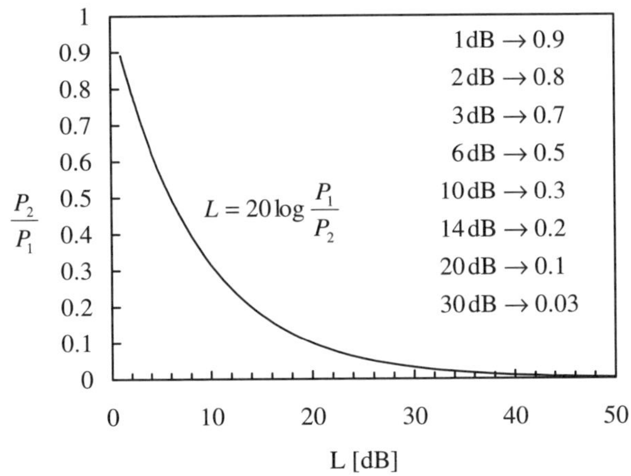

Fig. 8.1 Relationship between dB scale and linear scale.

8.2 Measurement of Vibration Attenuation for Soils

As the dynamic response of soils over a wide distance covers a rather large range, the attenuation of ground-borne vibration is often expressed by a logarithmic scale in dB. The following is the formula commonly used:

$$L[dB] = 20\log\frac{p_1}{p_2}, \tag{8.1}$$

where P_1 is the measured value of the soil response and P_2 a reference value. Figure 8.1 shows the relationship between the linear scale and dB scale of Eq. (8.1).

The vibration can be expressed either in terms of displacement, velocity, or acceleration, but more commonly in velocity. In this chapter, all of the three responses will be presented. When expressed in dB scale, a reference value should be selected. The reference values have not been internationally standardized. In this chapter, the reference values used by Esveld (1989) will be adopted, i.e., $a_0 = 10^{-5}\ \mathrm{m/s^2}$, $v_0 = 10^{-8}\ \mathrm{m/s}$, and $d_0 = 10^{-11}\ \mathrm{m}$, where a_0, v_0 and d_0 respectively denote the reference value for acceleration, velocity and displacement. By so setting, the relationship between the linear scale and dB scale in terms of the

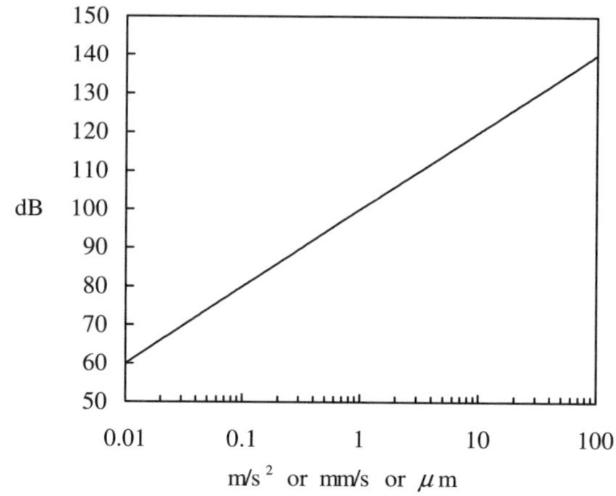

Fig. 8.2 Relationship between dB scale and linear scale for acceleration, velocity and displacement.

acceleration, velocity or displacement can be represented in a single plot as in Fig. 8.2.

8.3 Problem Description and Element Meshes

In this chapter, three types of soil geometry will be considered, i.e., a uniform half-space, as given in Fig. 8.3(a), a single soil stratum overlying a bedrock, as given in Fig. 8.3(b), and multi soil layers superimposed on a half-space, as given in Fig. 8.4. A moving load with a self oscillation frequency f_0 is assumed to travel at speed c along the z-axis on the surface, i.e.,

$$P(x=0, y=0, z, t) = \phi(z-ct)\exp(i2\pi f_0 t), \tag{8.2}$$

where the exponential term is introduced to account for the self oscillation of the moving load, which may be induced by the surface roughness of rails or mechanical system of the vehicle, and $\phi(z)$ is the *load distribution function*, as described in Chapter 2. In this chapter, the function given in Eq. (2.36) will be adopted to simulate the train-induced loads. Based on the definitions of Eq. (2.36) and Fig. 2.3(b), the following data are adopted: a = 2.56 m, b = 16.44 m and L = 25 m. The number of carriages of the train selected is N = 4, which has been

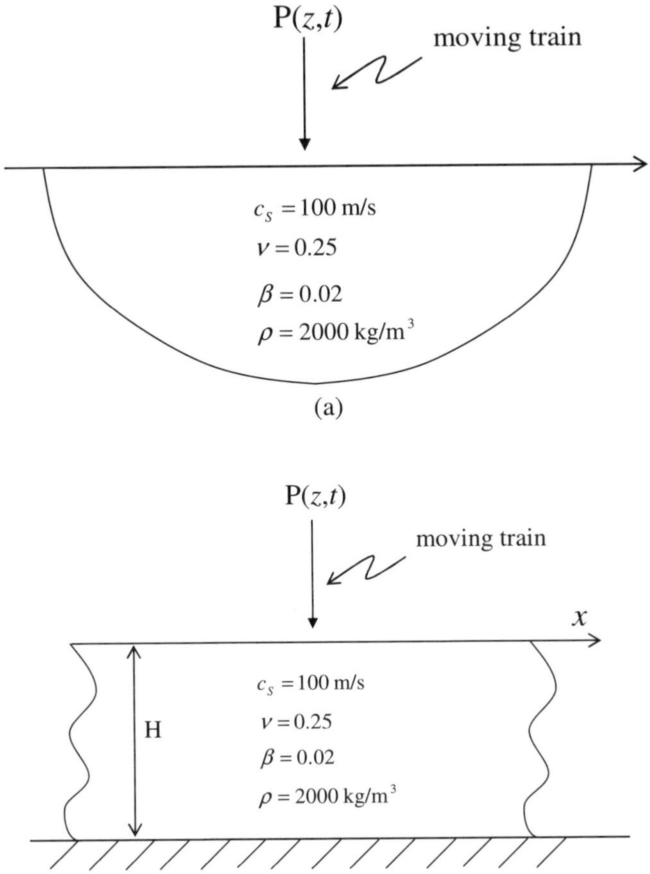

Fig. 8.3 Fundamental models: (a) homogeneous half-space; (b) single soil layer overlying a bedrock.

demonstrated to be enough for simulating the multi-wheel load effect. The wheel load is taken as T = 10 t and the characteristic length as $\alpha = 0.8$ m.

Since this chapter is concerned with wave propagation of the ground surface, a large mesh that covers a wide distance should be adopted. As shown in Fig. 8.5, the mesh to be used throughout this chapter covers a distance from x = 0 to 20 m, and a depth from y = 0 to 10 m, subjected to a half-unit load at the origin (x, y) = (0, 0), in which only half of the

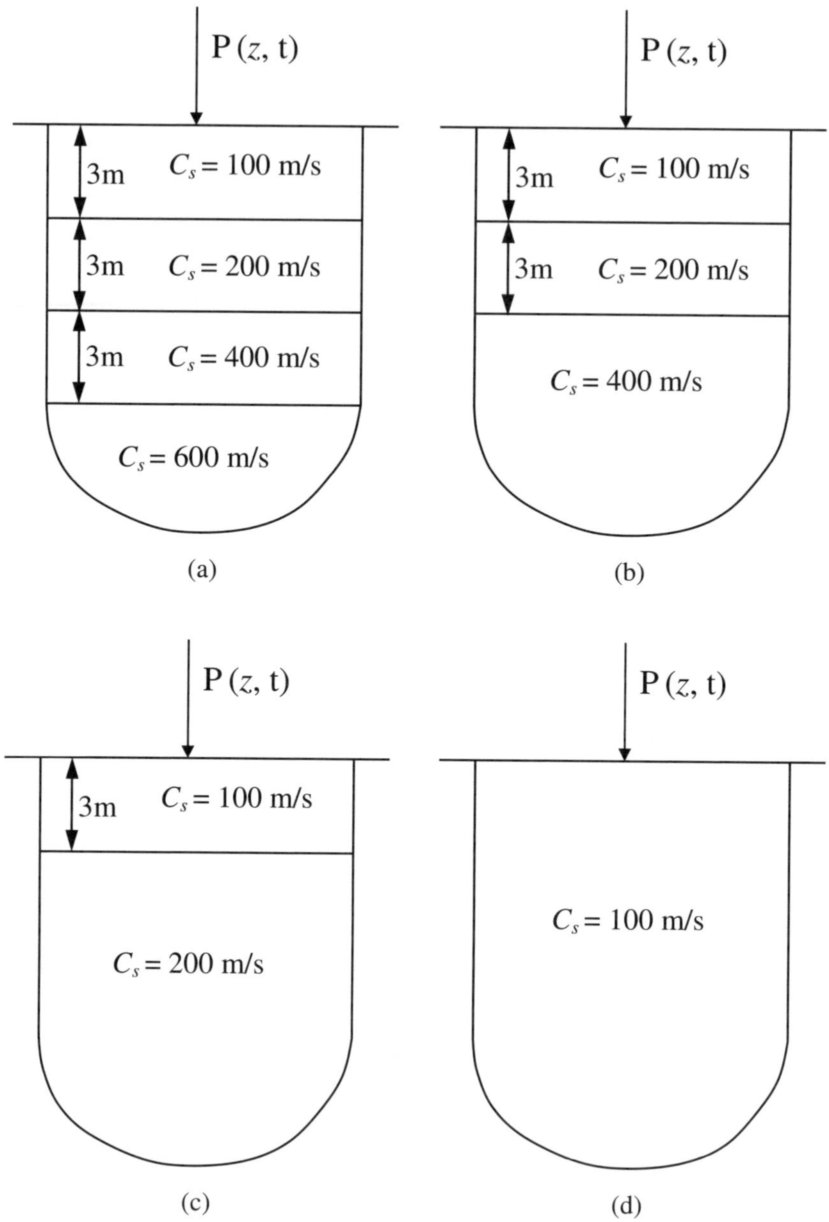

Fig. 8.4 Soil layers superimposed on a bedrock considered: (a) case 1; (b) case 2; (c) case 3; (d) case 4.

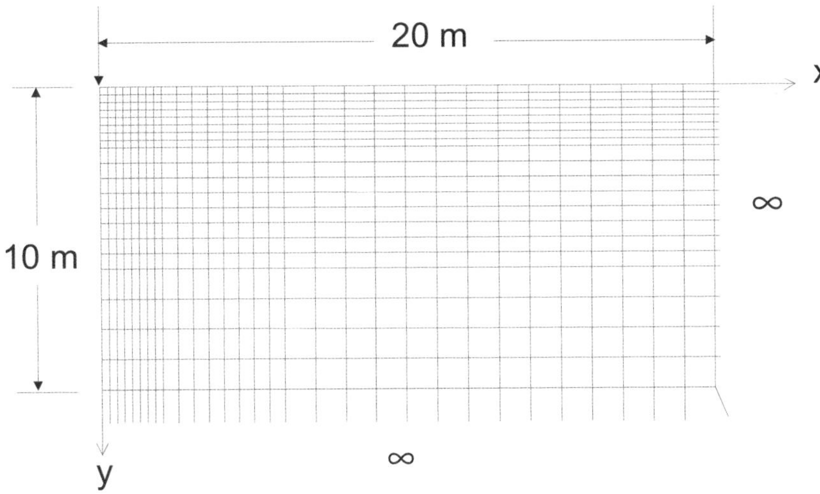

Fig. 8.5 Typical element mesh of the problem considered.

system is considered due to symmetry of the problem. The near field of the system is modeled by 2.5D Q8 elements and the far field by 2.5D infinite elements. By assigning different material properties and boundary conditions to this typical mesh, all the problems shown in Figs. 8.3 and 8.4 can be simulated.

The results obtained from the finite/infinite element model under a half-unit load do not represent directly the solution. They should be multiplied by the Fourier transform of Eq. (8.2), and then by an inverse Fourier transform to yield the final responses in time domain or space domain. Obviously, the contribution to the responses in frequency domain comes basically from two different sources. One is the Fourier transform of Eq. (8.2), which represents the effect of the moving load. And the other comes from the finite/infinite element analysis, which represents the *transfer function* of the system considered, similar to the role of a filter that produces an output response upon the receipt of an input signal from the moving loads. The transfer function determines how each frequency in the input signal is amplified or de-amplified by the soil model. In view of the fact that the transfer function can reflect the influence of soil properties, including the material properties and geometrical properties, in the present parametric study, the transfer functions will also be investigated in certain cases to explain some essential phenomena related to the effect of soil properties. It should be noted that for a plane strain problem, the transfer function depends solely on the soil properties, but for the present 2.5D problem, it also depends on the speed c and oscillation frequency f_0 of the moving loads. As a result, the transfer function appears to be much more complicated than the one for the case with no moving loads.

Of interest to note is that although the half-space is only discretized along a typical profile transverse to the moving load, i.e., on the plane $z = 0$, the vibration response can still obtained for the other parallel planes. This is owing to the fact that the 2.5D finite and infinite elements used have been defined in the wave number domain by k_z as well as in the frequency domain by ω. The effect of variable z has been included in the plane elements through addition of the term $\exp(-ik_z z)$ to the shape functions. As a result, the response for all values of z can be obtained simply by performing the inverse Fourier transform with respect to the wave number k_z.

Central to the present 2.5D finite/infinite element approach is calculation of the transfer function. To verify the adequacy of the mesh given in Fig. 8.5, the transfer function calculated by the 2.5D finite/infinite element procedure for a uniform half-space case will be investigated first and compared with the analytical solution $\tilde{v}_y$ given in Eq. (7.17). The transfer functions for the soils with S-wave speed $c_S = 100$ m/s, Poisson's ratio $\nu = 0.25$, mass density $\rho = 2000$ kg/m^3 and damping ratio $\beta = 0.02$ have been plotted in Figs. 8.6-8.9 respectively for the following four cases: (1) c = 70 m/s and f_0 = 5 Hz; (2) c = 70 m/s, f_0 = 30 Hz; (3) c = 100 m/s, f_0 = 5 Hz; and (4) c = 100 m/s and f_0 = 10 Hz.

In each of these figures, part (a) represents the transfer function plotted for the source at x = 0, and part (b) for a point near the mesh boundary at x = 18 m. As can be seen, no matter whether the location is at the source or near the mesh boundary, the transfer function obtained by the present 2.5D finite/infinite element method remains quite accurate over the range of frequencies considered. Moreover, for the cases with non-zero self-oscillation frequency f_0, the transfer functions are no longer symmetric with respect to the axis f = 0 Hz, unlike those for the

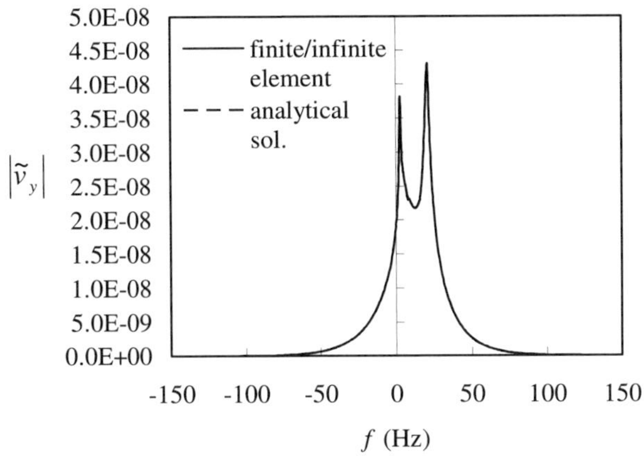

(a)

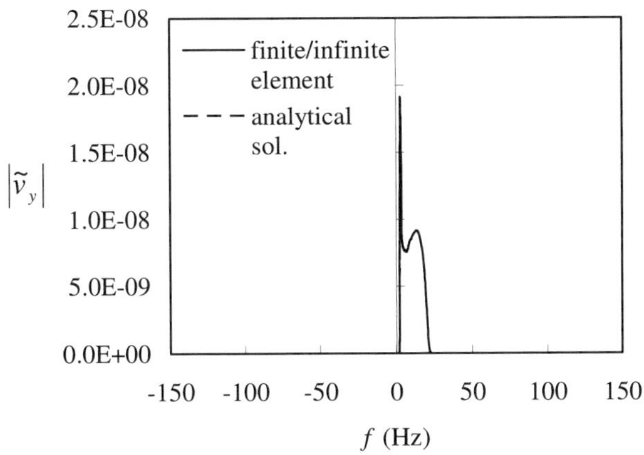

(b)

Fig. 8.6 Transfer function of a homogeneous half-space for c = 70 m/s, $f_0 = 5$ Hz at: (a) $(x, y) = (0, 1\text{ m})$; (b) $(x, y) = (18\text{ m}, 1\text{ m})$.

cases with zero self-oscillation frequency f_0. It follows that for the cases with non-zero self-oscillation frequency f_0, the time domain responses appear as complex numbers, rather than as real numbers. The real part of the responses can be interpreted as those induced by the real part of the moving load, i.e.,

$$P = \phi(z - ct)\cos(2\pi f_0 t), \tag{8.3}$$

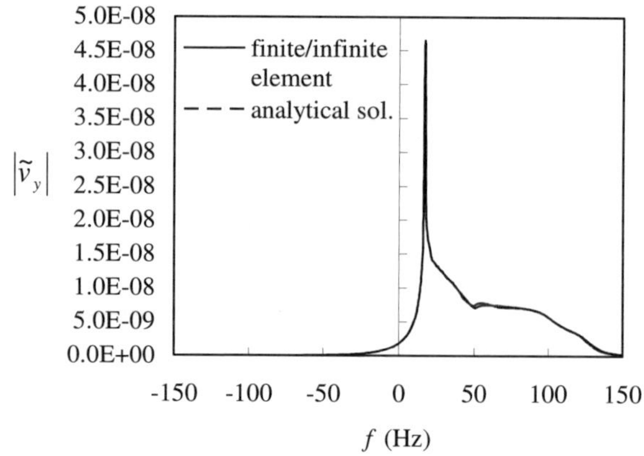

(a)

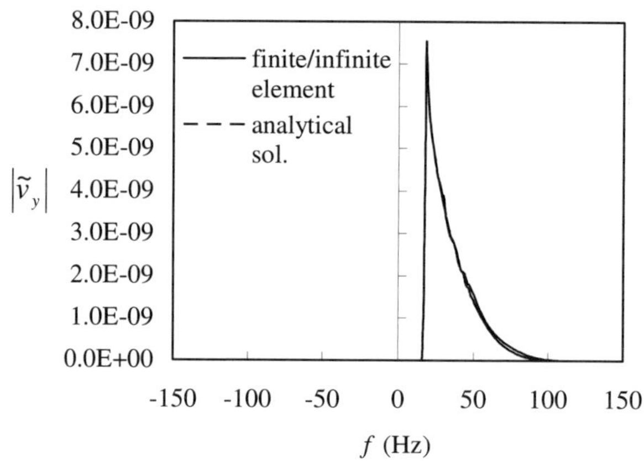

(b)

Fig. 8.7 Transfer function of a homogeneous half-space for c = 70 m/s, $f_0 = 30$ Hz at: (a) $(x, y) = (0, 1\text{ m})$; (b) $(x, y) = (18\text{ m}, 1\text{ m})$.

and the imaginary part by

$$P = \phi(z - ct)\sin(2\pi f_0 t). \tag{8.4}$$

Another feature that can be observed from the transfer functions is that the peaks shown in part (a) of each figure correspond to the critical frequencies given in Eq. (2.42) due to the *Doppler effect*.

Unless mentioned otherwise, the following soil properties will be adopted throughout the analyses in this chapter: S-wave speed

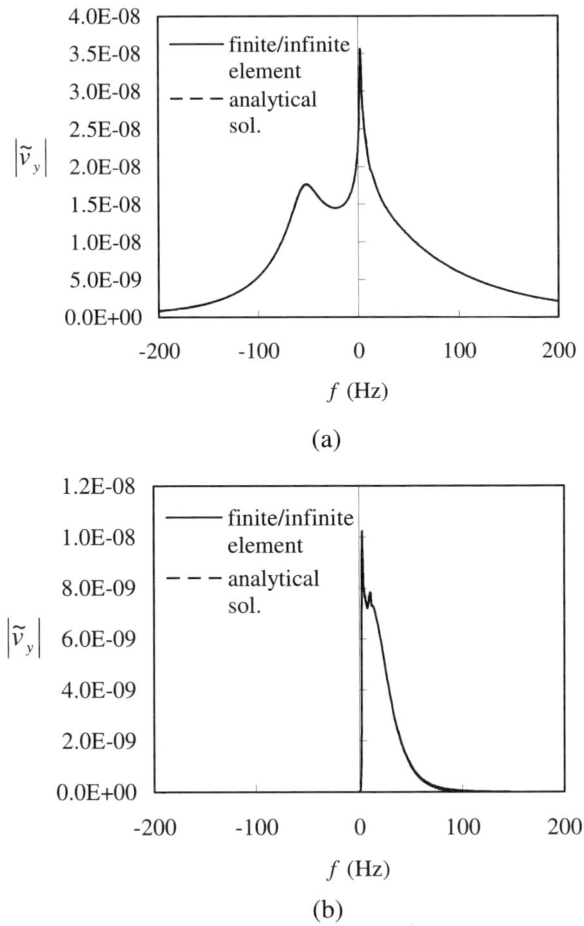

Fig. 8.8 Transfer function of a homogeneous half-space for c = 100 m/s, $f_0 = 5$ Hz at: (a) (x, y) = (0, 1 m); (b) (x, y) = (18 m, 1 m).

$c_S = 100$ m/s, Poisson's ratio $\nu = 0.25$, damping ratio $\beta = 0.02$, and mass density $\rho = 2000$ kg/m^3. The corresponding R-wave speed is 92.1 m/s. We recall from Chapter 2 that the ground vibrations induced by the moving loads with speeds lower than or greater than the R-wave speed can be quite different. Thus, except for the study focused exclusively on the effect of speeds, only two typical practical speeds, c = 70 m/s (252 km/hr) and c = 100 m/s (360 km/hr), will be considered, each to

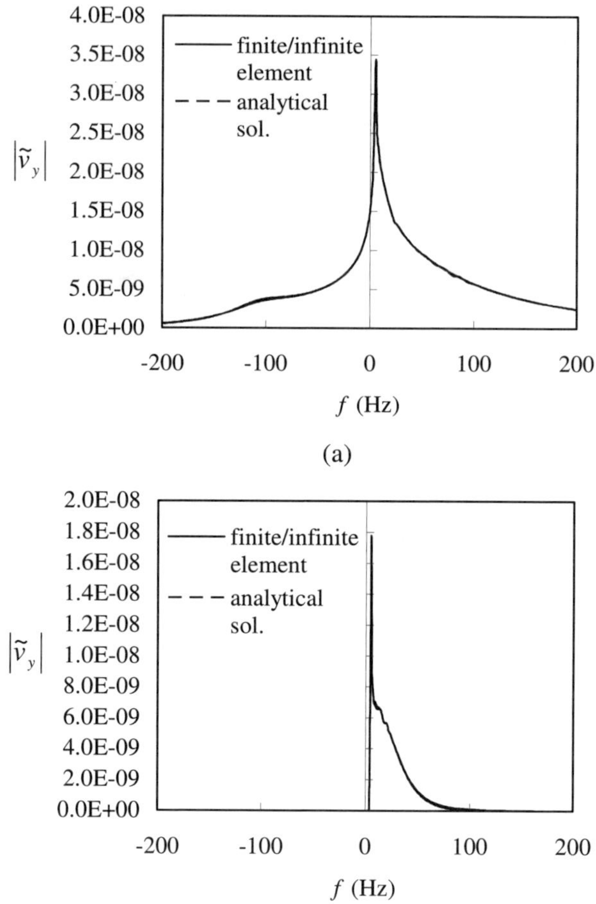

Fig. 8.9 Transfer function of a homogeneous half-space for c = 100 m/s, $f_0 = 10$ Hz at: (a) (x, y) = (0, 1 m); (b) (x, y) = (18 m, 1 m).

represent the range of speeds lower than and greater than the critical speed with respect to the R-waves, also referred to as the *sub-* and *super-critical speeds*. Note that the critical speed adopted herein is different from those used in Chapter 1, where the sub-, trans-, and super-critical speeds have been defined with respect to the shear and compressional wave speeds. In this chapter, care must be taken when talking about the reference of critical speed.

8.4 Parametric Study for a Uniform Half-Space

The model adopted in this section is the one plotted in Fig. 8.3(a). The effects of three parameters, i.e., S-wave speed, Poisson's ratio and damping ratio, will be investigated. Since an extensive study on the critical speed of a moving load on a uniform half-space has been conducted in Chapter 2 by an analytical approach, we shall not pursue the same subject in this section.

8.4.1 *Effect of shear wave speed*

By allowing the S-wave speed of the soil to vary from 100 to 400 m/s, the response attenuation in dB scale induced by a train with speed c = 70 and 100 m/s have been plotted in Figs. 8.10 and 8.11, respectively. In each figure, parts (a), (b) and (c) represent the vibration level of displacement, velocity and acceleration, respectively.

As can be seen, the variation of the S-wave speed affects the vibration level to a considerable extent. An increase in the S-wave speed implies a decrease in the *Mach number* according to the relation $M_2 = c/c_S$. We recall from Chapter 2 that as the speed is lower than the critical speed (i.e., the Rayleigh wave speed), a larger Mach number can result in greater vibration. It follows that for the load moving at the same speed c, *an increase in the S-wave speed can result in a decrease of the response*, which is exactly the phenomenon revealed in Figs. 8.10 and 8.11.

The other phenomenon observed from Fig. 8.11 is that the rates of attenuation for c_S = 200 and 400 m/s appear to be higher than that for the case with c_S = 100 m/s. The reason is that as the S-wave speed of soils becomes equal to 200 or 400 m/s, the load moving speed c = 100 m/s is no longer larger than the critical speed of the R-wave. For the case with moving speeds lower than the critical one, the attenuation rate of vibration is generally high.

8.4.2 *Effect of Poisson's ratio*

To investigate the influence of Poisson's ratio on the response of soils, four values of Poisson's ratio, $\nu = 0.25$, 0.33, 0.40, 0.48, are used. The

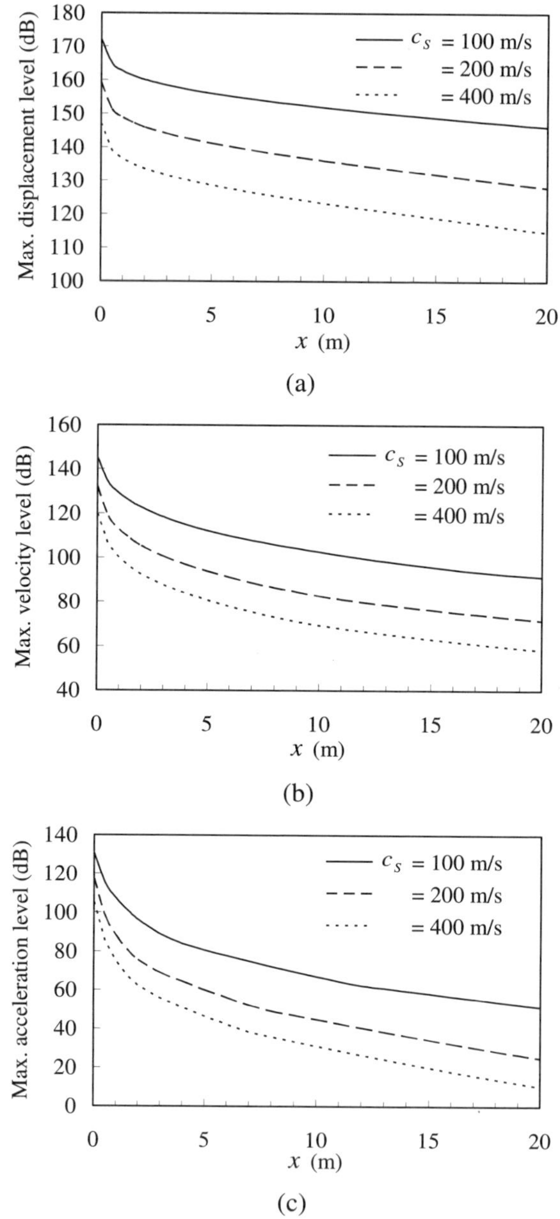

Fig. 8.10 Effect of S-wave speed on the vibration attenuation induced by a moving train with c = 70 m/s: (a) displacement; (b) velocity; (c) acceleration.

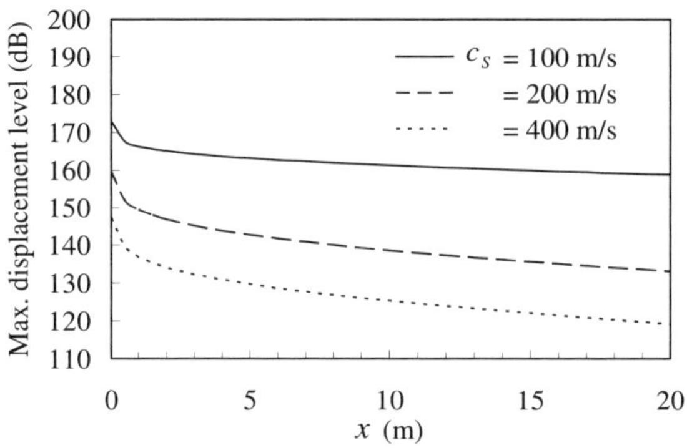

(a)

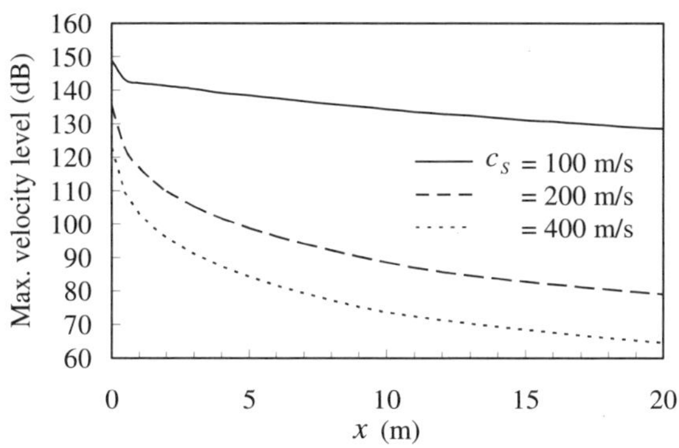

(b)

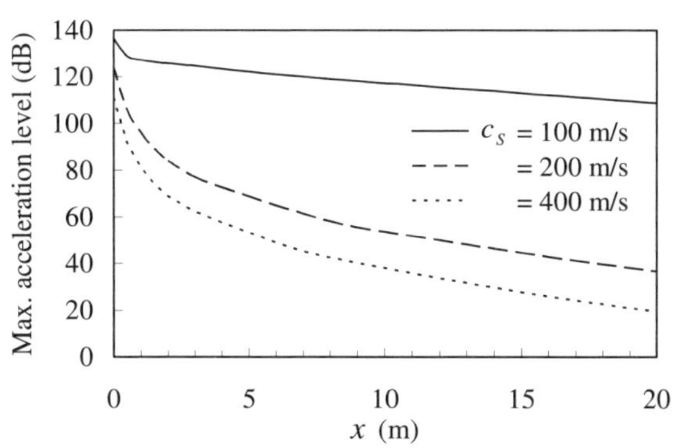

(c)

Fig. 8.11 Effect of S-wave speed on the vibration attenuation induced by a moving train with c = 100 m/s: (a) displacement; (b) velocity; (c) acceleration.

vibration response on the free surface solved for the sub- and super-critical speeds, c = 70 and 100 m/s, with respect to the Rayleigh wave speed has been plotted in Figs. 8.12 and 8.13, respectively. Again, parts (a), (b) and (c) of each figure represent the vibration level of displacement, velocity and acceleration, respectively.

From these figures, it is observed that an increase in Poisson's ratio can reduce the vibration level for both speeds in an even manner. For the case when the S-wave speed is fixed, an increase in Poisson's ratio implies an increase in the R- and P-wave speeds. According to Graff (1973), the relation between the R-wave and S-wave speeds can be approximately written as

$$c_R = \frac{0.87+1.12\nu}{1+\nu} c_S, \tag{8.5}$$

and the relation between the P- and S-wave speeds as

$$c_P = c_S \sqrt{\frac{2(1-\nu)}{1-2\nu}}, \tag{8.6}$$

both of which are functions of Poisson's ratio ν. Based on Eqs. (8.5) and (8.6), the R- and P-wave speeds computed for the four values of Poisson's ratio have been listed in Table 8.1. As can be seen, Poisson's ratio affects significantly the P-wave speed, but only slightly the R-wave speed. This observation suggests that the P-wave speed is also an important factor for vibration attenuation.

8.4.3 *Effect of damping ratio with no self oscillation*

To investigate the effect of damping ratio on vibrations, the response levels computed for five values of damping ratio, $\beta = 0.02$, 0.04, 0.06,

Table 8.1 Relations between Poisson's ratio and R- and P-wave speeds.

Poisson's ratio	c_R (m/s)	c_P (m/s)
0.25	92.0	173.2
0.33	93.2	198.5
0.40	94.1	250.0
0.48	95.1	510.0

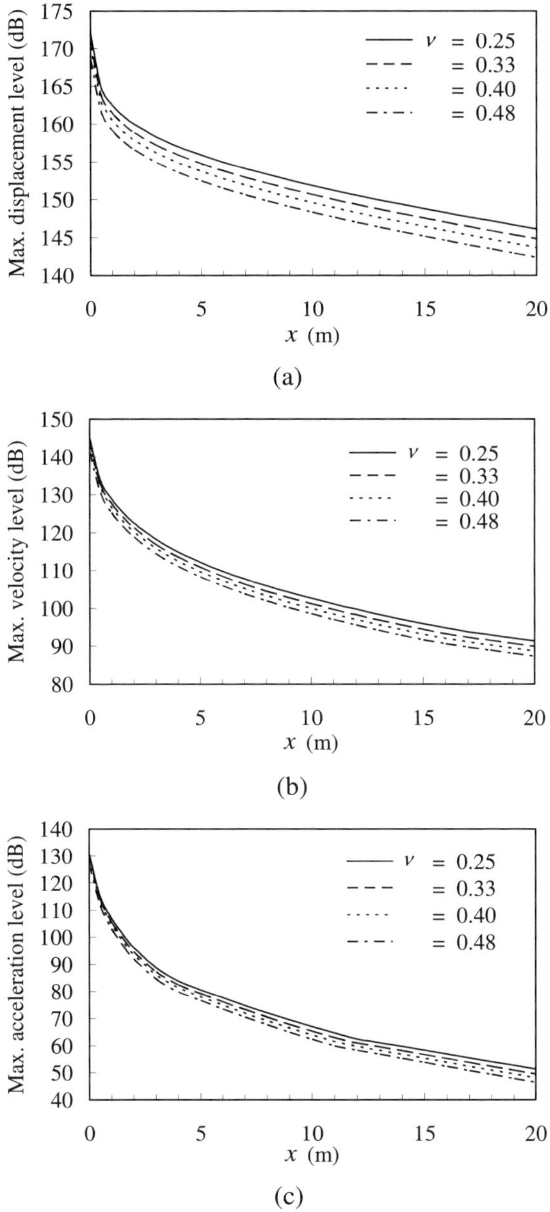

Fig. 8.12 Effect of Poisson's ratio on the vibration attenuation induced by a moving train with $c = 70$ m/s: (a) displacement; (b) velocity; (c) acceleration.

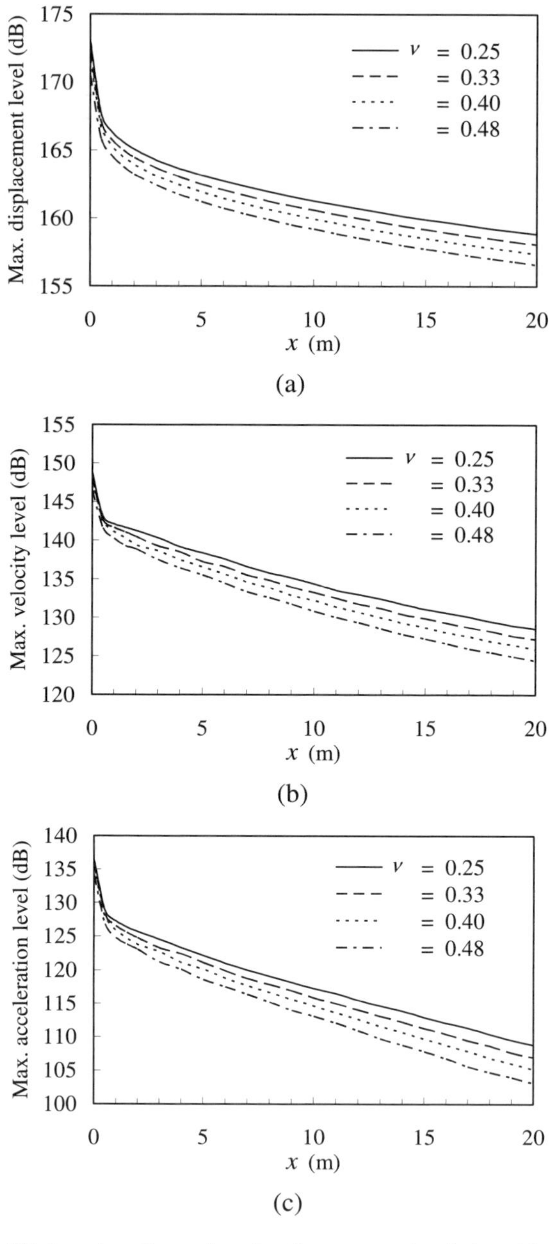

Fig. 8.13 Effect of Poisson's ratio on the vibration attenuation induced by a moving train with $c = 100$ m/s: (a) displacement; (b) velocity; (c) acceleration.

0.08 and 0.10 for the sub- and super-critical speeds, c = 70 and 100 m/s, with respect to the Rayleigh wave speed were plotted in Figs. 8.14 and 8.15 respectively. As can be seen, the effect of damping ratio on the vibration response for the sub-critical speed case (c = 70 m/s) is too small to be noted. However, the same is not true with the super-critical speed case (c = 100 m/s), in that the increase of damping ratio can result in a substantial reduction of vibrations, and that the amount of reduction in dB scale increases significantly with the distance.

8.4.4 *Effect of damping ratio for different oscillation frequencies*

To investigate the effect of damping ratio on the rate of attenuation for moving loads with different self oscillation frequencies f_0, the vibration responses of the soils with damping ratios β = 0.02 and 0.10 caused the passage of a moving train with the sub-critical speed c = 70 m/s (with respect to Rayleigh wave speed) and four different oscillation frequencies, f_0 = 5, 10, 20, 30 Hz, were plotted in Figs. 8.16 and 8.17, respectively. As can be seen, for the soils with a smaller damping ratio (β = 0.02), it is generally difficult to differentiate the rates of displacement attenuation for different oscillation frequencies f_0. But for the soils with a larger damping ratio (β = 0.10), the displacement responses induced by higher frequencies f_0 attenuate faster than those by the lower ones. The same phenomenon was noted by Jones (1991) for the plane strain problems.

A similar trend can also be observed for the moving train with the super-critical speed c = 100 m/s (with respect to the Rayleigh wave speed), as was depicted in Figs. 8.18 and 8.19 for the damping ratios of $\beta = 0.02$ and 0.10, respectively.

Again, it should be noted that for the case with non-zero self oscillation frequencies f_0, the final responses in time domain are complex numbers because the external load itself is a complex number. Thus, the maximum responses plotted in Figs. 8.16-8.19 have been obtained as the absolute values of the response in complex numbers. Another observation from these figures is that at locations closer to the moving train, i.e., the source of vibration, the velocity and acceleration

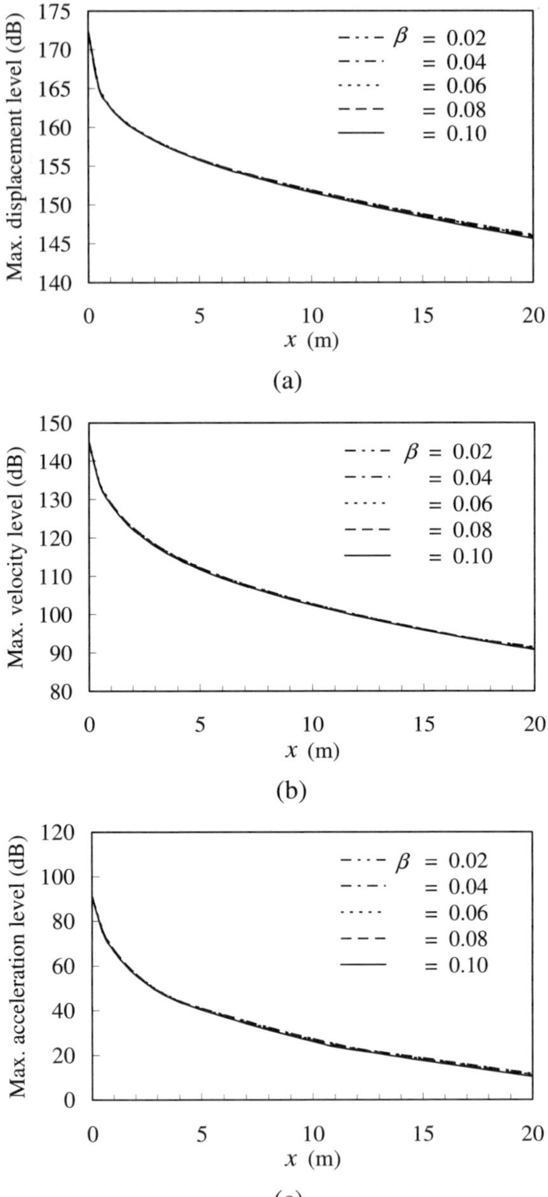

Fig. 8.14 Effect of damping ratio on the vibration attenuation induced by a moving train with $c = 70$ m/s: (a) displacement; (b) velocity; (c) acceleration.

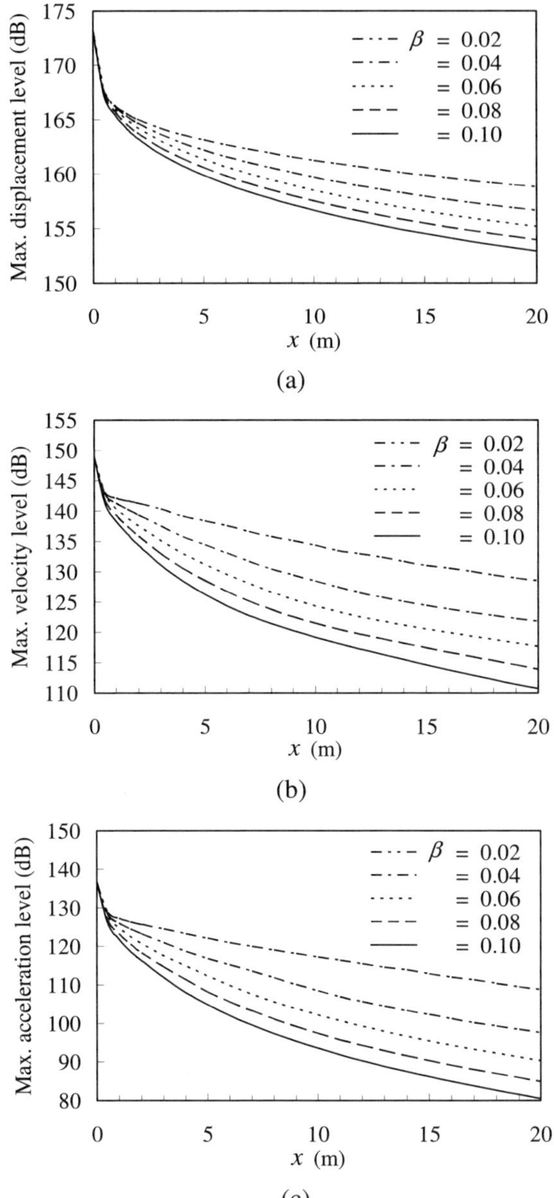

Fig. 8.15 Effect of damping ratio on the vibration attenuation induced by a moving train with $c = 100$ m/s: (a) displacement; (b) velocity; (c) acceleration.

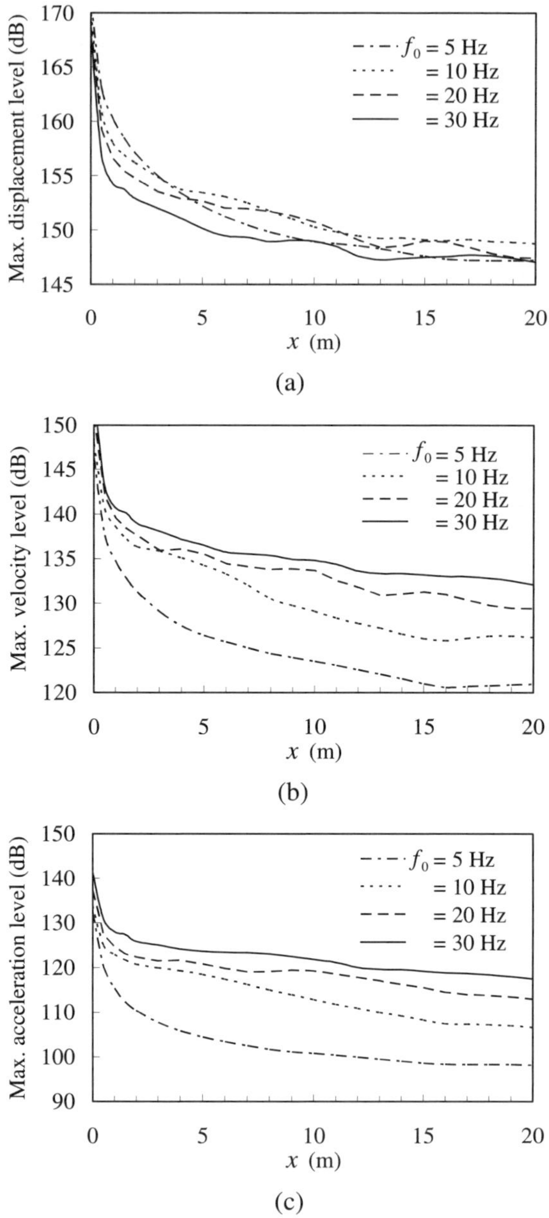

Fig. 8.16 Vibration attenuation for damping ratio $\beta = 0.02$ and a moving train with $c = 70$ m/s and different f_0: (a) displacement; (b) velocity; (c) acceleration.

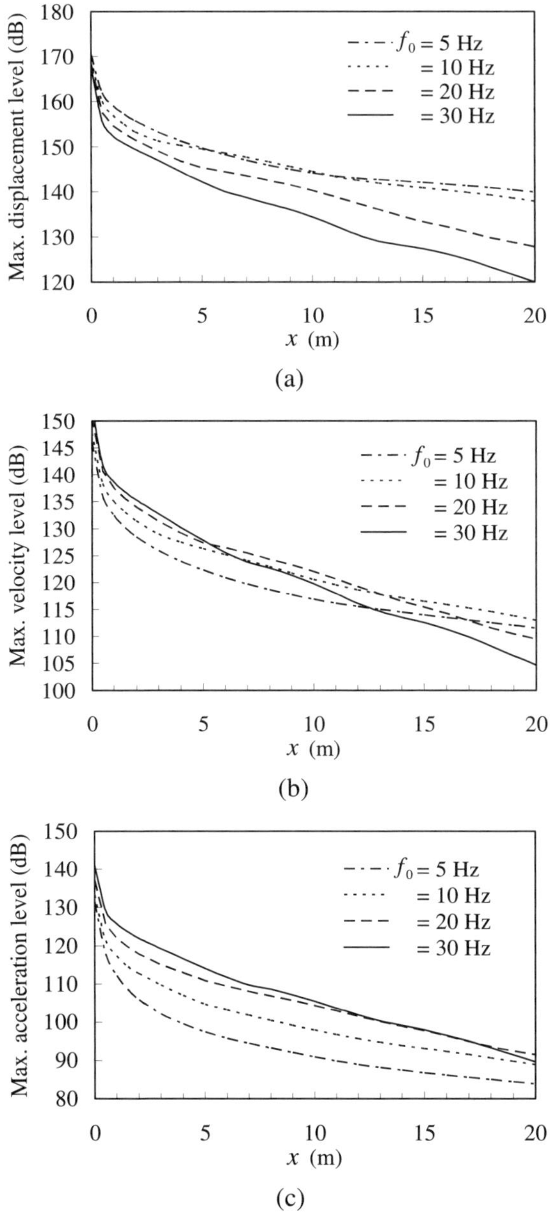

Fig. 8.17 Vibration attenuation for damping ratio $\beta = 0.1$ and a moving train with $c = 70$ m/s and different f_0: (a) displacement; (b) velocity; (c) acceleration.

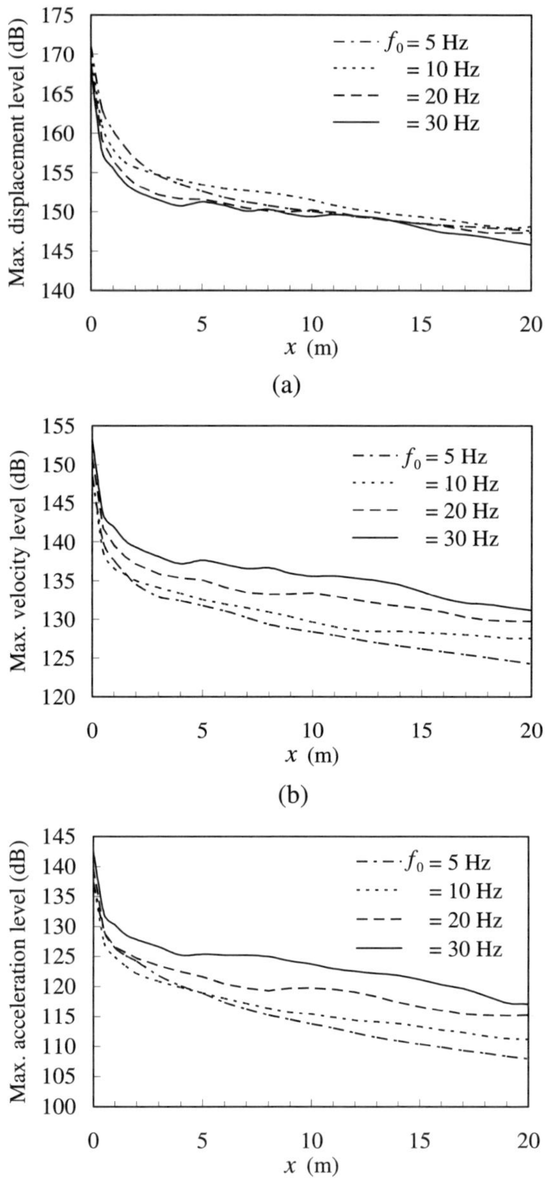

Fig. 8.18 Vibration attenuation for damping ratio $\beta = 0.02$ and a moving train with $c = 100$ m/s and different f_0: (a) displacement; (b) velocity; (c) acceleration.

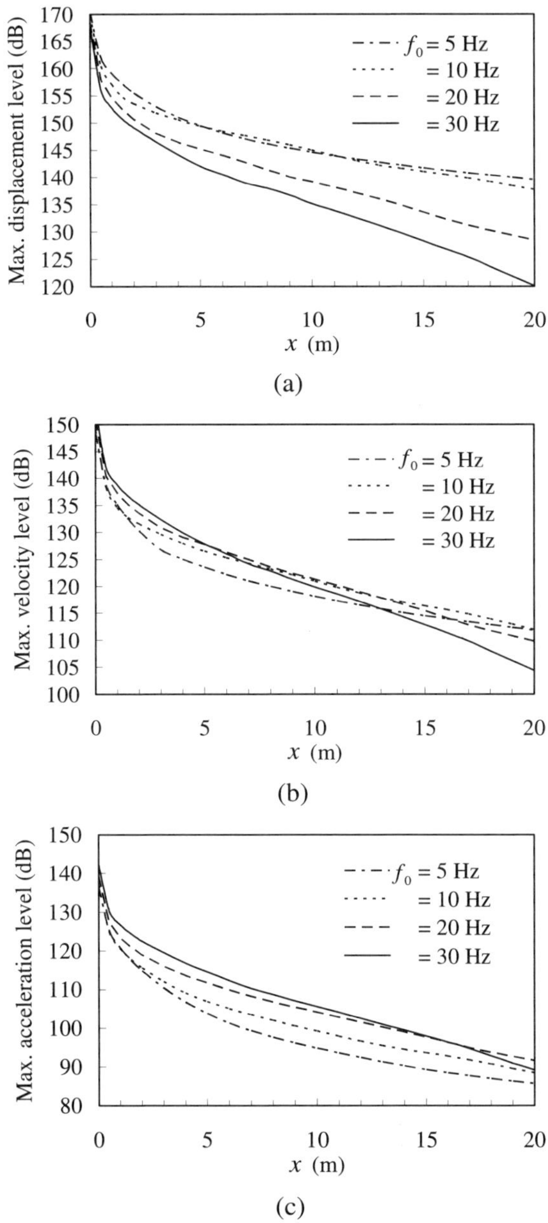

Fig. 8.19 Vibration attenuation for damping ratio $\beta = 0.1$ and a moving train with c = 100 m/s and different f_0: (a) displacement; (b) velocity; (c) acceleration.

responses are generally higher for higher self oscillation frequencies f_0, in consistence with the observation made in Chapter 2.

8.5 Parametric Study for Single Soil Layer Overlying a Bedrock

The analysis model used in this section is the one shown in Fig. 8.3(b) along with standard soil properties specified. For the uniform half-space model, the waves can be radiated from the source to the infinite boundary, but for the stratum models, because the existence of a rigid base, only waves with certain frequencies can transverse through the stratum. When the wavelength is greater than certain length implied by the soil deposit depth, the waves cannot propagate outward due to the restraining effect of the bedrock. In this regard, the important factor is the *cut-off frequency*, which can be obtained from the frequency equation for waves traveling over a single soil layer.

For a simple out-of-plane model (with respect to the soil profile), the cut-off frequency equals the fundamental vibration frequency induced by the shear waves propagating out of plane (also known as *SH waves*) of speed c_S (Wolf, 1985), i.e.,

$$f = \frac{c_S}{4H}(2n-1), \quad \text{for } n = 1,\ 2,\ 3..., \tag{8.7}$$

where H is the depth of the bedrock. The fundamental vibration frequency is the one obtained by letting n = 1. Below this fundamental or cut-off frequency, no radiation damping can occur and no waves can propagate outward.

For an in-plane plane-strain problem, because of the involvement of both the P-waves and shear waves propagating in plane (also known as *SV waves*), another set of vibration frequencies will be induced, i.e.,

$$f = \frac{c_P}{4H}(2n-1), \quad n = 1,\ 2,\ 3... \tag{8.8}$$

For the case with vertically applied loadings, the peak frequencies of the vertical response are close to the ones induced by the P-waves given in Eq. (8.8).

In this section, three-dimensional plots of displacement on the ground surface will be given, which allow us to capture the wave propagation properties of soil layers in a more direct way. In order to highlight the effects of soil layers, only a single wheel load is considered, that is,

$$P(z,t) = q_0(z-ct)\cos(2\pi f_0 t), \tag{8.9}$$

where $q_0(z)$ indicates the elastically distributed wheel load, as given in Eq. (2.31). Although there exists no difficulty in dealing with the more general case of a train consisting of a sequence of wheel loads, a complex loading function can simply make the displacement graphs more complicated for understanding of the soil effects. As a result, we insist on the use of a single wheel load in this part of study.

8.5.1 *Effect of stratum depth for a quasi-static moving load*

To study the influence of the soil stratum depth H, the response of the soil surface for different stratum depths have been plotted in Figs. 8.20 and 8.21 for a load moving at speeds c = 70 and 100 m/s, respectively. As can be seen, for the speed lower than the R-wave speed, i.e., with c = 70 m/s, a shallower soil stratum can result in greater reduction of vibrations, including the displacement, velocity and acceleration. On the other hand, for the speed greater than the R-wave speed, i.e., with c = 100 m/s, only the soil stratum with depth H = 1 m shows a reduced of vibration. As for thicker strata, the effect of layer depth is rather small, except for the displacement. Such phenomena can be explained by the transfer functions of the models considered, as will be given below.

The transfer functions for the speeds c = 70 and 100 m/s have been plotted in Figs. 8.22 and 8.23, respectively, in which parts (a) and (b) represent the transfer function at the locations x = 0 and 10 m, respectively. Great difference exists between the two figures. Firstly, by observing the results for $H \to \infty$, i.e., for the uniform half-space, we find that for speed c = 70 m/s, the vibrations of high frequencies decay quite fast with respect to the distance. However, the same is not true for the case with speed c = 100 m/s. For instance, from the results for c = 70 m/s in Fig. 8.22(b), we observe that at distance x = 10 m, the amplitude of the transfer function is localized in the range of low frequencies,

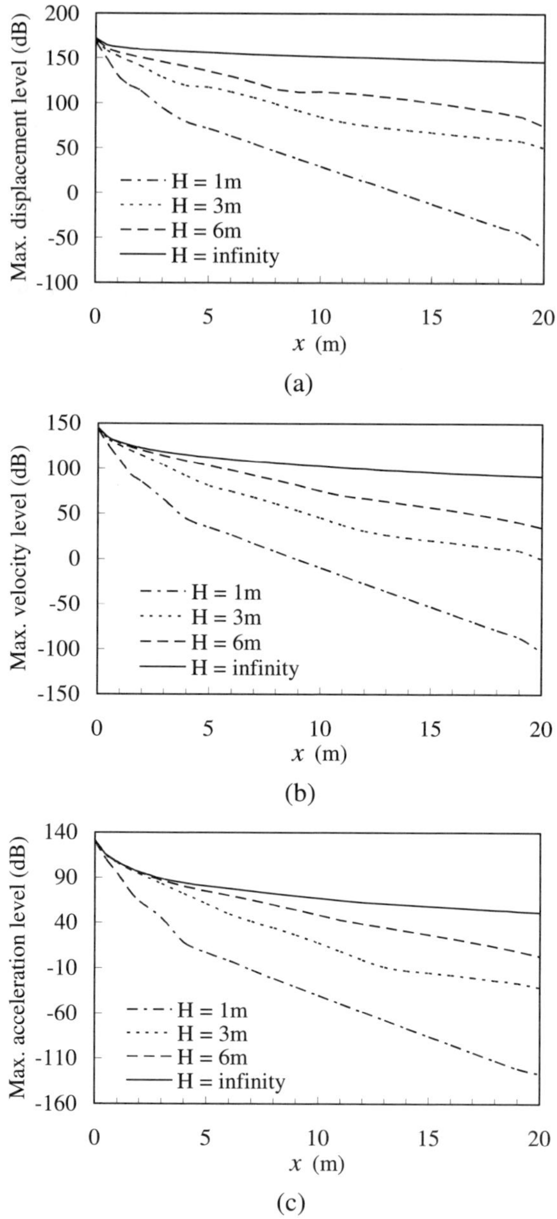

Fig. 8.20 Effect of stratum depth on vibration attenuation induced by a moving wheel load with c = 70 m/s: (a) displacement; (b) velocity; (c) acceleration.

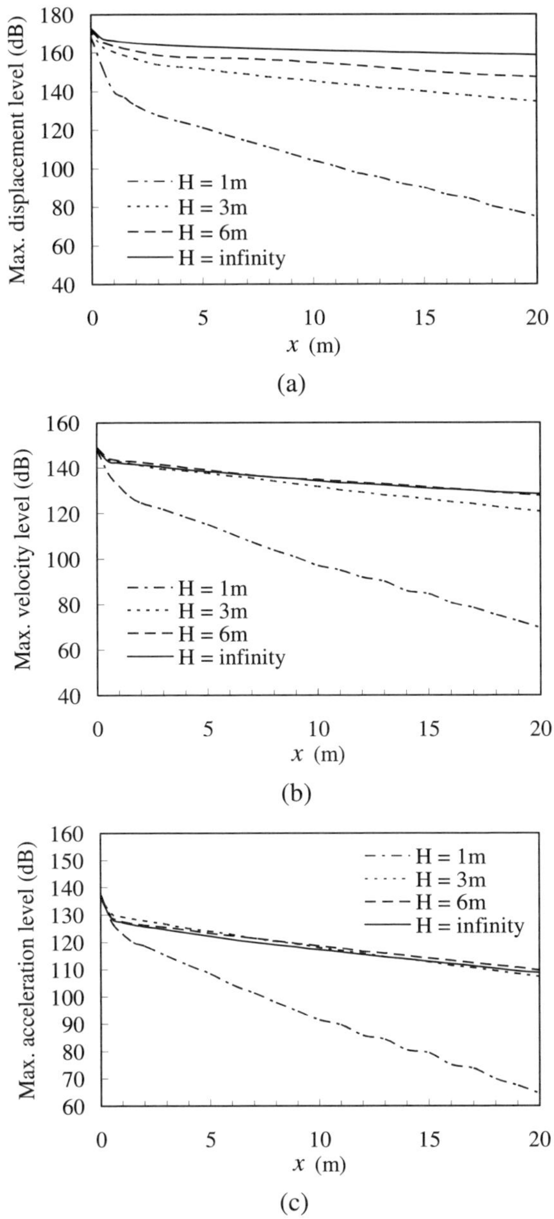

Fig. 8.21 Effect of stratum depth on vibration attenuation induced by a moving wheel load with $c = 100$ m/s: (a) displacement; (b) velocity; (c) acceleration.

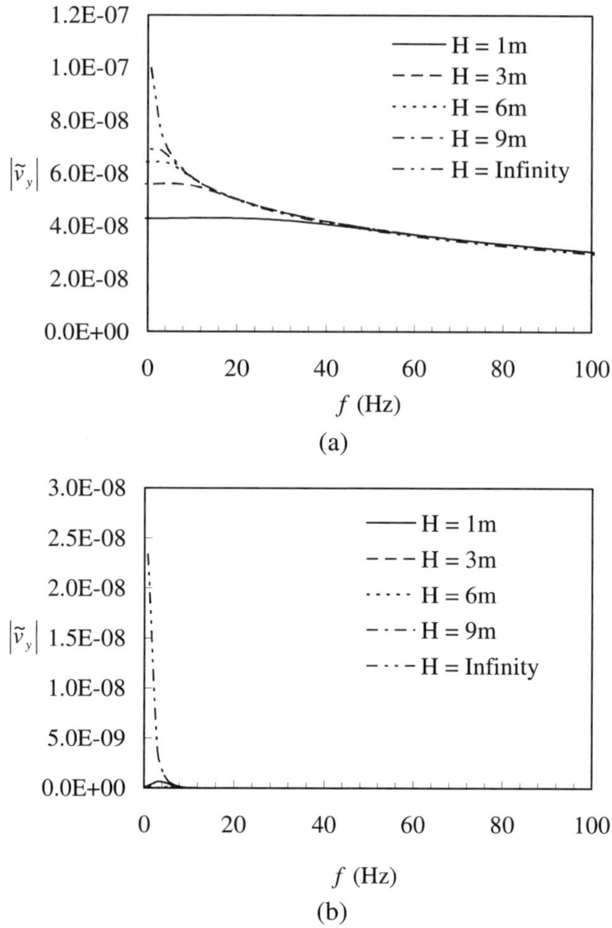

Fig. 8.22 Effect of stratum depth on the transfer function induced by a moving wheel load with $c = 70$ m/s observed at: (a) origin; (b) $(x, y) = (10$ m, 0 m$)$.

which happens to be the range of frequencies mostly affected by the soil stratum. Thus the effect of soil depth for c = 70 m/s is not negligible.

The other feature that can be observed from the figures is that for the soil depth $H \neq$ infinity, the transfer function for speed $c = 70$ m/s does not show any peak in the frequency domain, implying that the load did not resonate with the fundamental frequency of the soil layer. However, for speed $c = 100$ m/s, distinct peak frequencies can be observed, which

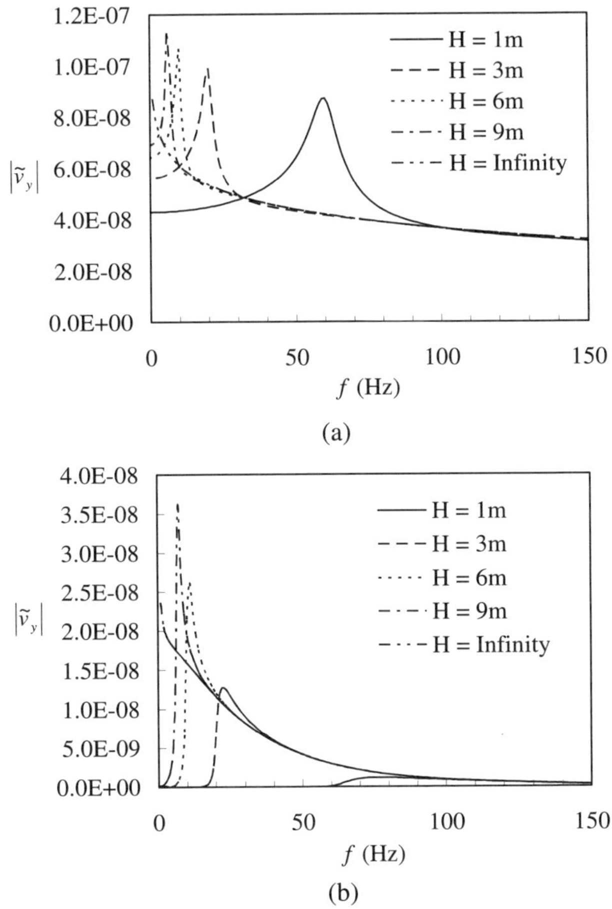

Fig. 8.23 Effect of stratum depth on the transfer function induced by a moving wheel load with $c = 100$ m/s observed at: (a) origin; (b) $(x, y) = (10$ m, 0 m).

reveals a trend similar to the vibration frequencies given by Eq. (8.8), but with slightly higher values. These peak frequencies also correspond to the *cut-off frequencies* of the soil layer. As can be seen from Fig. 8.23(b), virtually no response exists for the frequencies lower than the peak frequencies.

Due to the presence of distinct resonance peaks at different frequencies in Fig. 8.23, it is observed that the influence of layer depth for speed $c = 100$ m/s on wave propagation is rather significant. As a matter of fact, it affects greatly the mode shapes of waves propagating through the layers (as will be illustrated in the three-dimensional plots to follow), but *not* the maximum responses, especially for the velocity and acceleration. As far as the velocity and acceleration are concerned, the corresponding transfer functions are not the ones shown in Figs. 8.22 and 8.23, but are the ones further multiplied by $i\omega$ and $-\omega^2$, respectively. Thus the amplitude of peaks for the case with a lower H, which implies higher resonant frequency, can be much larger than the one shown in Fig. 8.23. Consequently, the decrease in soil depth may not always result in reduction of vibration amplitudes.

One exception is the case with H = 1 m, for which the cut-off frequency is so large, nearly 60 Hz, compared with those of the other depths. Thus, most of the frequency content that can be induced by the moving train has been cut off, thereby resulting in much lower responses shown in Fig. 8.21. From Figs. 8.22 and 8.23, we also observe that for the range with larger frequencies f, the layer solution is almost the same as the uniform half-space solution, i.e., with H = infinity. This is not surprising, if one realizes that the boundary condition at the bottom of the layer remains practically irrelevant to short wavelengths.

To have a clear picture on the effect of soil layer depth, the displacement of the ground surface at the instant when the wheel moves over the origin, i.e., $(x, z) = (0, 0$ m), were plotted in Figs. 8.24 and 8.25 for the speed c = 70 and 100 m/s, respectively, in which parts (a), (b), (c) and (d) represent the soil stratum with depth H = 1, 3, 6 and 9 m, respectively. As can be seen from Fig. 8.24, for the case with a speed lower than the R-wave speed, i.e., with c = 70 m/s, the decrease in stratum depth can result in a much more localized vibration on the ground surface. Because the displacements are localized around the source even for the uniform half-space case, the effect of stratum depth for such a speed is mainly on the response amplitude, rather than on the shape of the displacement field. For this reason, the effect of soil layer depth can hardly be distinguished from a spatial graph.

On the other hand, for the speed greater than the R-wave speed (Fig. 8.25), a distinct wave shape can be observed behind the moving load, which should be interpreted as the accumulation of waves reflected back and forth from the bedrock. Moreover, with the increase in layer

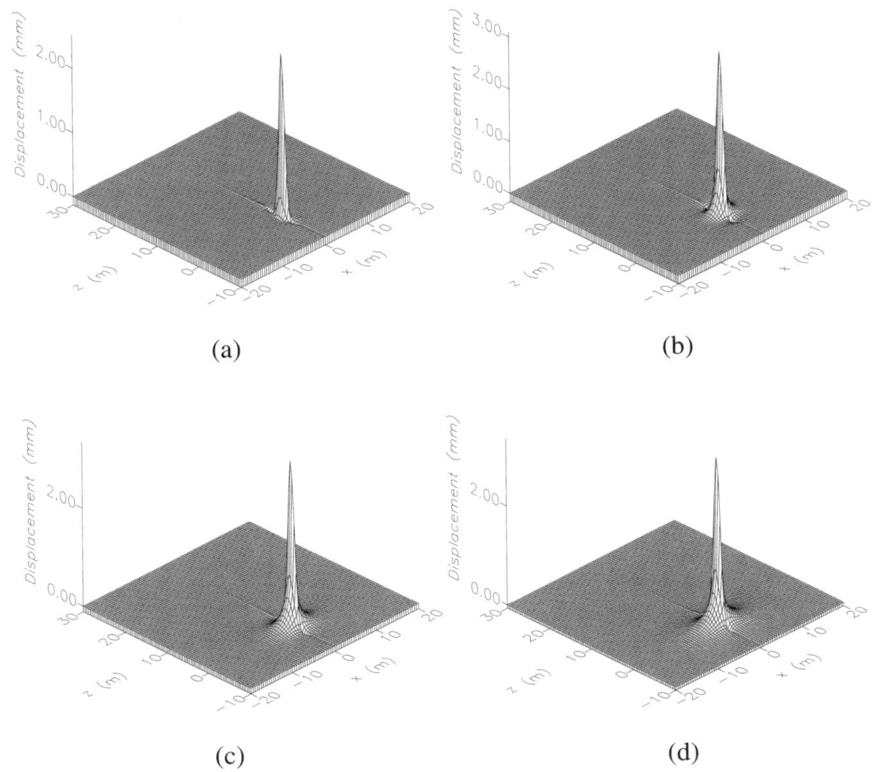

Fig. 8.24 Displacements caused by a moving wheel load with c = 70 m/s: (a) H = 1 m; (b) H = 3 m; (c) H = 6 m; (d) H = 9 m.

depth, the wavelength increases. Of interest is the result shown for the soil depth H = 1 m in Fig. 8.25(a), where the powerful effect of cut-off frequency is clearly revealed. Almost all waves have been suppressed by the bedrock, except those with very high frequencies.

8.5.2 *Effect of stratum depth for a moving load with self oscillation*

For the moving load with a self oscillation frequency of f_0 = 10 Hz, Figs. 8.26 and 8.27 depict the effect of soil layer depth on the response attenuation for the speed c = 70 and 100 m/s, respectively. From these figures, it is clear that the attenuation of vibration for depth H = 1 m is

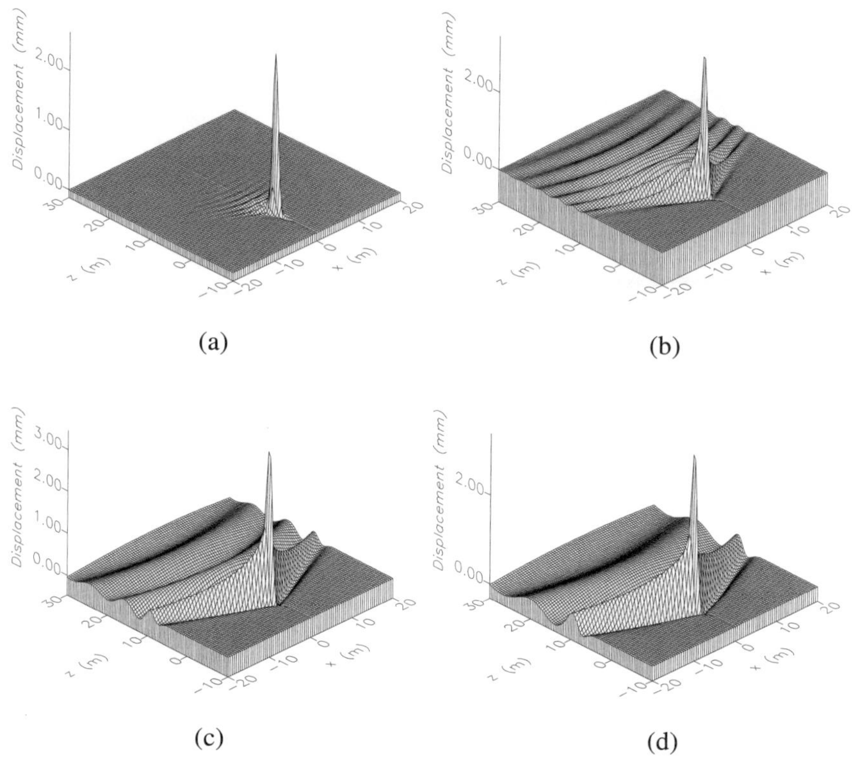

Fig. 8.25 Displacements caused by a moving wheel load with c = 100 m/s: (a) H = 1 m; (b) H = 3 m; (c) H = 6 m; (d) H = 9 m.

much faster than the others for both speeds. Such a result can also be explained by the transfer functions obtained for f_0 = 10 Hz in Figs. 8.28 and 8.29 for the two speeds c = 70 and 100 m/s, respectively. By examining the results for c = 70 m/s and $H \to \infty$ at a distance 10 m away from the source, i.e., Fig. 8.28(b), we observe that the main frequency content of the transfer function ranges from 5.68 to 41.67 Hz, corresponding to the critical frequencies $f_{cr} = f_0 /(1 \pm c / c_R)$ caused by the *Doppler effect* mentioned in Chapter 2. However, for the soil stratum with depth H = 1 m, due to the cut-off frequency effect, the amplitudes for frequencies lower than 50 Hz decay very fast, implying that almost all frequencies have been cut off and no waves can propagate outward.

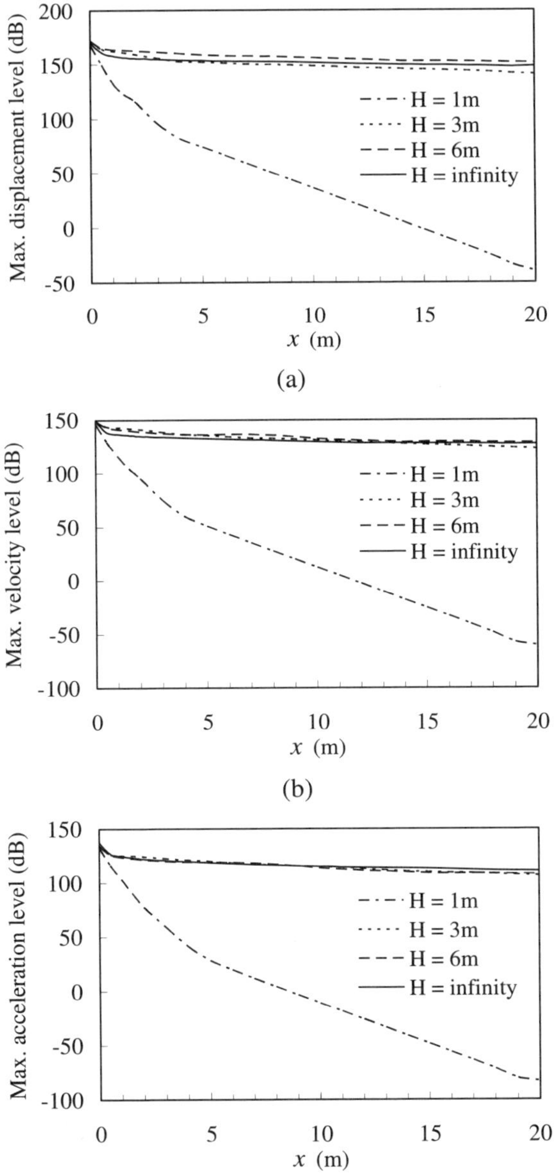

Fig. 8.26 Effect of stratum depth on vibration attenuation induced by a moving wheel load with $c = 70$ m/s and $f_0 = 10$ Hz: (a) displacement; (b) velocity; (c) acceleration.

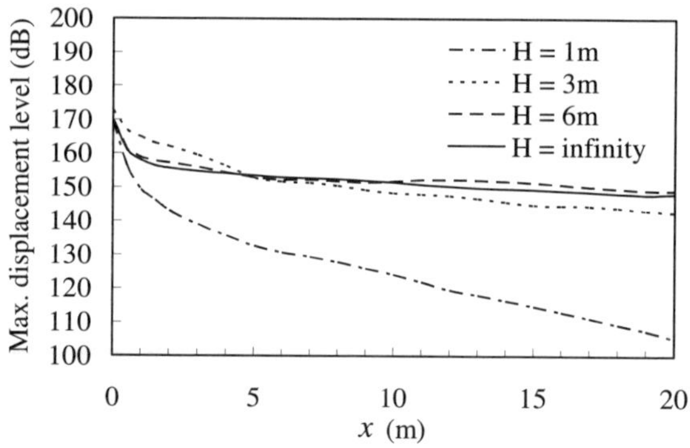

(a)

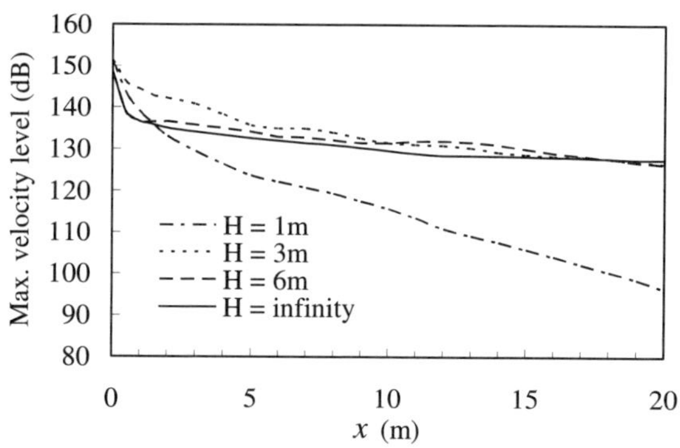

(b)

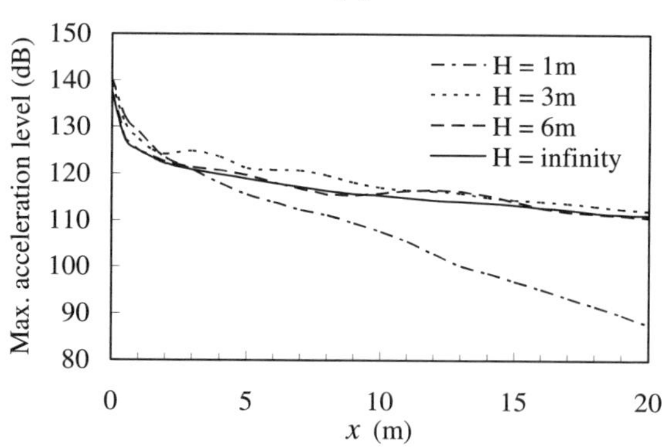

(c)

Fig. 8.27 Effect of stratum depth on vibration attenuation induced by a moving wheel load with c = 100 m/s and $f_0 = 10$ Hz: (a) displacement; (b) velocity; (c) acceleration.

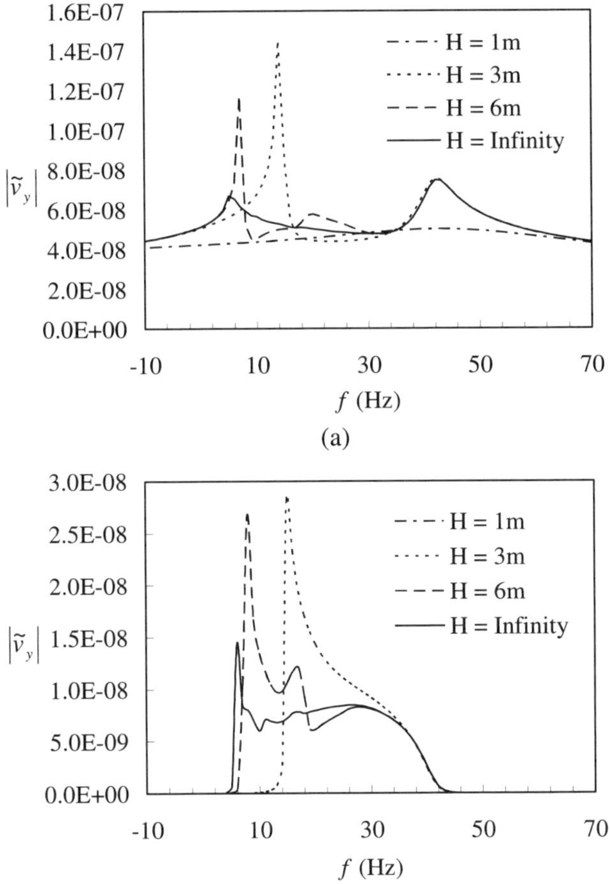

(b)

Fig. 8.28 Effect of stratum depth on the transfer function induced by a moving wheel load with c = 70 m/s and $f_0 = 10$ Hz observed at: (a) origin; (b) (x, y) = (10 m, 0 m).

Similarly, for a moving wheel load with c = 100 m/s and f_0 = 10 Hz (Fig. 8.29), the cut-off frequency for soil layer with H = 1 m has a value much higher than the others with greater depths, which implies a much smaller response for the ground surface, as indicated in Fig. 8.27.

For a moving load with a self oscillation frequency f_0 = 10 Hz, Figs. 8.30 and 8.31 show the real-part displacements in spatial domain for speeds c = 70 and 100 m/s, respectively. As revealed by Fig. 8.30 for

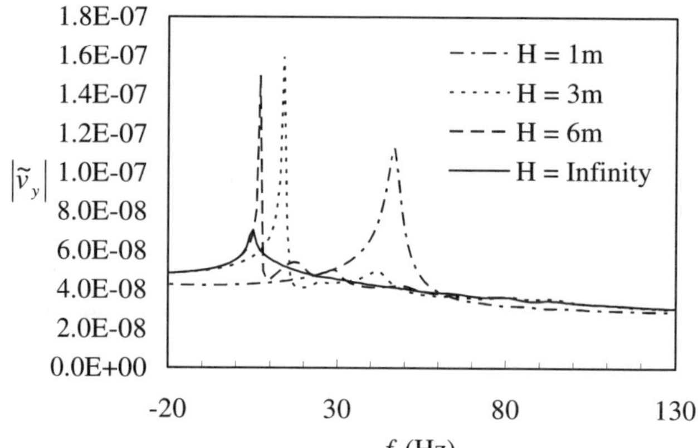

(a)

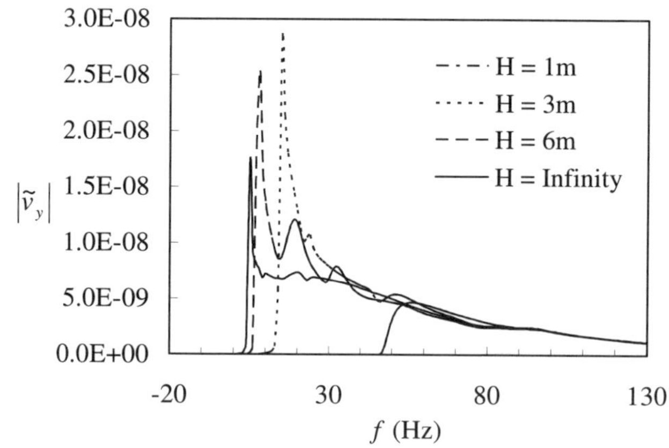

(b)

Fig. 8.29 Effect of stratum depth on the transfer function induced by a moving wheel load with $c = 100$ m/s and $f_0 = 10$ Hz observed at: (a) origin; (b) $(x, y) = (10$ m, 0 m).

speed $c = 70$ m/s, no waves can propagate outward from the source for the case with $H = 1$ m, because all the vibration modes have been restrained by the bedrock. In comparison, for speed $c = 100$ m/s shown in Fig. 8.31, the soil stratum with $H = 1$ m allows only a small amount of vibrations of high frequencies to travel outward. These figures also indicate that as the soil depth increases, its influence becomes less noticeable.

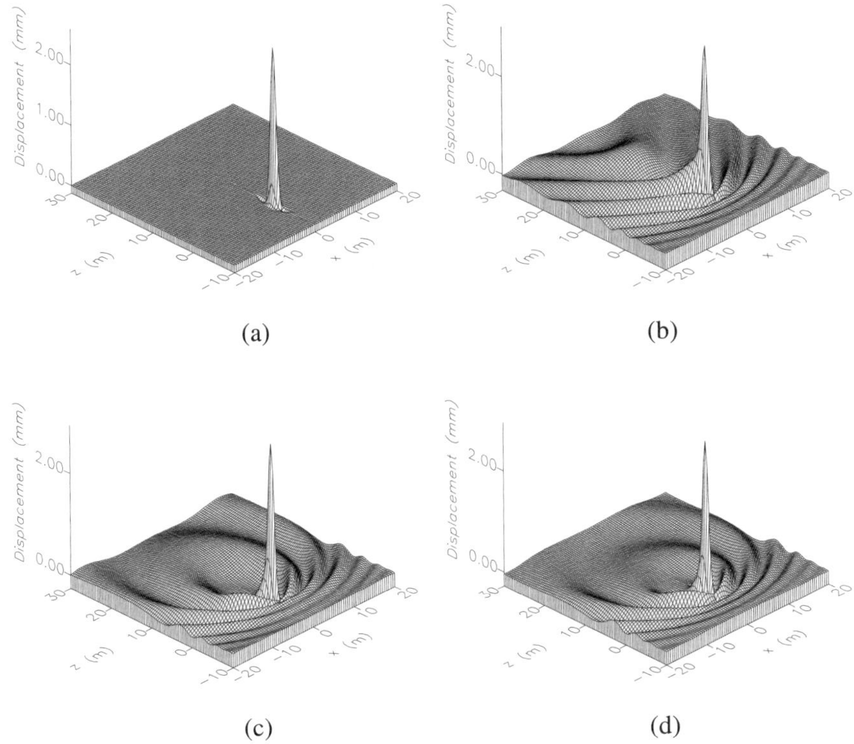

Fig. 8.30 Real-part displacements caused by a moving wheel load with c = 70 m/s and f_0 = 10 Hz: (a) H = 1 m; (b) H = 3 m; (c) H = 6 m; (d) H = 9 m..

8.5.3 *Effect of self oscillation frequency*

The effect of soil depth can be considerably different for different oscillation frequencies f_0. Consider the case for a soil layer with depth H = 3 m. As can be seen from Fig. 8.32 for speed c = 70 m/s and different values of f_0, the vibration induced by f_0 = 5 Hz seems to be trapped around the source and prevented from traveling outward. From the corresponding transfer functions plotted in Fig. 8.33, we find that the variation of the oscillation frequency f_0 affects the peak frequency to a large extent. In particular, for the transfer function shown in Fig. 8.33(a) for the origin (x = 0 m), two clear peaks can be found each for the cases

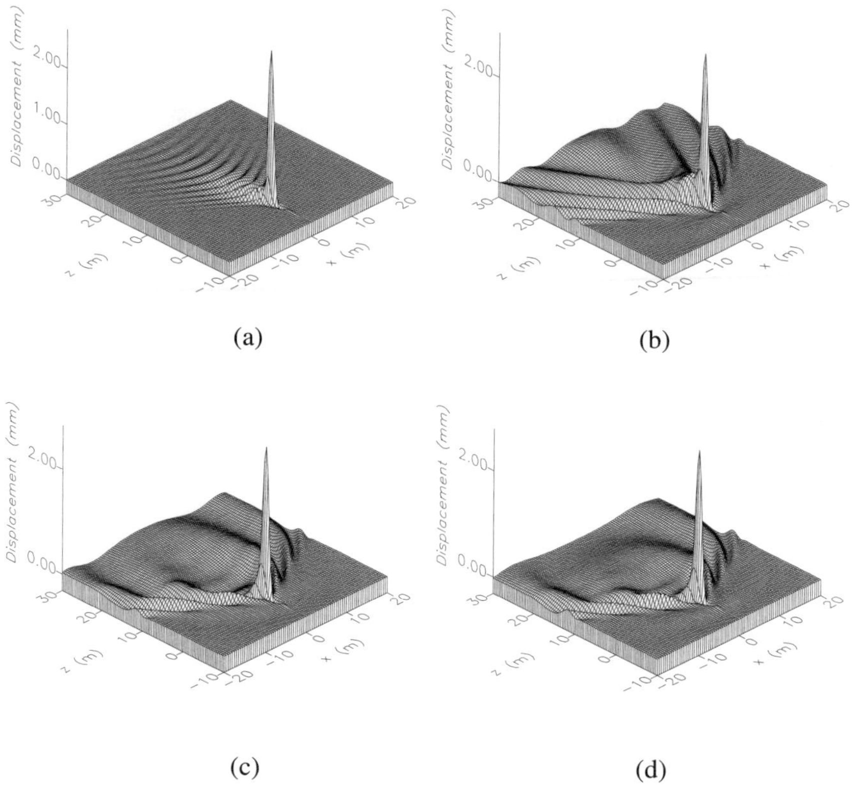

Fig. 8.31 Real-part displacements caused by a moving wheel load with c = 100 m/s and f_0 = 10 Hz: (a) H = 1 m; (b) H = 3 m; (c) H = 6 m; (d) H = 9 m.

with f_0 = 10, 20, 30 Hz, while only one peak exists for the case with f_0 = 5 Hz, due to the coupling of the *Doppler effect* with the *cut-off frequency effect*. For a uniform half-space, the two critical speeds computed from the equation $f_{cr} = f_0 /(1 \pm c / c_R)$ for the oscillation frequencies f_0 = 5, 10, 20, 30 Hz are (2.84, 20.84), (5.68, 41.67), (11.36, 83.35), and (17.05, 125.02) Hz, respectively. But for the soil stratum with a bedrock at depth H = 3 m considered herein, due to the cut-off effect of the bedrock, the first peak frequencies revealed by Fig. 8.33(a) for f_0 = 10 and 20 Hz are all clustered around the frequency of 15 Hz, which is close to the frequency of the stratum with depth H = 3 m, as given by Eq. (8.8).

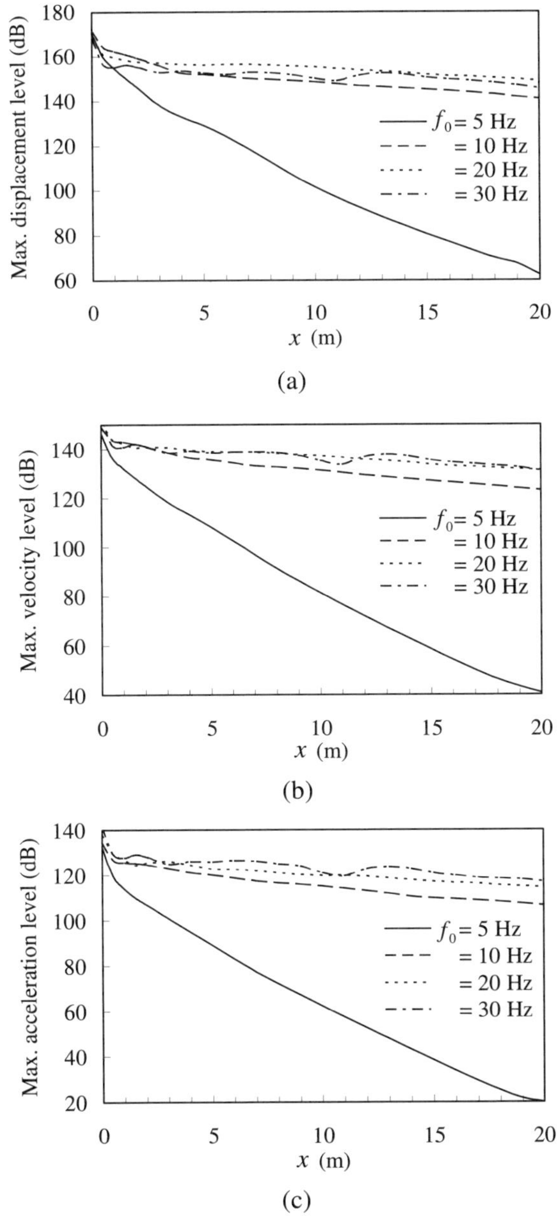

Fig. 8.32 Vibration attenuation for stratum with depth H = 3 m due to a moving wheel load with c = 70 m/s and different f_0: (a) displacement; (b) velocity; (c) acceleration.

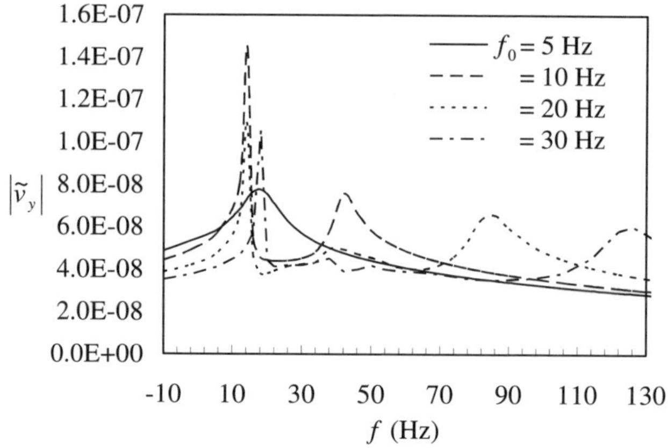

(a)

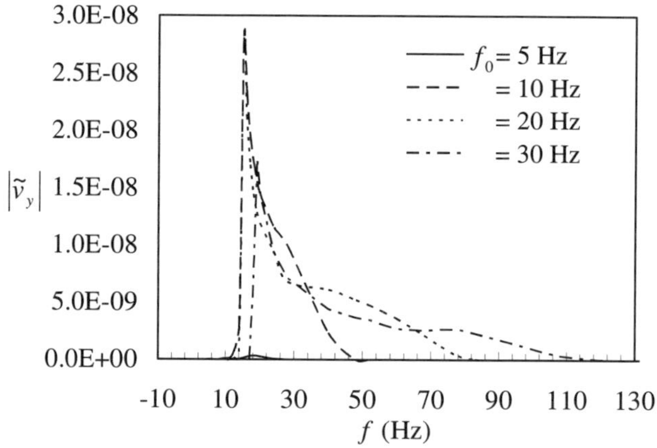

(b)

Fig. 8.33 Transfer function for stratum with H = 3 m induced by a moving wheel load with c = 70 m/s and different f_0 observed at: (a) origin; (b) (x, y) = (10 m, 0 m).

As for the self oscillation frequency of f_0 = 5 Hz, the calculated second critical frequency is 20.84 Hz, only slightly larger than the fundamental frequency of the soil stratum 15 Hz. Thus, two peaks mixed together and only one peak frequency with flat amplitude around 18 Hz can be observed in Fig. 8.33(a).

As mentioned previously, for the case with speed c lower than the R-wave speed, only the vibration with frequencies located between two critical frequencies $f_{cr} = f_0 /(1 \pm c / c_R)$ can propagate outward as a result of the Doppler effect. In addition, due to the cut-off effect of the soil stratum, the vibration with frequencies lower than the fundamental frequency of the soil stratum will be suppressed and cannot travel outward, either. Thus, due to the coupling effect, as can be seen from the transfer function given in Fig. 8.33(b) for location x = 10 m, the amplitudes for frequencies lower than the first critical frequencies or the fundamental frequency 15 Hz of the soil stratum almost reduce to zero. As for the special case of f_0 = 5 Hz, because a large portion of the frequency content, 2.84-20.84 Hz, due to the Doppler effect is lower than the fundamental frequency of the soil stratum; therefore, most vibrations have been cut off and virtually no waves are allowed to propagate outward.

The influence of stratum depth can be visualized from the spatial graph given in Fig. 8.34. In part (a) for f_0 = 5 Hz, the vibration seems to be suppressed by the underlying bedrock because the wavelength induced by 5 Hz self oscillation is too long to propagate via the stratum of depth H = 3 m. In parts (b), (c), (d) for f_0 = 10, 20 and 30 Hz, we can still observe the effect of stratum depth, but with the increase of f_0, such an effect becomes less pronounced.

8.5.4 *Effect of load-moving speed*

In Fig. 8.35, the vibration levels computed for four different locations, x = 0, 1, 5, 10 m, for a soil stratum of H = 3 m were plotted with respect to the S-wave Mach number, defined as $M_2 = c / c_S$. Here, parts (a), (b) and (c) represent the displacement, velocity and acceleration, respectively. At the location right under the railway, i.e., at x = 0 m, the critical speed occurs at M_2 = 0.92, which corresponds to the R-wave speed. Such a result is exactly the same as the one obtained for the uniform half-space in Chapter 2. However, for the present case, the critical speed shifts to a higher value as the distance x increases, which can also be verified from the vibration attenuation displayed in Fig. 8.36 for different speeds. As can be seen, the rate of attenuation decreases with the increase of speed. In other words, the maximum response no longer occurs at the R-wave speed at points away from the source.

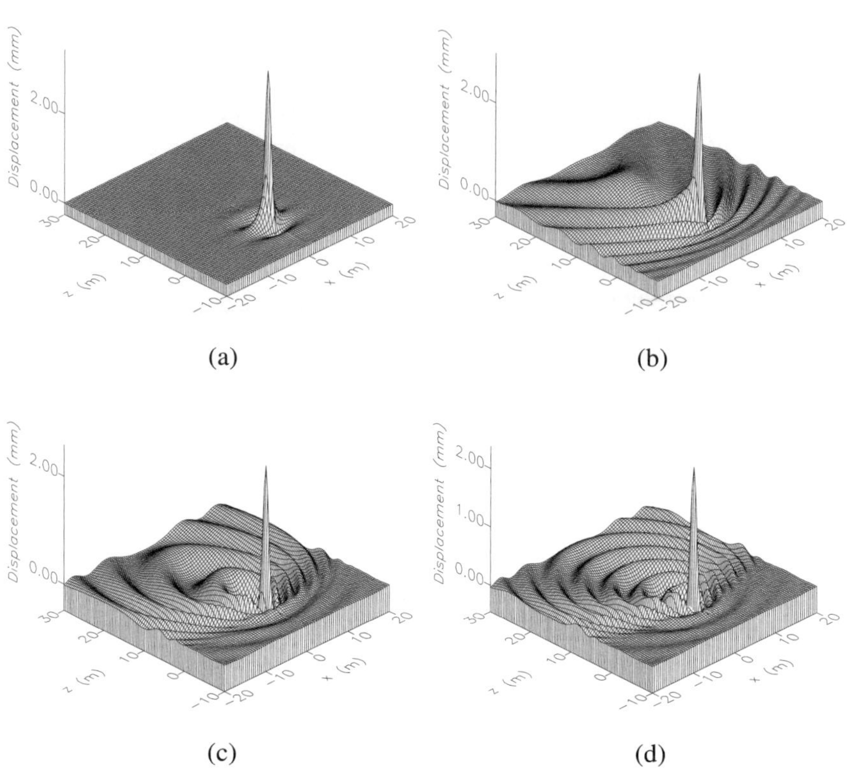

Fig. 8.34 Real-part displacements for stratum with H = 3 m caused by a moving wheel load with c = 70 m/s: (a) f_0 = 5 Hz; (b) f_0 = 10 Hz; (c) f_0 = 20 Hz; (d) f_0 = 30 Hz.

Correspondingly, Figs. 8.37(a) and (b) show the transfer functions at the locations x = 0 and 10 m, respectively. As can be seen, with the increase of speed, the resonance peaks become narrower and steeper, while the resonant frequencies shift to lower values. It is expected that as the speed approaches infinity, the resonant frequency will be the same as that for the plane strain case under a vertical line load, which should be around 14.4 Hz. Figure 8.37 also indicates that for speeds lower than the R-wave speed, such as $M_2 = 0.8$, no distinct resonance peaks can be observed under the source [see part (a)], and the amplitude of the transfer function reduces virtually to zero at locations away from the source, compared with those cases with speeds greater than the R-wave speed [see part (b)].

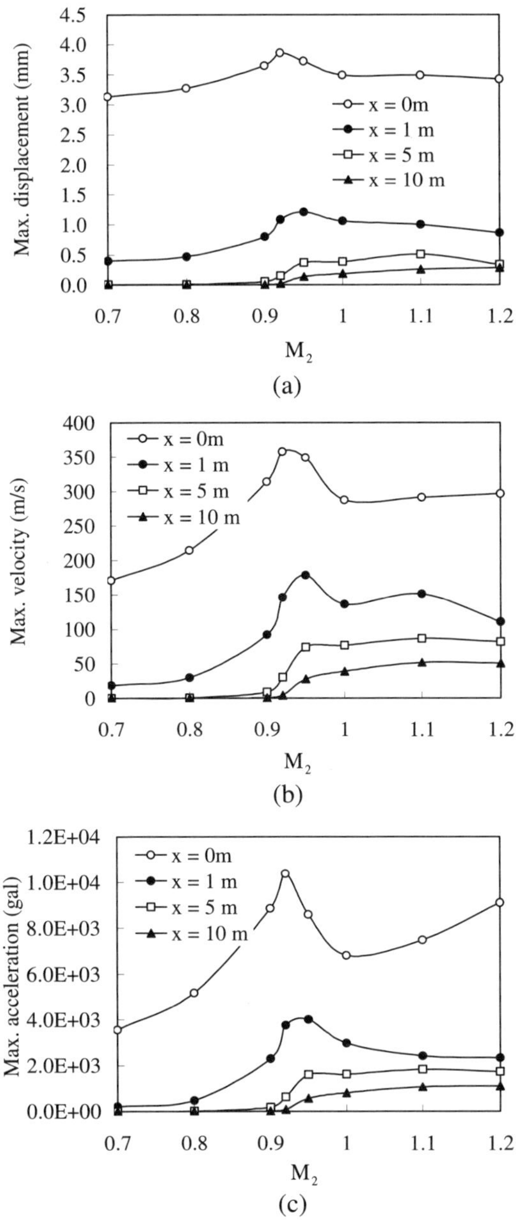

Fig. 8.35 Effect of speed on the maximum responses of a stratum with H = 3 m due to a moving wheel load: (a) displacement; (b) velocity; (c) acceleration.

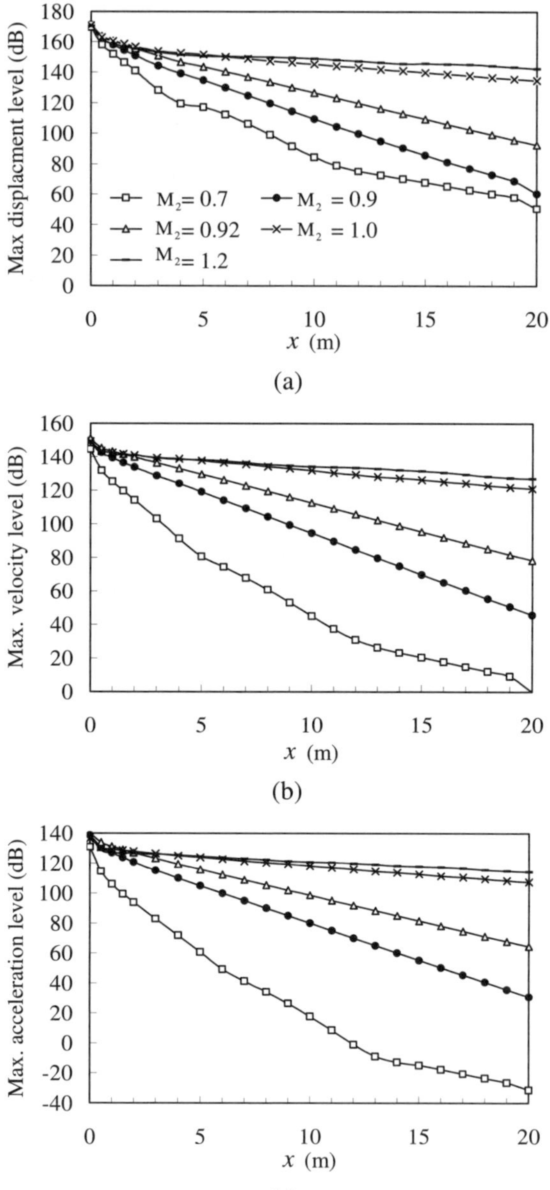

Fig. 8.36 Effect of speed on the response attenuation of a stratum with $H = 3$ m due to a moving wheel load: (a) displacement; (b) velocity; (c) acceleration.

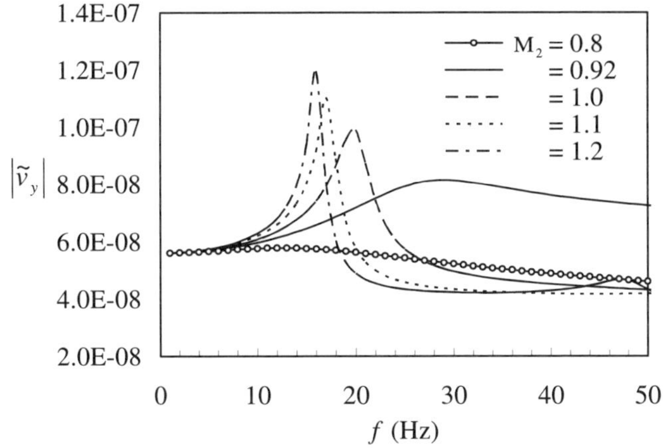

(a)

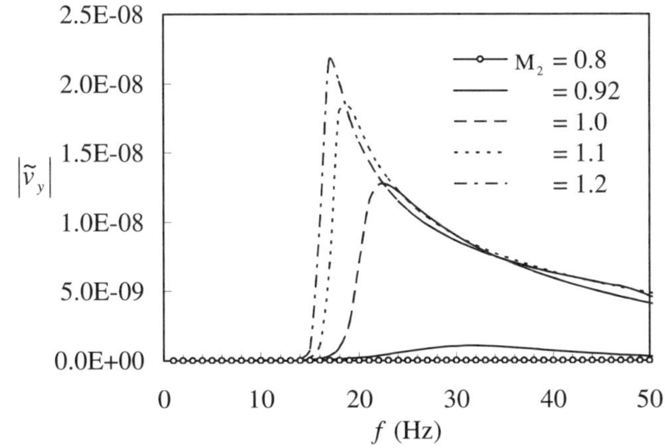

(b)

Fig. 8.37 Effect of speed on transfer function for stratum with H = 3 m induced by a moving wheel load observed at: (a) origin; (b) (x, y) = (10 m, 0 m).

8.6 Parametric Study for Multi Soil Layers

In general, the S-wave speed of soils at a greater depth is larger than that near the surface. In the section, a model with larger S-wave speeds for soil layers at greater depths is denoted as Case 1, as depicted in Fig. 8.4(a). The other cases shown in Figs. 8.4(b)-(d) will be denoted as

Cases 2-4. In their sequence, the four cases differ in that the lowest layer of soil is ignored one by one. The S-wave speeds for each case are given in Fig. 8.4. In what follows, a parametric study will be performed to investigate the consequence of approximating Case 1 by Cases 2, 3 or 4, with less and less soil layers included in the simulation.

8.6.1 *Effect of soil layers for a quasi-static moving load*

Figures 8.38 and 8.39 show the vibration attenuation for a moving wheel load with speeds c = 70 and 100 m/s, respectively. For the load with the sub-critical speed c = 70 m/s (with respect to the Rayleigh wave speed), great differences can be observed between different soil profiles. The less the number of layers included in the simulation, the larger the response is. On the other hand, for the wheel load with the super-critical speed c = 100 m/s, the difference between the cases considered is less pronounced except for Case 4, which implies a uniform half-space. Basically, larger displacement response can be observed for a small number of layers, as revealed by Fig. 8.39(a). But for the velocity and acceleration, the responses are almost the same for Cases 1-3, as can be seen from Figs. 8.39(b) and (c). These results share the same feature as that of Figs. 8.20 and 8.21 in that the influence of layer depth is limited to low frequencies.

For speeds lower than the R-wave speed, most contribution to the response at a distance away from the source comes from low frequencies, as indicated by the transfer function for c = 70 m/s in Fig. 8.40(b). Consequently, the effect of soil layering for such a speed may affect the response significantly.

In contrast, for the wheel load with speed c = 100 m/s, which is larger than the R-wave speed, the contribution to the response comes from a wide range of frequencies. Thus, the difference of the transfer functions in the low frequency range, as shown in Fig. 8.41, due to consideration of different numbers of soil layers does not make the maximum responses in time domain much different. This has the implication that for trains moving at high speeds, the influence of soil layers is relatively small and can be neglected.

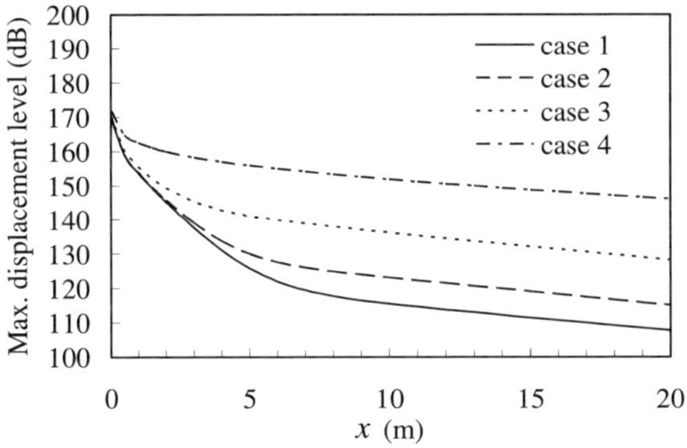

(a)

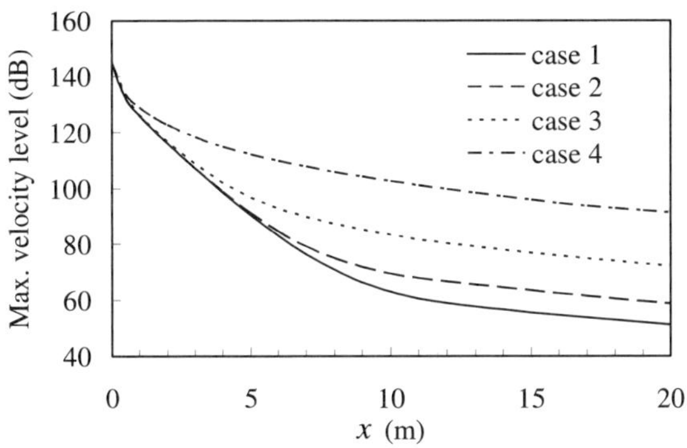

(b)

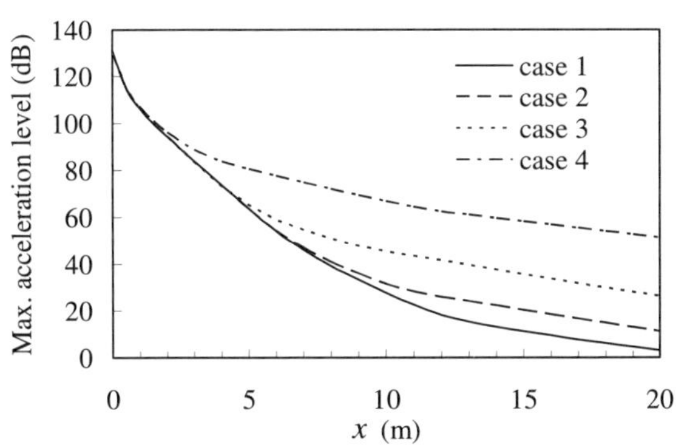

(c)

Fig. 8.38 Response attenuation of different soil profiles for a moving wheel load with $c = 70$ m/s: (a) displacement; (b) velocity; (c) acceleration.

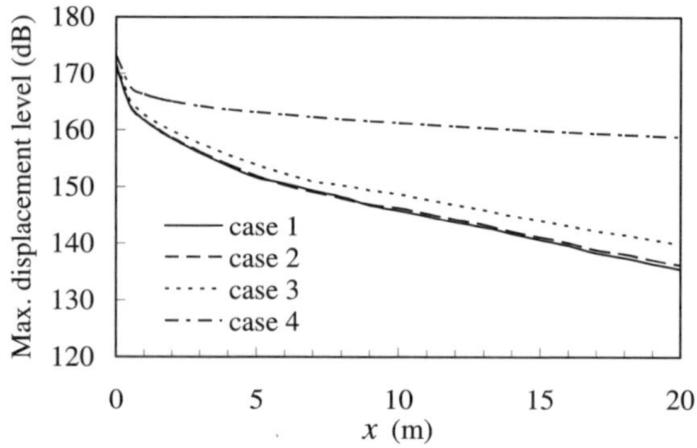

(a)

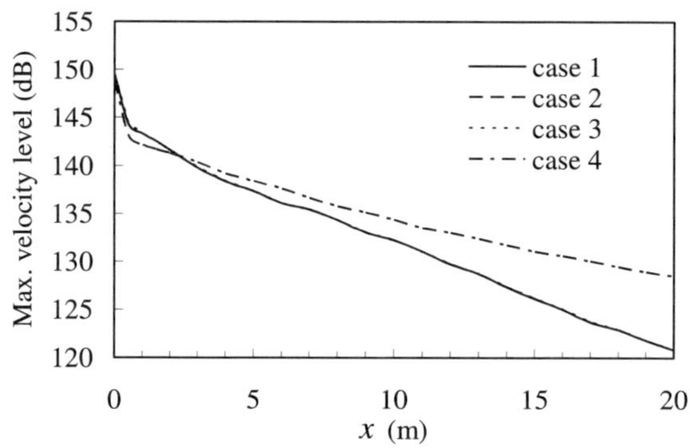

(b)

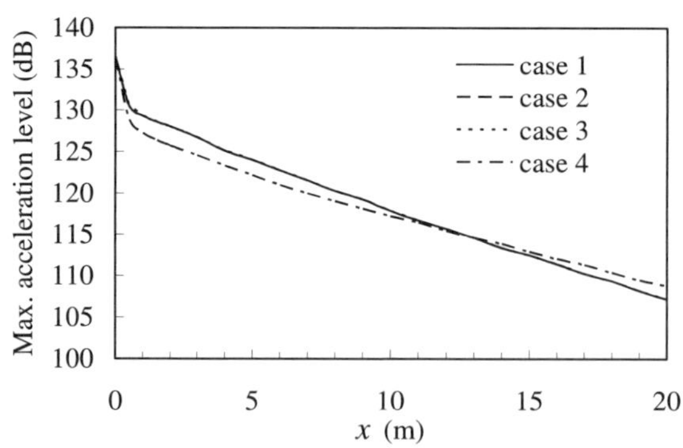

(c)

Fig. 8.39 Response attenuation of different soil profiles for a moving wheel load with $c = 100$ m/s: (a) displacement; (b) velocity; (c) acceleration.

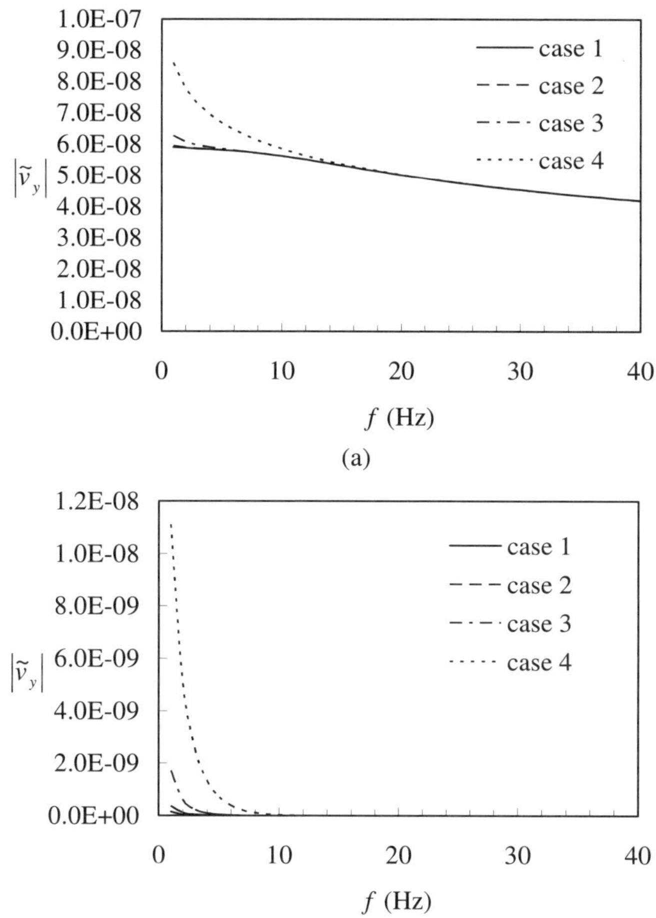

Fig. 8.40 Transfer function of different soil profiles for a wheel load with c = 70 m/s observed at: (a) origin; (b) (x, y) = (10 m, 0 m).

8.6.2 *Effect of soil layers for a moving load with self oscillation*

For a wheel load with a self oscillation frequency of f_0 = 10 Hz moving at the sub- and super-critical speeds c = 70 and 100 m/s with respect to the Rayleigh wave speed, the vibration responses of the ground surface for the four different soil profiles in Fig. 8.4 were plotted in Figs. 8.42 and 8.43, respectively. Clearly, for both wheel load speeds,

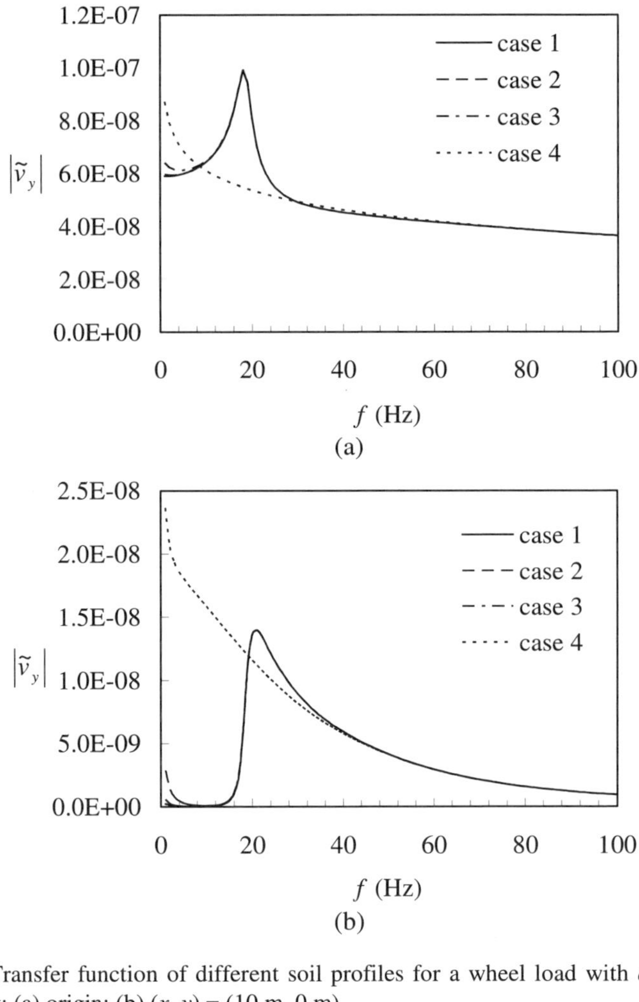

Fig. 8.41 Transfer function of different soil profiles for a wheel load with c = 100 m/s observed at: (a) origin; (b) (x, y) = (10 m, 0 m).

Case 4 shows the smallest response and Case 1 the greatest. This result is contrary to the one obtained in the preceding section for the case with no self oscillation frequency, i.e., with f_0 = 0 Hz, primarily due to the coupling effect between the input vibration frequency f_0 = 10 Hz and the layered soil frequencies.

As was expected, the transfer functions computed of the four soil models in Figs. 8.44 and 8.45 for the two speeds reveal rather

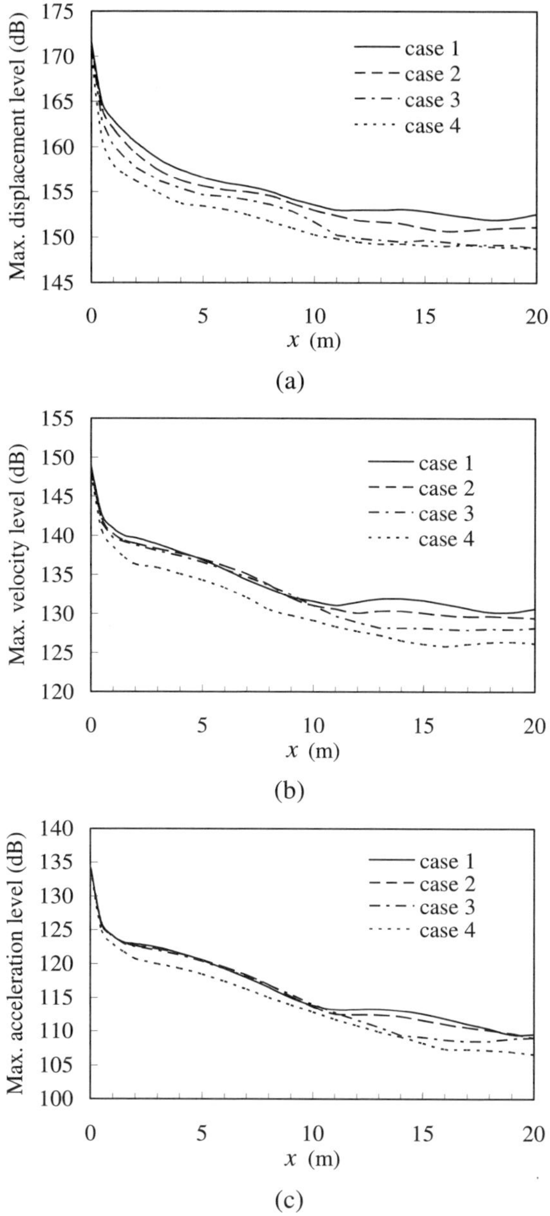

Fig. 8.42 Response attenuation of different soil profiles for a moving wheel load with $c = 70$ m/s and $f_0 = 10$ Hz: (a) displacement; (b) velocity; (c) acceleration.

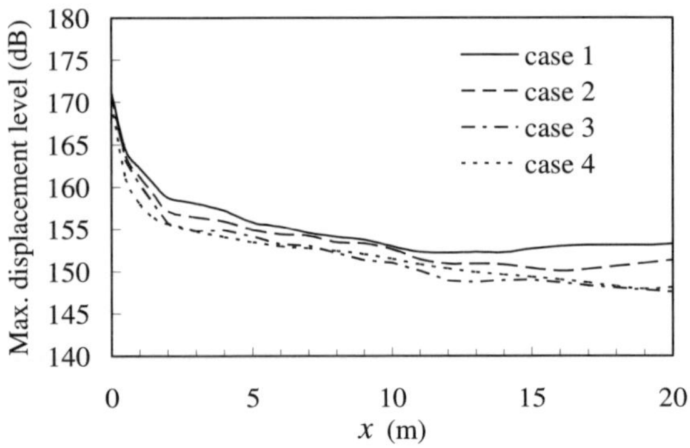

(a)

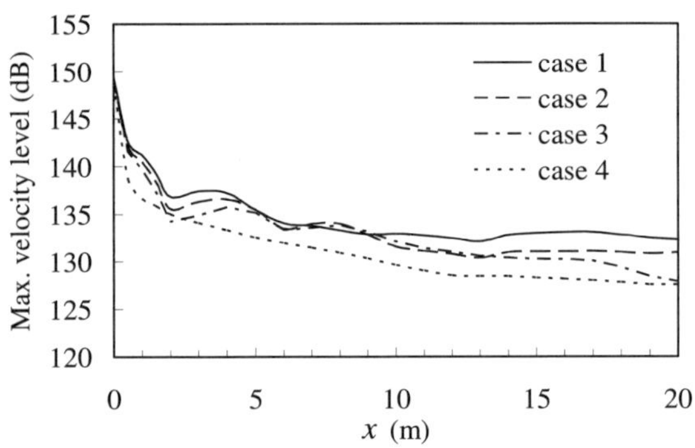

(b)

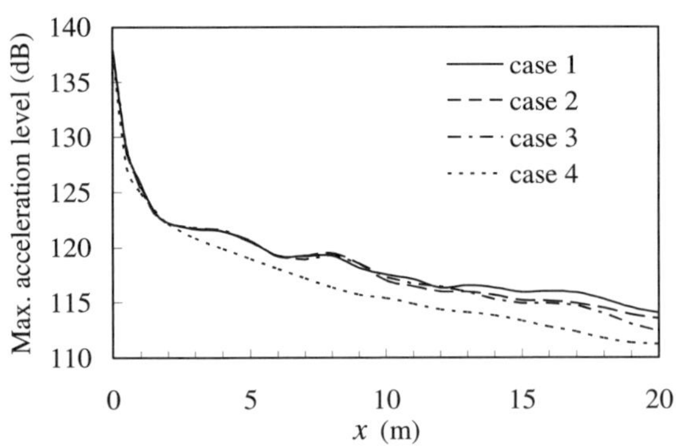

(c)

Fig. 8.43 Response attenuation of different soil profiles for a moving wheel load with $c = 100$ m/s and $f_0 = 10$ Hz: (a) displacement; (b) velocity; (c) acceleration.

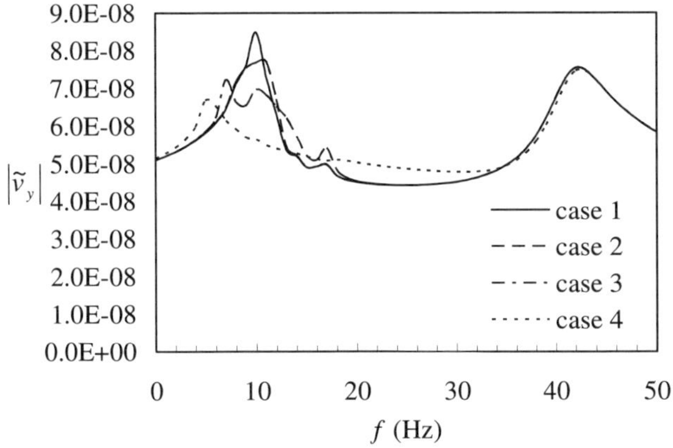

(a)

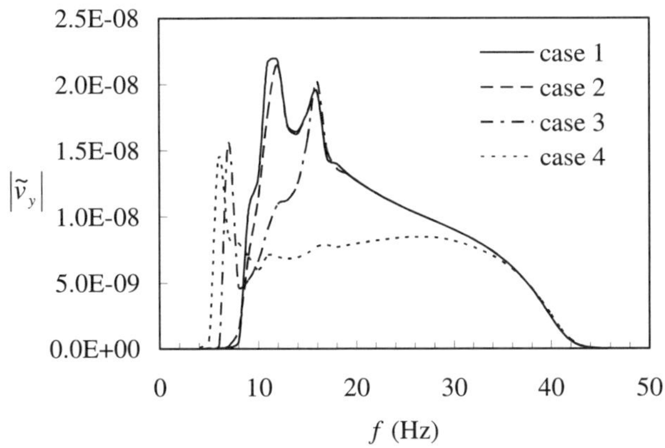

(b)

Fig. 8.44 Transfer function of different soil profiles for a wheel load with c = 70 m/s and f_0 = 10 Hz observed at: (a) origin; (b) (x, y) = (10 m, 0 m).

complicated patterns. In particular, multi peaks occur in each figure, due to coupling of the oscillation frequency f_0 of the moving load with those of the soil layers.

8.6.3 *Effect of load-moving speed for multi-layered soils*

Consider the multi-layered soil shown as Case 1 in Fig. 8.4(a). Figure 8.46 shows the effect of load-moving speed on the maximum responses

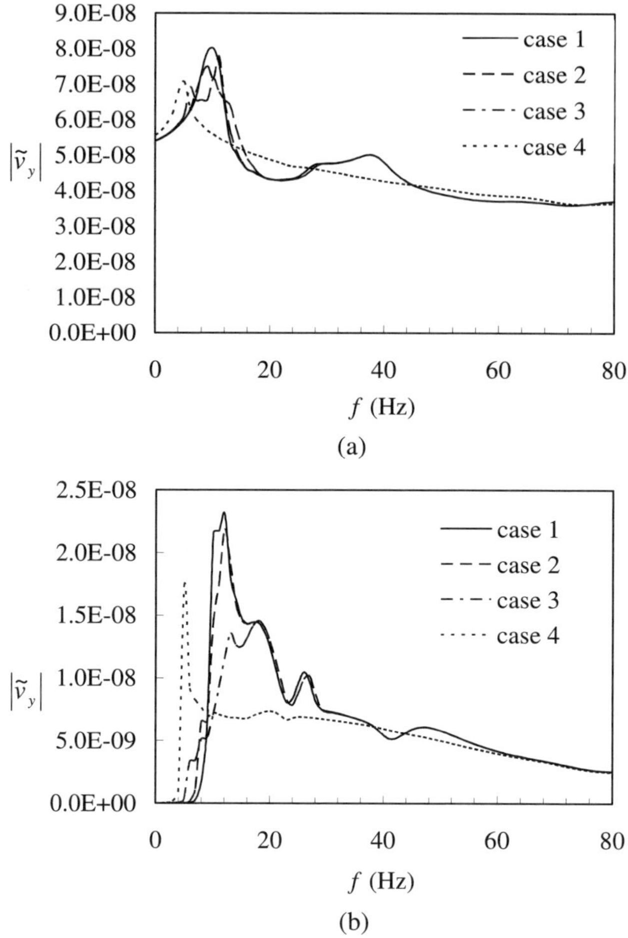

Fig. 8.45 Transfer function of different soil profiles for a wheel load with c = 100 m/s and f_0 = 10 Hz observed at: (a) origin; (b) (x, y) = (10 m, 0 m).

at four different locations, i.e., at x = 0, 1, 5 and 10 m, away from the source. As can be seen, at the location right under the source (x = 0 m), the first critical speed occurs at $M_2 = 0.92$, which corresponds to the R-wave speed of the top layer. This result indicates that for a multi-layered soil, the critical speed at the location under the moving load can be determined primarily by the R-wave speed of the top layer. But with the increase in distance from the source, the maximum responses tend to

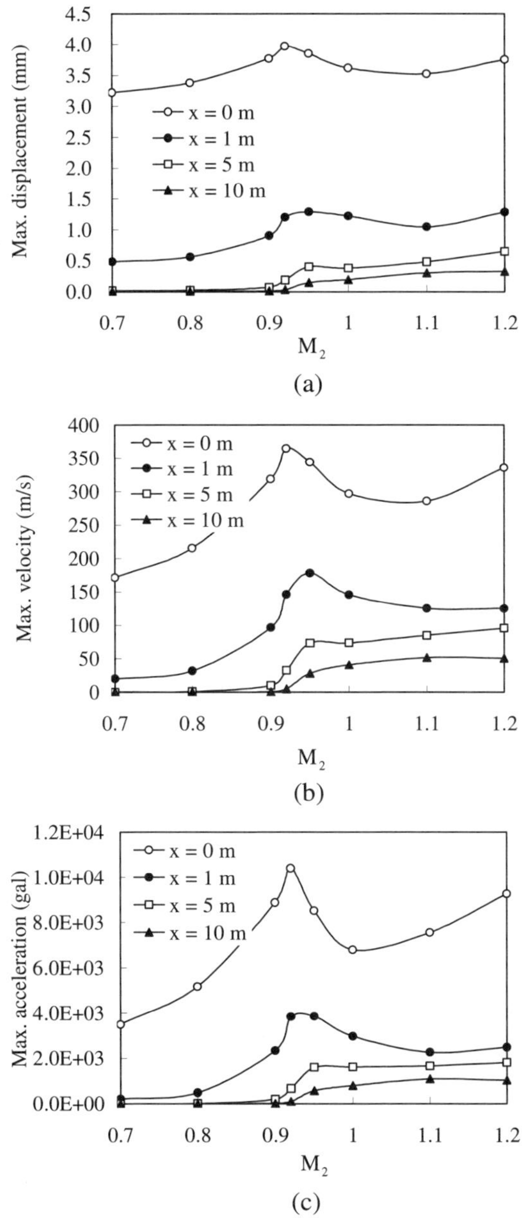

Fig. 8.46 Effect of speed on the maximum responses for soil profile Case 1: (a) displacement; (b) velocity; (c) acceleration.

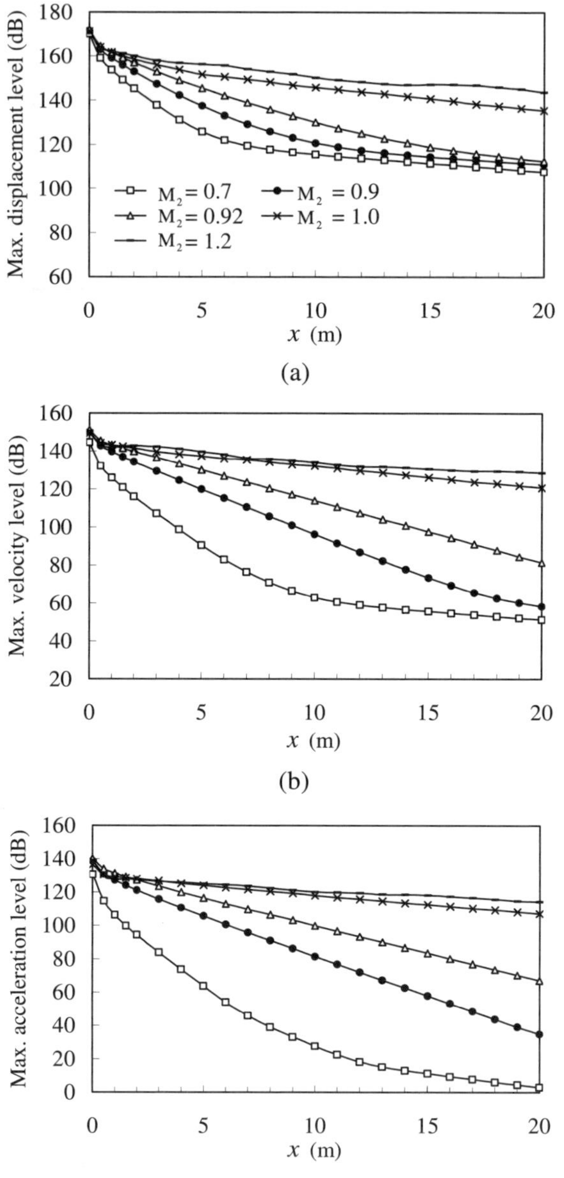

Fig. 8.47 Effect of speed on the response attenuation for soil profile Case 1: (a) displacement; (b) velocity; (c) acceleration.

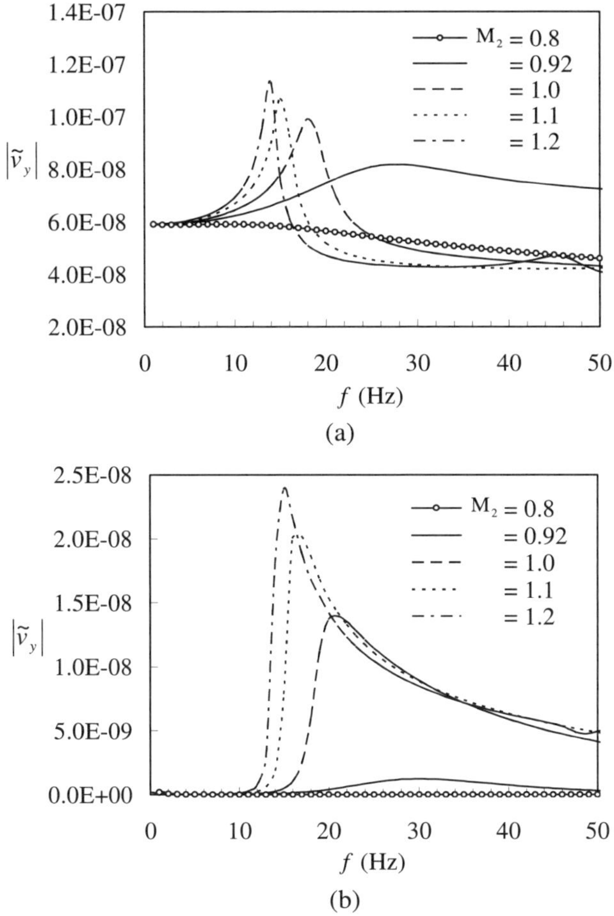

Fig. 8.48 Effect of speed on the transfer function for soil profile Case 1 observed at: (a) origin; (b) $(x, y) = (10\text{ m}, 0\text{ m})$.

occur at higher speeds, as a result of contribution from the underlying layers. This phenomenon is similar to that of Fig. 8.35 for the case with bedrock located underneath a soil deposit of depth $H = 3$ m.

Figure 8.47 shows the vibration attenuation for different load speeds, which, again, reveals a pattern similar to the case of a soil deposit resting on bedrock, as shown in Fig. 8.36, but on a diminutive scale, mainly due to the fact that the soils below the top layer of the present case is much softer than the bedrock. Corresponding to this case, the transfer function was plotted in Fig. 8.48. As can be seen, with the increase of the load moving speed, the resonant frequencies shift to lower values. Such a trend is similar to that observed from Fig. 8.37, but with lower resonant frequencies.

8.7 Concluding Remarks

The 2.5D finite/infinite element procedure presented in Chapter 7 has been adopted in the parametric study for layered soils subjected to high-speed moving loads in this chapter. Many factors are considered, including the S-wave speed, damping ratio, Poisson's ratio, stratum depth, etc. Basically, only two train speeds are considered, i.e., c = 70 and 100 m/s, each to represent the effect of sub- and super-critical speeds with respect to the Rayleigh waves. It is realized that different moving speeds may produce quantitatively different results. Nevertheless, the results obtained for these two specific speeds are believed to be capable of capturing the basic, qualitative trend of the soil effects for the loads moving in the two ranges of speeds, i.e., sub- and super-critical speeds.

The following are the conclusions extracted from the parametric studies performed in this chapter:

(1) For moving load with no self oscillation, i.e., with $f_0 = 0$, the increase of the S-wave speed and Poisson's ratio of soils can result in reduction of the responses for both of the speeds considered, while the increase of damping ratio can only decrease the responses for the super-critical speed with respect to the Rayleigh waves.
(2) The soil damping has an important effect on the level of attenuation, particularly at higher frequencies f_0. In general, the existence of damping will make the response obtained for a moving load with higher frequency f_0 attenuate faster than that for a lower frequency f_0. However, for lower values of damping ratio, a longer distance is needed to demonstrate the full effect of damping.
(3) The stratum depth is a key factor for vibration propagation due to its inherent cut-off frequency. Below the cut-off frequency, no waves can propagate outward. The cut-off frequency differs for different

load speeds c, self oscillation frequencies f_0, as well as for different layer depths. For $f_0 = 0$, with the increase of load-moving speed, the cut-off frequency tends to approach the one obtained from the plane-strain case. For $f_0 \neq 0$, the vibration induced by lower values of self oscillation frequencies f_0 can be easily suppressed by the bedrock, because shallow layers will prohibit the propagation of waves of long wavelengths.

(4) For a multi-layered soils subjected to non-pulsating moving loads, i.e., with $f_0 = 0$, the critical speed occurring at the location near the source can be determined from the R-wave speed of the top layer soil. But at locations away from the moving load, the maximum responses may occur at a higher speed due to contribution of the underlying soil layers.

(5) As the self oscillation frequency f_0 of the moving load and the fundamental frequencies of the multi-layered soils are concerned, the transfer function becomes quite complicated due to the coupling of these frequencies. The results obtained differ from case to case. No simple rule can be established to predict the response. The only tool we can count on for such a case is by field experiment or by a proper numerical analysis that takes into account the variations of soil properties.

Chapter 9

Wave Barriers for Reduction of Train-Induced Vibrations: Parametric Study

The 2.5D finite/infinite element procedure presented in Chapter 7 is adopted to study the effectiveness of three vibration barriers in isolating the ground vibrations induced by trains moving at sub- and super-critical speeds, with respect to the Rayleigh wave speed of the supporting soils. The vibration barriers considered include the open trenches, in-filled trenches and wave impeding blocks (WIB). By the 2.5D finite/infinite element approach, the effect of load-moving can be duly taken into account, while three-dimensional results can be obtained using only the two-dimensional profile. The moving train is simulated as a sequence of moving wheel loads with self oscillation frequencies. The performance of the three types of wave barriers in isolating soil vibrations for trains moving at sub- and super-critical speeds with various self oscillation frequencies is evaluated with respect to the key parameters, along with suggestions made for enhancing the efficiency of isolation.

9.1 Introduction

The problem of vibration isolation for soils induced by various sources has been the focus of a great deal of research since the mid-twentieth century. Common vibration countermeasures that have been adopted include the installation of *open trenches*, *in-filled trenches* and *wave impeding blocks* (WIBs). Previous works related to trenches that may be cited include Woods (1968), Beskos *et al.* (1986), Ahmad *et al.* (1996), and Yang and Hung (1997), among others. As for the works on WIB, the following may be cited: Schmid *et al.* (1991), Antes and von Estorff (1994), and Takemiya and Fujiwara (1994).

As revealed by the results of the works cited above, the most important requirement for a trench to achieve a good effect of isolation is that the trench should have a depth of an order comparable to that the surface wavelength. Consequently, the isolation of ground-born vibrations by trenches is effective only for moderate to high frequency vibrations. On the other hand, the WIB is not good at reducing vibrations with very high frequencies, since the basic idea of using the WIBs comes from the observation on the vibration transmission behavior of a soil stratum lying over a bedrock. According to Wolf (1985), no vibration eigenmodes can be induced below the *cut-off frequency* of the soil stratum, which equals $c/(4H)$, with H denoting the depth of the soil stratum and c the compressional or shear wave speed of the soil stratum. Thus, it is expected that an artificial solid plate constructed underneath the soil can exhibit the same cut-off effect to some extent, thereby impeding the spreading of vibrations with longer wavelengths, i.e., low-frequency waves.

A review of all the aforementioned works indicates that the isolation performance of these wave barriers depends mainly on the frequency range of vibrations. Thus, for vibrations induced by machine foundations with a single dominant frequency, rather good performance of isolation can be achieved by simply adjusting the dimension of wave barriers to an optimized size. However, for vibrations induced by moving trains, it is not as easy to achieve the same effect of isolation, due to the fact that train-induced vibrations usually involve a wide range of frequencies, depending on the speed of the train. Thus, as far as the performance of wave barriers in reducing train-induced vibrations is concerned, further investigations in time domain that takes into account the effects of speed and self oscillation of the train should be conducted.

In the literature, a great deal of research has been carried out on the ground vibrations induced by high-speed trains, including the works of Krylov (1995), Degrande and Lombaert (2000), Hung and Yang (2001), Paolucci *et al.* (2003), and so on. However, relatively few studies were focused on the performance of wave barriers in reducing train-induced vibrations, especially those induced by high-speed trains. In this chapter, the 2.5D finite/infinite element scheme previously presented in Chapter 7 will be employed to investigate the effectiveness of the three barriers, i.e., the open trenches, in-filled trenches, and WIBs, in reducing the train-induced vibrations on soils. Using such an approach, the irregularities in geometry and material of the 2D profile can be dealt with in a relatively easy way, as is with the traditional finite element method.

In this chapter, comprehensive parametric studies will be conducted for the three wave barriers. In particular, focus will be placed on the effects of the moving speed and self oscillation frequencies of the train on the isolation efficiency of the three wave barriers. Such effects were not covered by any previous studies using the 2D approaches, because of theoretical handicap. Owing to inclusion of the third degree of freedom along the load-moving direction (i.e., perpendicular to the soil profile considered), the 2.5D finite/infinite approach offers a convenient means for investigating such effects.

9.2 Major Considerations in Parametric Studies

For the purpose of wave reduction, three types of wave barriers will be considered. The models to be investigated in this chapter have been schematically shown in Fig. 9.1, where parts (a), (b) and (c) represent the open trench, in-filled trench, and wave impedance block (WIB), respectively. Either the open trench or in-filled trench is constructed between the railway and the target structure to be protected. The WIB is installed at a certain depth underneath the railway. The 2.5D finite/infinite element method presented in Chapter 7 will be employed to investigate the influence of various parameters upon the screening effect of these three wave barriers at different train speeds, including the sub-critical and super-critical speeds with respect to the Rayleigh waves.

The same element mesh as the one depicted in Fig. 8.5 is adopted in this chapter, in which only half of the soil profile is modeled due to symmetry consideration. The soil profile considered is a uniform half-space with the material properties listed in Table 9.1. Correspondingly, the shear and Rayleigh wave speeds are $c_S = 100$ m/s and $c_R = 93.2$ m/s. Obviously, the latter is the *critical speed*.

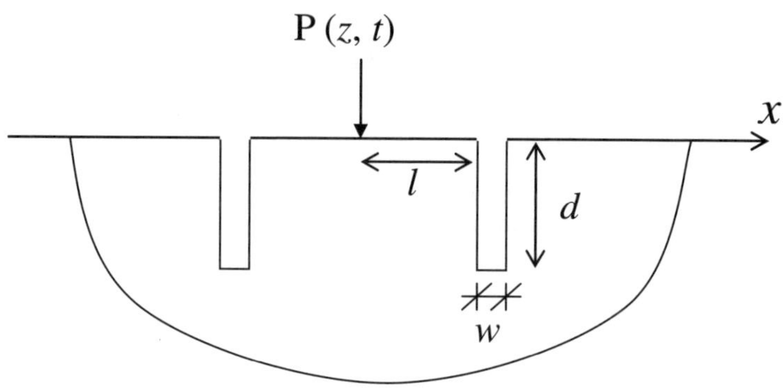

(a)

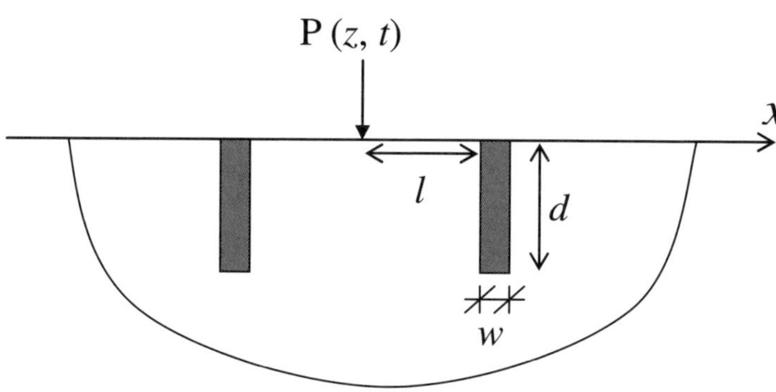

(b)

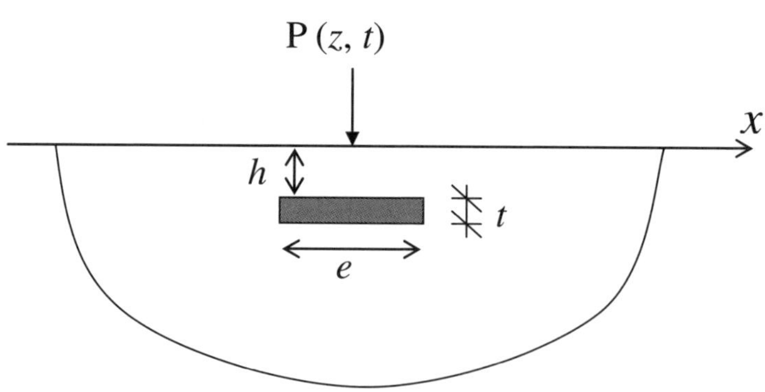

(c)

Fig. 9.1 Typical model of the problem: (a) open trench; (b) in-filled trench; (c) wave impedance block (WIB).

Table 9.1 Material properties.

Material	Young's modulus E (MPa)	Poisson's ratio ν	Density ρ (kg/m^3)	Damping ratio β
Soil	53.2	0.33	2,000	0.05
In-filled trench	11,760	0.25	2,400	0.05
WIB	11,760	0.25	2,400	0.05

Similar to what is done in Chapter 7, the moving train is simulated by a sequence of wheel loads traveling on the ground surface along the z-axis, which can be written as:

$$P(x=0, y=0, z, t) = \phi(z-ct)\exp(i2\pi f_0 t), \tag{9.1}$$

where the exponential term is to account for the dynamic effect that may arise from the rail irregularity or mechanical system of the vehicles, the function $\phi(z)$ has been given in Eq. (2.36). Based on the definitions given in Eq. (2.36) and Fig. 2.3(b), the following parameters are adopted in this chapter: a = 2.56 m, b = 16.44 m and L = 25 m. The number of carriages of the train is selected to be N = 4, for the reason stated previously. The wheel load is T = 10 t and the characteristic length is $\alpha = 0.8$ m.

The screening effect of a wave barrier can be evaluated using the *reduction of vibration level* with the unit dB, i.e.,

$$\text{Reduction of vibration level [dB]} = L_b - L_a, \tag{9.2}$$

where L_a denotes the vibration level obtained after installation of wave barriers and L_b the vibration level with no isolation. The definition of vibration level L[dB] has been given in Eq. (8.1). Based on Eq. (8.1), the preceding equation can be rewritten as:

$$\text{Reduction of vibration level [dB]} = -20\log\frac{P_1}{P_2}. \tag{9.3}$$

Here, P_1 denotes the measured value when the wave barrier is present, P_2 the reference value obtained from an associated analysis excluding the wave barrier. The vibration level of interest can be either displacement, velocity or acceleration level. For brevity, only the reduction of velocity level will be considered in this chapter. Another parameter previously used to evaluate the isolation efficiency of wave barriers is the *amplitude reduction ratio* A_r, which is defined as:

$$A_r = \frac{P_1}{P_2}, \tag{9.4}$$

where P_1 and P_2 have the same definition as those given in Eq. (9.3). The relation between the amplitude reduction ratio A_r and the reduction of vibration level with the unit dB in Eq. (9.3) can be referred to Fig. 8.1. For instance, a reduction of vibration level of 10 dB amounts to an amplitude reduction ratio of $A_r = 0.3$.

By the present approach, the responses along all the three directions can be obtained at the same time. For the sake of brevity, however, only the isolation efficiency in reducing the vertical and horizontal velocities along the load-moving direction will be presented in the following analysis.

A review of the related literature indicates that the effect of isolation generally depends on the exciting frequencies. However, as was mentioned previously, a moving train in time domain can produce vibrations that are of a wide range of frequencies. Thus, in the following section, the transfer function in frequency domain will be presented as well for certain cases to highlight the frequency dependent characteristics of the problems considered.

9.3 Vibration Reduction by Open Trenches

The analytical model for this section has been given in Fig. 9.1(a), where two open trenches are placed each at a distance l away from the railway center on each side. Unless noted otherwise, the following dimensions are adopted for the open trench: distance l = 5 m, depth d = 4 m, and width w = 1 m.

9.3.1 *Moving loads with no self oscillation*

In this subsection, the effectiveness of open trenches in reducing the vibrations induced by moving loads with no self oscillation, i.e., with $f_0 = 0\text{ Hz}$, will be investigated. For this case, the external load function

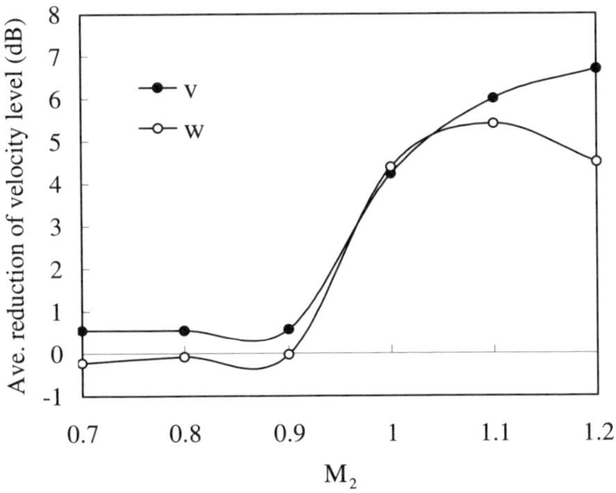

Fig. 9.2 Effect of train speed on the average reduction of velocity level (open trenches).

is simply $P(z,t) = \phi(z - ct)$, where c is the load moving speed, following the definition of Chapter 2.

9.3.1.1 *Effect of load-moving speed*

To investigate the screening effectiveness of open trenches for different train speeds, the reduction of velocity level for six different train speeds ranging from $M_2 = 0.7$ to 1.2 has been computed and plotted in Fig. 9.2, in which the Mach number M_2 is defined as $M_2 = c/c_S$, v represents the vertical response and w the horizontal response along the load-moving direction. Since the S-wave speed of the underlying soils is $c_S = 100$ m/s, the corresponding train speeds considered in Fig. 9.2 range from $c = 70$ m/s to 120 m/s. The average reduction of velocity level shown in Fig. 9.2 is computed as the average of vibration reduction over the distance from $x = 6$ m to 20 m. Evidently, for a train speed lower than the critical speed (i.e., the Rayleigh wave speed with $M_2 = 0.932$), the effectiveness of the open trench is pretty poor. As for the horizontal response w, the installation of open trenches even adversely amplifies the vibration. On the other hand, for speeds greater than the critical speed (i.e., the Rayleigh wave speed with $M_2 = 0.932$), the screening efficiency of the open trenches appears to be rather good. Another trend that can be

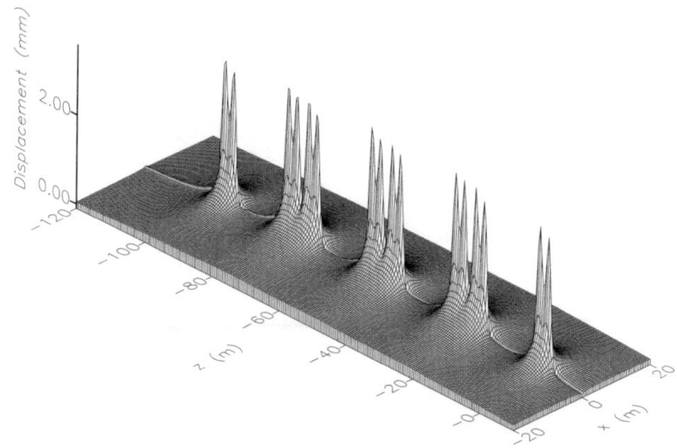

(a)

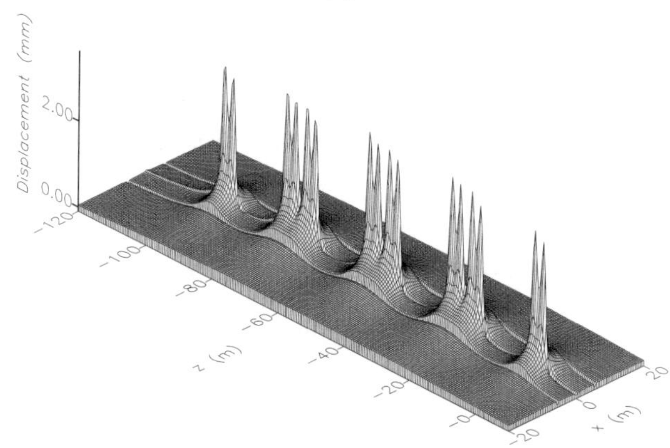

(b)

Fig. 9.3 Displacement field for a moving train with speed c = 70 m/s and f_0 = 0 with: (a) no isolation; (b) open trenches.

observed from Fig. 9.2 is that the open trenches tend to be more effective in isolating the vertical than the horizontal vibrations.

As can be seen from Fig. 9.2, the effectiveness of open trenches is quite different for the loads moving at the sub- and super-critical speed ranges, as divided by M_2 = 0.932. For illustration, we shall select only one speed from each of two speed ranges in the following study, i.e., c = 70 m/s (= 252 km/hr) and c = 100 m/s (= 360 km/hr).

Before we proceed with parametric study of the trench dimensions, the responses due to the installation of an open trench with the typical dimensions mentioned in the beginning of this section will be studied at first, in order to get an overall view of the isolation effect of the open trench. Figures 9.3(a) and (b) respectively depict the displacement field before and after installation of the open trenches for the train speed $c = 70$ m/s. Clearly, the existence of open trenches has almost no influence on the ground surface displacement.

The corresponding time history responses for the vertical displacement, velocity and acceleration at the location $x = 10$ m from the centerline of the railway were shown in Figs. 9.4(a)-(c). Again, it is confirmed that the effect of open trenches in reducing the vibration induced by the train with a speed of $c = 70$ m/s in the sub-critical range (with respect to the Rayleigh wave speed) is rather small. Such a result is in consistence with that observed by Yeh *et al.* (1997). One possible explanation for this is that the open trenches are only effective for isolating high-frequency vibrations, as has been indicated in Chapter 5 or Yang and Hung (1997). However, for a static moving load with speeds lower than the R-wave speed, the vibrations of high frequencies decay rather fast against distance. For instance, from the transfer function plotted in Fig. 9.5 for the location $x = 10$ m, we observe that the frequency content of the transfer function is only localized for a small range of frequencies, i.e., 0 ~ 10 Hz, for which the effect of open trenches can basically be neglected.

As for the super-critical case (with reference to the Rayleigh wave speed), i.e., for $c = 100$ m/s, the displacement on the ground surface with and without open trenches has been plotted in Fig. 9.6, As can be seen, with the presence of open trenches, the displacements of the ground surface on the two outer regions of the trenches become smoother, but with little change in amplitude. The same phenomenon can also be observed from the time history of the vertical displacement at the location $x = 10$ m shown in Fig. 9.7(a). The other message from Fig. 9.7 is that open trenches are generally effective for reducing the vertical acceleration, but only moderately effective for reducing the vertical velocity.

The above phenomenon can be explained using the transfer function computed for the vertical response of the half-space with open trenches in Fig. 9.8. The frequency content of the transfer function for the train

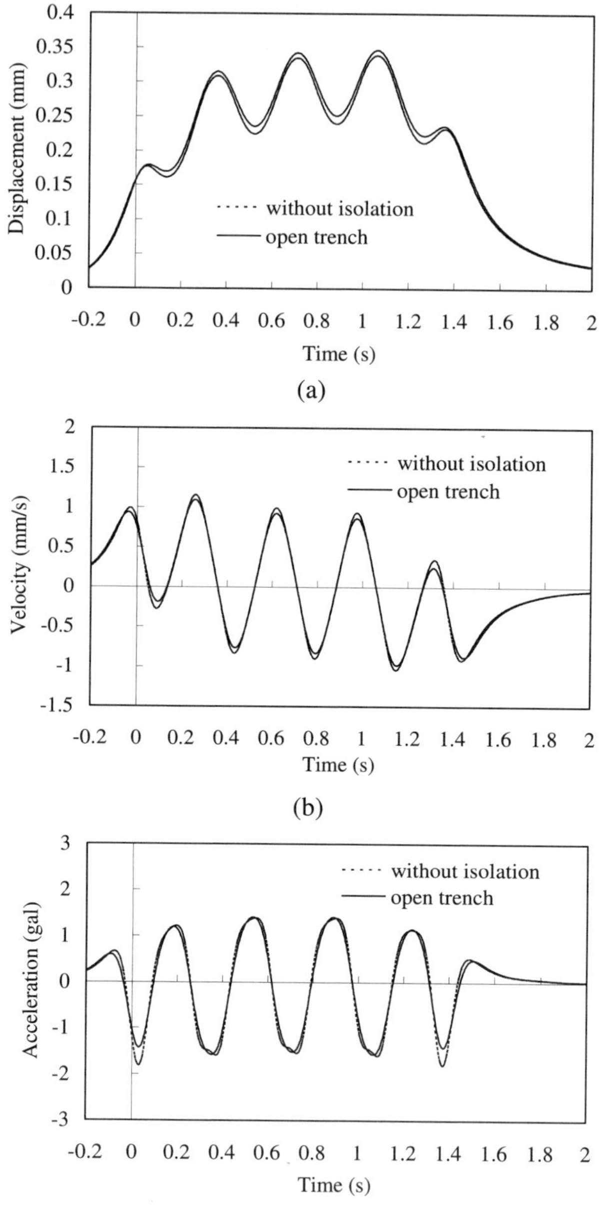

Fig. 9.4 Screening efficiency of open trenches on vertical responses for x = 10 m and c = 70 m/s: (a) displacement; (b) velocity; (c) acceleration.

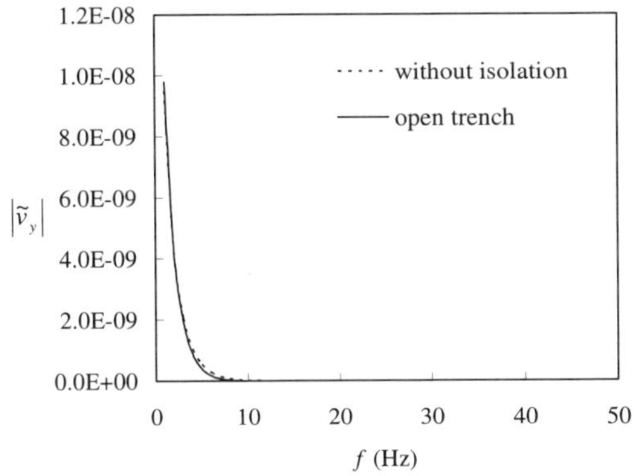

Fig. 9.5 Influence of open trenches on the transfer function of vertical response for $x = 10$ m and $c = 70$ m/s.

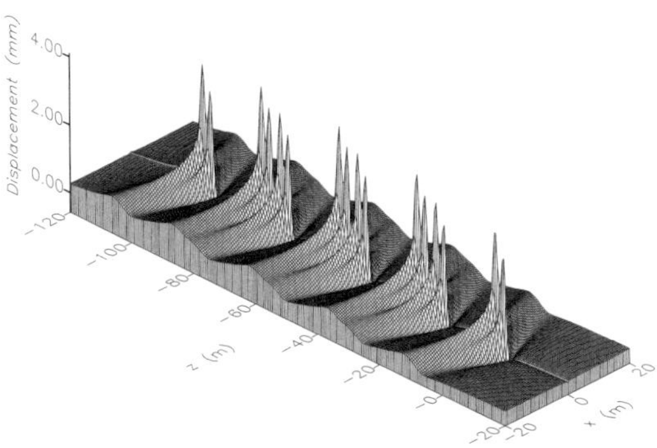

(a)

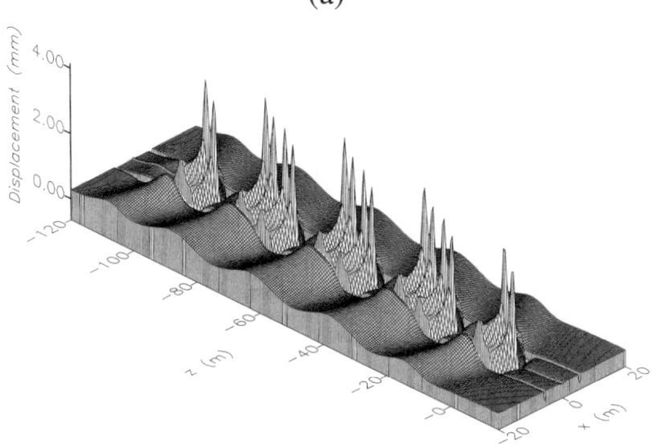

(b)

Fig. 9.6 Displacement field for a moving train with speed $c = 100$ m/s and $f_0 = 0$ with: (a) no isolation; (b) open trenches.

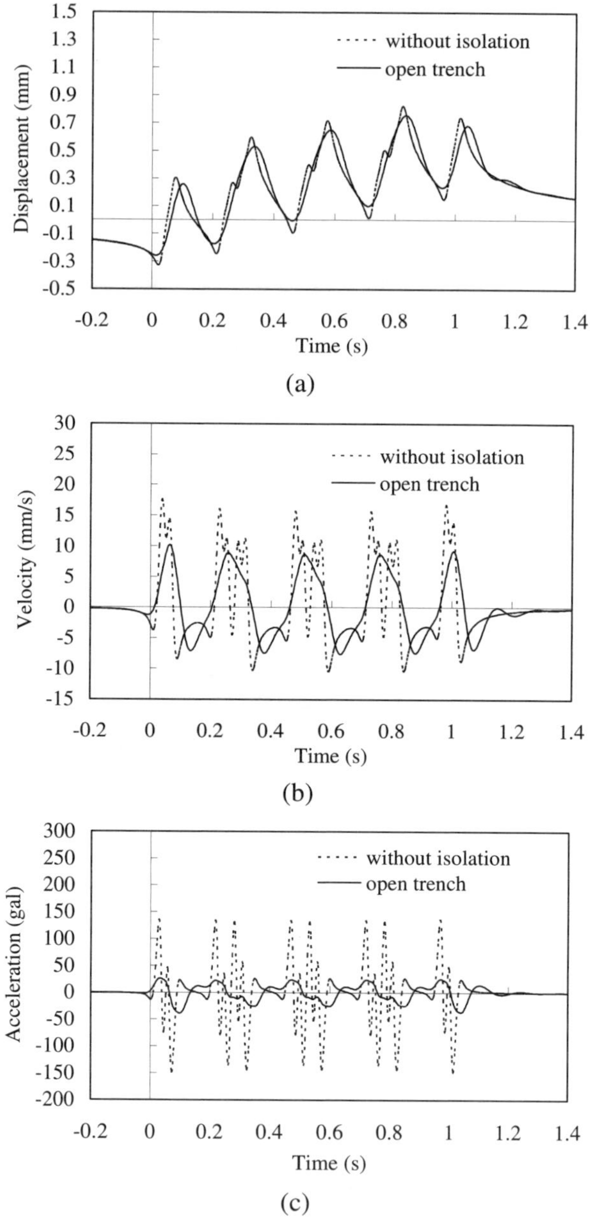

Fig. 9.7 Screening efficiency of open trenches on vertical responses for x = 10 m and c = 100 m/s: (a) displacement; (b) velocity; (c) acceleration.

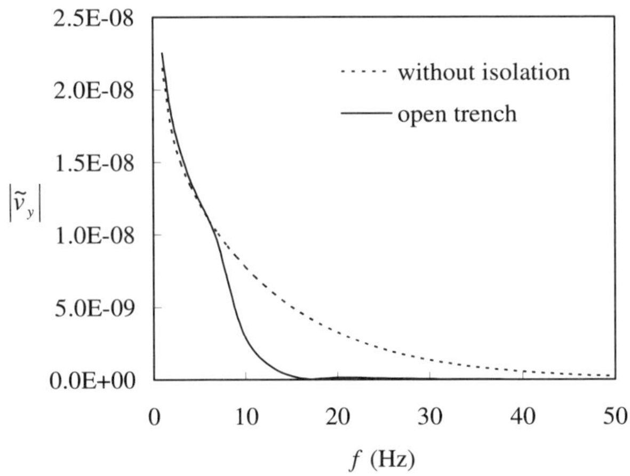

Fig. 9.8 Influence of open trenches on the transfer function of vertical response for $x = 10$ m and $c = 100$ m/s.

at speed $c = 100$ m/s (super-critical case) is much wider than that at $c = 70$ m/s (sub-critical case). Thus, the effectiveness of open trenches in reducing the high-frequency vibrations caused by trains moving in the super-critical speed range is confirmed.

9.3.1.2 *Effect of trench depth*

By varying the depth of the open trench, the results computed for the reduction of velocity level for the train moving at speed $c = 70$ m/s (sub-critical speed) have been plotted against the distance x in Fig. 9.9, where parts (a) and (b) respectively represent the reduction of the vertical and horizontal velocity level of the ground surface. Obviously, an increase in the trench depth can improve the efficiency of isolation. However, the influence is mainly concentrated in the regions outside the two trenches, but not far away from the trenches. In other words, the efficiency of isolation declines with the increase in distance from the railway. Another phenomenon that can be observed from this figure is that in the region inside the trenches, i.e., at any location from $x = 0$ to $x = \pm 5$ m, the vibration is adversely amplified due to installation of the open trenches, and the degree of amplification increases with the increase in trench depth, too. It should be noted that the installation of open trenches may

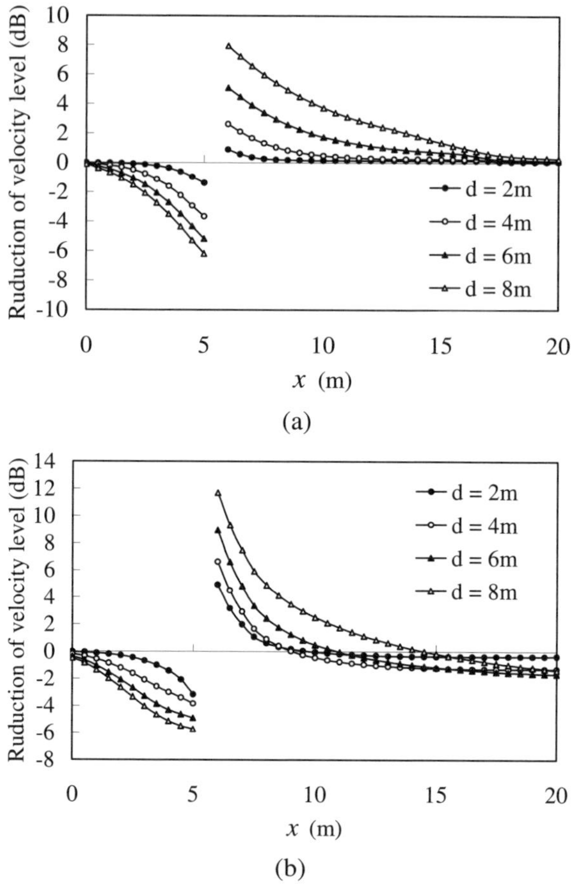

Fig. 9.9 Effect of trench depth on velocity reduction for c = 70 m/s (open trenches): (a) vertical; (b) horizontal.

adversely affect the horizontal responses at a certain distance outside the trenches, as can be seen from Fig. 9.9(b).

Correspondingly, the velocity reduction for the ground surface under the moving loads at the super-critical speed c = 100 m/s has been plotted in Fig. 9.10. For this case, the efficiency of isolation increases as the trench depth increases. However, for $d \geq 4$ m, little improvement can be made by increasing the trench depth. The efficiency of velocity reduction is generally not good for open trenches with smaller depths d, say, less than 2 m.

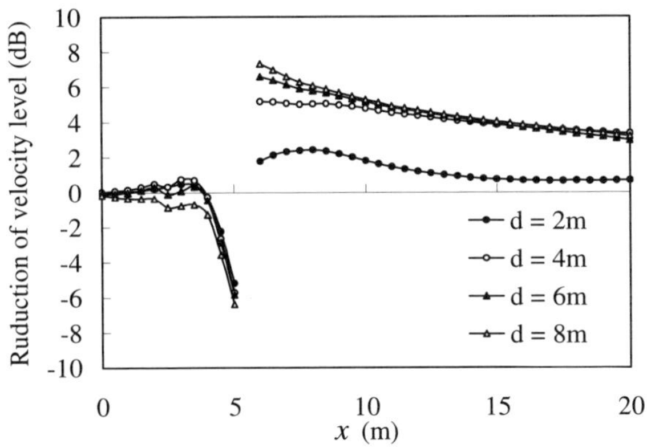

(a)

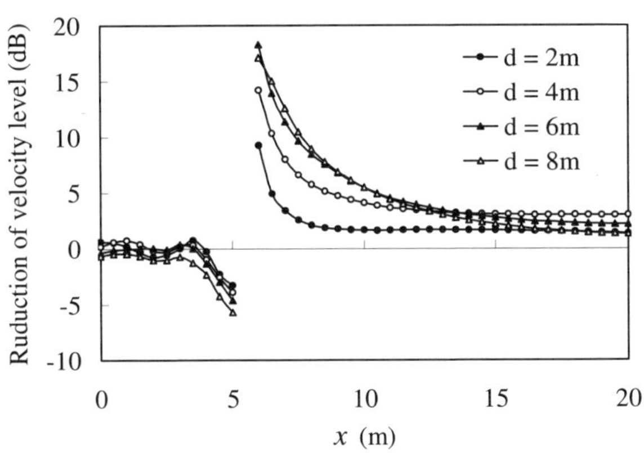

(b)

Fig. 9.10 Effect of trench depth on velocity reduction for c = 100 m/s (open trenches): (a) vertical; (b) horizontal.

9.3.1.3 *Effect of trench width*

The reduction of velocity level for both the vertical and horizontal response has been plotted for various trench widths with respect to the distance x for the sub-critical speed c = 70 m/s in Fig. 9.11. As can be seen, wider open trenches can improve the isolation efficiency in some regions outside the trenches, but the amount of increase is generally insignificant. Moreover, the average reduction of the response outside the trenches does not change too much. The improvement of isolation effect

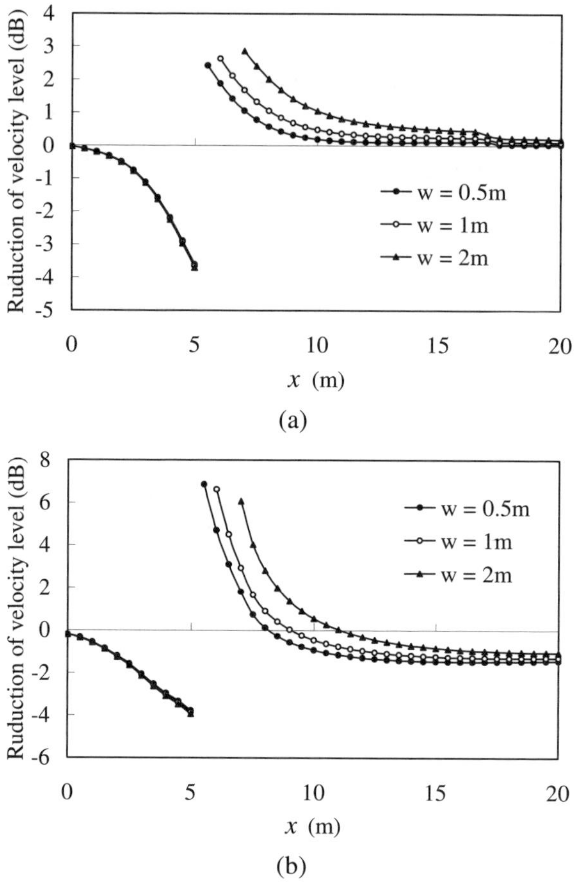

Fig. 9.11 Effect of trench width on velocity reduction for c = 70 m/s (open trenches): (a) vertical; (b) horizontal.

at some location can be interpreted as the shifting of the same reduction pattern to a farther distance simply because the trench wall moves to a farther distance, too. The same phenomenon can also be observed from the results computed for the case with the super-critical speed c = 100 m/s in Fig. 9.12. Moreover, both Figs. 9.11 and 9.12 indicate that the variation of trench width does not affect the response inside the trenches. Based on the above analyses, the width of open trenches is not considered a very effective parameter in isolation of the train-induced vibrations.

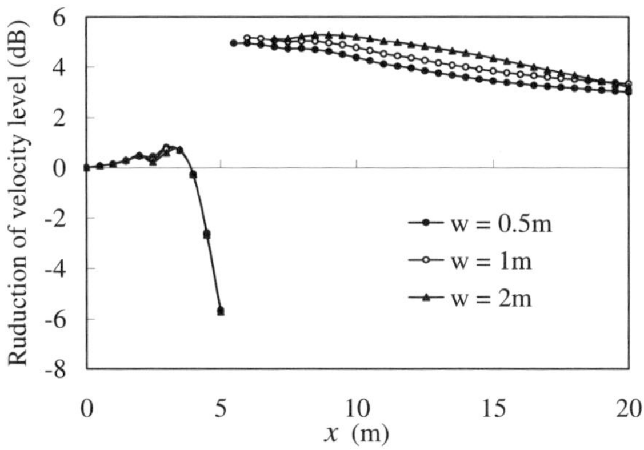

(a)

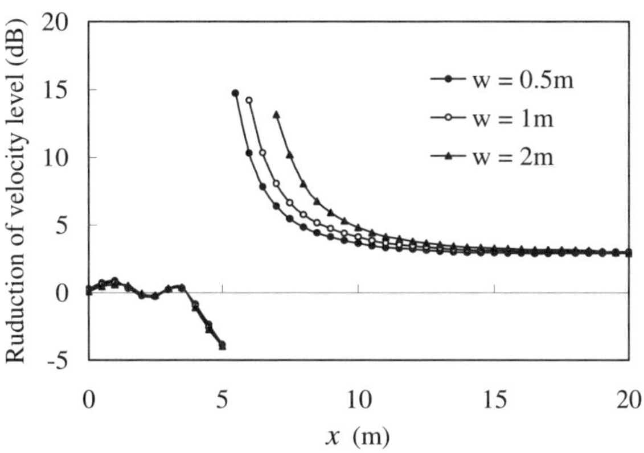

(b)

Fig. 9.12 Effect of trench width on velocity reduction for c = 100 m/s (open trenches): (a) vertical; (b) horizontal.

9.3.2 *Moving loads with self oscillation*

By assuming that the self oscillation frequency of the moving loads is $f_0 = 20$ Hz and the train speed is c = 70 m/s, i.e., in the sub-critical range with respect to the Rayleigh wave speed, the real-part vertical displacements along the ground surface before and after installation of the open trenches were plotted in Fig. 9.13. A comparison between Figs. 9.13(a) and (b) indicates that the open trenches are very effective in reducing the displacement induced by the moving loads with self

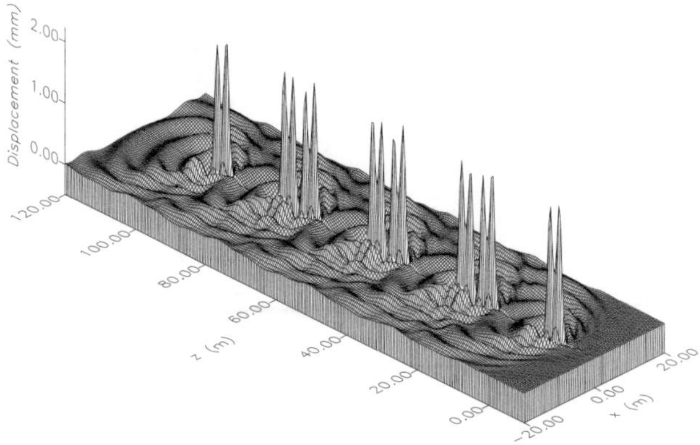

(a)

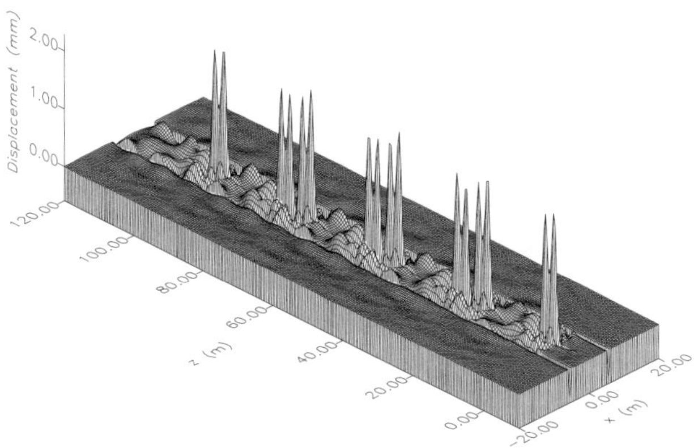

(b)

Fig. 9.13 Displacement field for a moving train with speed c = 70 m/s and f_0 = 20 Hz with: (a) no isolation; (b) open trenches.

oscillation, as the regions outside the trenches have become much smoother. To further explore this effect, the corresponding time history responses of the real-part vertical displacement, velocity and acceleration for the location x = 10 m were plotted in Figs. 9.14(a)-(c), respectively. Clearly, the presence of open trenches has resulted in a significant reduction of all the responses of the ground surface. Corresponding to Fig. 9.14, the effect of open trenches on the transfer function at the

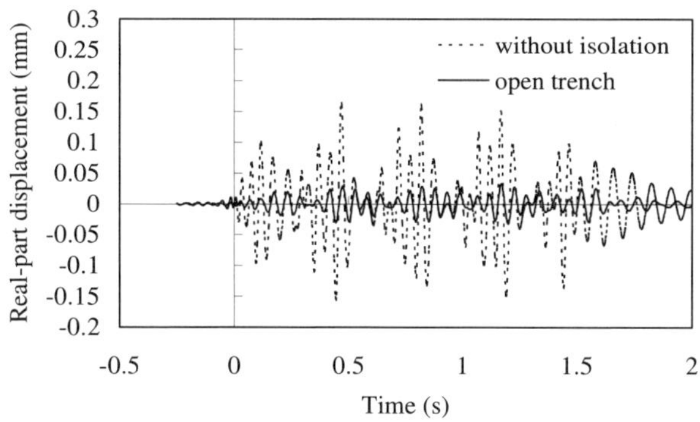

(a)

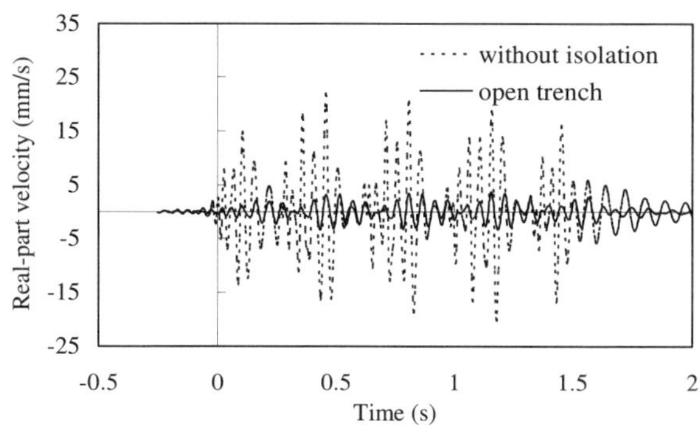

(b)

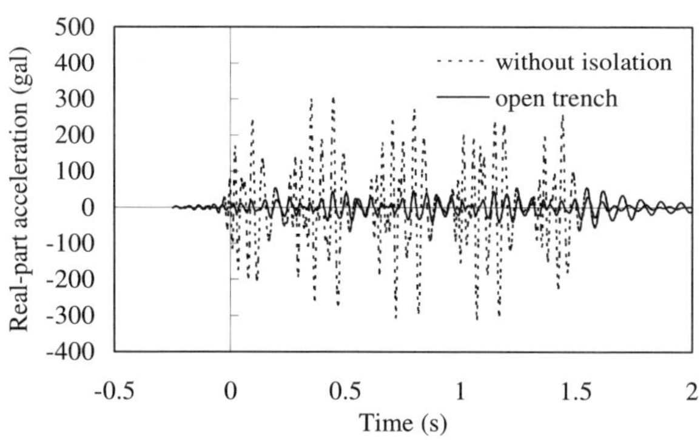

(c)

Fig. 9.14 Screening efficiency of open trenches on vertical responses for x = 10 m, c = 70 m/s and f_0 = 20 Hz: (a) displacement; (b) velocity; (c) acceleration.

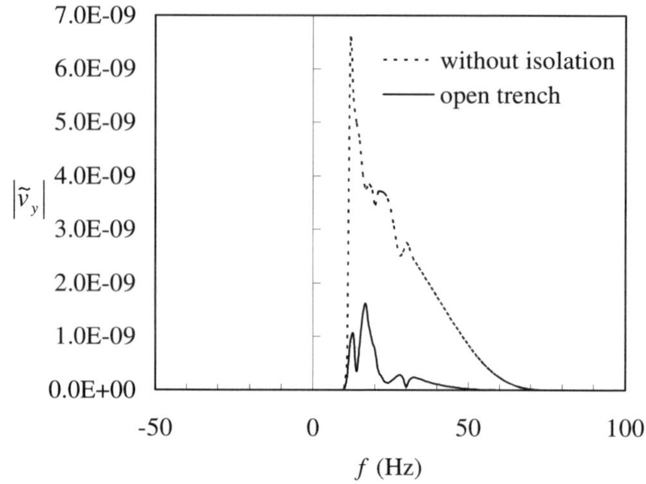

Fig. 9.15 Influence of open trenches on the transfer function of vertical response for $x = 10$ m, $c = 70$ m/s, and $f_0 = 20$ Hz .

location $x = 10$ m on the ground was given in Fig. 9.15, which shows a drastic decrease in the amplitude of the transfer function due to the presence of open trenches.

To further investigate the screening efficiency of open trenches for different wheel oscillation frequencies f_0, four different oscillation frequencies, i.e., $f_0 = 5$, 10, 20, and 30 Hz, are considered for the moving loads with $c = 70$ m/s. The corresponding results for reduction of velocity level along the ground surface of the profile were solved and plotted in Fig. 9.16. As can be seen, higher efficiency of isolation can be achieved for moving loads with higher self oscillation frequencies f_0. Moreover, the efficiency of isolation remains almost the same over a longer distance than the case with zero f_0.

The average reduction levels computed over the distance from $x = 6$ m to $x = 20$ m for the vertical response (v) and horizontal response (w) against the self oscillation frequency f_0 have been plotted in Fig. 9.17. Clearly, the isolation efficiency of open trenches improves significantly with the increase in the oscillation frequency f_0 of the moving loads. Besides, the open trenches are more effective in reducing the vertical response than the horizontal response of the ground.

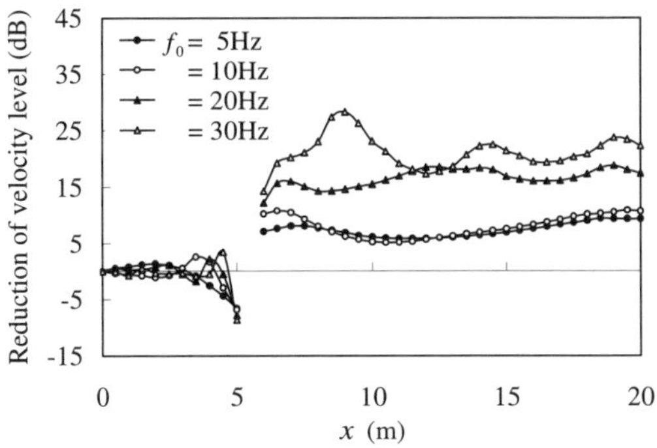

(a)

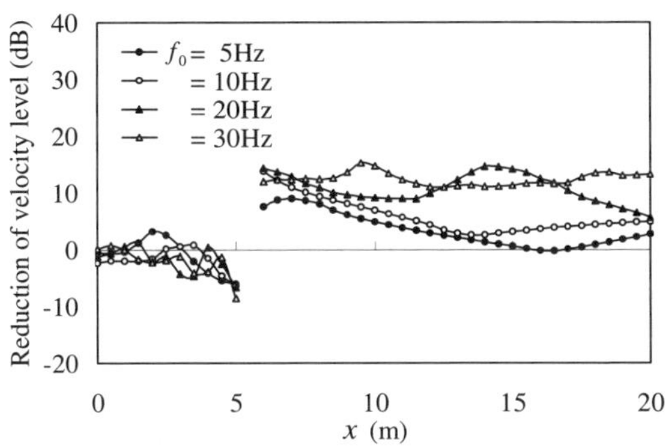

(b)

Fig. 9.16 Effect of excitation frequency f_0 on velocity reduction for c = 70 m/s (open trenches): (a) vertical; (b) horizontal.

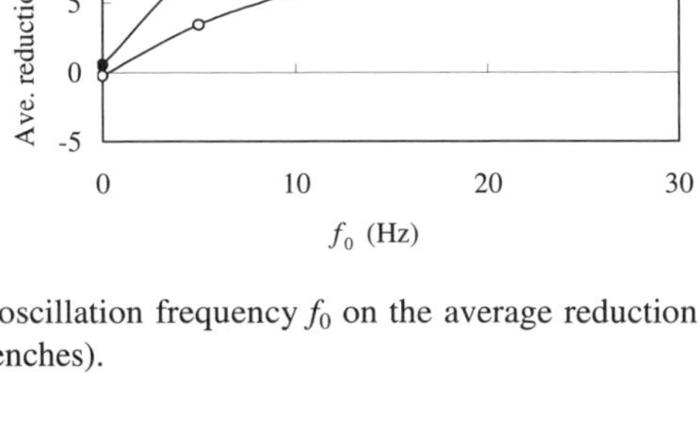

Fig. 9.17 Effect of oscillation frequency f_0 on the average reduction of velocity level for c = 70 m/s (open trenches).

9.4 Vibration Reduction by In-Filled Trenches

The analytical model adopted in this section is the one shown in Fig. 9.1(b), where two in-filled trenches are constructed at a distance l away from the centerline of the railway on each side. Unless otherwise noted, the geometric parameters for the in-filled trench considered are the same as the ones used previously for the open trenches, i.e., distance $l = 5$ m, depth $d = 4$ m, and width $w = 1$ m. The material used to fill the trenches is concrete, of which the properties are listed in Table 9.1.

9.4.1 *Moving loads with no self oscillation*

By assuming that the moving wheel loads do not oscillate by themselves, the effects of various parameters, i.e., the train speed, the dimensions and S-wave speed of the trench, on the wave screening effectiveness will be investigated. Both the damping ratio and Poisson's ratio of the trenches will be excluded from the analysis, since their effects on vibration reduction are known to be negligible from the literature.

9.4.1.1 *Effect of load-moving speed*

The isolation efficiency of in-filled trenches for different train speeds has been shown in Fig. 9.18. As can be seen, the performance of in-filled trenches attains its maximum when $M_2 = 1.0$ or when $c = 100$ m/s for the soil considered. Besides, in-filled trenches are more effective in reducing the horizontal than vertical responses. Similar to the previous section, only two train speeds will be considered in the following analyses, i.e., $c = 70$ m/s and $c = 100$ m/s, to represent the effect of trains moving in sub-critical and super-critical speeds, respectively, with respect to the Rayleigh wave speed.

For the sub-critical speed $c = 70$ m/s, the displacements of the ground surface with and without installation of the in-filled trenches have been plotted in Figs. 9.19(a) and (b), respectively. As can be seen, the displacement outside the trenches shows a trend much smoother than the case with no trenches, which clearly indicates the effect of isolation of the in-filled trenches. The same phenomenon can also be observed from

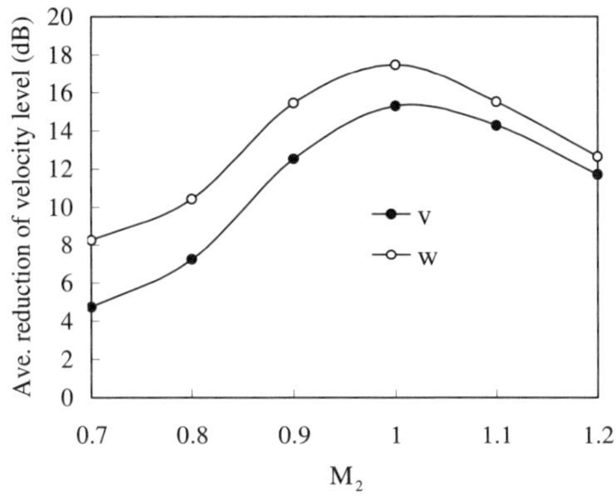

Fig. 9.18 Effect of train speed on the average reduction of velocity level (in-filled trenches).

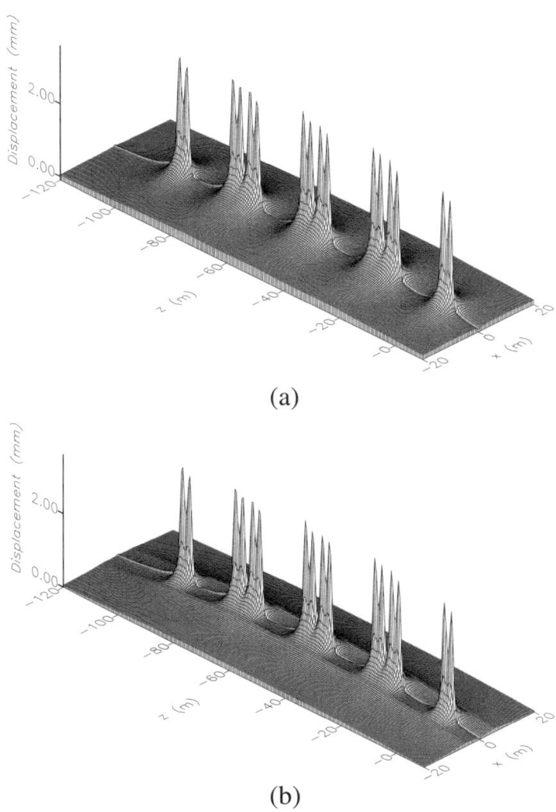

Fig. 9.19 Displacement field for a moving train with speed c = 70 m/s and f_0 = 0 with: (a) no isolation; (b) in-filled trenches.

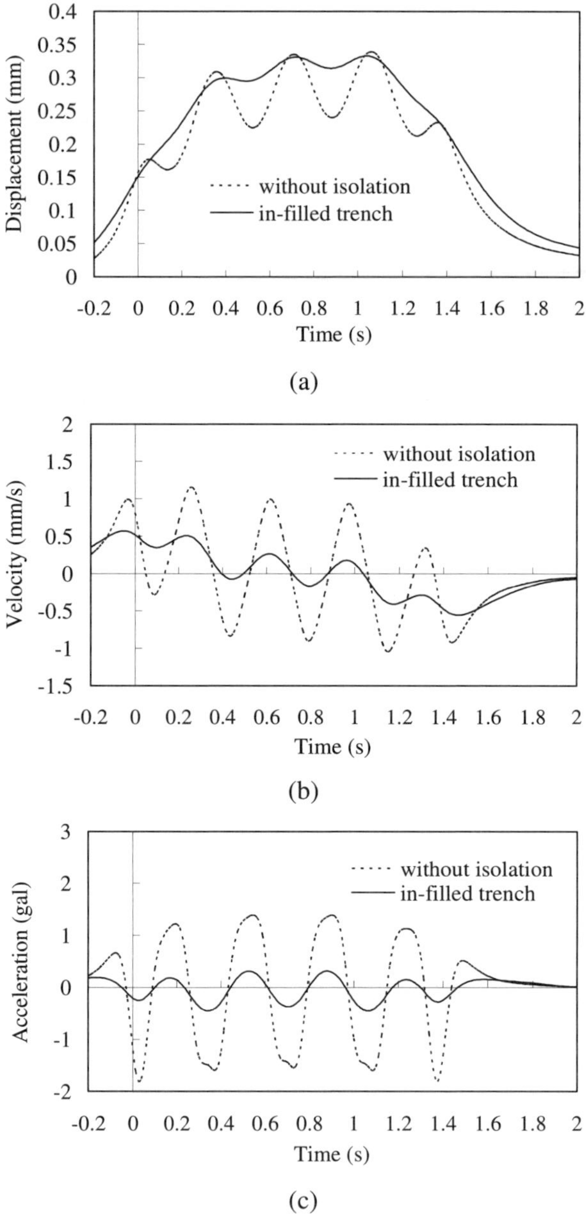

Fig. 9.20 Screening efficiency of in-filled trenches vertical responses for $x = 10$ m and $c = 70$ m/s: (a) displacement; (b) velocity; (c) acceleration.

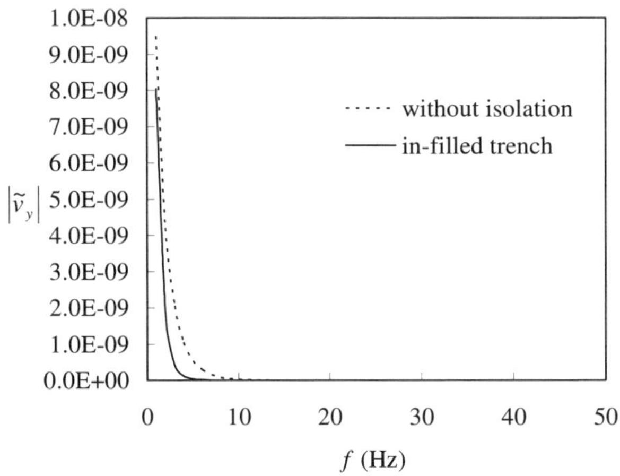

Fig. 9.21 Influence of in-filled trenches on the transfer function of vertical response for $x = 10$ m and $c = 70$ m/s (in-filled trenches).

the time history responses plotted for the location $x = 10$ m in Fig. 9.20. As can be seen from Fig. 9.20(a), the existence of in-filled trenches can help smooth out the displacement field, but with basically no change on the amplitude. As for the velocity and acceleration (see Figs. 9.20(b)-(c)), the screening efficiency of in-filled trenches appears to be much more pronounced. This observation can be explained by the corresponding transfer function given in Fig. 9.21, where the reduction in amplitude of the transfer function increases with the increase in the frequency f, implying a better performance in reducing the velocity and acceleration than the displacement.

In comparison, the displacement fields caused by the train loads moving at the super-critical speed $c = 100$ m/s for the cases without or with in-filled trenches have been plotted in Figs. 9.22(a) and (b), respectively. Correspondingly, the time history responses of the displacement, velocity, and acceleration for the location $x = 10$ m have been plotted in Figs. 9.23(a)-(c), respectively. From Fig. 9.23, one observes that in-filled trenches are very effective for reducing the vibrations, especially for acceleration. Such a significant effect can also be observed from the transfer function given in Fig. 9.24, where a substantial reduction of amplitude can be observed, especially within the high frequency range.

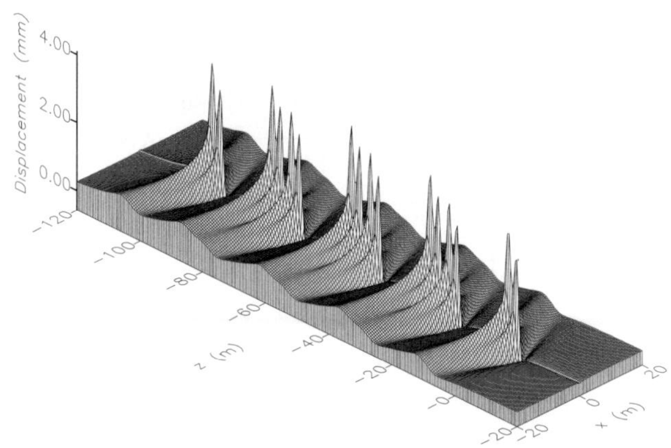

(a)

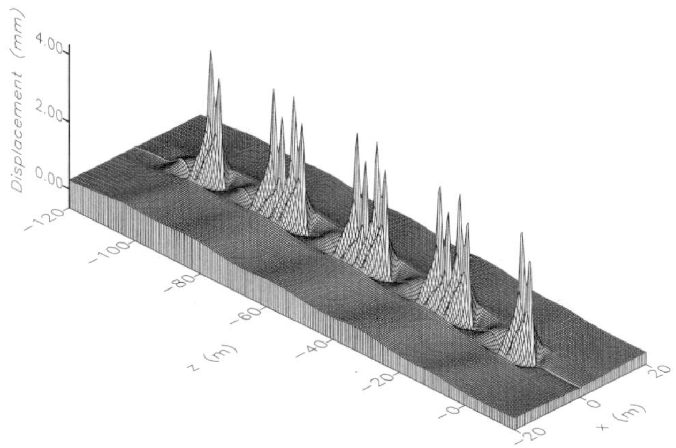

(b)

Fig. 9.22 Displacement field for a moving train with speed c = 100 m/s and f_0 = 0 with: (a) no isolation; (b) in-filled trenches.

9.4.1.2 *Effect of trench depth*

The influence of trench depth on the isolation efficiency of in-filled trenches for the sub-critical and super-critical speeds, i.e., c = 70 m/s and c = 100 m/s, has been plotted in Figs. 9.25 and 9.26, respectively, where part (a) denotes the reduction of the vertical response and (b) the horizontal response. As for c = 70 m/s, the increase of trench depth results in a significant improvement of the isolation effect. Noteworthy is the fact that the installation of in-filled trenches does not adversely

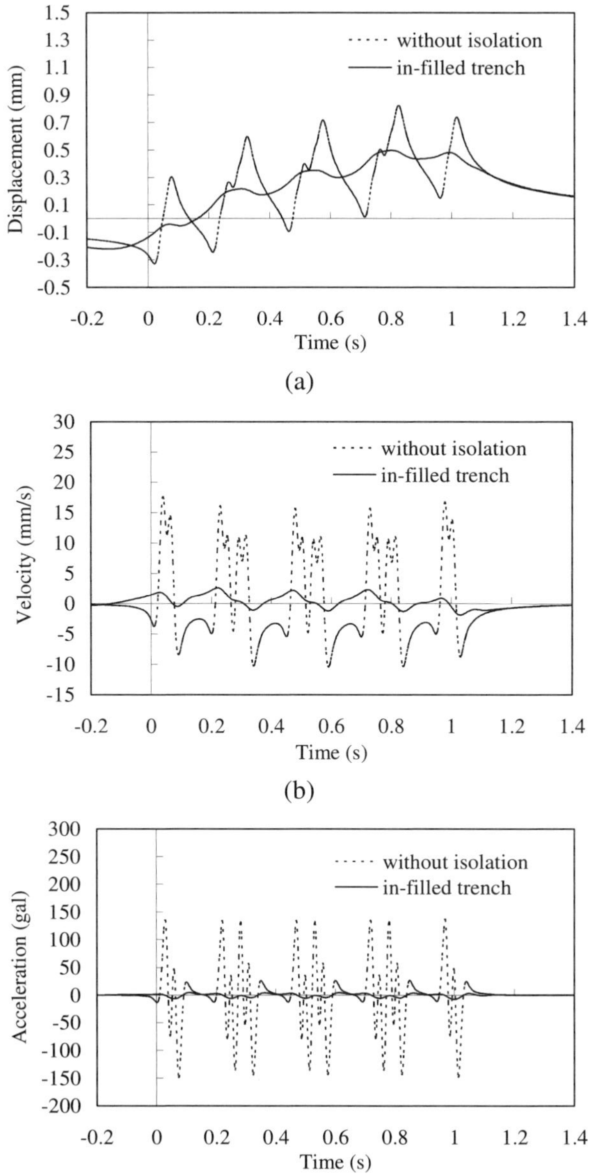

Fig. 9.23 Screening efficiency of in-filled trenches on vertical responses for $x = 10$ m and $c = 100$ m/s: (a) displacement; (b) velocity; (c) acceleration.

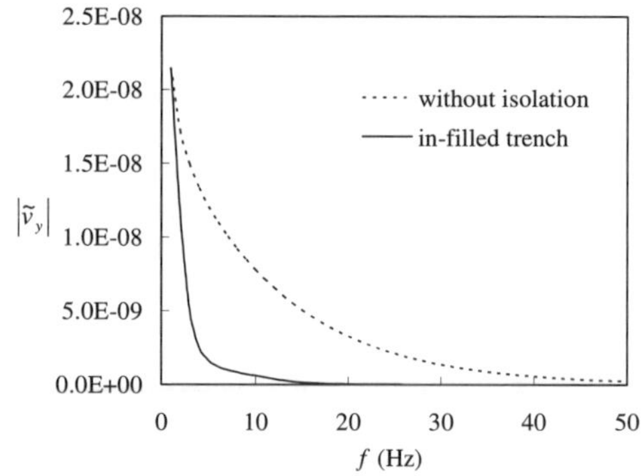

Fig. 9.24 Influence of in-filled trenches on the transfer function of vertical response for $x = 10$ m and $c = 100$ m/s (in-filled trenches).

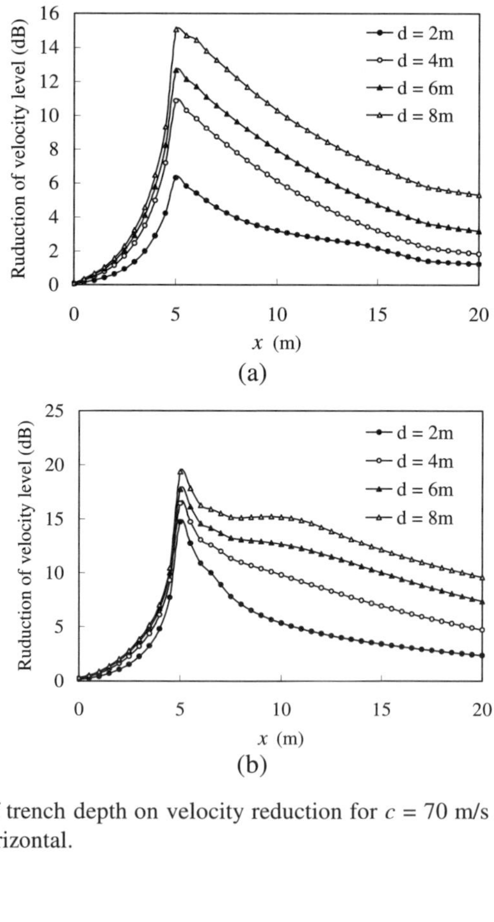

Fig. 9.25 Effect of trench depth on velocity reduction for $c = 70$ m/s (in-filled trenches): (a) vertical; (b) horizontal.

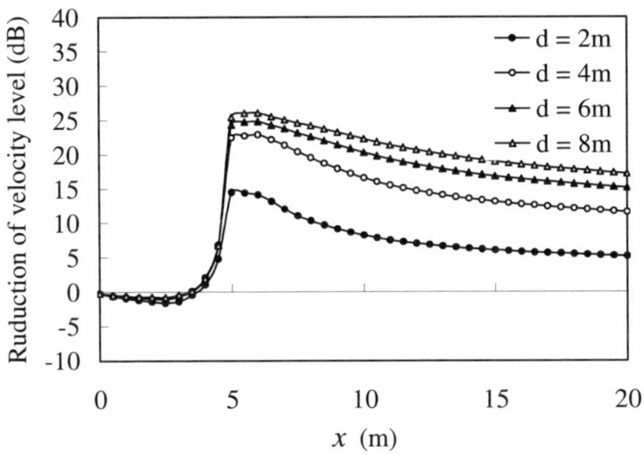

(a)

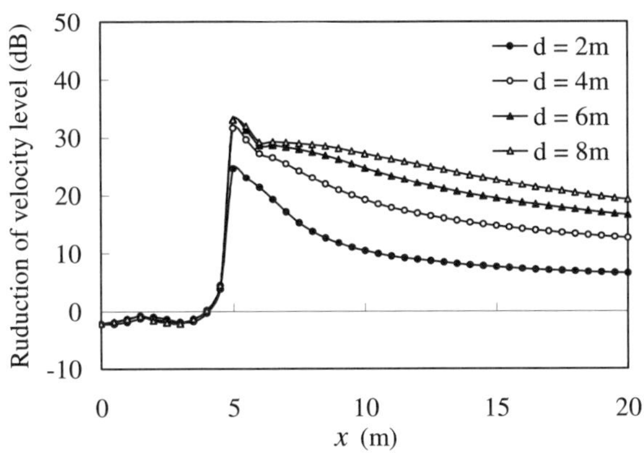

(b)

Fig. 9.26 Effect of trench depth on velocity reduction for c = 100 m/s (in-filled trenches): (a) vertical (b) horizontal.

amplify the vibration inside the trenches as the open trench does (see Fig. 9.9). The difference between open and in-filled trenches can be attributed to the instability or lack of restraint for the soil bodies located inside the open trenches. It is for this reason that open trenches are not considered really practical.

For c = 100 m/s, the influence of trench depth is generally similar to the case for c = 70 m/s. However, some minor amplification may occur in the region between the two trenches, as was revealed by the results given in Fig. 9.26.

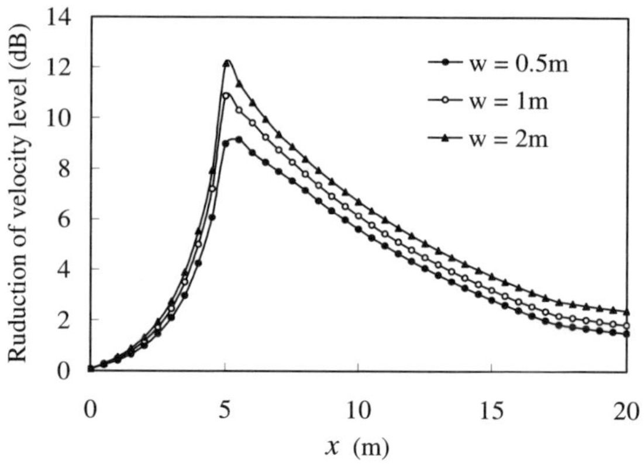

(a)

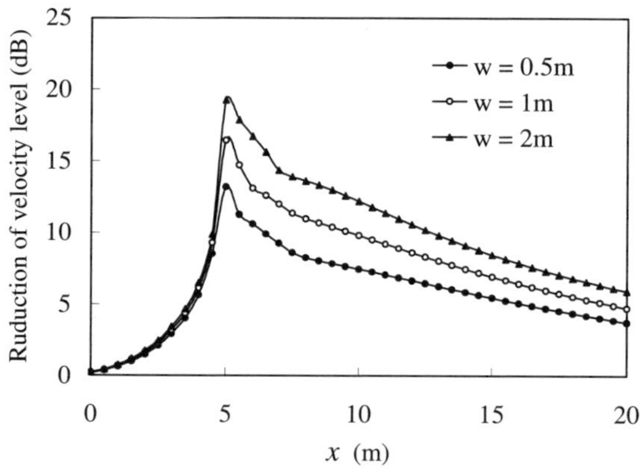

(b)

Fig. 9.27 Effect of trench width on velocity reduction for c = 70 m/s (in-filled trenches): (a) vertical; (b) horizontal.

9.4.1.3 *Effect of trench width*

To investigate the influence of trench width on the isolation efficiency of in-filled trenches, three widths are adopted to evaluate the reduction of velocity level on the ground surface. The results obtained for the two ranges of speeds, i.e., c = 70 m/s and c = 100 m/s, have been plotted in Figs. 9.27 and 9.28, respectively. As can be seen, for both speeds, an increase in the trench width leads to improvement of isolation efficiency, in a manner better than that of the open trenches. But for the response

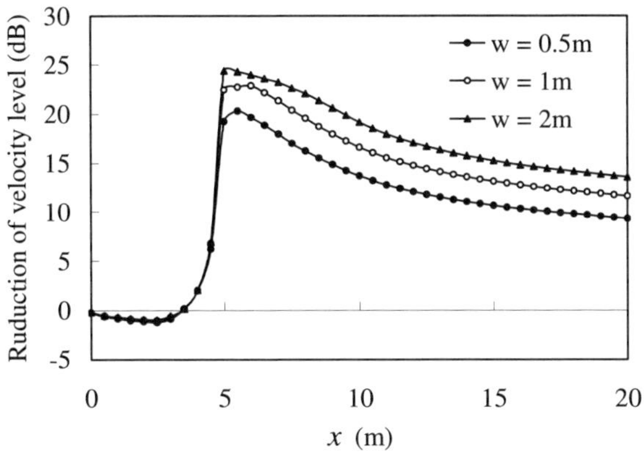

(a)

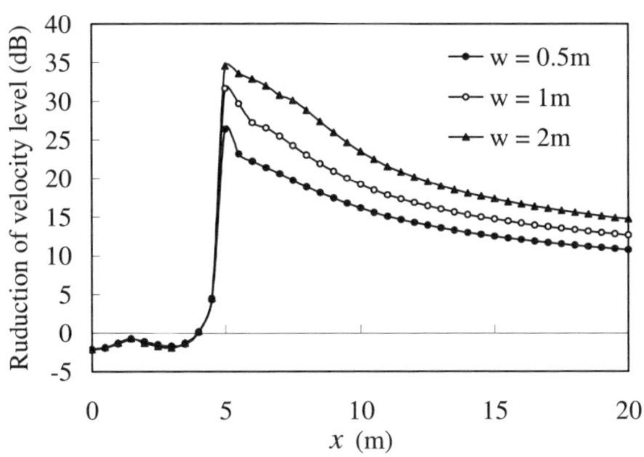

(b)

Fig. 9.28 Effect of trench width on velocity reduction for c = 100 m/s (in-filled trenches): (a) vertical; (b) horizontal.

inside the trenches, the level of improvement made by increasing the trench width is generally negligible for both ranges of speed. In particular, the response is adversely amplified inside the trenches for the case with c = 100 m/s.

9.4.1.4 *Effect of shear wave speed of trenches*

Let us define the ratio of shear wave speeds as $\overline{c}_S = c'_S / c_S$, where c'_S and c_S denotes the S-wave speed of the in-filled trenches and surrounding

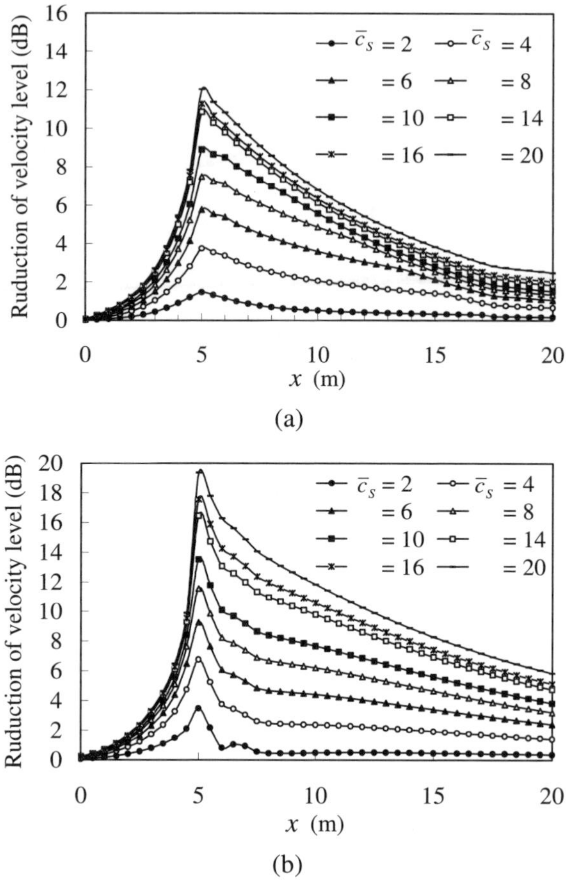

Fig. 9.29 Effect of ratio of S-wave speeds $\overline{c}_s$ on velocity reduction for c = 70 m/s (in-filled trenches with $\overline{c}_s > 1$): (a) vertical; (b) horizontal.

soils, respectively. The ratio of shear wave speeds $\overline{c}_S$ for the typical in-filled trenches considered herein is taken as 14. Consider first the case in which the S-wave speed of the in-filled trenches is greater than that of the surrounding soil ($\overline{c}_S > 1$), meaning that the in-filled trenches are stiffer or harder than the surrounding soils. By varying the S-wave speed of the trenches, the reduction of velocity level computed for different $\overline{c}_S$ values has been given in Figs. 9.29 and 9.30 for c = 70 m/s and c = 100 m/s, respectively. These figures indicate that for the case of stiffer trenches, i.e., with $\overline{c}_S > 1$, an increase in the shear wave speed can result

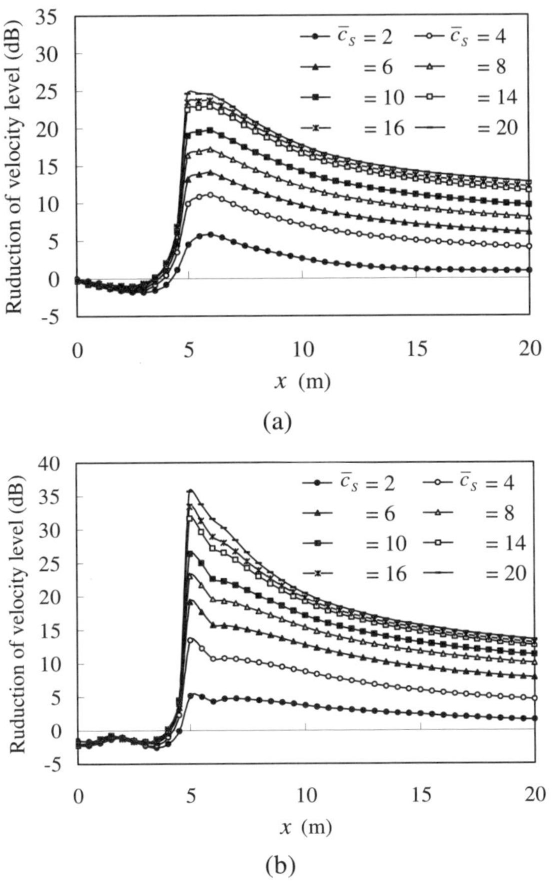

Fig. 9.30 Effect of ratio of S-wave speeds $\bar{c}_s$ on velocity reduction for c = 100 m/s (in-filled trenches with $\bar{c}_s > 1$): (a) vertical; (b) horizontal.

in improvement of the isolation performance of the in-filled trenches for both train speeds, primarily due to the restraining effect of the in-filled trenches.

Another possible choice for the fill material is soft materials, i.e., materials with shear wave speeds lower than the surrounding soil ($\bar{c}_S < 1$). Since a softer material generally has a lower mass density, a small mass density of $\rho = 150\,\text{kg/m}^3$ is assumed for the fill material herein. By varying the shear wave speed of the fill material, the reduction of velocity level against distance x for the sub-critical speed c = 70 m/s has

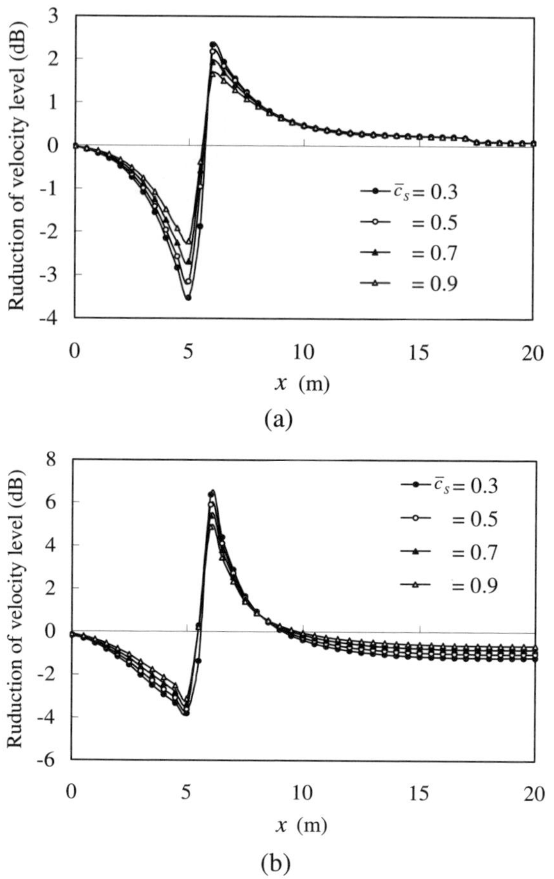

Fig. 9.31 Effect of ratio of S-wave speeds $\bar{c}_s$ on velocity reduction for $c = 70$ m/s (in-filled trenches with $\bar{c}_s < 1$): (a) vertical; (b) horizontal.

been plotted in Fig. 9.31. As can be seen, the isolation performance of in-filled trenches with a softer material is generally poor.

For the case with wheel loads moving in the super-critical range, i.e., with $c = 100$ m/s (Fig. 9.32), the performance appears to be much better, but still not so good as that for trenches with stiffer in-filled materials. By carefully examining the results in Figs. 9.31 and 9.32 with the corresponding ones for the open trench with $d = 4$ m in Figs. 9.9 and 9.10, one finds that the influence patterns for these two cases are quite similar. From the point of construction, an in-filled trench with soft material

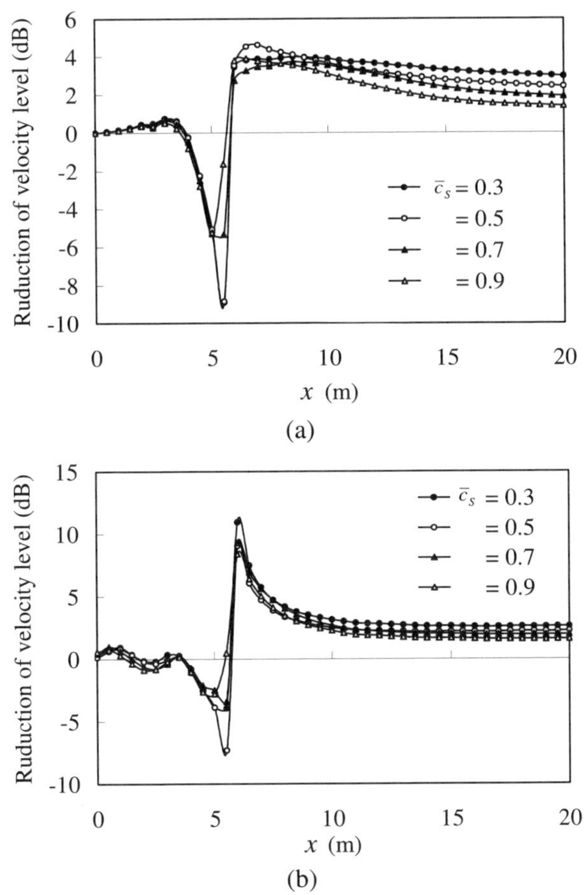

Fig. 9.32 Effect of ratio of S-wave speeds $\bar{c}_s$ on velocity reduction for c = 100 m/s (in-filled trenches with $\bar{c}_s < 1$): (a) vertical; (b) horizontal.

can be achieved by using properly designed soil-bentonite mix as the fill material.

The *impedance ratio* (IR) is a factor widely used by geotechnical engineers to distinguish whether a wave barrier is soft or hard with respect to the surrounding soil, which is defined as

$$\mathrm{IR} = \frac{\rho' c_S'}{\rho\, c_S} \tag{9.5}$$

where ρ' and ρ denote the mass density of the barrier and the soil,

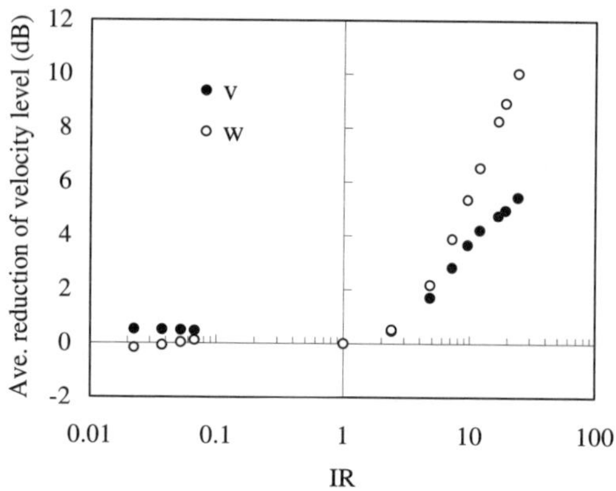

Fig. 9.33 Effect of impedance ratio on the average reduction of velocity level for $c = 70$ m/s (in-filled trenches).

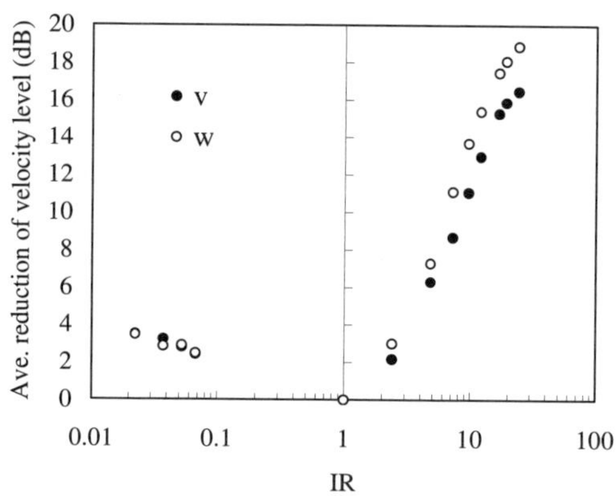

Fig. 9.34 Effect of impedance ratio on the average reduction of velocity level for $c = 100$ m/s (in-filled trenches).

respectively, and c_S' and c_S the S-wave speed of the two. By averaging the reduction of velocity level from distance $x = 6$ m to $x = 20$ m given in Figs. 9.29-9.32, the influence of trench impedance ratio, for both IR > 1 and IR < 1, can be plotted in the same figures as in Figs. 9.33 and 9.34 for the two train speeds $c = 70$ m/s and $c = 100$ m/s, respectively. Evidently, the in-filled trench is effective only for stiff fill materials, i.e., with IR > 1. Within this region, the stiffer the trench, the better the isolation efficiency is. The other phenomenon observed from Figs. 9.33 and 9.34 is that, for IR > 1, the performance of in-filled trenches is better

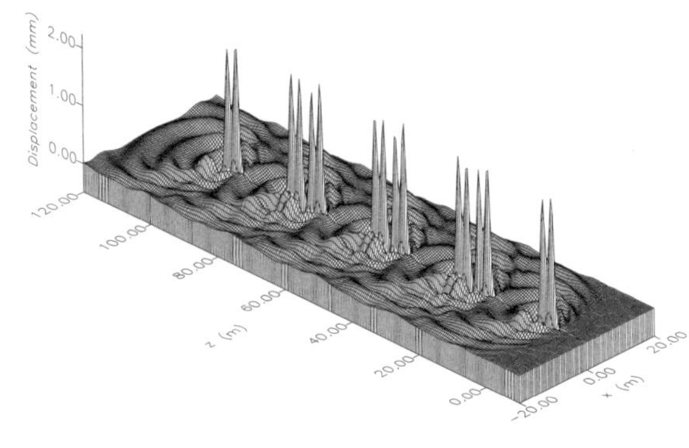

(a)

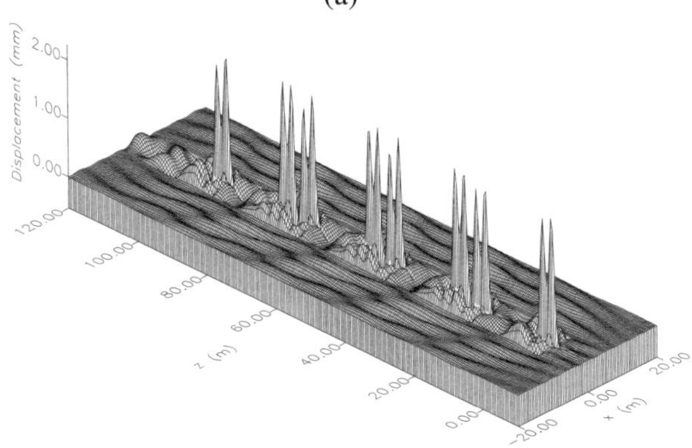

(b)

Fig. 9.35 Displacement field for a moving train with speed c = 70 m/s and f_0 = 20 Hz with: (a) no isolation; (b) in-filled trenches.

for reducing the horizontal than vertical responses, whereas for IR < 1, the reverse is true.

9.4.2 *Moving loads with self oscillation*

Let us turn to the influence of in-filled trenches in reducing the vibration caused by moving loads with self oscillation. By assuming the oscillation frequency to be f_0 = 20 Hz and the train speed to be c = 70 m/s, the displacement field computed for the case with and with no installation of the in-filled trenches have been depicted in Figs. 9.35(a) and (b), respectively. Clearly, the in-filled trenches have changed the wave shape

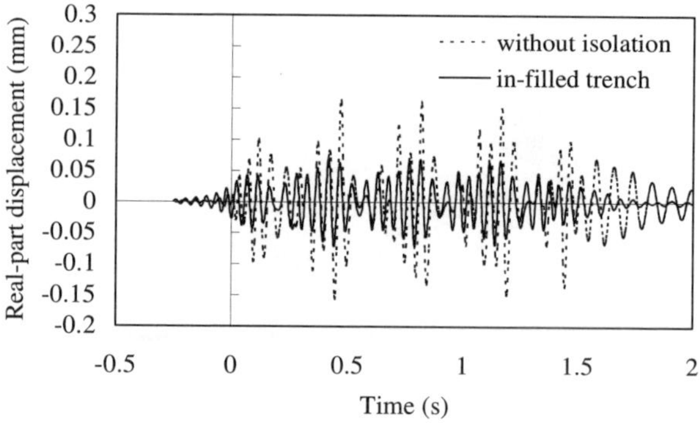

(a)

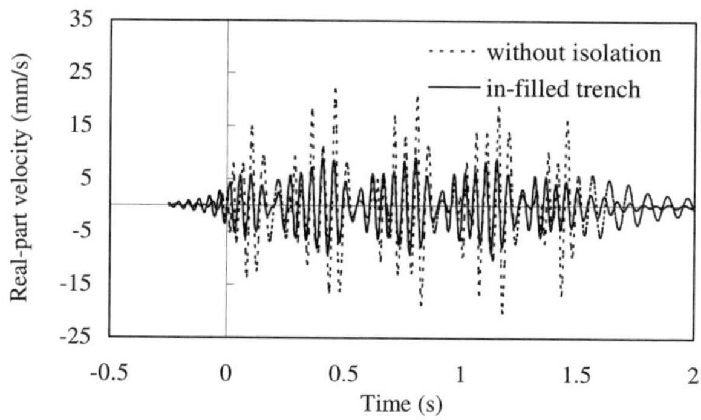

(b)

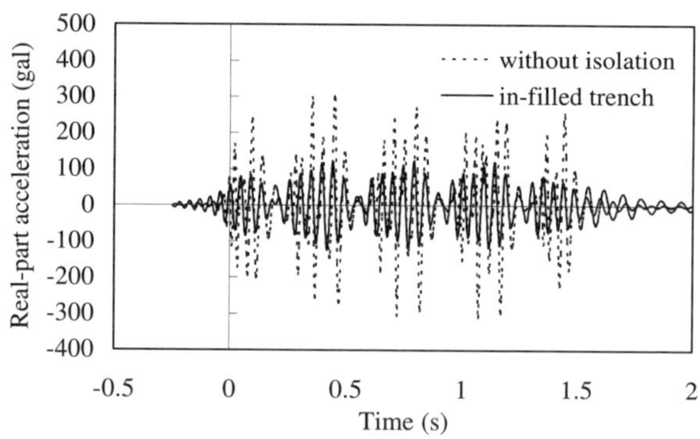

(c)

Fig. 9.36 Screening efficiency of in-filled trenches on vertical responses for x = 10 m, c = 70 m/s and f_0 = 20 Hz: (a) displacement; (b) velocity; (c) acceleration.

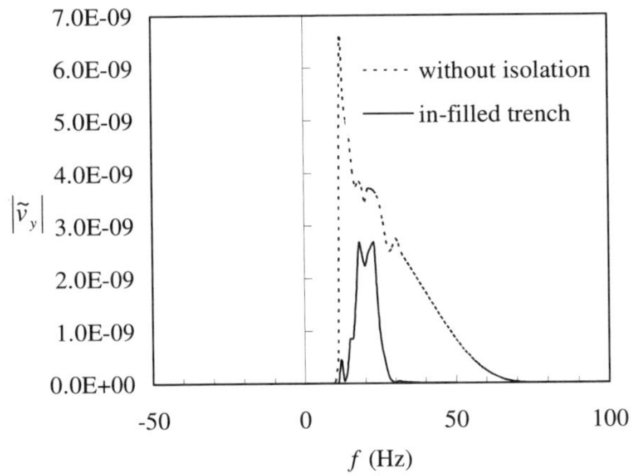

Fig. 9.37 Influence of in-filled trenches on the transfer function of vertical response for $x = 10$ m, $c = 70$ m/s, and $f_0 = 20$ Hz (in-filled trenches).

on the ground surface. But it is not easy to distinguish the variation in vibration amplitudes. To better understand this problem, the real-part time history responses of the displacement, velocity and acceleration for the location $x = 10$ m have been plotted in Figs. 9.36(a)-(c). As can be seen, the in-filled trenches are generally effective for reducing the response, but not as effective as the open trenches are, as compared with Fig. 9.14. The effect of in-filled trenches in the frequency domain was given in Fig. 9.37, which indicates a decrease in magnitude of the transfer function due to installation of the in-filled trenches.

To further investigate the screening efficiency of in-filled trenches for different oscillation frequencies f_0, four different frequencies f_0, i.e., $f_0 = 5$, 10, 20, 30 Hz, are considered for $c = 70$ m/s. The reduction levels of the vertical and horizontal velocities along the x-axis were plotted in Figs. 9.38(a) and (b), respectively. Because of the fluctuation nature of the results shown, it is not easy to distinguish the effect of different frequencies f_0 in the figures. By averaging the reduction value over the distance from $x = 6$ to 20 m, the average reduction of velocity level for the vertical (v) and horizontal (w) responses with respect to the frequency f_0 was plotted in Fig. 9.39. Although the isolation efficiency of the in-filled trenches tends to improve with the increase of f_0, the effect is not as obvious as that of the open trenches.

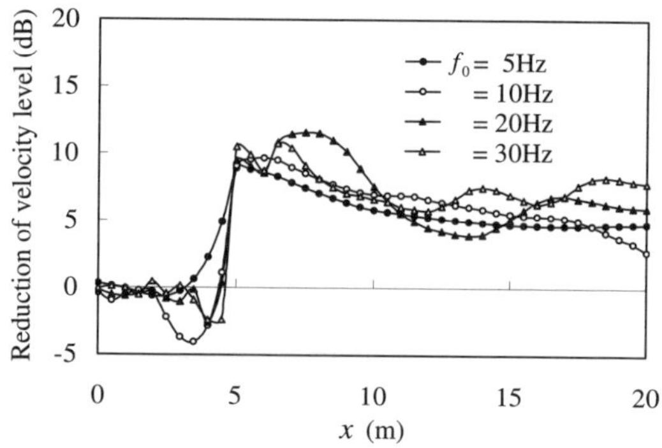

(a)

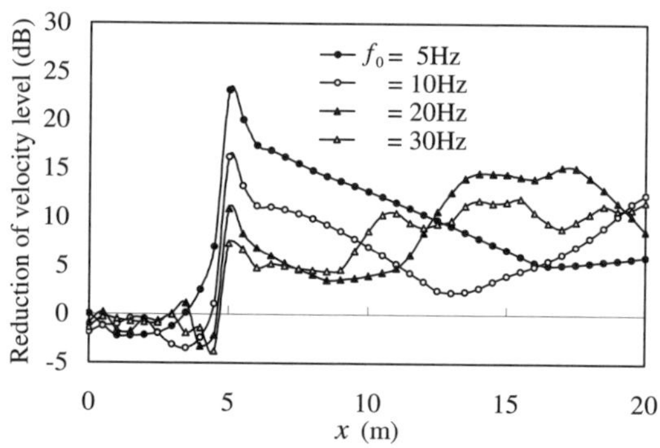

(b)

Fig. 9.38 Effect of oscillation frequency f_0 on velocity reduction for c = 70 m/s (in-filled trenches): (a) vertical; (b) horizontal.

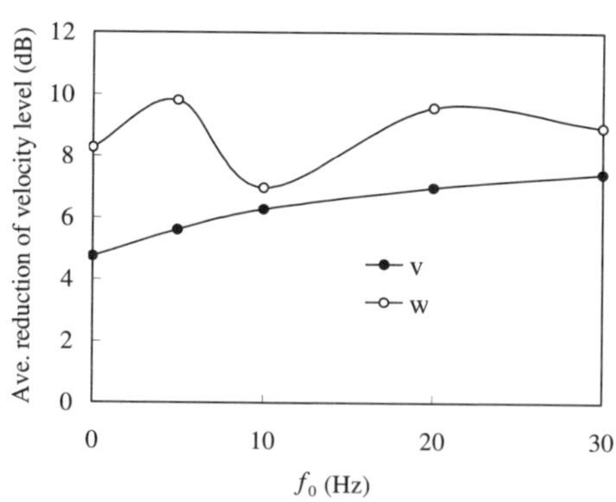

Fig. 9.39 Effect of oscillation frequency f_0 on the average reduction of velocity level for c = 70 m/s (in-filled trenches).

9.5 Vibration Reduction by Wave Impeding Block

The model of the wave impeding block (WIB) studied in this section was plotted in Fig. 9.1(c), where a block with thickness t and width e is placed underneath the railway at a depth h. Unless mentioned otherwise, the following data will be adopted in all analyses in this section: thickness $t = 1$ m, width $e = 4$ m, and depth $h = 1$ m. It should be noted that the dimensions of the cross section of the WIB are made equal to those of the open and in-filled trenches, so as to provide an equal basis for comparison. The material properties for the WIB considered have been listed in Table 9.1, which are the same as those of the in-filled trenches considered. The basic idea of using wave impeding blocks comes from the bedrock underlying a soil layer. As we have observed in the preceding chapter, the existence of a bedrock underneath the soils can suppress the waves with longer wavelengths, i.e., low-frequency waves, from propagating outward. Thus, it is expected that a solid block constructed underneath the soil can reveal the effect of suppression to some degree.

9.5.1 *Moving loads with no self oscillation*

Similar to the study for the open and in-filled trenches, the static moving loads will be considered first in studying the effectiveness of the WIB for vibration reduction. In what follows, the depth and thickness of the WIB, shear wave speed of the WIB, as well as the train speed, will be investigated parametrically.

9.5.1.1 *Effect of load-moving speed*

The results computed by varying the train speed have been plotted in Fig. 9.40, where v and w represent respectively the vertical and horizontal responses. The average reduction of velocity level is obtained by averaging the reduction of velocity level over the distance from $x = 2$ m to 20 m. It is observed that the performance of the WIB reaches its maximum when $M_2 = 1.0$, i.e., when $c = 100$ m/s. In general, the screening effect of the WIB is better for trains moving at speeds higher

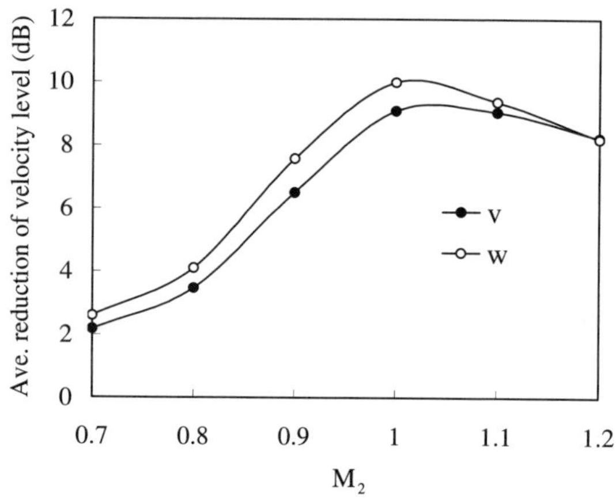

Fig. 9.40 Effect of train speed on the average reduction of velocity level (WIB).

than the critical speed, i.e., the Rayleigh wave speed. The results also indicate that the WIB tends to reduce the horizontal vibration more effectively than the vertical one.

If we compare Fig. 9.40 with Fig. 9.2 for open trenches and Fig. 9.18 for in-filled trenches, which have the same size as the WIB, we find that among these three wave barriers, in-filled trenches are most effective for reducing the vibration induced by the static moving loads, while open trenches are the worst.

A general view of the effect of the WIB can be appreciated from Figs. 9.41(a) and (b) for the ground displacement field before and after installation of the WIB for a train speed of c = 70 m/s. Obviously, the installation of the WIB does not affect the displacement too much. The same result can also be observed from Fig. 9.42 for the time history responses at location x = 10 m, which indicates that the WIB can reduce only slightly the displacement, velocity and acceleration responses.

Again, if we compare the results in Fig. 9.42 with Fig. 9.4 for the open trenches and Fig. 9.20 for the in-filled trenches, we find that only in-filled trenches are effective for vibration reduction the sub-critical speed case with c = 70 m/s. From the transfer function plotted in Fig. 9.43 for the location x = 10 m, one observes that the WIB is rather ineffective for reducing the amplitude of the transfer function for waves of low frequencies. Such a result is not surprising, if one realizes that

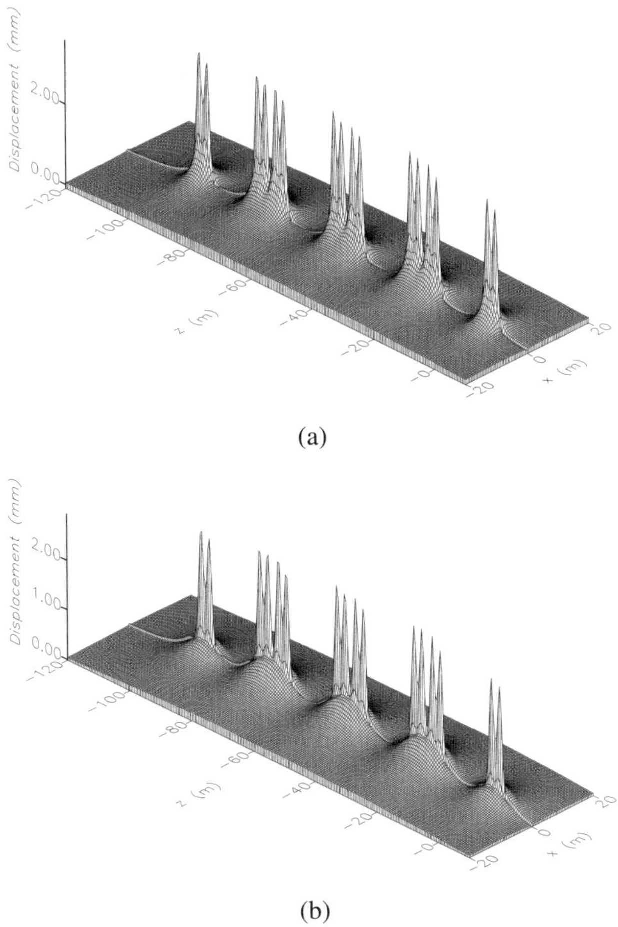

Fig. 9.41 Displacement field for a moving train with speed c = 70 m/s and f_0 = 0 with: (a) no isolation; (b) WIB.

the WIB is not a real bedrock, it behaves like a bedrock only for restraining waves of which the wavelengths are within a specific range. For a real bedrock, it is well known that all the waves with wavelengths longer than that of the cut-off frequency, as determined by the stratum depth in Eqs. (8.7) and (8.8), will be suppressed. Owing to its limited length, the WIB is effective not for waves with wavelengths longer than its length. Clearly, the depth and width of a WIB determine the range of

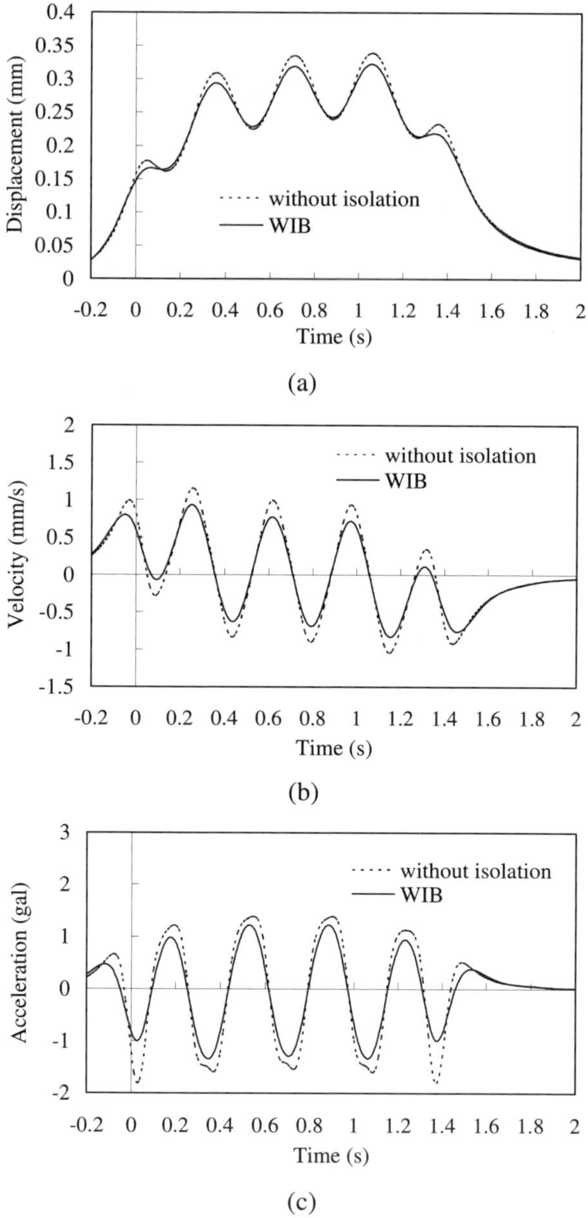

Fig. 9.42 Screening efficiency of WIB on vertical responses for $x = 10$ m and $c = 70$ m/s: (a) displacement; (b) velocity; (c) acceleration.

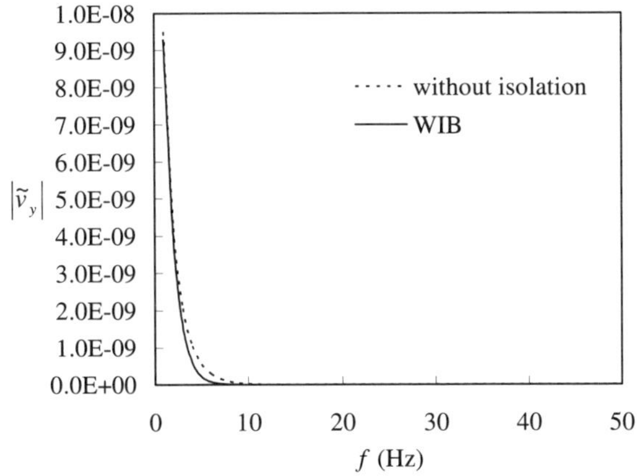

Fig. 9.43 Influence of WIB on the transfer function of vertical response for $x = 10$ m and $c = 70$ m/s.

wavelengths that will be suppressed by the WIB, if its thickness is large enough, as will be discussed in Section 9.5.1.3. Another observation from Fig. 9.43 is that the transfer function is concentrated mainly in a limited range of low frequencies. By comparing the transfer function in Fig. 9.43 with the curve with depth $H = 1$ m in Fig. 8.22(b), one observes that some *leaking problem* exists with the WIB for the range of low frequencies down to zero. As a result, waves of rather long wavelengths, particularly, those longer than the length of the WIB, can still propagate through the soil stratum, thereby causing a leaking problem on the WIB concerning vibration suppression.

As for the case of a train moving at the super-critical speed of $c = 100$ m/s, the displacement field has been given in Fig. 9.44. Here, one observes that the installation of the WIB can make the displacement field much smoother, but not very effective in reducing the amplitude of vibrations. Similar observation can also be obtained from the displacement response plotted in Fig. 9.45(a) for the point at a distance of $x = 10$ m from the source. As for the velocity and acceleration responses shown in Figs. 9.45(b) and (c), the effectiveness of the WIB is demonstrated to be much better.

By comparing Fig. 9.45 with Fig. 9.7 for open trenches and Fig. 9.23 for in-filled trenches, one observes that in-filled trenches are most

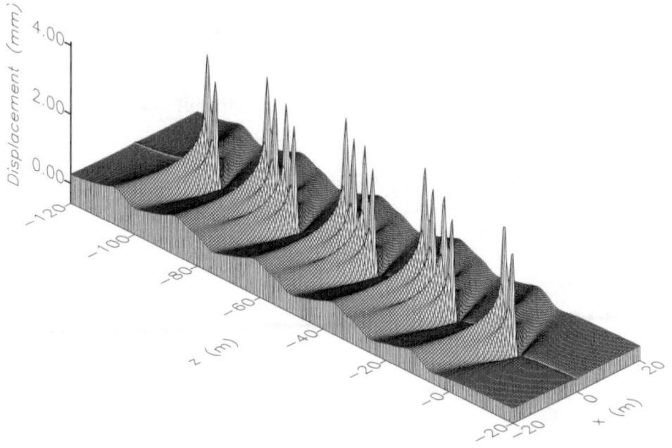

(a)

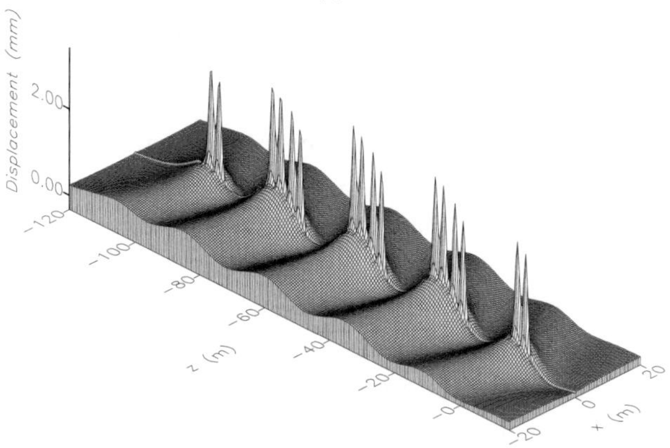

(b)

Fig. 9.44 Displacement field for a moving train with speed c = 100 m/s and $f_0 = 0$ with: (a) no isolation; (b) WIB.

effective among the three. From the transfer function plotted for the location x = 10 m in Fig. 9.46, it is evident that the WIB is rather ineffective in reducing the amplitude in the low frequency range, due to the fact that the dimensions of the WIB are too small compared with the wavelengths of low-frequency vibrations. Note that for isolating the vibrations between f = 10 Hz and 50 Hz, the WIB is quite effective, due to the fact that for frequencies in this range, the WIB behaves like an artificial bedrock installed at the depth of H = 1 m, of which the cut-off frequency is around 50 Hz for compressional waves.

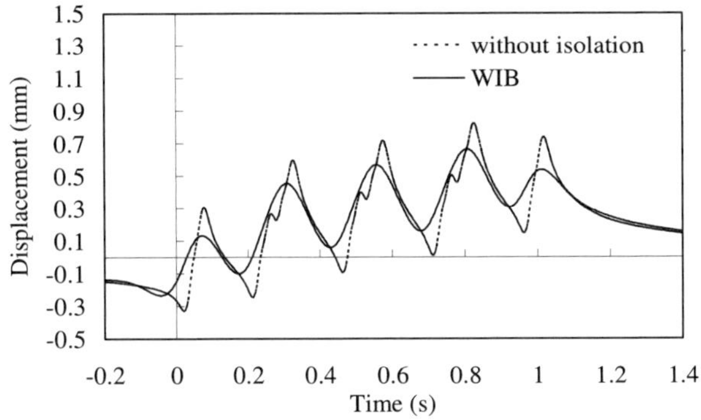

(a)

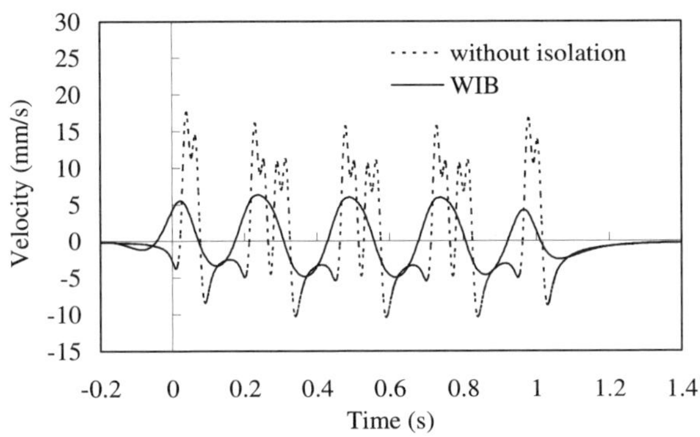

(b)

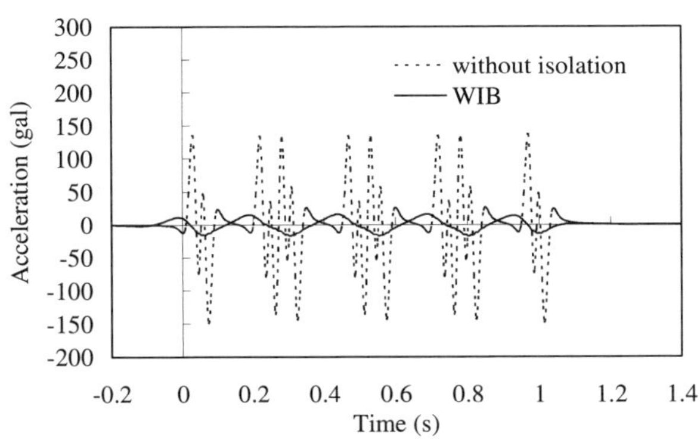

(c)

Fig. 9.45 Screening efficiency of WIB on vertical responses for $x = 10$ m and $c = 100$ m/s: (a) displacement; (b) velocity; (c) acceleration.

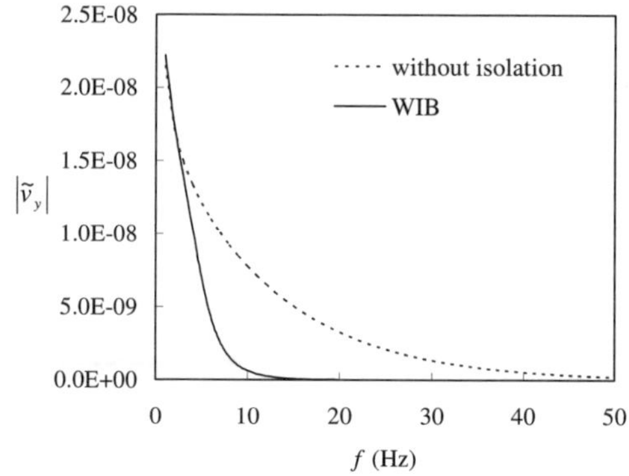

Fig. 9.46 Influence of WIB on the transfer function of vertical response for x = 10 m and c = 100 m/s.

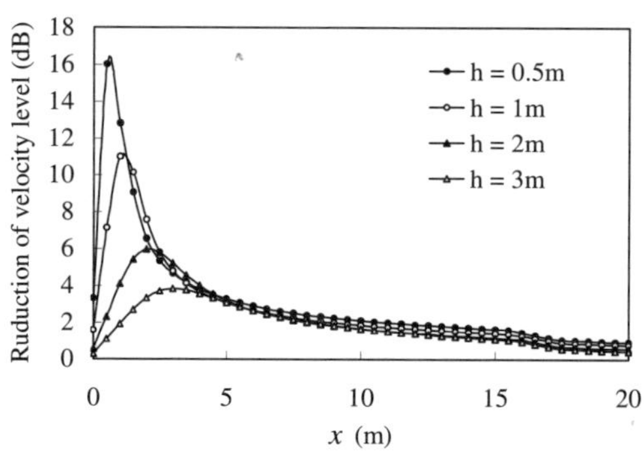

(a)

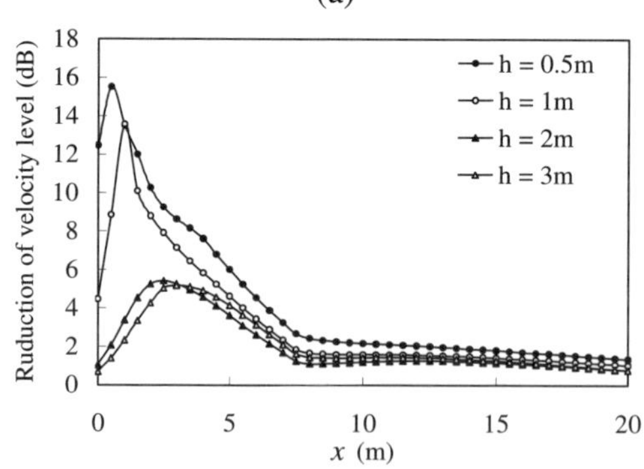

(b)

Fig. 9.47 Effect of block depth on velocity reduction for c = 70 m/s (WIB): (a) vertical; (b) horizontal.

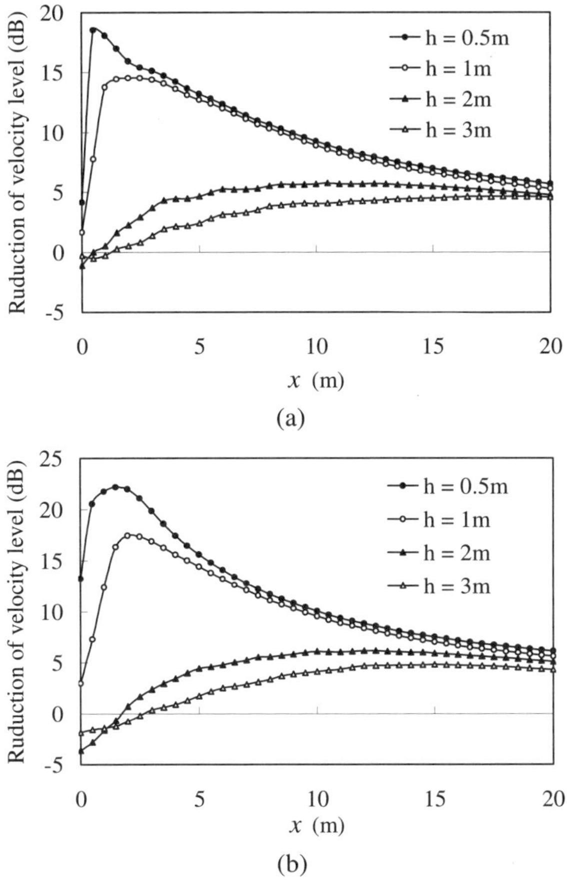

Fig. 9.48 Effect of block depth on velocity reduction for c = 100 m/s (WIB): (a) vertical; (b) horizontal.

9.5.1.2 *Effect of depth of WIB*

By changing the block depth h, the reduction of velocity level computed for the vertical and horizontal responses of the ground surface has been plotted in Figs. 9.47(a) and (b), respectively, for the train speed c = 70 m/s. Clearly, at locations near the source, the shallower the block, the better the effect of isolation is. The same effect remains generally true, but less pronounced, for locations at a distance away from the source, say, for regions with x > 10 m, which are too far for the WIB (with width e = 4 m)

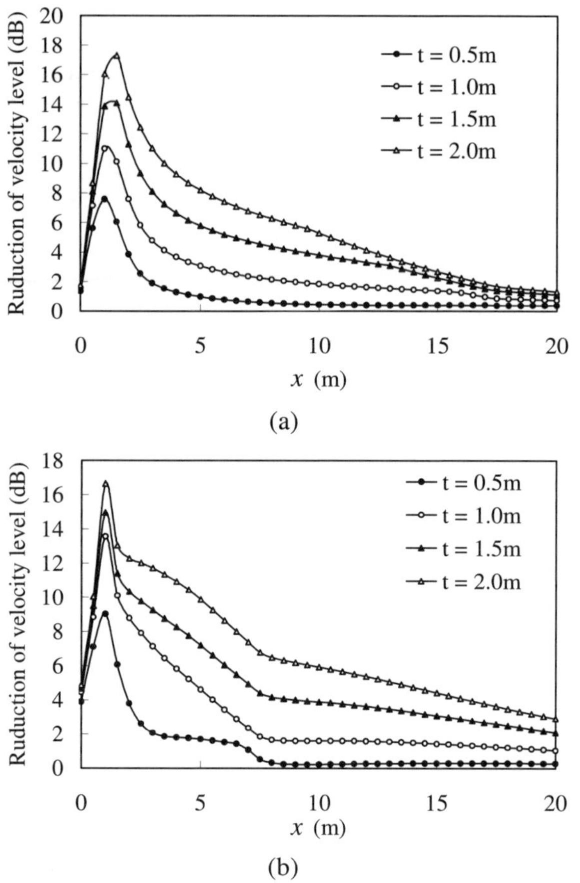

Fig. 9.49 Effect of block thickness on velocity reduction for c = 70 m/s (WIB): (a) vertical; (b) horizontal.

to be effective. For a speed greater than the critical speed, i.e., for c = 100 m/s, the effect of block depth was given in Fig. 9.48, which reveals a trend similar to that for c = 70 m/s, except that the influence of block depth is much more pronounced for the speed c = 100 m/s.

9.5.1.3 *Effect of thickness of WIB*

The velocity reductions for the vertical and horizontal responses of the ground surface computed for block thickness varying from t = 0.5 m to

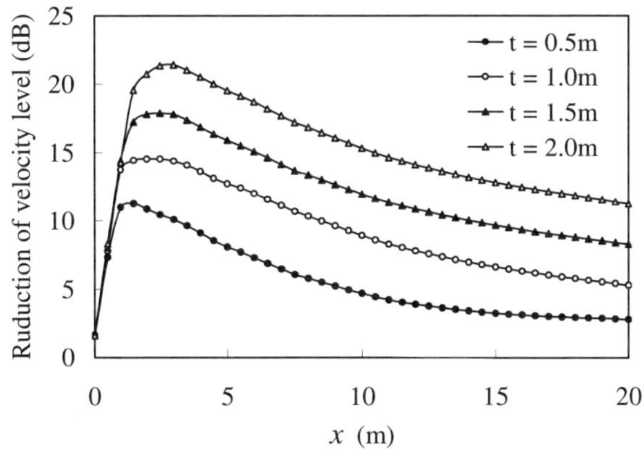

(a)

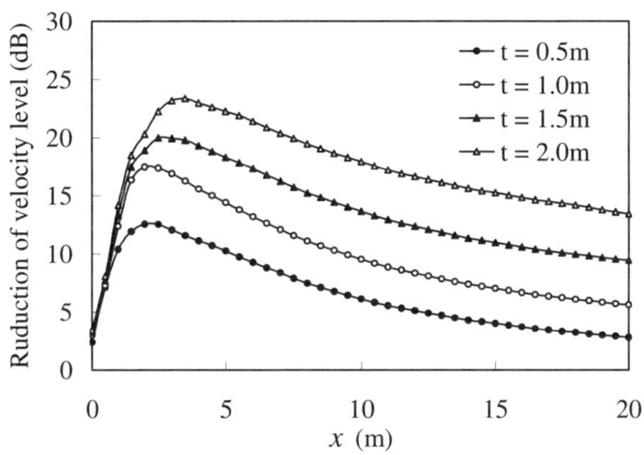

(b)

Fig. 9.50 Effect of block thickness on velocity reduction for c = 100 m/s (WIB): (a) vertical; (b) horizontal.

2.0 m, has been plotted in Figs. 9.49(a) and (b), respectively, for c = 70 m/s. Evidently, the influence of block thickness is rather significant. An increase in the block thickness can result in drastic improvement of the efficiency of isolation, which is especially true for c = 100 m/s, as can be seen from Fig. 9.50. Such a result is not surprising, if one realizes that the purpose of installing a WIB is to produce the effect of an artificial bedrock. Thus, a thicker block will make the WIB perform much like the real bedrock and become more effective in restraining the waves from traveling outward.

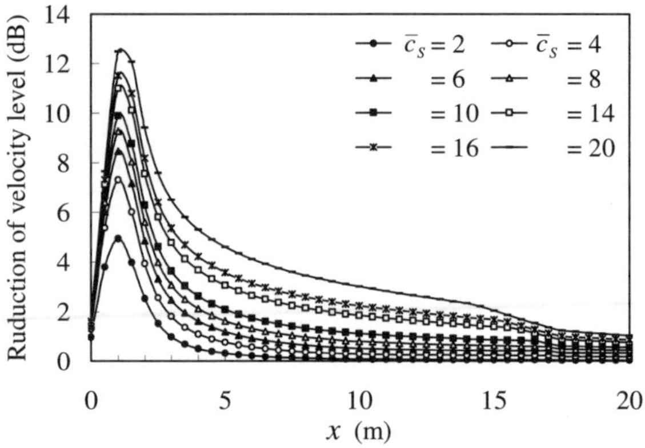

(a)

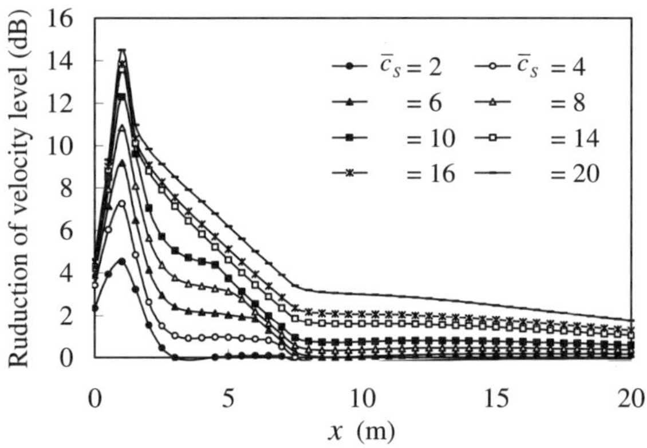

(b)

Fig. 9.51 Effect of ratio of S-wave speeds $\overline{c}_s$ on velocity reduction for c = 70 m/s (WIB with $\overline{c}_s > 1$): (a) vertical; (b) horizontal.

9.5.1.4 *Effect of shear wave speed of WIB*

Let c_S' denote the S-wave speed of the WIB, the reduction of velocity level calculated for different ratios of S-wave speeds $\overline{c}_S = c_S' / c_S$ has been plotted in Figs. 9.51 and 9.52 for the two speeds c = 70 m/s and 100 m/s, respectively. As can be seen, for a stiffer WIB ($\overline{c}_S$ > 1), an increase in the speed ratio $\overline{c}_S$ can result in better efficiency of isolation for trains moving at both speeds, which is similar to that observed for the in-filled trenches. It should be noted that because of its limited length, the WIB

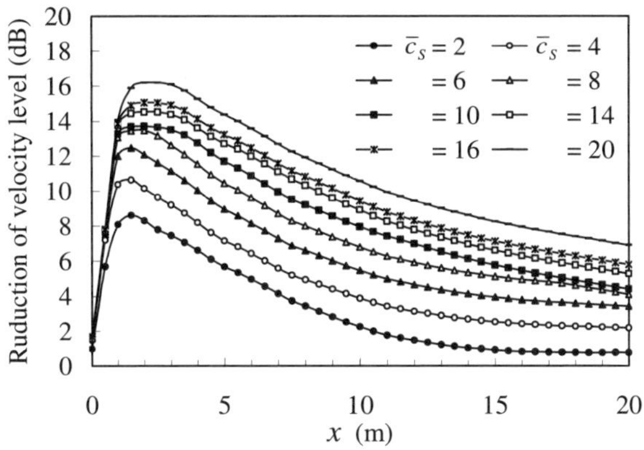

(a)

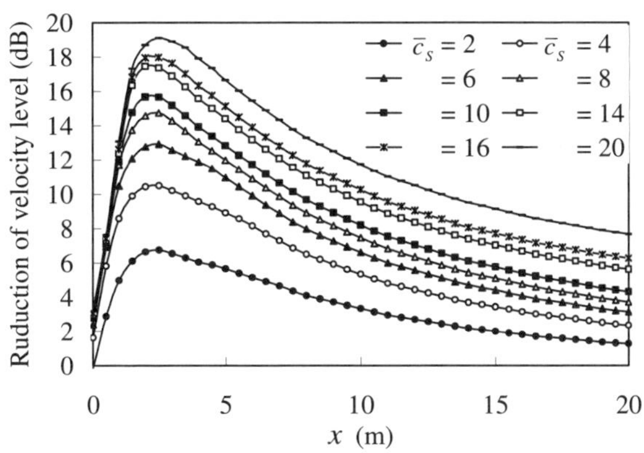

(b)

Fig. 9.52 Effect of ratio of S-wave speeds $\bar{c}_s$ on velocity reduction for c = 100 m/s (WIB with $\bar{c}_s > 1$): (a) vertical; (b) horizontal.

shows a decline in its shielding effect for regions away from the source, e.g., with a distance of over 8 m, for the sub-critical speed c = 70 m/s, as was revealed by Fig. 9.51(a).

On the other hand, for a softer WIB ($\bar{c}_S < 1$), say, with a mass density ρ of 150 kg/m^3, the effect of $\bar{c}_S$ has been given in Figs. 9.53 and 9.54 for the two speeds. Obviously, the inclusion of a softer WIB is not good at all for the purpose of reducing vibrations, which therefore should always be avoided. By averaging the results presented in Figs. 9.51-9.54 over the distance from x = 2 m to 20 m, the average reduction of velocity

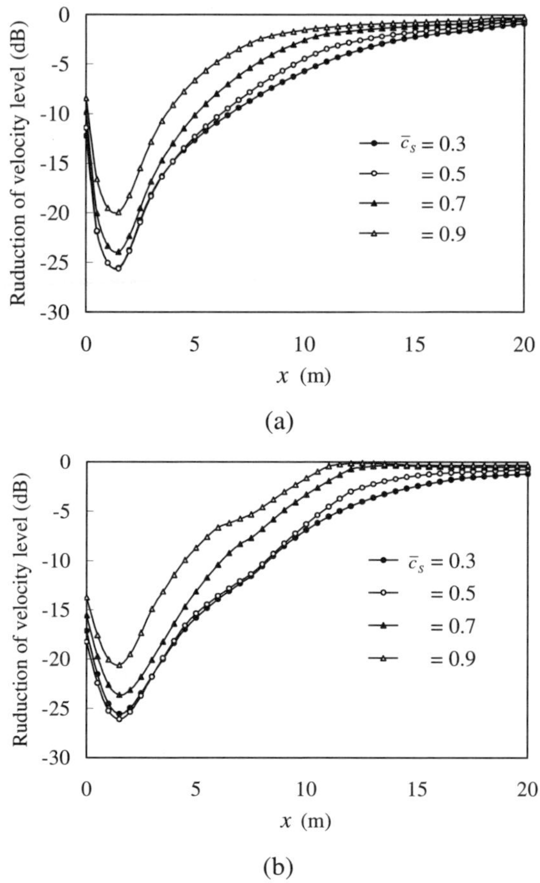

Fig. 9.53 Effect of ratio of S-wave speeds $\overline{c}_s$ on velocity reduction for c = 70 m/s (WIB with $\overline{c}_s < 1$): (a) vertical; (b) horizontal.

level was plotted against the impedance ratio IR in Figs. 9.55 and 9.56 for the two speeds. Evidently, the WIB is effective only when the material used is stiffer than the surrounding soils, i.e., with IR > 1.

9.5.2 *Moving loads with self oscillation*

Assume that the wheel loads have a self oscillation frequency of $f_0 = 20\,\text{Hz}$ and the train speed is c = 70 m/s. The displacement field computed for this case before and after installation of the WIB has been

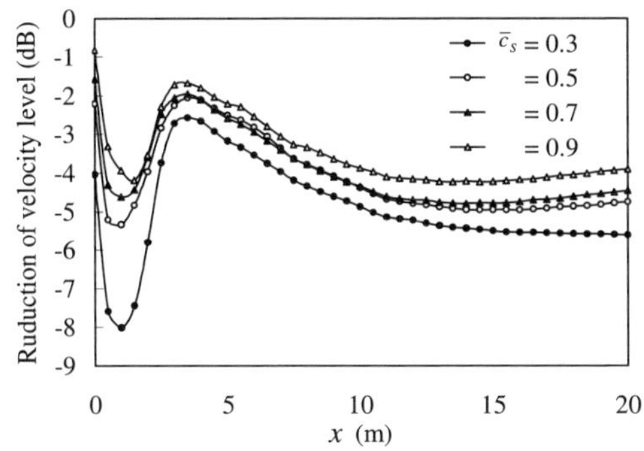

(a)

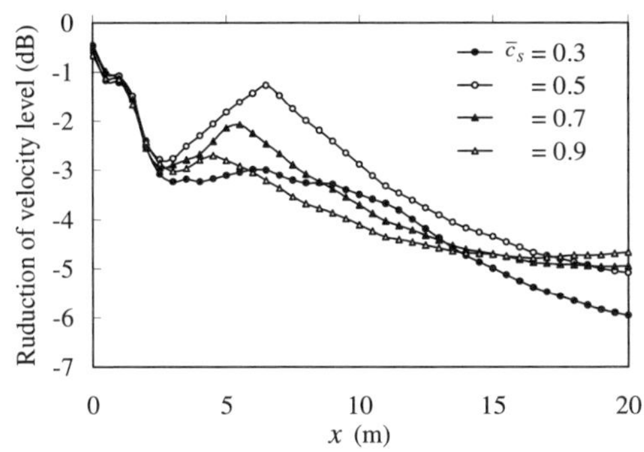

(b)

Fig. 9.54 Effect of ratio of S-wave speeds $\bar{c}_s$ on velocity reduction for c = 100 m/s (WIB with $\bar{c}_s$ < 1): (a) vertical; (b) horizontal.

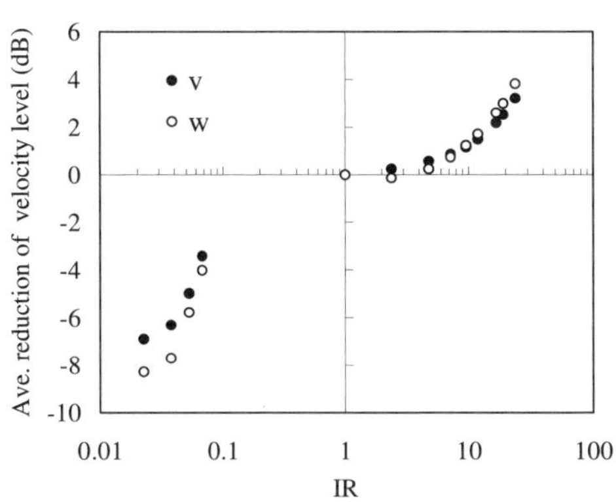

Fig. 9.55 Effect of impedance ratio on average reduction of velocity level for c = 70 m/s (WIB).

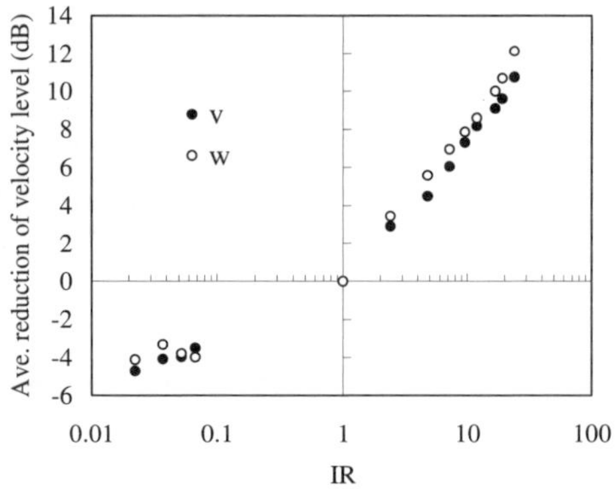

Fig. 9.56 Effect of impedance ratio on average reduction of velocity level for c = 100 m/s (WIB).

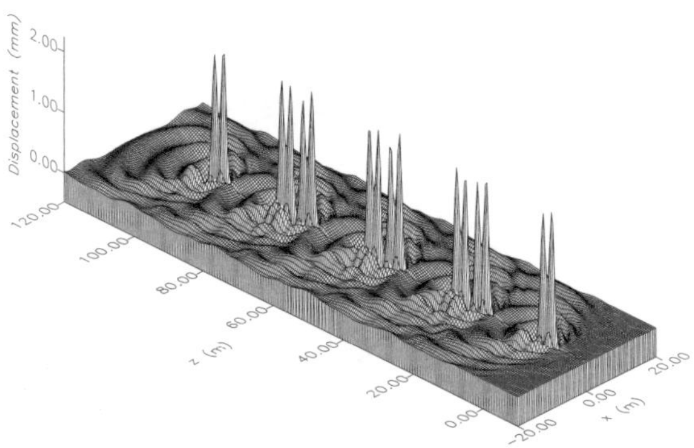

(a)

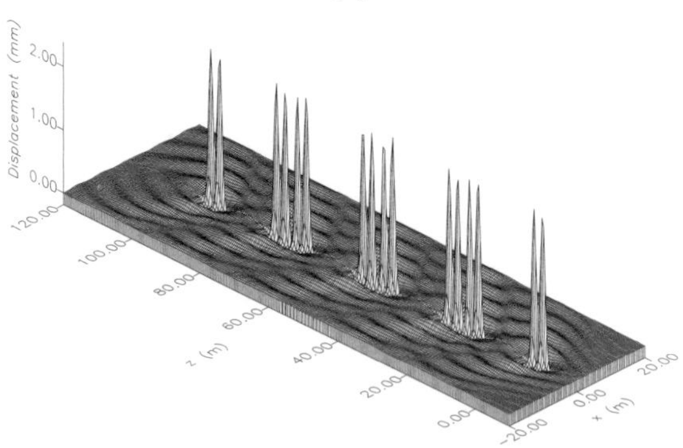

(b)

Fig. 9.57 Displacement field for a moving train with speed c = 70 m/s and f_0 = 20 Hz with: (a) no isolation; (b) WIB.

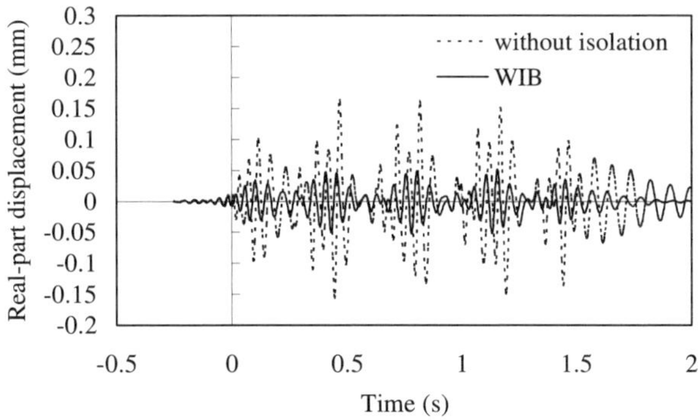

(a)

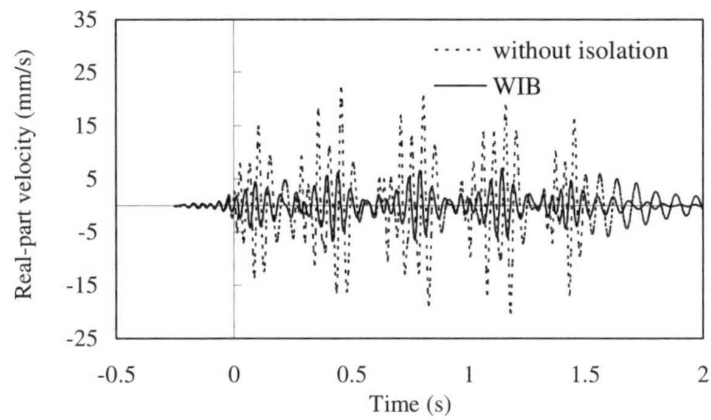

(b)

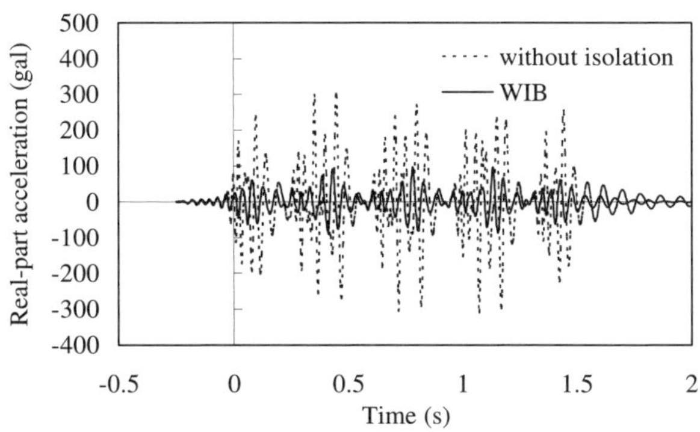

(c)

Fig. 9.58 Screening efficiency of WIB on vertical responses for x = 10 m, c = 70 m/s and f_0 = 20 Hz: (a) displacement; (b) velocity; (b) acceleration.

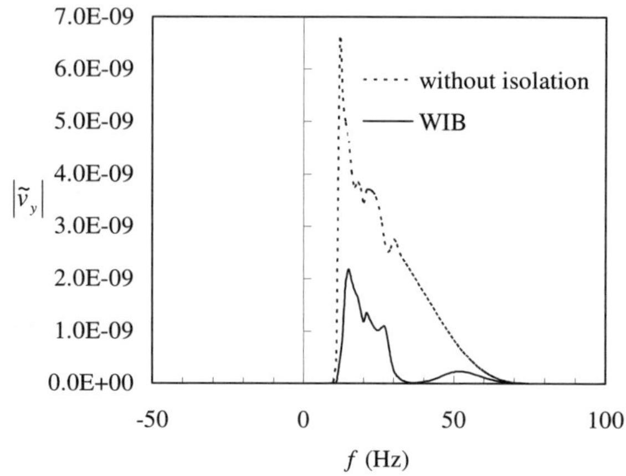

Fig. 9.59 Influence of WIB on the transfer function of vertical response for x = 10 m, c = 70 m/s, and f_0 = 20 Hz.

given in Figs. 9.57(a) and (b), respectively. As can be seen, the presence of the WIB can only reduce the displacement response slightly. The effect of the WIB can also be examined from the responses plotted for the location at x = 10 m in Fig. 9.58, which indicates that the reduction effect of the WIB is about the same for the displacement, velocity and acceleration. From the transfer function plotted in Fig. 9.59 for the location x = 10 m, one observes that the WIB can significantly reduce the response below f = 50 Hz. However, with the decrease of f, the screening effect declines. This phenomenon can be easily conceived, if one realizes that as the frequency becomes smaller and the wavelength becomes longer, the dimensions of the WIB become too small to trap the outgoing waves.

In order to investigate the effect of self oscillation frequencies f_0 of the train loads on the screening efficiency of the WIB, four different self oscillation frequencies f_0, i.e., f_0 = 5, 10, 20, 30 Hz, are considered for the train moving at c = 70 m/s. Correspondingly, the vertical and horizontal velocity reductions of the ground surface have been plotted in Figs. 9.60(a) and (b), respectively. By averaging the reduction values over the distance from x = 2 m to 20 m, the average reduction of velocity level for the vertical (v) and horizontal response (w) against the self oscillation frequency f_0 has been plotted in Fig. 9.61. From Figs. 9.60

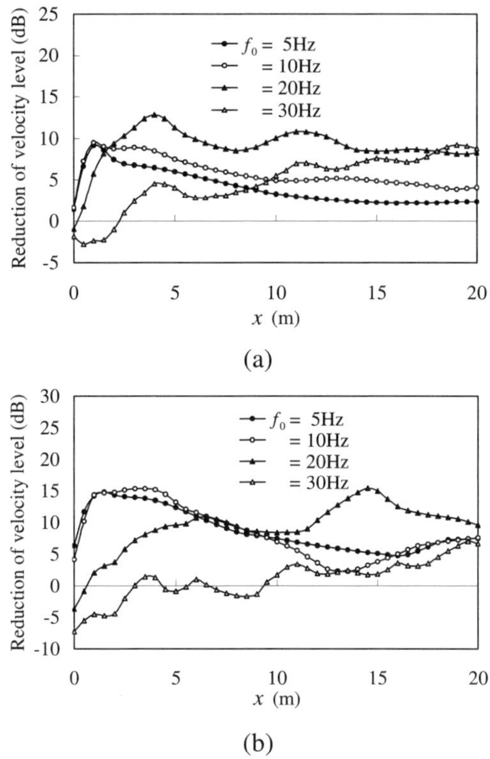

Fig. 9.60 Effect of self oscillation frequency f_0 on velocity reduction for c = 70 m/s (WIB): (a) vertical; (b) horizontal.

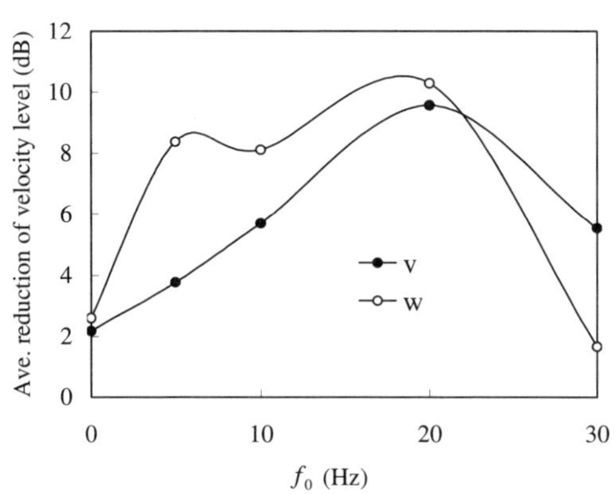

Fig. 9.61 Effect of self oscillation frequency f_0 on the average reduction of velocity level for c = 70 m/s (WIB).

and 9.61, one observes that the performance of the WIB reaches its maximum when $f_0 = 20$ Hz, below which a decrease of f_0 is accompanied by a reduction in the efficiency of isolation. The reason is that the amplitude of the transfer function in the low-frequency range can hardly be reduced by the WIB, as was explained previously. The performance of the WIB appears to be quite poor at $f_0 = 30$ Hz. This is primarily due to the fact that the main frequency content induced by $f_0 = 30$ Hz and $c = 70$ m/s, i.e., 17-120 Hz, goes generally beyond the cut-off frequency of a bedrock with depth $H = 1$ m, i.e., around 50 Hz.

The performance of the three wave barriers in reducing train-induced vibrations with non-zero oscillating f_0 for a train moving at the sub-critical speed $c = 70$ m/s can be evaluated through comparison of Fig. 9.17 for the open trenches, Fig. 9.39 for the in-filled trenches, and Fig. 9.61 for the WIB. It is easy to see that the isolation efficiency of open trenches is the best among the three wave barriers, especially for reducing the vertical responses.

9.6 Comparison and Discussion

In the preceding sections, parametric studies for the effectiveness of the open trenches, in-filled trenches and WIB have been performed, assuming either non-pulsating moving loads or moving loads with a single oscillation frequency f_0. In this section, a more realistic representation of the train loads moving over the railway that includes both the static and dynamic terms will be assumed and adopted to investigate the screening efficiency of the three barriers. The following is a general expression for the moving train loads:

$$P(x=0, y=0, z, t) = \phi(z-ct) f(t) \tag{9.6}$$

where the function $\phi(z)$ accounting for the real distribution of the wheel loads is the same as the one used in the previous sections; the train speed c is taken to be sub-critical, i.e., 70 m/s (252 km/hr), for the present case; and the function $f(t)$ represents the contact forces between the wheels and rails, which move with speed c.

Instead of treating $f(t)$ as a static or a dynamic term with a single frequency, as was done previously, we shall assume the oscillation function $f(t)$ to be composed of multi frequencies, given in the following form:

$$f(t) = 1 + \frac{1}{10}\left[\cos(10\pi t) + \cos(20\pi t) + \cos(40\pi t) + \cos(60\pi t)\right] \quad (9.7)$$

which includes a static term contributed mainly by the wheel load and four dynamic terms with the vibration frequencies: 5 Hz, 10 Hz, 20 Hz and 30 Hz. In (6.7), the weight of the dynamic terms is assumed to be 1/10, which is usually smaller than the static term.

For the train loads with the multi-frequency oscillation function given in Eq. (9.7), by adopting the typical parameters specified in the preceding three sections for the open trenches, in-filled trenches and WIB, the effectiveness of the three wave barriers in reducing the displacement, velocity and acceleration at location x = 10 m has been presented in Figs. 9.62-9.64. As can be seen from Fig. 9.62, all the three barriers are not effective for reducing the displacement, due to the fact that the contribution to the displacement comes mainly from the static term, i.e., the vehicle weight. This is consistent with what we observed from the previous parametric studies, that is, for train speeds lower than the critical speed, all the three barriers are not effective for reducing the displacement.

From Figs. 9.63 and 9.64, one observes that all the three barriers are generally effective for reducing the velocity and acceleration responses for the present train loads with the multi-frequency oscillation as given in Eq. (9.7). Moreover, the open trenches appear to be most effective among the three barriers. Such a result is consistent with our previous finding that for the cases with single oscillation frequencies f_0, the contribution by the moving static term to the velocity and acceleration responses is rather small, compared with that by the moving dynamic term. As such, the responses of velocity and acceleration are dominated by the dynamic term for the present load case considered.

Another trend revealed by these figures is that open trenches isolate mainly waves of the high-frequency components. Because of this, distinct low-frequency waves are left in the velocity response of Fig. 9.63(a). In contrast, the WIB is effective for isolating waves of the middle-frequency range, i.e., from 10 Hz to 20 Hz. As a result, both the

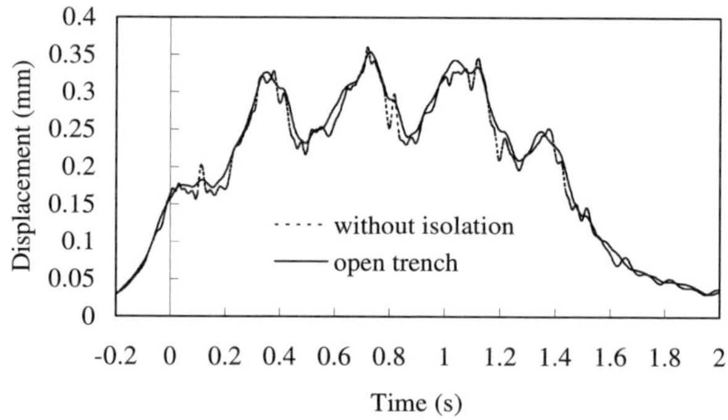

(a)

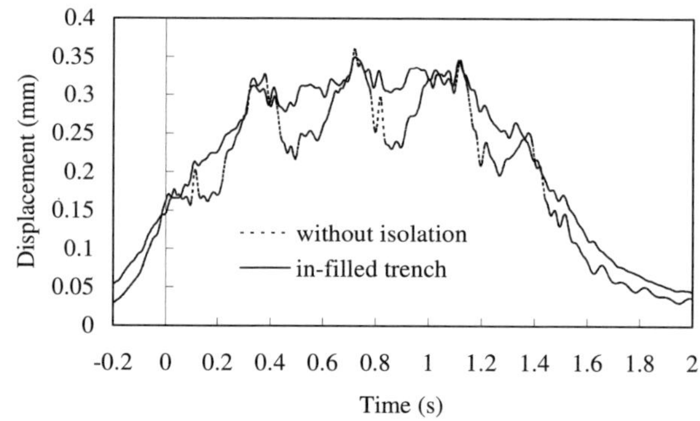

(b)

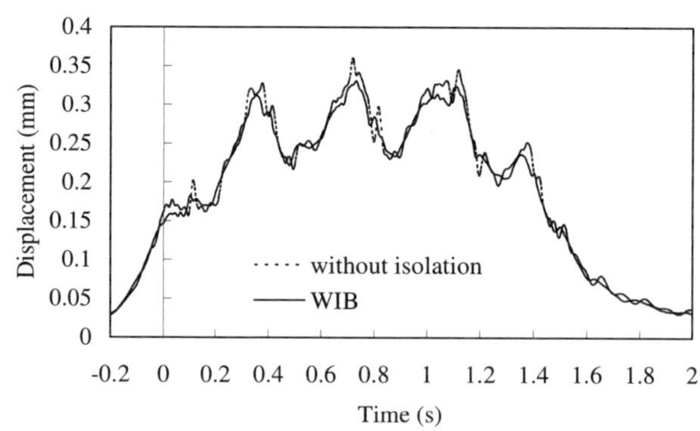

(c)

Fig. 9.62 Effectiveness of the three wave barriers in reducing displacement at x = 10 m: (a) open trenches; (b) in-filled trenches; (c) WIB.

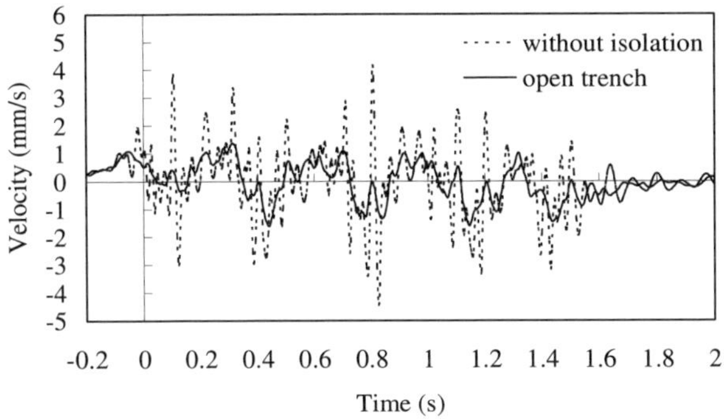

(a)

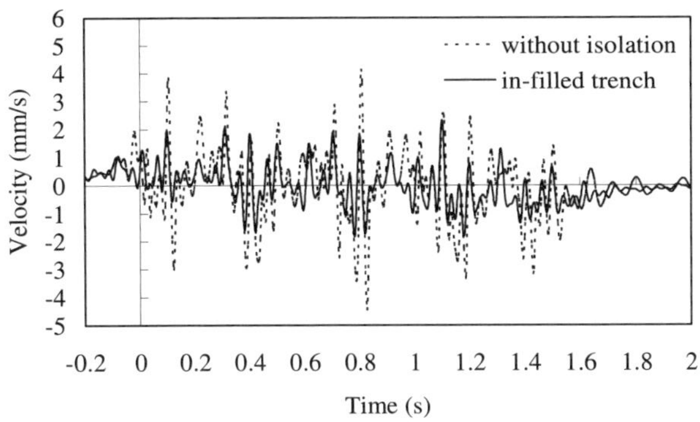

(b)

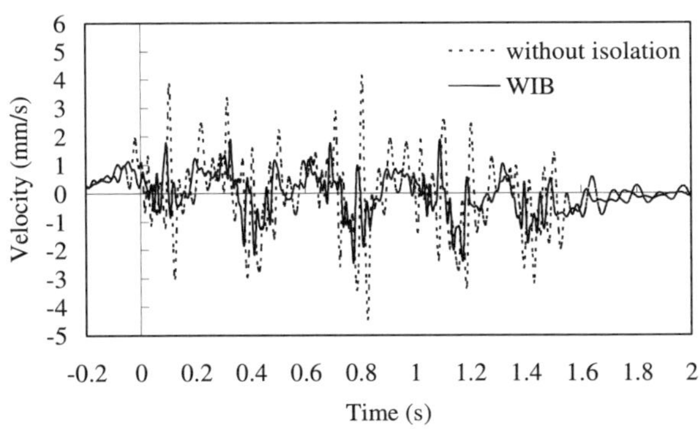

(c)

Fig. 9.63 Effectiveness of the three wave barriers in reducing velocity at x = 10 m: (a) open trenches; (b) in-filled trenches; (c) WIB.

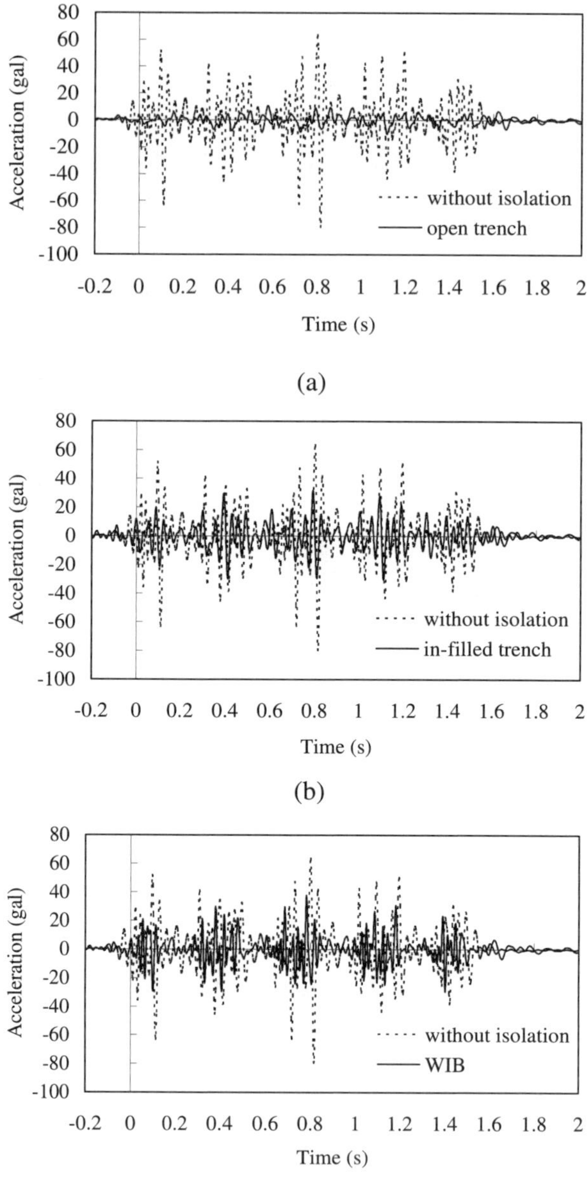

Fig. 9.64 Effectiveness of the three wave barriers in reducing acceleration at $x = 10$ m: (a) open trenches; (b) in-filled trenches; (c) WIB.

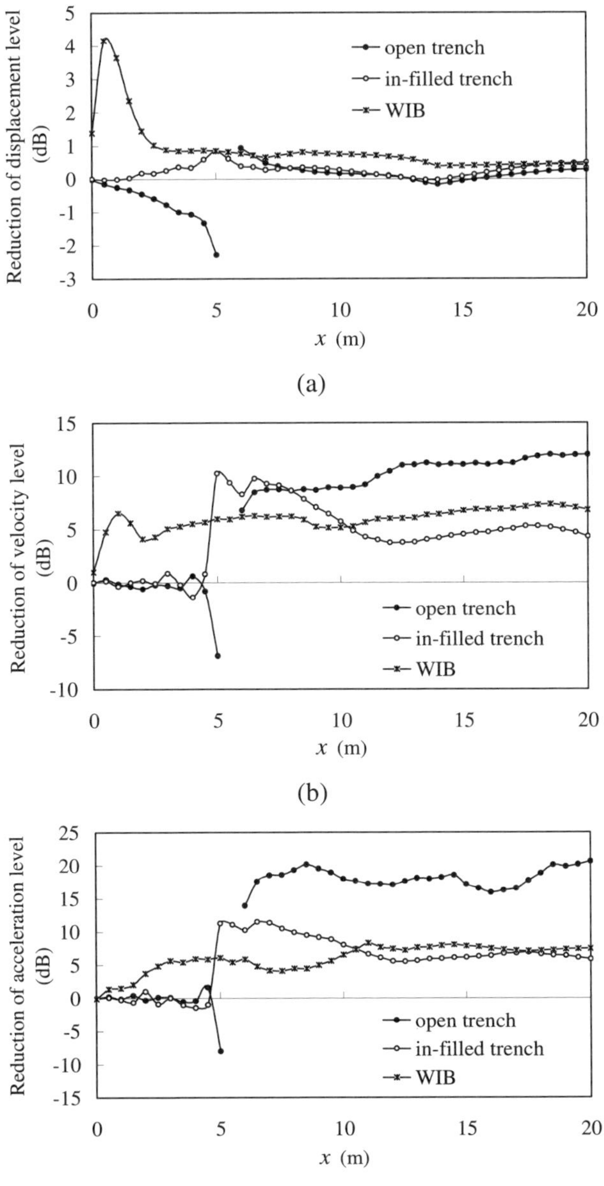

Fig. 9.65 Comparison between the isolation efficiency of three wave barriers: (a) vertical displacement; (b) vertical velocity; (c) vertical acceleration.

high-frequency and low-frequency waves are left in the velocity response of Fig 9.63(c).

Finally, the reduction of vibration level for the displacement, velocity and acceleration level was computed for all the three barriers and plotted with respect to the distance x in Fig. 9.65. Evidently, for the present train loads with the multi-frequency oscillation function given Eq. (9.7), the open trench is most effective among the three wave barriers, especially for reducing the acceleration response.

9.7 Concluding Remarks

In this chapter, a parametric study for investigating the major factors affecting the effectiveness of three wave barriers, i.e., the open trenches, in-filled trenches, and wave impeding block, in isolating the train-induced vibrations has been performed. The conclusions acquired from the parametric study can be summarized as follows:

(1) For isolating the vibrations induced by moving static loads (with $f_0 = 0\,\text{Hz}$), in-filled trenches appear to be the best choice. For isolating vibrations induced by moving dynamic loads (with $f_0 \neq 0\,\text{Hz}$), the performance of open trenches perform is the best. When both the static and dynamic terms of moving loads are taken into account, open trenches remain the most effective in reducing the velocity and acceleration responses.

(2) For the case with moving static loads, all the three wave barriers perform better in reducing the vibrations induced by a train moving at super-critical speeds than at sub-critical speeds, with respect to the Rayleigh wave speed.

(3) For the case with moving static loads, the key geometric parameter for both the open and in-filled trenches is depth, for the WIB it is thickness. The efficiency of isolation can be improved through an increase of the trench depth or WIB thickness.

(4) For the case with moving static loads, stiffer in-filled trenches or WIB with respect to the surrounding soil perform better than softer ones.

(5) For the case with moving dynamic loads, the performance of both the in-filled and open trenches improves with the increase of the self oscillation frequency f_0, but this phenomenon is more pronounced for open trenches. The WIB is only effective in isolating the vibrations with frequencies below the cut-off frequency, as determined by its depth, and the vibrations with wavelengths not longer than the length of the WIB itself.

Chapter 10

Soil Vibrations Caused by Underground Moving Trains

The wave propagation problems caused by the underground moving trains are analyzed by the 2.5D finite/infinite element approach. The near field of the half-space, including the tunnel and parts of the soil, is simulated by finite elements, and the far field extending to infinity by infinite elements. The train is simulated as a sequence of wheel loads moving at constant speeds. Using the present approach, a 2D profile with three degrees per node is used to simulate the 3D behavior of the half-space, which is valid for the case when the material and geometry of the system can be regarded as invariant along the tunnel direction. The factors considered in the analysis of ground-borne vibrations include the damping ratio and stratum depth of the supporting soils, the depth and thickness of the tunnel, and the speed and excitation frequency of the moving loads. It was found that the moving loads with non-zero self oscillation frequencies can induce significantly higher vibrations than the static moving loads. The effect of stratum depth depends highly on the self oscillation frequency of the moving loads. It is concluded that the ground surface vibrations can be greatly reduced if the tunnels are constructed of stiff soils. Other conclusions useful to practical engineering design are contained in the parametric study.

10.1 Introduction

In metropolitan areas where the road traffic suffers seriously from daily congestion, underground trains have emerged as a more efficient way for mass transportation, as revealed by the increasing popularity of subway systems in major cities all over the world. However, ground-borne vibrations due to subway trains have sometimes reached the level that cannot be tolerated by residents living in buildings located alongside the subways (Shyu *et al.* 2002). To address such a problem, the project CONVURT was launched by the European Union, with the objective of predicting and controlling the noise and vibration in buildings induced by the underground rail traffic. As part of this effort, in-situ vibration measurements in Paris and London were carried out (Chatterjee *et al.* 2003; Degrande *et al.* 2006a). Also, approaches for predicting vibrations caused by metro trains moving through the tunnel were developed (Gupta *et al.* 2007), e.g., using a semi-analytical pipe-in-pipe model (Forrest and Hunt 2006a,b) and a coupled periodic finite element-boundary element model (Clouteau *et al.* 2005; Degrande *et al.* 2006b). Clearly, ground-borne vibrations have become an issue of great concern, which will continuously attract the attention of researchers and engineers worldwide.

The ground vibration induced by underground moving trains is a complicated dynamic problem due to the involvement of a number of factors along the paths of wave propagation, including the load generation mechanism of the train-track system, the geometry and location of the tunnel structure, and the irregularity of soil layers, etc. Previously, numerous researches on the ground-borne vibrations due to subway trains were conducted by field measurement (Vadillo *et al.* 1996; Degrande *et al.* 2006a) and empirical or semi-empirical prediction models were proposed (Kurzweil 1979; Trochides 1991; Melke 1998). These studies provide practical reference for estimation of relevant wave propagation problems. However, most of these studies were performed for a specific condition, thereby suffering from the lack of generality. On the other hand, concerning the techniques of simulation, most previous works have been based on the 2D models (Balendra *et al.* 1991; Yun *et al.* 2000; Metrikine and Vrouwenvelder 2000a; Yang *et al.* 2008). The results obtained by the 2D models are approximate and qualitative in nature, but have been frequently adopted by designers. One drawback with the 2D models is that the effect of train speeds was overlooked.

Enhanced by the advent of high-performance computers, 3D simulation methods were also used by researchers, e.g., Gardien and

Stuit (2003), Park *et al.* (2004); Forrest and Hunt (2006), Andersen and Jones (2006), etc. However, a full 3D modeling is still too time-consuming using the high-performance computers available nowadays. In fact, by considering the invariance of soil-tunnel system along the load-moving direction, a 3D response can be obtained from the 2D profile by the Fourier transformation technique. By such an idea, Yang and Hung (2001) proposed the 2.5D finite/infinite element approach for modeling the visco-elastic bodies subjected to moving loads, as was presented in Chapter 7. The wavenumber finite/boundary element methods was proposed by Sheng *et al.* (2005) for predicting the vibration spectra for circular tunnel structures. By assuming the geometry to be periodic in the tunnel direction, the coupled finite/boundary element model was proposed by Degrande *et al.* (2006b) using the Floquet transform, instead of the Fourier transform.

Evidently, there is a tendency of using the 2D profile to generate the 3D response, for its relative efficiency and accuracy, by taking advantage of the invariance or periodicity of the geometry along the load-moving direction. However, most of these studies were carried out for a specific case and none has performed a parametric study. To fill such a gap, a parametric study will be performed in this chapter, using the 2.5D finite/infinite element approach presented in Chapter 7. By such an approach, the 3D time-history response of the soil-tunnel system to the underground moving trains can be investigated. The parameters to be considered include the damping ratio and stratum depth of the supporting soils, the depth and thickness of the tunnel, and the speed and self oscillation frequency of the moving loads. The paper by Yang and Hung (2008) has been included as parts of the material presented in this Chapter. It should be noted that to reduce the amount of citations to previous chapters, while improving the logistics of presentation, some of the materials presented in previous chapters will just be repeated.

10.2 Problem Formulation and Basic Assumptions

The problem to be considered is schematically shown in Fig. 10.1, where a train with speed c is traveling in the underground tunnel along the

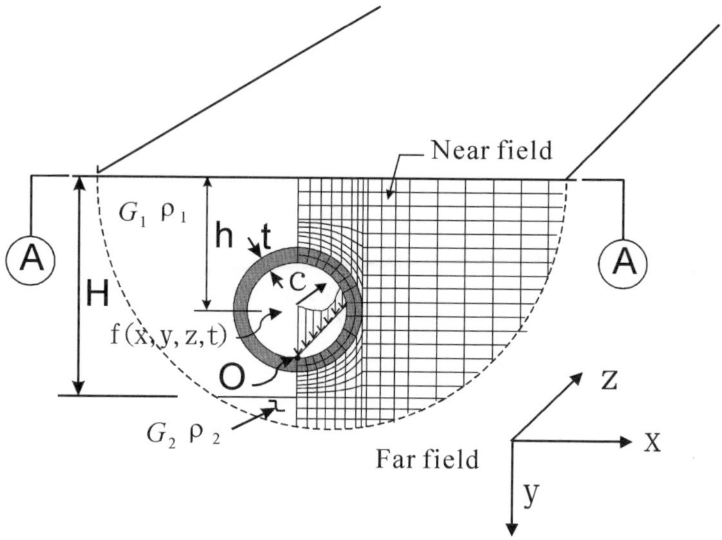

Fig. 10.1 Schematic representation of the soil-tunnel interaction system.

z-direction. To study the effect of soil layers, two configurations with and without bedrock underlying the top soil layer are considered. By assuming the material and geometry properties of the soil-tunnel system to be uniform along the load-moving direction, i.e., the z-axis, one may simulate the whole system by considering only a profile perpendicular to the z-axis, i.e., the profile A-A in Fig. 10.1. The profile A-A contains a near field of finite irregular region, as enclosed by the circular boundary line, and a semi-infinite far field. In this chapter, the near field containing the train loads, soils and tunnel structure is simulated by finite elements, while the far field with unbounded domain by infinite elements, as will be summarized in the section to follow.

As shown in Fig. 10.2, the moving train traveling underground along the z-direction can be simulated as a sequence of wheel loads, expressed as follows:

$$f(x,y,z,t) = \psi(x,y)\phi(z-ct)\exp(i2\pi f_0 t), \tag{10.1}$$

where $\phi(z)$ is the *load distribution function* of the train along the z-axis, as was given in Eq. (2.36), $\psi(x,y)$ represents the influence function of the moving loads on the xy plane, and the term $\exp(i2\pi f_0 t)$ is to account

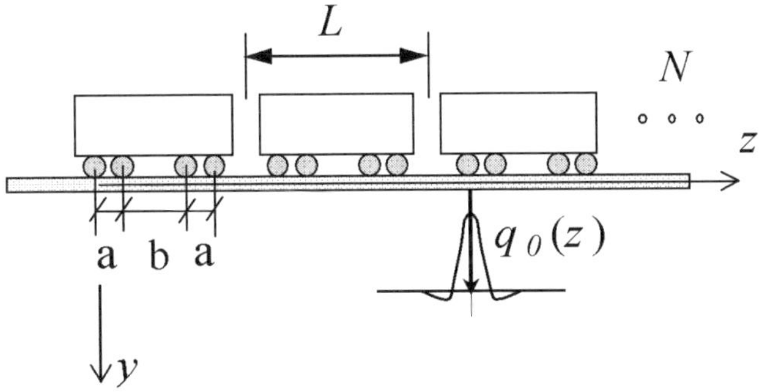

Fig. 10.2 Schematic representation of train induced loadings.

for the *self oscillation* of the moving loads, arising from the suspension system of train cars and rail unevenness. By and large, the term $\exp(i2\pi f_0 t)$ should be replaced by a general term $p(t)$ to account for the interaction between the wheels and rails. However, due to the complicated nature of interaction between the train, rails, sleepers, subgrades, and soils, such a function cannot be readily obtained. In this chapter, the self oscillation frequency f_0 is used to indicate the inherent frequency of the sprung mass or suspension system of train cars.

As indicated in Fig. 10.2, a train is assumed to contain a total of N identical cars, each of which has four axle loads of magnitude T. The load distribution function $\phi(z)$ of the train involved in Eq. (10.1) and explicitly given in Eq. (2.36) has been obtained as the superposition of the distribution function $q_0(z)$ associated with each of the axle loads, determined as the deflection curve of an infinite elastically supported beam subjected to the axle load T (Esveld 1989):

$$q_0(z) = \frac{T}{2\alpha}\exp\left(\frac{-|z|}{\alpha}\right)\left(\cos\frac{|z|}{\alpha} + \sin\frac{|z|}{\alpha}\right), \tag{10.2}$$

where α is the *characteristic length* of the beam, $\alpha = \left(4EI/s\right)^{1/4}$, with E denoting Young's modulus, I the moment of inertia of the beam, and s the spring constant of the elastic foundation on which the infinite beam is supported.

By performing the Fourier transformation to Eq. (10.1), one can express the external loads in frequency domain, $\tilde{f}(x, y, z, \omega)$, as

$$\tilde{f}(x,y,z,\omega)=\psi(x,y)\frac{1}{c}\exp(-ikz)\tilde{\phi}(-k), \tag{10.3}$$

in which $k=(\omega-2\pi f_0)/c$ and $\tilde{\phi}(k)$ is the Fourier transform of $\phi(z)$. On the other hand, by the inverse Fourier transformation, the external loading in time domain $f(x,y,z,t)$ can be recovered as

$$f(x,y,z,t)=\psi(x,y)\int_{-\infty}^{\infty}\frac{1}{c}\tilde{\phi}(-k)\exp(-ikz)\exp(i\omega t)d\omega. \tag{10.4}$$

Equation (10.4) shows that the external loading can be expressed as the summation of a series of harmonic components.

For a linear system, the final steady-state response in time domain can be obtained by superposing the response generated by each of the harmonic components. Let $H(i\omega)$ denote the response generated by the harmonic component $\psi(x,y)\exp(-ikz)\exp(i\omega t)$ at location (x',y'). The total response of the system in time domain can be written as

$$d(x',y',z,t)=\int_{-\infty}^{\infty}\frac{1}{c}\tilde{\phi}(-k)H(i\omega)\exp(-ikz)\exp(i\omega t)d\omega, \tag{10.5}$$

which contains information not only for the 2D profile, but also along the load-moving direction. In this chapter, the *complex response function* $H(i\omega)$ involved in Eq. (10.5) will be computed by the 2.5D finite/infinite element approach to be summarized below.

10.3 Formulation of 2.5D Finite/Infinite Element Method

Corresponding to Eqs. (10.4) and (10.5) for a harmonic load component $\psi(x,y)\exp(-ikz)\exp(i\omega t)$, the 3D time-history displacements of the system can be related to the displacements of the 2D profile as

$$\mathbf{u}(x,y,z,t)=\hat{\mathbf{u}}(x,y)\exp(-ikz)\exp(i\omega t), \tag{10.6}$$

where $k=(\omega-2\pi f_0)/c$ and $\hat{\mathbf{u}}(x,y)$ represents the displacement field of the 2D profile and $\mathbf{u}(x,y,z,t)$ the time-history displacements along the three axes. As can be seen, the term $\exp(-ikz)$ involving the variable z has been separated from the displacements $\hat{\mathbf{u}}(x,y)$. It follows that the originally 3D continuous solid can be represented by elements on the xy plane via the function $\hat{\mathbf{u}}(x,y)$.

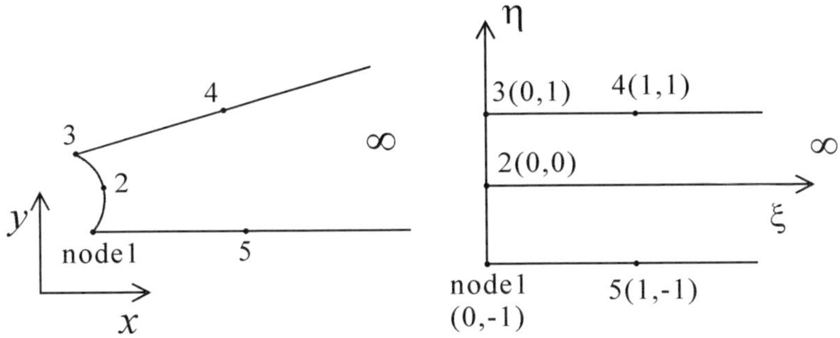

(a) (b)

Fig. 10.3 Infinite element: (a) global coordinates; (b) local coordinates.

The displacement $\hat{\mathbf{u}}(x, y)$ within each element can be interpolated as follows:

$$\hat{\mathbf{u}}(x, y) = \sum_{i=1}^{n} N_i \mathbf{u}_i, \tag{10.7}$$

in which N_i are the shape functions for the displacement of a plane element, n the number of nodes for each element, e.g., $n = 8$ for a quadratic 8-node (Q8) element. The coordinates x and y within the element can be expressed as

$$\begin{aligned} x &= \sum_{i=1}^{n} M_i x_i, \\ y &= \sum_{i=1}^{n} M_i y_i, \end{aligned} \tag{10.8}$$

where M_i are the shape functions for the coordinates. The shape functions in Eqs. (10.7) and (10.8) may be selected to be identical to those of the conventional plane finite/infinite elements.

As was stated previously, the near field of the soil-tunnel system, including the tunnel structure, will be modeled by finite elements and the far field with unbounded domain by infinite elements. The finite element adopted herein is the Q8 element, and the infinite element adopted is the one, shown in Fig. 10.3, all based on the 2.5D formulation. As for the infinite element, a *propagation function* $P(\xi)$ was included in the shape function along the infinite axis to account for the wave number and amplitude attenuation due to wave dispersion. Though the shape functions are the same for the 2D and 2.5D elements, the size of the element matrices is enlarged by 1.5 × 1.5 times due to inclusion of the third degree per node by the 2.5D approach. Besides, the stiffness matrices derived are complex and asymmetric, because of the existence of the term $\exp(-ikz)$ in the displacement field in Eq. (10.6) (see Chapter 7).

Substituting the displacement field in Eq. (10.6) into the equation of virtual work for each element, and summing the contribution of all the finite and infinite elements, the equation of motion for the system in the frequency domain can be obtained,

$$([K]-\omega^2[M])\{D\}=\{F\}, \tag{10.9}$$

in which $\{D\}$ and $\{F\}$ denote the amplitudes of the nodal displacements and applied loads, respectively, and $[M]$ and $[K]$ the global mass and stiffness matrices, both of which are obtained by assembly of their corresponding element matrices. From the structural equation in Eq. (10.9), the displacements $\{D\}$ in frequency domain can be solved. Next, by setting the transfer function $H(i\omega)$ equal to $\{D\}$, the time-domain response in Eq. (10.5) can be computed using the fast Fourier transformation. Meanwhile, by setting the transfer function $H(i\omega)$ in Eq. (10.5) equal to $i\omega\{D\}$, the velocities in time domain can be computed, too. Other details concerning application of the finite/infinite element approach used herein, including the guidelines for determining the amplitude decay factor and wave number for infinite elements used in different regions of the half-space, as well as the automatic mesh generation procedure, are available in Chapters 3 and 7.

10.4 Verification of the Present Approach

The finite/infinite element approach has been successfully applied to solving ground vibration problems due to surface moving loads in the previous chapters. To verify its applicability to solution of problems with underground moving trains, the steady-state response for the uniform elastic *infinite space* in Fig. 10.4(a), referred to as Case 1, and uniform

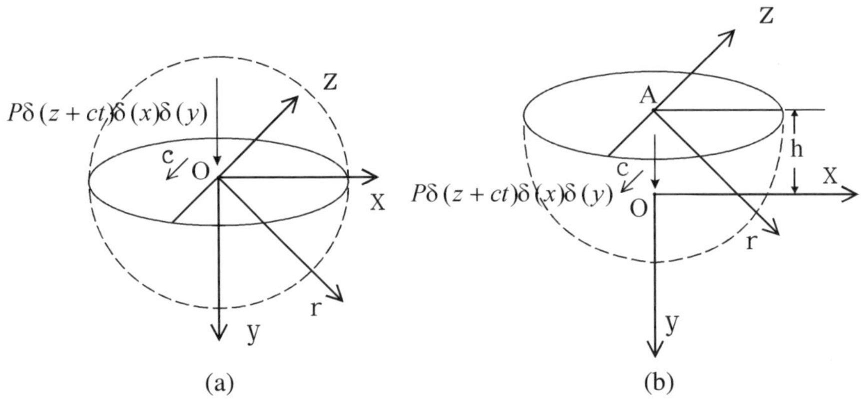

Fig. 10.4 Two cases for verification: (a) Case 1: uniform infinite space; (b) Case 2: uniform half-space.

visco-elastic half-space in Fig. 10.4(b), referred to as Case 2, subjected to a vertical buried point load P traveling at a sub-critical speed c will be studied and compared with the analytical solutions. Particularly, the capability of the infinite elements in dealing with the infinite space, rather than, half-space, will be demonstrated in Case 1.

The following properties are adopted for the two cases shown in Figs. 10.4(a) and (b): shear modulus $G = 1.154 \times 10^7$ N/m^2, Poisson's ratio $\nu = 0.3$, density $\rho_{\text{soil}} = 1900$ kg/m^3, and load moving speed $c = 40$ m/s. Accordingly, the P- and S-wave speeds are computed as 145.79 m/s and 77.93 m/s, and the Mach numbers related to the P- and S-waves are $M_1 = 0.274$ and $M_2 = 0.513$, respectively. The meshes adopted for the two cases were plotted in Fig. 10.5, where only half of the system is modeled due to symmetry considerations, and the far and near fields are separated by dotted lines. The far field is modeled by infinite elements, and the near field with a range of 4 m from the loading point O by finite elements.

For the uniform infinite space with no material damping, i.e., Case 1 in Fig. 10.4(a), the displacements u, v and w computed for point B ($x = 0$ m, $y = -0.5$ m, $z = 0$ m) and point C ($x = 3.5$ m, $y = -3.5$ m, $z = 0$ m) have been normalized and compared with the theoretical solutions by Frýba (1972) in Fig. 10.6, using the relations $U = (4\pi G/P)u$, $V = (4\pi G/P)v$ and $W = (4\pi G/P)w$. The displacement v denotes the vertical response

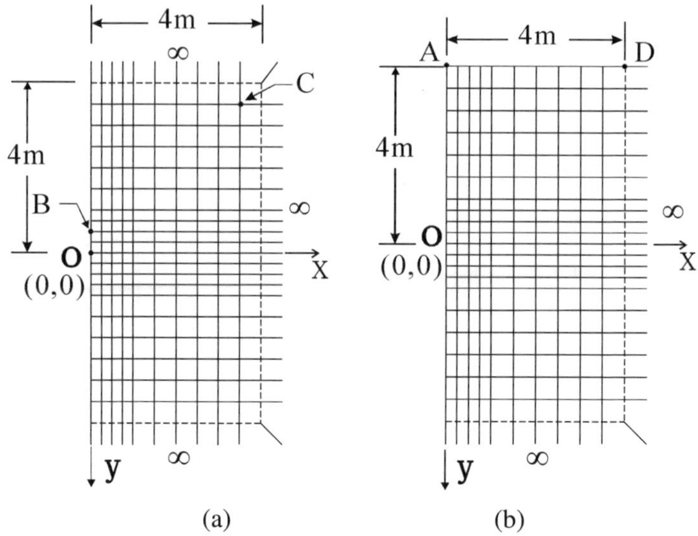

Fig. 10.5 Meshes for two cases of verification: (a) Case 1: uniform infinite space; (b) Case 2: uniform half-space.

along the y-axis, and (w, u) the horizontal responses along the z- and x-axis. The theoretical solution for the displacement v was also given in Eq. (1.18a) provided that z is replaced by z-ct. In Fig. 10.6, the time $t = 0$ corresponds to the instant at which the load passes through the point of the profile with $z = 0$. Evidently, the results computed agree excellently with the theoretical ones for both the locations near the source (point B) and near the boundary (point C).

By the reciprocity theorem, the responses obtained for an observation point (point A or point D) of the elastic half-space induced by a load applied at the source (point O) should be equal when the source and observation points are exchanged. Therefore, the ground surface response in time domain for Case 2 in Fig. 10.4(b) can be obtained from the analytical solution derived in Chapter 2 or Hung and Yang (2001), by which the underground responses due to a surface point load moving at a constant speed can be calculated analytically. For a vertical point load traveling at a depth of 4 m (point O), which is the case of concern here, the normalized displacements V_y and W_y at the ground surface point A should be equal to the vertical responses V_y and V_z at point O when

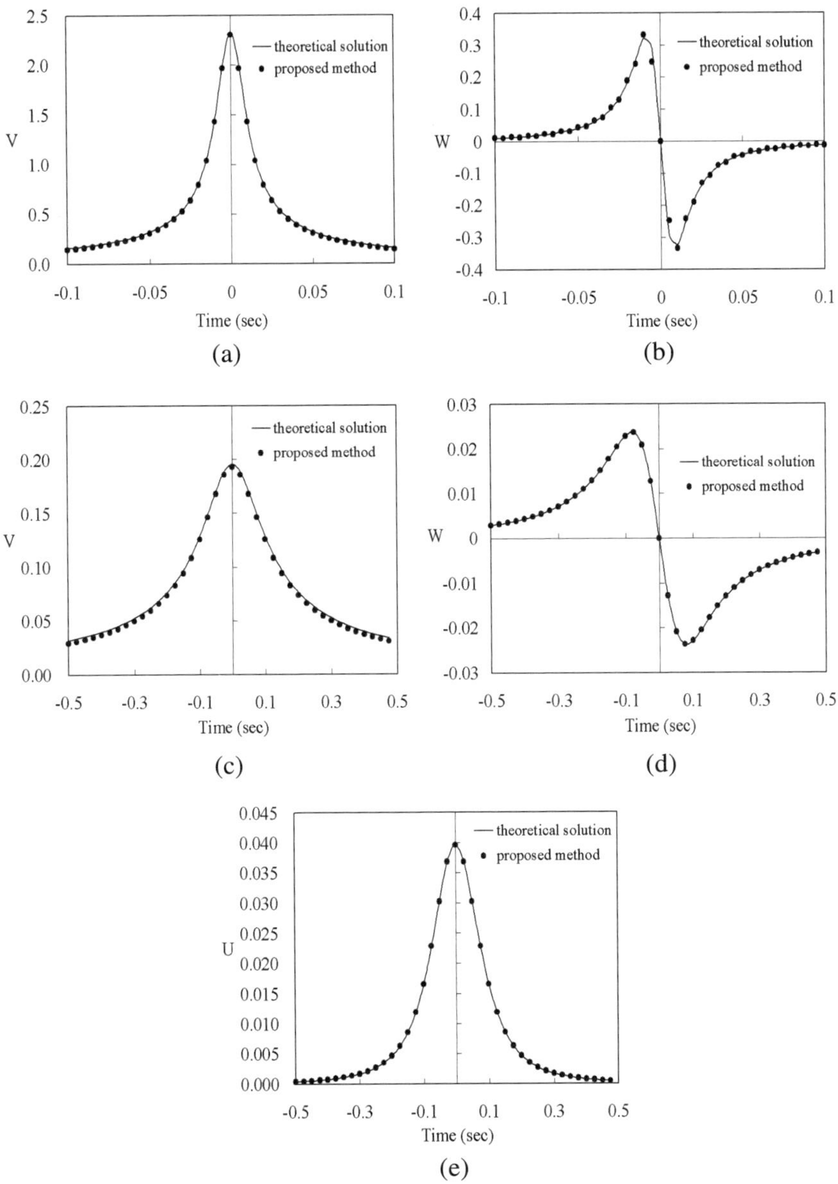

Fig. 10.6 Comparison of present results with theoretical ones for normalized displacements (Case 1): (a) V at $x = 0$ m, $y = -0.5$ m; (b) W at $x = 0$ m, $y = -0.5$ m; (c) V at $x = 3.5$ m, $y = -3.5$ m; (d) W at $x = 3.5$ m, $y = -3.5$ m; (e) U at $x = 3.5$ m, $y = -3.5$ m.

subjected to a point load at point A traveling along the z-axis, but directed along the y- and z-axes. Similarly, the normalized displacements V_y, W_y and U_y at the ground surface point D should be equal to the vertical responses V_y, V_z and V_x at point O when subjected to a point load at point D traveling along the z-axis, but directed along the y-, z- and x-axes, respectively.

For the purpose of verification, by adopting the material damping ratio as $\beta_{soil} = 0.02$, the normalized displacements V_y and W_y calculated for point A and V_y, W_y and U_y for point D using the mesh in Fig. 10.5(b) have been compared with the analytical ones for Case 2 in Fig. 10.7 using the reciprocity theorem. Again, the present results agree excellently with the analytical ones for both the locations near the source (point A) and on the boundary (point D). Thus, it is confirmed that the present finite/infinite element approach can be reliably used in the solution of related soil vibration problems.

10.5 Numerical Modeling and Related Considerations

Since this chapter is concerned with the wave propagation behavior of the half-space, a mesh with a proper range should be used, according to Chapter 2. The mesh to be used herein is the one shown in Fig. 10.8 which has a width of 50 m and depth of 32 m. Again, only half of the system is considered due to symmetry considerations, and the near and far fields are separated by dotted lines. In this section, the train loads are assumed to move through the invert of the tunnel (point O). The train is simulated by a sequence of moving loads with self oscillation frequency f_0 at speed c. According to Esveld (1989), the following are the ranges of frequencies that may be induced by a moving train:

(a) sprung mass: 0 - 20 Hz;

(b) unsprung mass: 0 - 125 Hz;

(c) corrugations, welds, and wheel flats: 0 - 2,000 Hz.

Focus is placed on inclusion of the frequencies f_0 of the moving loads in the range 0 - 20 Hz, arising mainly from the sprung mass.

As for the train, the following data for the Taipei rapid transit system are adopted: $a = 2.3$ m, $b = 14.2$ m, $L = 23.5$ m, wheel load $T = 16$ tf, and

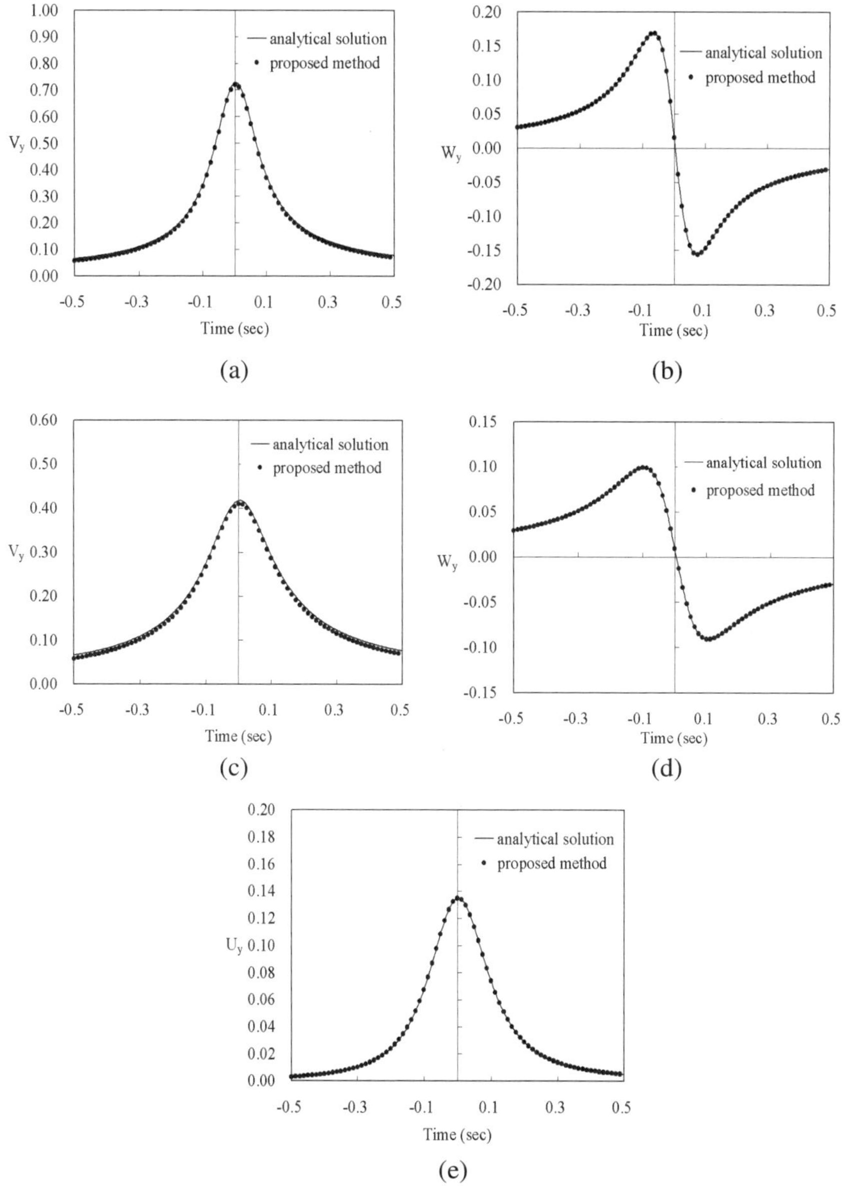

Fig. 10.7 Comparison of present results with theoretical ones for normalized displacements (Case 2): (a) V at point A; (b) W at point A; (c) V at point D; (d) W at point D; (c) U at point D.

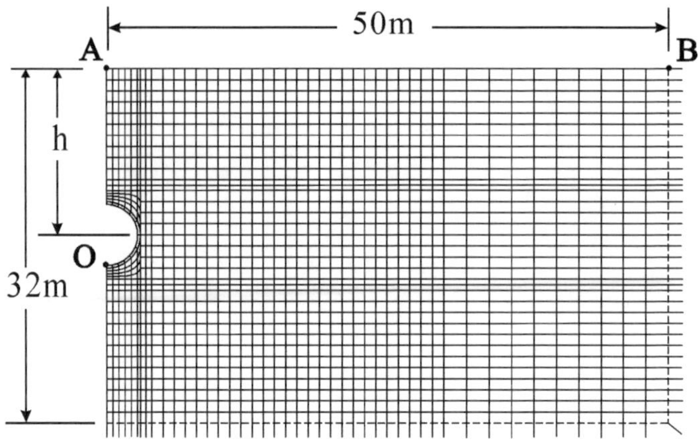

Fig. 10.8 Finite/infinite element mesh for a half-space.

characteristic length $\alpha = 0.8$ m, based on the definitions of Fig. 10.2 and Eq. (2.36). The load distribution functions for the train with N (= 6) cars in the spatial and transformed domains were plotted in Figs. 10.9(a) and (b), respectively. In Fig. 10.9(b), the load distribution function in the transformed domain appears to be periodically distributed, with each peak indicating one passing frequency implied by the distances between different wheels, especially for the distance L between neighboring carts. If we redraw the Fourier transform of the load distribution function with respect to $f/(c/L)$ instead of wave number k in Fig. 10.9(b), we can obtain Fig. 10.10, where clear peaks can be observed at frequencies corresponding to cart traveling frequencies, i.e., at $f/(c/L) = 1, 2, 3, 4, 6, 7, 8$ and 9, etc.

The problem considered herein is a tunnel embedded in a uniform visco-elastic half-space, with or without bedrock, subjected to an underground moving train as depicted in Fig. 10.1. Unless mentioned otherwise, the following data are adopted for the soil: Young's modulus $E_{soil} = 3 \times 10^7$ N/m^2, Poisson's ratio $\upsilon_{soil} = 0.3$, density $\rho_{soil} = 1900$ kg/m^3, material damping ratio $\beta_{soil} = 0.05$. Correspondingly, the Rayleigh, shear, and compressional wave velocities are $C_R = 72.294$ m/s, $C_S = 77.929$ m/s, $C_P = 145.791$ m/s, respectively. The material properties for the concrete tunnel are: Young's modulus $E_{concrete} = 2.5 \times 10^{10}$ N/m^2, Poisson's ratio

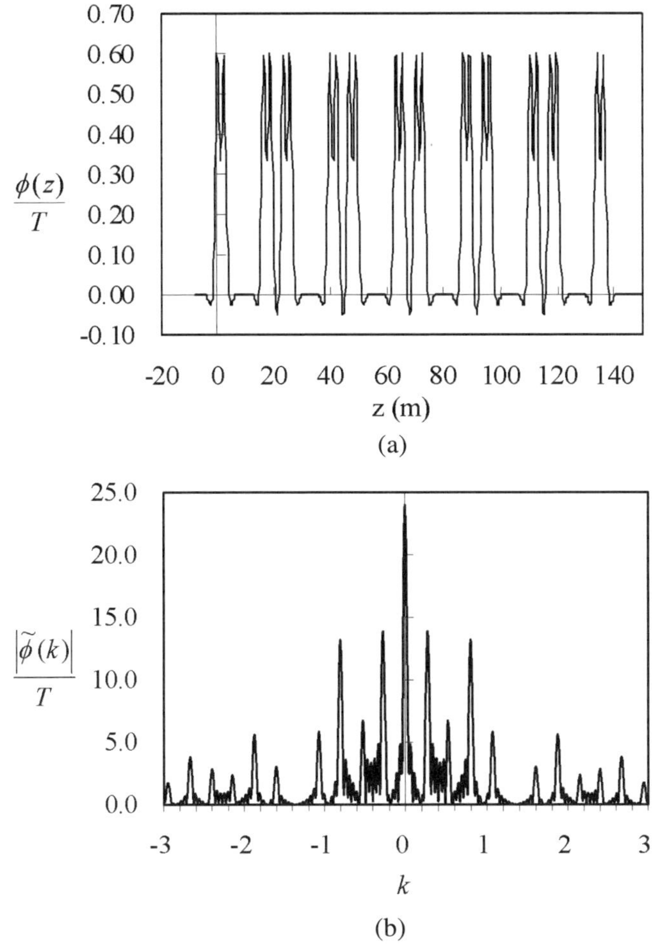

Fig. 10.9 A sequence of wheel loads: (a) load distribution function; (b) Fourier transform of load distribution function.

$\upsilon_{\text{concrete}} = 0.2$, density $\rho_{\text{concrete}} = 2400$ kg/m^3, and damping ratio $\beta_{\text{concrete}} = 0.02$. The centroid of the tunnel is located at a depth of $h = 15$ m beneath the ground surface, the inner diameter of the tunnel is 5.5 m, and the wall thickness of the tunnel is $t = 25$ cm.

To visualize the ground vibration due to the underground moving loads, the spatial distribution of the ground displacement generated by a single moving wheel load in the underground tunnel was studied first.

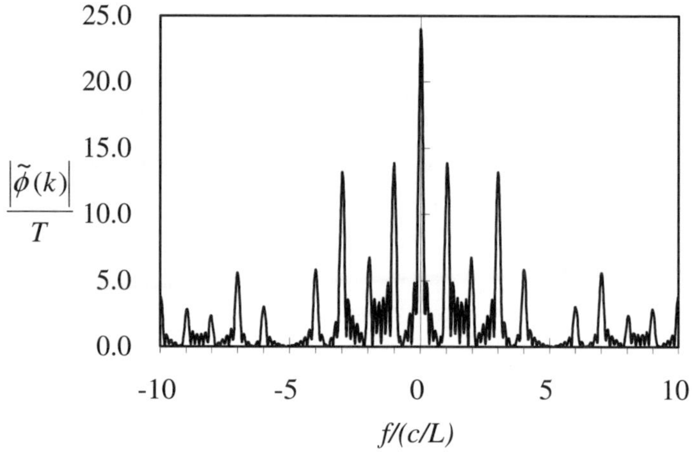

Fig. 10.10 Fourier transform of load distribution function with respect to $f/(c/L)$.

For the case with no self oscillation, i.e., with $f_0 = 0$, the ground displacements computed for the speeds c = 30 m/s and 80 m/s at the instant when the wheel moves to point O (Fig. 10.8) were plotted in Figs. 10.11(a) and (b), respectively. As can be seen, for a load moving in the tunnel with a speed lower than the shear wave velocity (c = 30 m/s), only localized quasi-static behavior above the source can be observed. However, for the case with a speed larger than the shear wave velocity (c = 80 m/s), a Mach cone can be observed for the displacement field, which shows a magnitude much larger than the one for c = 30 m/s. Notwithstanding the above observation, we shall use the train speed of c = 30 m/s (108 km/hr) in most of the following studies, since the operating speed of an underground train, which is the major concern of this study, is usually below 100 km/hr, unless noted otherwise.

Consider the case where the wheel load vibrates by itself with a self oscillation frequency ($f_0 \neq 0$). The real-part displacement field for the cases with f_0 = 1 Hz and f_0 = 5 Hz under the speed c = 30 m/s were computed and plotted in Figs. 10.11(c) and (d), respectively, from which distinct fluctuating vibrations can be observed on the ground surface. However, the frequencies observed are not exactly equal to those given by $f_{cr} = f_0/(1 \pm c/C_R)$, as is the case encountered by surface moving loads in Chapter 8, but with a tendency of being localized around the

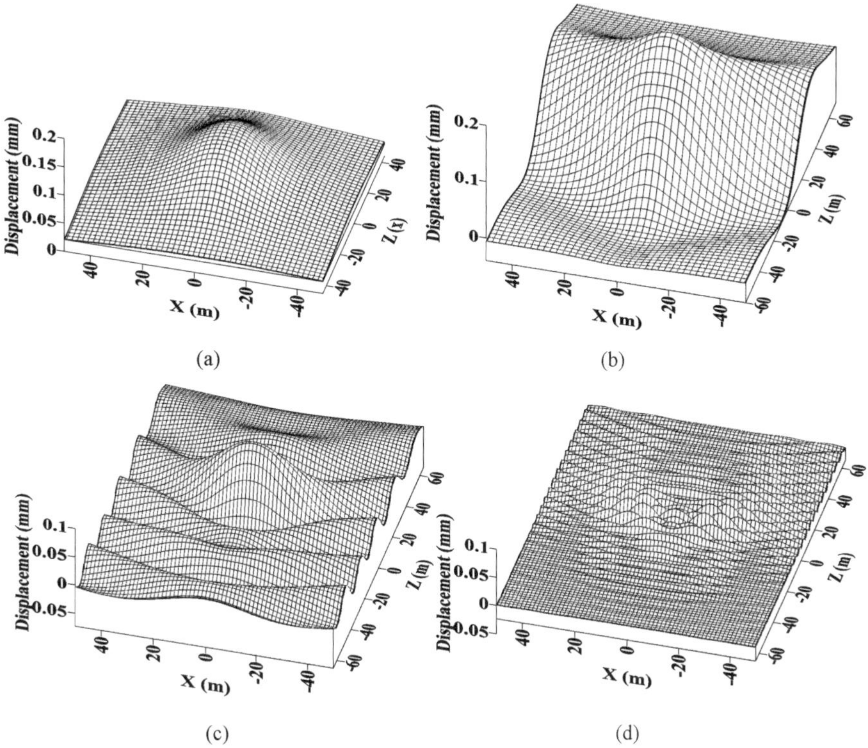

Fig. 10.11 Displacements caused by a moving wheel load with speed: (a) $c = 30$ m/s, $f_0 = 0$ Hz; (b) $c = 80$ m/s, $f_0 = 0$ Hz; (c) $c = 30$ m/s, $f_0 = 1$ Hz; (d) $c = 30$ m/s, $f_0 = 5$ Hz.

oscillation frequency f_0. The reason is as follows. For the case with a surface moving load, the surface responses are mostly generated by the vibration traveling along the same route as that of the moving load on the surface. In contrast, for the case with an underground moving load, the surface responses are generated as a superposition of waves from different routes, including those through the tunnel structure, which travel faster than those in the soils. For the case with $f_0 = 5$ Hz in Fig. 10.11(d), the phase velocities observed on the free surface above the tunnel are higher along the tunnel (z) axis than in the direction perpendicular to the tunnel, resulting in an elliptical wavefront. Similar phenomenon was observed by Clouteau *et al.* (2005) for a moving point force applied at the invert of a shallow cut-and-cover tunnel.

Theoretically, the vibration can be expressed either in terms of displacement, velocity or acceleration. However, displacement was rarely used for describing ground-borne vibrations, simply because the response of humans, buildings, and equipment to vibrations can be more accurately described in terms of velocity or acceleration. As such, the ground-borne vibrations to be studied herein will be represented in velocity on a logarithmic scale in dB:

$$L(\mathrm{dB}) = 20 \log \frac{P_1}{P_2} \tag{10.10}$$

where P_1 is the computed velocity amplitude and P_2 is a reference value. In this study, the reference value is adopted as $P_2 = 10^{-8}$ m/s.

10.6 Parametric Study for an Underground Moving Train

The problem considered herein is a tunnel embedded in a half-space subjected to an underground moving train (Fig. 10.1). The vibration energy induced by a moving train starts with the rolling action of the car wheels over the rails, which transmits through the track system into the soil and nearby structures. For car wheels that are perfectly round, the axle inputs can be assumed to be moving loads of the quasi-static type ($f_0 = 0$), which differ by a time delay only. In reality, however, the wheels may not be perfectly round. The responses are generated as the supposition of quasi-static moving loads and oscillations associated with the frequencies ($f_0 \neq 0$) of the sprung mass (of concern in this chapter), wheel flats or rail unevenness.

For comparison, both the above two cases will be studied in the following. Factors such as the number and speed of carriage, damping ratio, and stratum depth of the soil layer, as well as the depth and thickness of the tunnel structure, will be studied as well.

10.6.1 *Effect of number of carriages*

For a uniform half-space, i.e., with $H = \infty$, and for a train speed of $c = 30$ m/s, the effect of the number of carriages is studied in Fig. 10.12, in which the maximum velocity levels for the responses along the x-, y-,

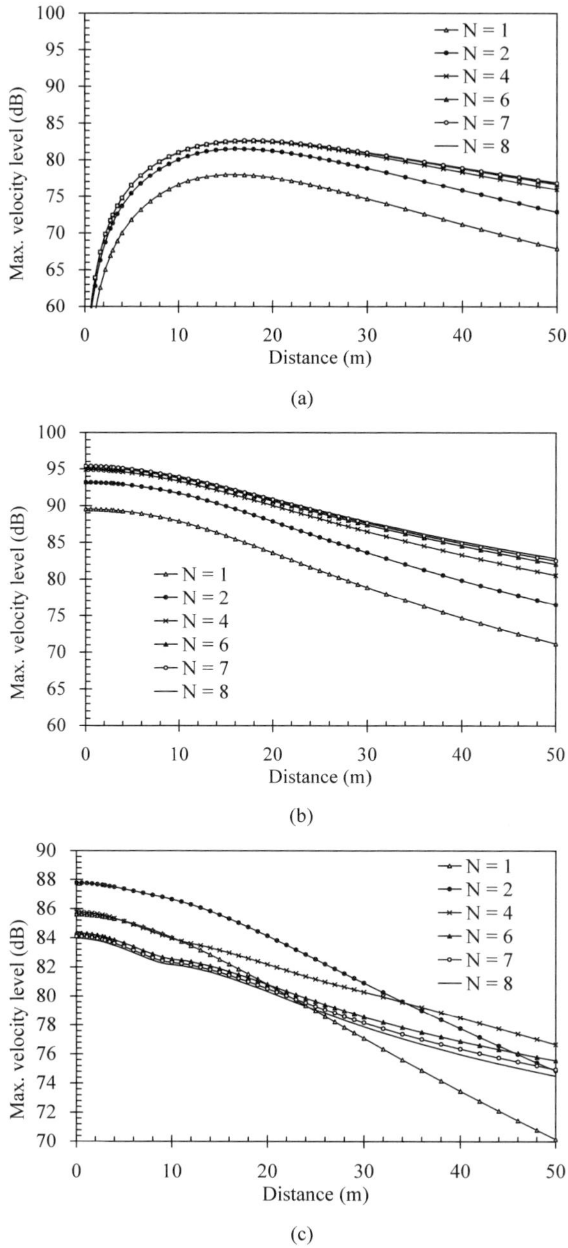

Fig. 10.12 Effect of carriage number N on ground vibration attenuation induced by a moving train with speed $c = 30$ m/s: (a) $\dot{u}$; (b) $\dot{v}$; (c) $\dot{w}$.

and z-axes, i.e., $\dot{u}$, $\dot{v}$ and $\dot{w}$, have been plotted against the distance on the surface (as indicated by point A to B in Fig. 10.8) for the carriage number $N = 1$ to 8. Clearly, the velocity levels of $\dot{u}$ and $\dot{v}$ increase with the increase in carriage number, but little difference can be observed for the cases with $N = 6$ and $N = 8$. In contrast, the velocity level of $\dot{w}$ along the load-moving direction does no show a clear trend of increase following the increase in the carriage number. This phenomenon can be explained by the fact that the soil particles along the loading-moving direction, i.e., the z-axis, before and after the moving load's arrival move in opposite directions, as shown in Figs. 10.7(b) and (d), in which the responses for $t < 0$ are positive, while those for $t > 0$ are negative. As a result, the cancelling effect between different carts may result in the decrease of vibration with the increase of cart number. For the sake of simplicity, the carriage number N will be set to 6 in the following parametric studies.

Another feature can be observed from Fig. 10.12 is that the ground vibration attenuation for the vertical velocity ($\dot{v}$) and two horizontal velocities ($\dot{u}$ and $\dot{w}$) are quite different. To further gain some insight about their differences, the velocity attenuations of the three quantities $\dot{u}$, $\dot{v}$ and $\dot{w}$ along the ground surface of the profile have been plotted for a uniform half-space ($H \to \infty$) and for a stratum case (H = 30 m) in Figs. 10.13(a) and (b), respectively. As can be seen, the horizontal velocities $\dot{u}$ and $\dot{w}$ are generally lower than vertical velocity $\dot{v}$ for a uniform half-space. However, for a stratum case, the vertical velocity decays faster than the other two horizontal velocities. Thus, horizontal velocities are likely to be dominant at locations away from the source. All the responses along the three axes will be studied in the following parametric studies.

10.6.2 *Effect of load-moving speed*

For the case of quasi-static moving loads (i.e., with $f_0 = 0$), the velocity response attenuation along the ground surface of the profile in dB scale was plotted for various train speeds (i.e., c = 10 - 80 m/s or c = 36 - 288 km/hr) for a uniform half-space ($H \to \infty$) and a stratum case (H = 30 m) in Figs. 10.14 and 10.15, respectively, in which parts (a), (b) and (c)

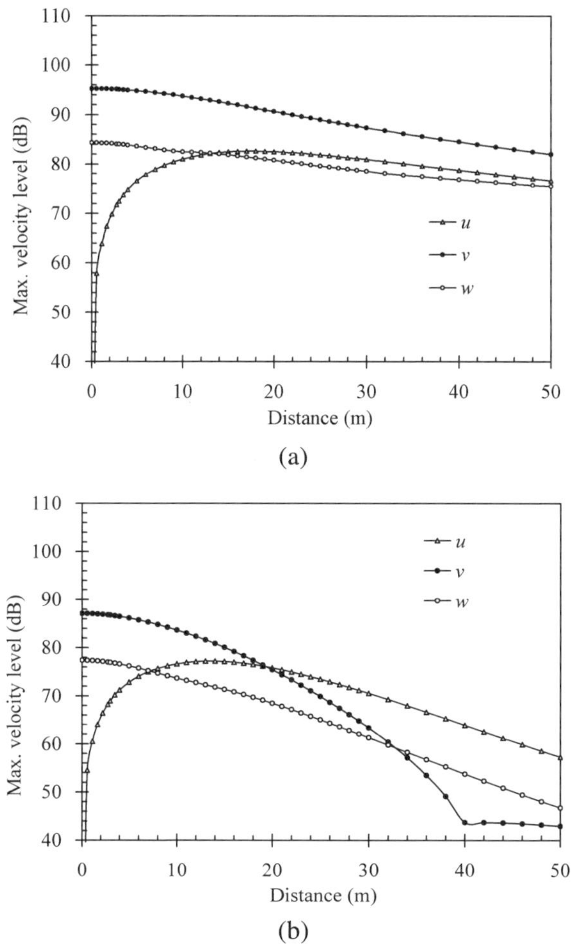

Fig. 10.13 Vibration attenuation of $\dot{u}$ and $\dot{w}$ (horizontal) and $\dot{v}$ (vertical) due to an underground moving train with speed $c = 30$ m/s and $f_0 = 0$: (a) $H \to \infty$; (b) $H = 30$ m.

represent the velocity level of $\dot{u}$, $\dot{v}$ and $\dot{w}$, respectively. Evidently, an increase in the train speed c is accompanied by an increase in the vibration level for the velocities along all the three directions. For the uniform half-space, the attenuation rates for $\dot{v}$ and $\dot{w}$ remain almost constant on the surface. Such a phenomenon is different from the one observed for a surface moving train in Chapter 8, where larger decaying rate exists for observation points closer to the source.

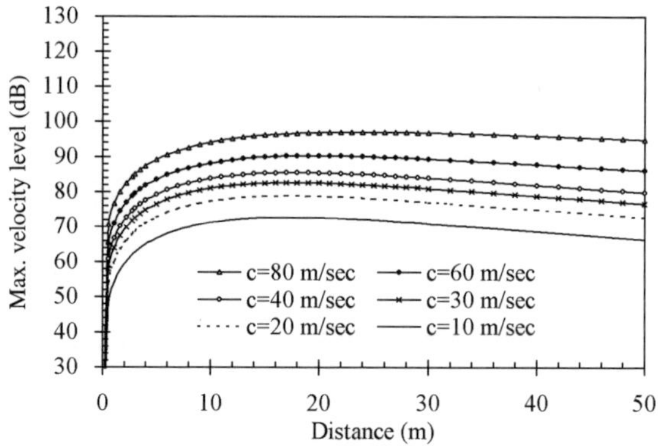

(a)

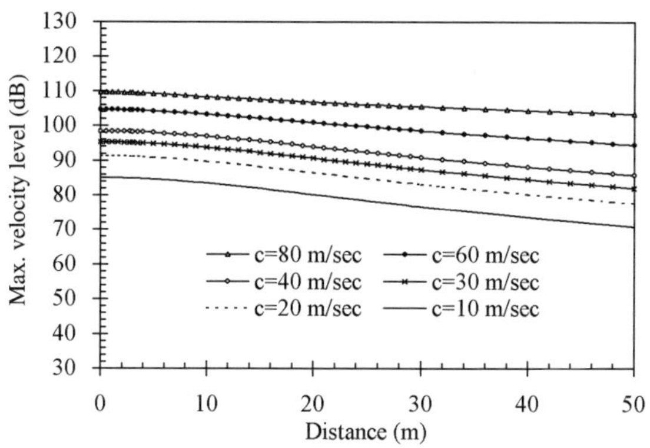

(b)

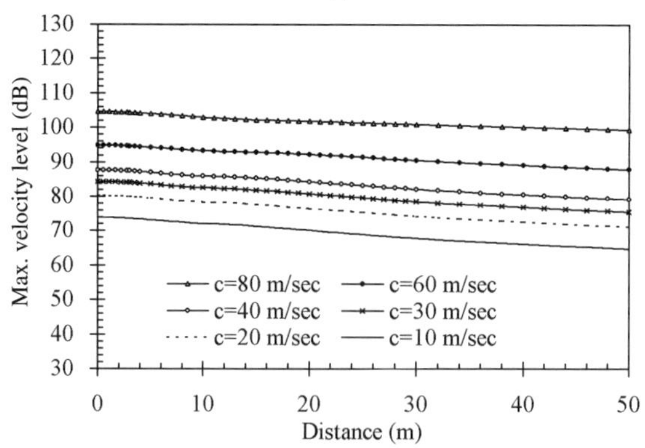

(c)

Fig. 10.14 Effect of train speed on vibration attenuation induced by an underground moving train for $H \to \infty$ ($f_0 = 0$): (a) $\dot{u}$; (b) $\dot{v}$; (c) $\dot{w}$.

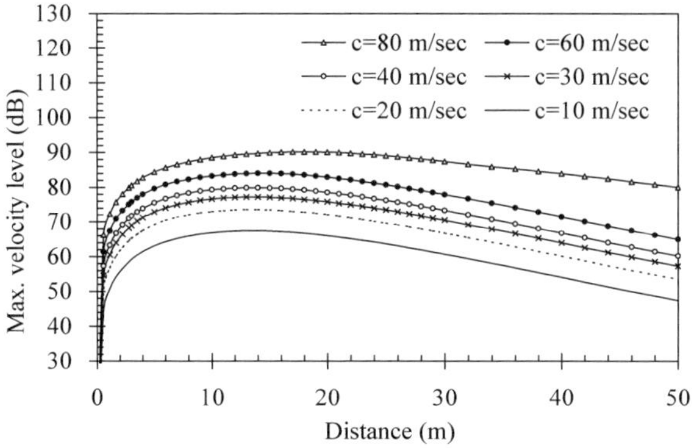

(a)

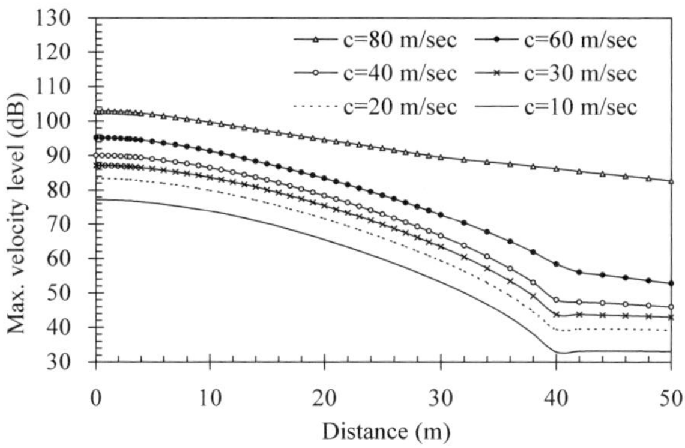

(b)

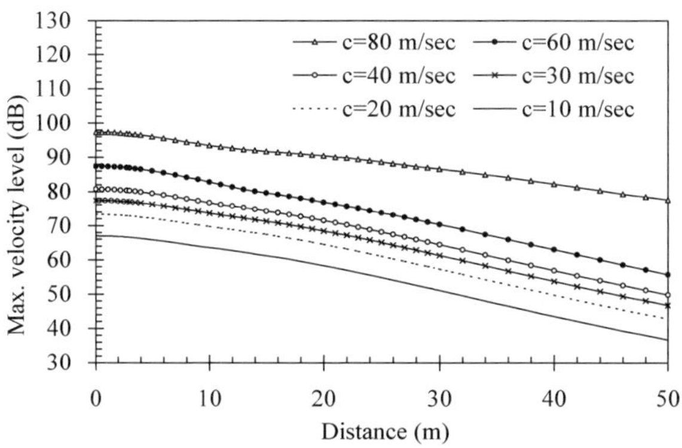

(c)

Fig. 10.15 Effect of train speed on vibration attenuation induced by an underground moving train for $H = 30$ m ($f_0 = 0$): (a) $\dot{u}$; (b) $\dot{v}$; (c) $\dot{w}$.

As for the case with bedrock at the depth $H = 30$ m in Fig. 10.15, the decay rate is generally higher, compared with the uniform half-space case in Fig. 10.14. The decay rate of the vertical vibration stagnates at the distance around $x = 40$ m, except for the case with transonic speed ($c = 80$ m/s). Besides, the attenuation rates for both the horizontal velocities $\dot{u}$ and $\dot{w}$ appear to be slower than that of the vertical velocity.

Now let us investigate the effects of train speed and self oscillation frequency f_0 all together. Consider a train moving at speeds in the range from 10 to 40 m/s (36 - 144 km/hr) with self oscillation frequencies f_0 in the range from 0 to 20 Hz. The average velocity level versus the train speed c for different self oscillation frequencies f_0 have been plotted in Figs. 10.16 and 10.17 for the uniform half-space ($H \rightarrow \infty$) and a soil deposit ($H = 30$ m), respectively, in which parts (a), (b) and (c) represent the velocity levels of $\dot{u}$, $\dot{v}$ and $\dot{w}$, respectively. The average velocity level shown in the figures was obtained by averaging the vibration levels computed over the distance from $x = 0$ to 50 m along the ground surface of the profile.

As can be observed from these figures, except for the case with no self oscillation (i.e., with $f_0 = 0$), the effect of train speed c on the average vibration level is small for all the cases. In addition, the velocity level induced by a moving train with self oscillation (i.e., $f_0 \neq 0$) is appreciably higher than that for the case with no self oscillation (i.e., $f_0 = 0$). The implication here is that *the self oscillation frequencies f_0 resulting from the mechanical system of the train cars or rail unevenness should be considered in the study of ground-borne vibrations*, although the weighting of this term is usually lower than the static term in practice. It can also be observed that the average velocity level for $\dot{u}$ (horizontal) is almost equal to that of $\dot{v}$ (vertical) for the case with $f_0 \neq 0$. This observation further implies the need to investigate the horizontal responses in addition to the vertical responses.

It should be noted that for the case with non-zero self oscillation frequencies f_0, the final responses in time domain are complex because the external load by itself is complex. Thus, the maximum velocity level of concern has been obtained as the absolute value of the computed results.

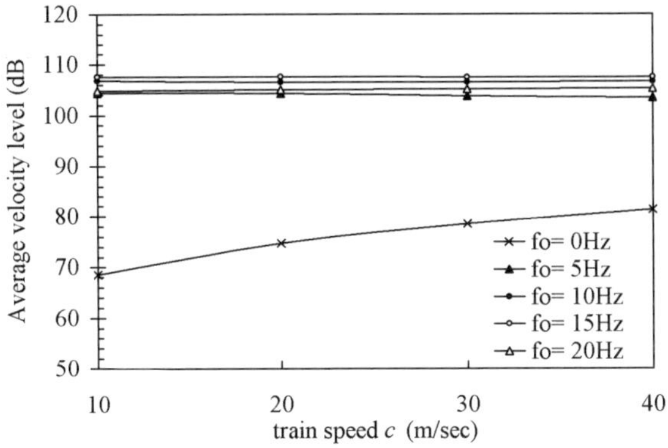

(a)

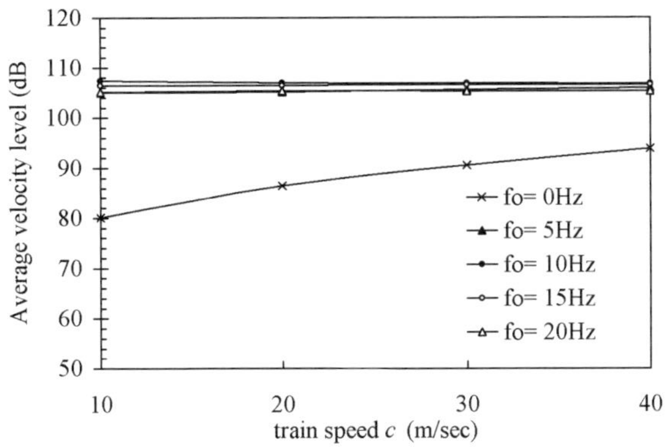

(b)

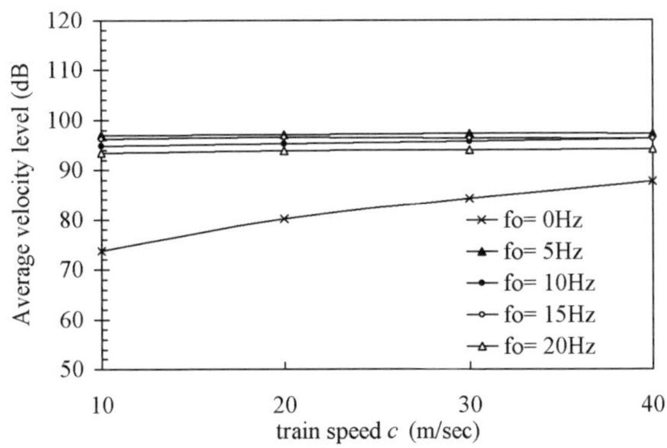

(c)

Fig. 10.16 Effect of train speed and frequency f_0 on vibration attenuation at different depths induced by an underground moving train for $H \rightarrow \infty$: (a) $\dot{u}$; (b) $\dot{v}$; (c) $\dot{w}$.

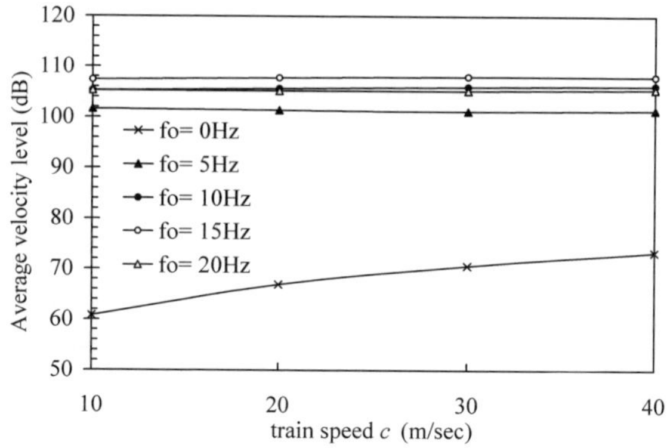

(a)

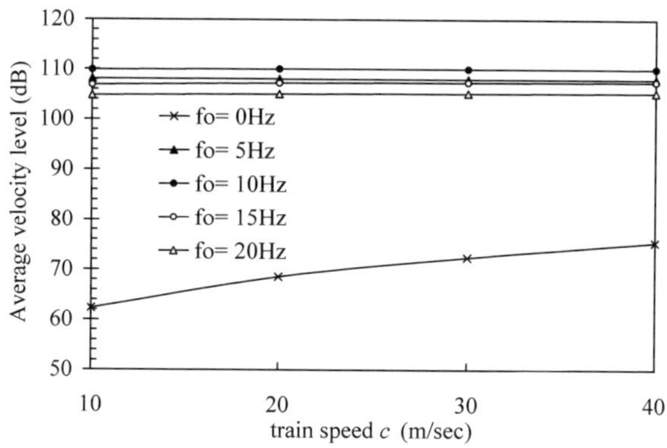

(b)

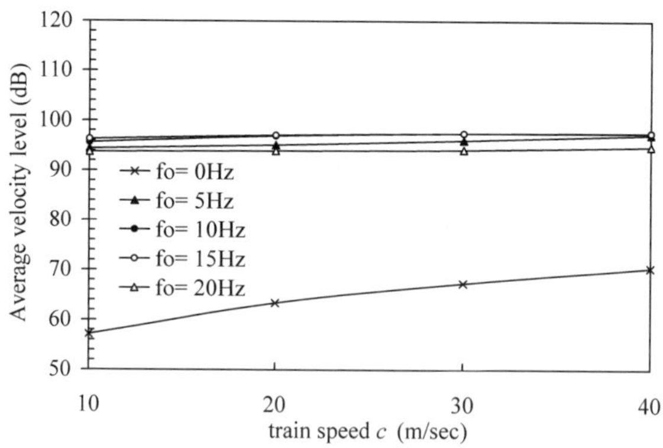

(c)

Fig. 10.17 Effect of train speed and frequency f_0 on vibration attenuation at different depths induced by an underground moving train for $H = 30$ m: (a) $\dot{u}$; (b) $\dot{v}$; (c) $\dot{w}$.

10.6.3 *Effect of bedrock depth H*

For a uniform half-space, all the waves will be radiated from the source to infinity. But for a soil stratum case, no waves can propagate outward for vibrations with frequencies lower than the cut-off frequency due to existence of the rigid base (Wolf 1985). For the vertically applied loads considered, the peak frequencies of the vertical response are close to the one induced by $f_P = (C_P / 4H)$. To investigate the influence of the soil stratum depth H, the three velocity responses $\dot{u}$, $\dot{v}$ and $\dot{w}$ respectively along the soil surface of the profile with different stratum depths, i.e., H = 15, 20, 30 m and ∞, have been plotted in Figs. 10.18-10.20 for train speed c = 30 m/s. In these figures, parts (a), (b), (c) and (d) represent the case for self oscillation frequency f_0 equal to 0, 1, 5, and 15 Hz, respectively. Obviously, for all the three responses, *the effect of stratum depth depends highly on the self oscillation frequency of the moving loads*.

For the case with static moving loads (i.e., with $f_0 = 0$), a shallower soil stratum reveals smaller levels of vibration. In contrast, for the case with $f_0 = 5$ and 15 Hz, only the soil stratum with H = 15 m, for which half of the tunnel is embedded in the bedrock, experiences rather small levels of vibration. As for thicker strata (e.g., H = 20 or 30 m), the effect of bedrock is not obvious as compared with the case without bedrock. Finally, for the case with $f_0 = 1$ Hz, a shallower soil stratum experiences a smaller level of vibration, too, except for H = 30 m. Such phenomena can be explained using the concept of cut-off frequency below.

According to Chapter 8, for a static moving load ($f_0 = 0$) with speeds lower than the critical one, the vibrations of high frequencies decay rather fast with respect to distance from the source. Thus, the frequency content is concentrated mainly in the low frequency range, for which the effect of bedrock is obvious. For the present case, the primary resonance frequencies or cut-off frequencies for H = 15, 20, and 30 m are 2.4, 1.8, and 1.2 Hz, respectively. Thus, with the decrease in stratum depth, more frequency content can be cut off. On the other hand, for the case with $f_0 = 5$ and 15 Hz, the frequency contents of their response are generally concentrated around the self oscillation frequency f_0, which already exceed the cut-off frequency range of 1.8 and 1.2 Hz. As such, the

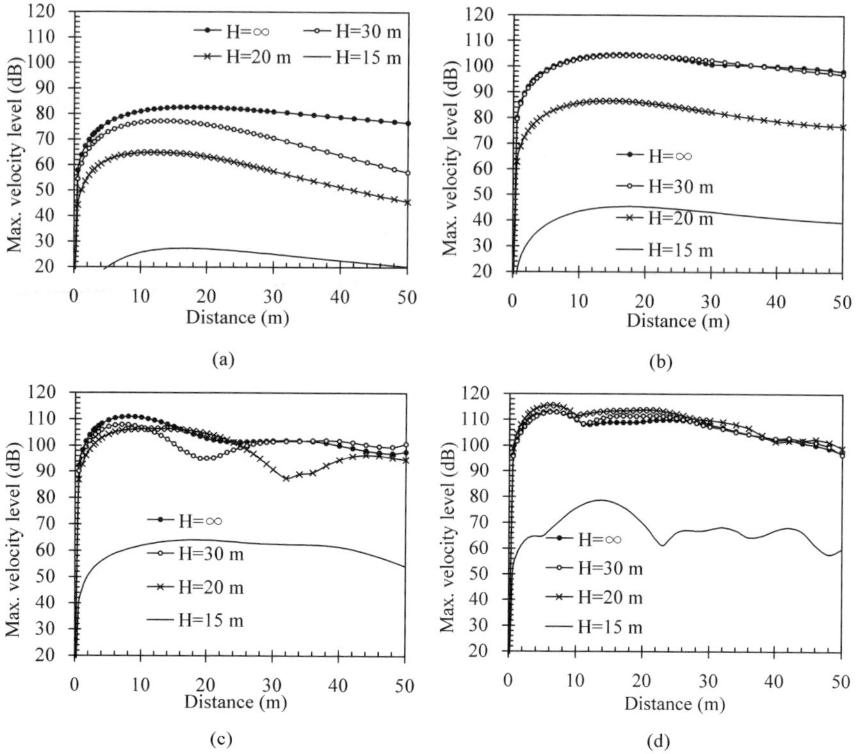

Fig. 10.18 Effect of stratum depth on ground vibration attenuation of $\dot{u}$ induced by an underground moving train with speed c = 30 m/s: (a) f_0 = 0 Hz; (b) f_0 = 1 Hz; (c) f_0 = 5 Hz; (d) f_0 = 15 Hz.

difference in the responses between the cases of H = 20 m, H = 30 m, and $H \to \infty$ appears to be invisible.

For the case with self oscillation frequency of f_0 = 1 Hz, which is close to the natural or cut-off frequency of the soil layer with H = 30 m, i.e., 1.2 Hz, a slight increase in response is observed as compared with the case of $H \to \infty$ in the region near the source in Figs. 10.19 and 10.20 for the responses $\dot{v}$ and $\dot{w}$. But as the distance increases, a slight decrease in response can be observed due to the cut-off effect of the bedrock. Noteworthy is that for the case with a soil stratum of thickness H = 15 m, where half of the tunnel structure is embedded in the bedrock, the velocity levels observed for all the four frequencies are substantially

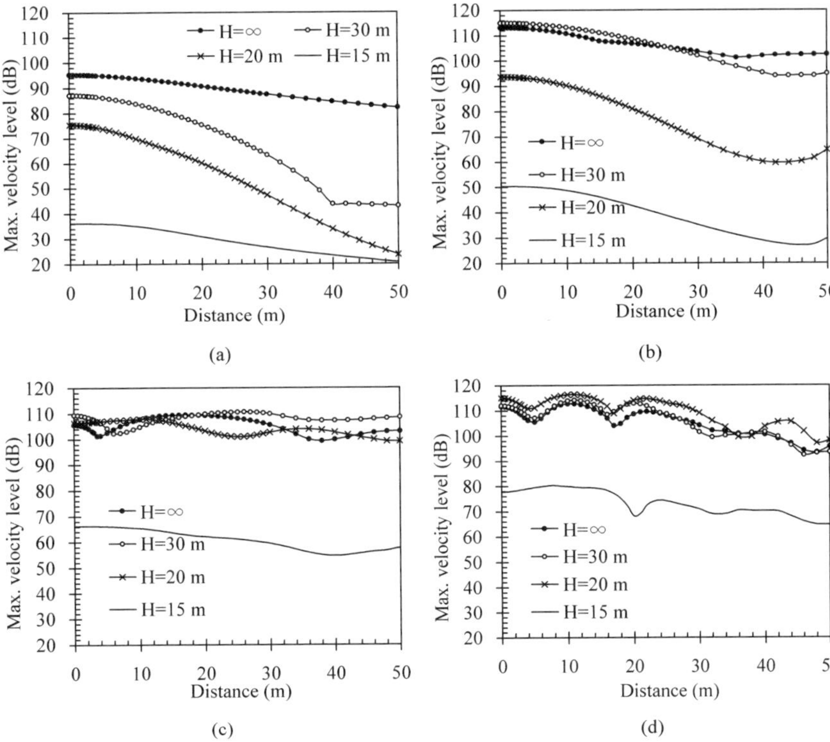

Fig. 10.19 Effect of stratum depth on ground vibration attenuation of $\dot{v}$ induced by an underground moving train with speed c = 30 m/s: (a) f_0 = 0 Hz; (b) f_0 = 1 Hz; (c) f_0 = 5 Hz; (d) f_0 = 15 Hz.

lower than those with deeper bedrocks, which imply no direct contact between the tunnel structure and bedrock. Such a result indicates that *for a tunnel constructed on a bedrock, the ground surface vibrations can be greatly reduced, since the vibration energy brought by the moving trains can be transmitted directly to the bedrock and dissipated.*

10.6.4 *Effect of damping ratio*

The effect of damping ratio on the vibration level for the case of static moving loads (i.e., with f_0 = 0) is studied at first. By assuming the loads to move at speed c = 30 m/s, the velocity levels computed for five values of damping ratios, i.e., β = 0.02, 0.04, 0.06, 0.08 and 0.1, were plotted for

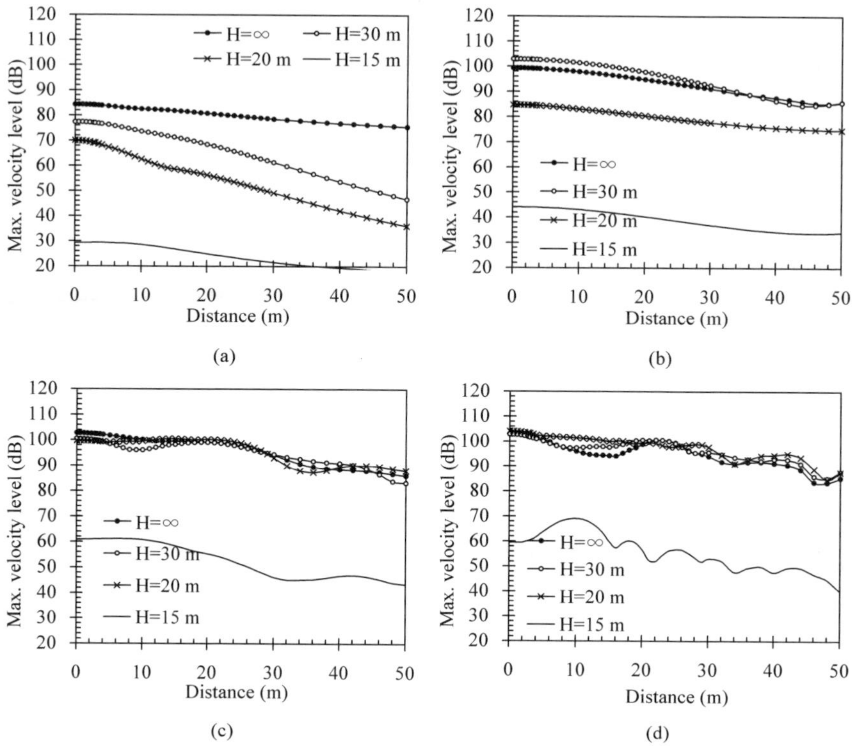

Fig. 10.20 Effect of stratum depth on ground vibration attenuation of $\dot{w}$ induced by an underground moving train with speed c = 30 m/s: (a) f_0 = 0 Hz; (b) f_0 = 1 Hz; (c) f_0 = 5 Hz; (d) f_0 = 15 Hz.

a uniform half-space (with $H \to \infty$) in Fig. 10.21, where parts (a), (b) and (c) represent the responses of $\dot{u}$, $\dot{v}$ and $\dot{w}$, respectively. As can be seen, *no matter the response of which direction is considered, the effect of damping ratio for the sub-critical speed case is too small to become distinguishable for the quasi-static moving loads case.*

Let us consider the same uniform half-space subjected to underground moving loads with self oscillation, i.e., with $f_0 \neq 0$. As can be seen from Figs. 10.22 and 10.23 *for the cases with* f_0 = 5 *and* 15 Hz, *respectively, the influence of damping ratio on the responses is generally significant*, as an increase in damping ratio can result in a substantial reduction of vibration, and the amount of reduction in dB scale increases significantly with the distance. The other trend revealed by Figs. 10.22

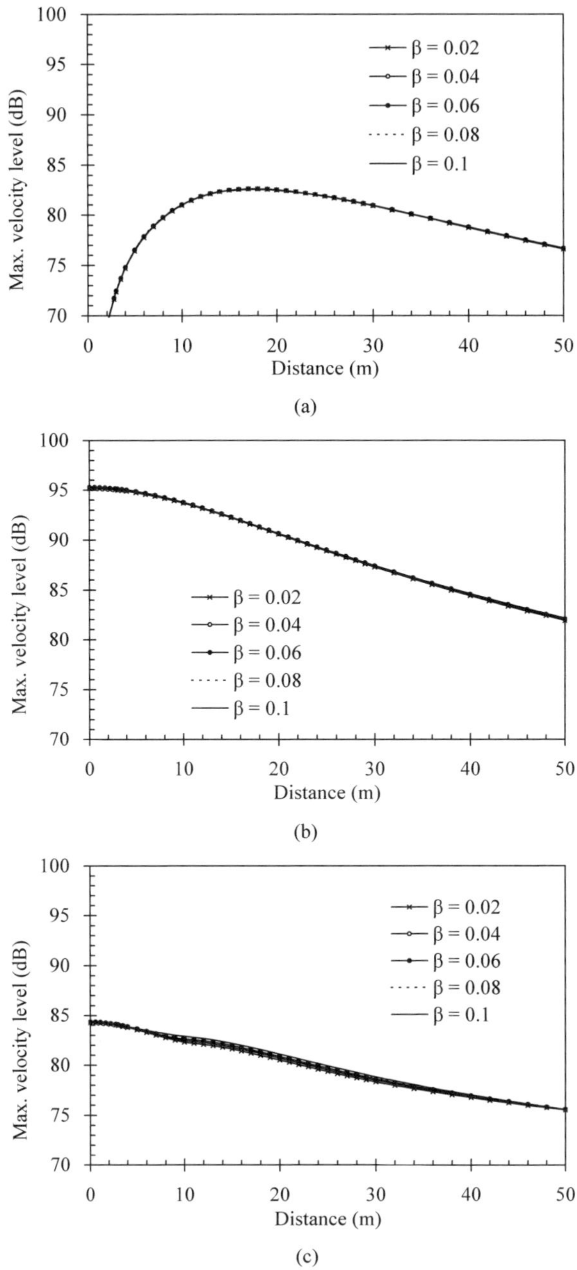

Fig. 10.21 Effect of damping ratio on ground vibration attenuation induced by an underground moving train for $f_0 = 0$ Hz (c = 30 m/s): (a) $\dot{u}$; (b) $\dot{v}$; (c) $\dot{w}$.

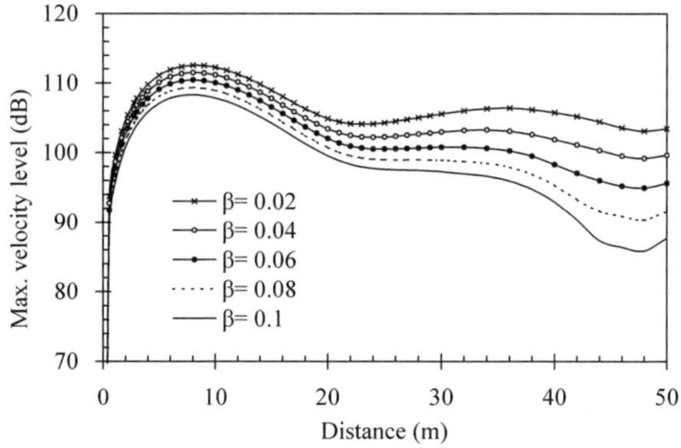

(a)

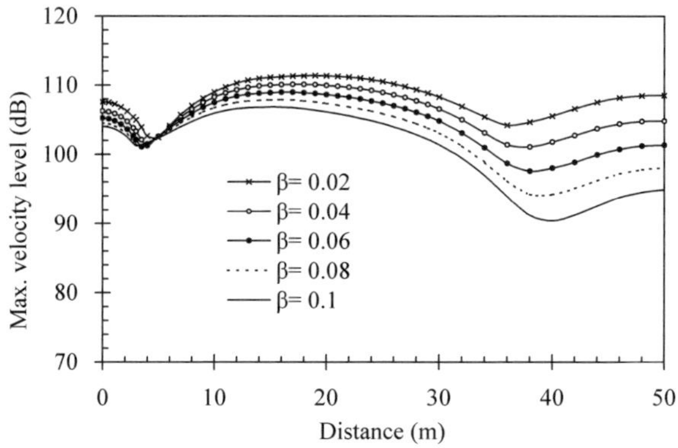

(b)

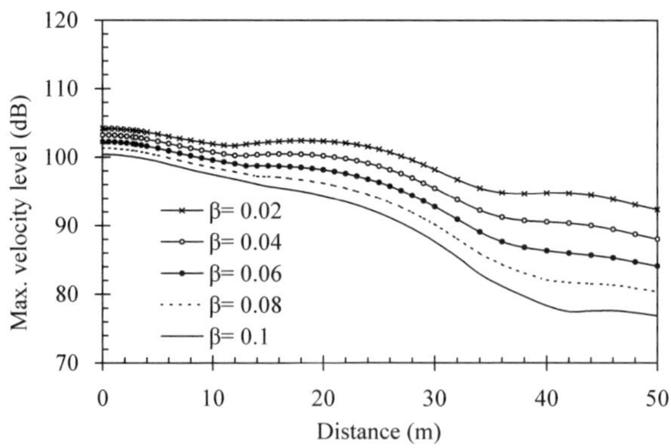

(c)

Fig. 10.22 Effect of damping ratio on ground vibration attenuation induced by an underground moving train for $f_0 = 5$ Hz ($c = 30$ m/s): (a) $\dot{u}$; (b) $\dot{v}$; (c) $\dot{w}$.

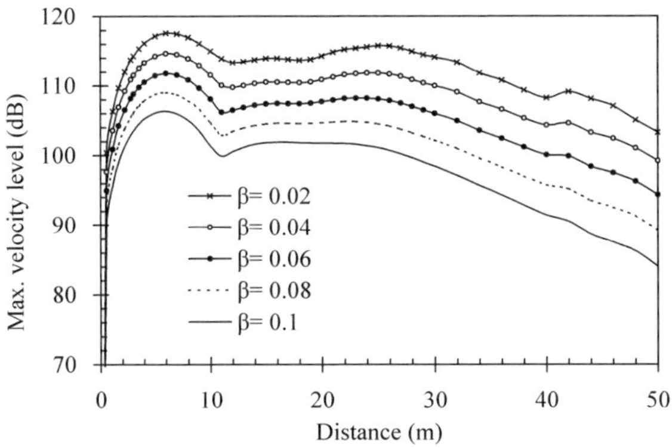

(a)

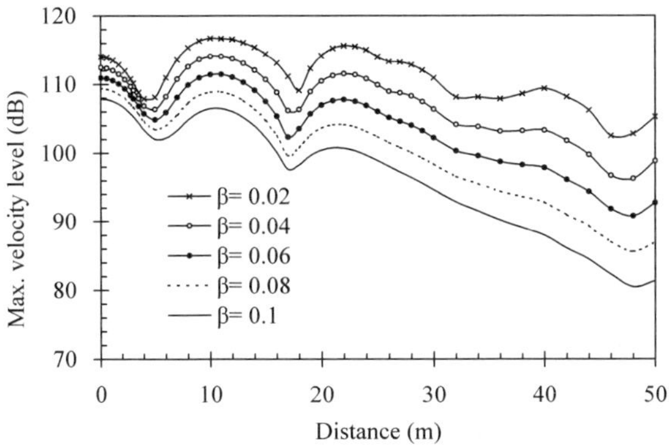

(b)

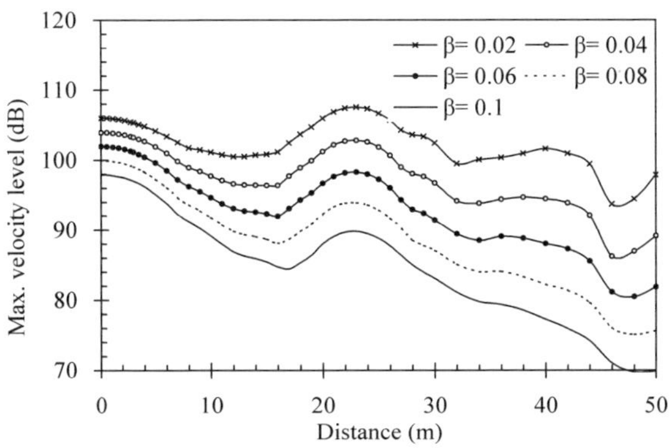

(c)

Fig. 10.23 Effect of damping ratio on ground vibration attenuation induced by an underground moving train for $f_0 = 15$ Hz ($c = 30$ m/s): (a) $\dot{u}$; (b) $\dot{v}$; (c) $\dot{w}$.

and 10.23 is that the velocities induced by higher frequencies ($f_0 = 15$ Hz) attenuate faster than those by the lower ones ($f_0 = 5$ Hz).

10.6.5 *Effect of tunnel lining thickness*

By assuming the moving speed to be c = 30 m/s and by varying the lining thickness of the tunnel from t = 12.5 to 50 cm, the maximum velocity levels computed along the ground surface of the profile were plotted for the three oscillation frequencies f_0 = 0, 5, and 15 Hz each in Figs. 10.24-10.26, where parts (a), (b) and (c) represent the velocity levels of $\dot{u}$, $\dot{v}$ and $\dot{w}$, respectively. From Fig. 10.24(a), one observes that for velocity $\dot{u}$, an increase in the wall thickness leads to slight decrease of surface vibration for distance greater than 20 m. However, it is generally concluded that *for the case with no self oscillation*, i.e., with f_0 = 0, *the effect of wall thickness of the tunnel on the surface vibration along each direction is small and can be neglected.*

For the case with non-zero self oscillation frequencies (i.e., with $f_0 \neq 0$), the responses along different directions show different trends. For the velocity $\dot{w}$ along the load-moving direction, the surface responses predicted for thicker tunnel walls are generally smaller than those for thinner tunnel walls.

For the horizontal response $\dot{u}$, the wall thickness does not show an absolute effect on the magnitude of vibration. For instance, for the case of f_0 = 5 Hz, the velocity $\dot{u}$ is higher for the case with thicker walls at locations with a distance less than 46 m. However, at location with a distance greater than 46 m, the influence of wall thickness is the other way around.

For the vertical velocity $\dot{v}$, although the surface responses show a trend of phase shift due to variation of the wall thickness, the effect of wall thickness t on the average response amplitudes for both cases with f_0 = 5 and 15 Hz is generally insignificant. However, if the invert of the tunnel, i.e., point O, is of concern, the increase in tunnel thickness is generally accompanied by a large reduction in the maximum vertical velocity $\dot{v}$ for different train speeds from c = 10 to 40 m/s, as shown in Fig. 10.27, and for different self oscillation frequencies, as shown in Fig. 10.28.

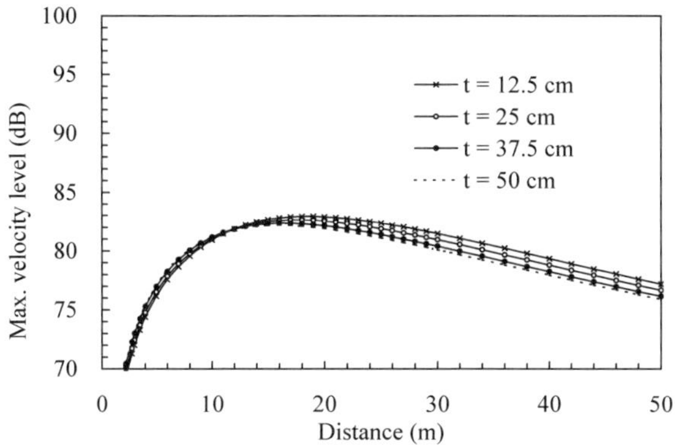

(a)

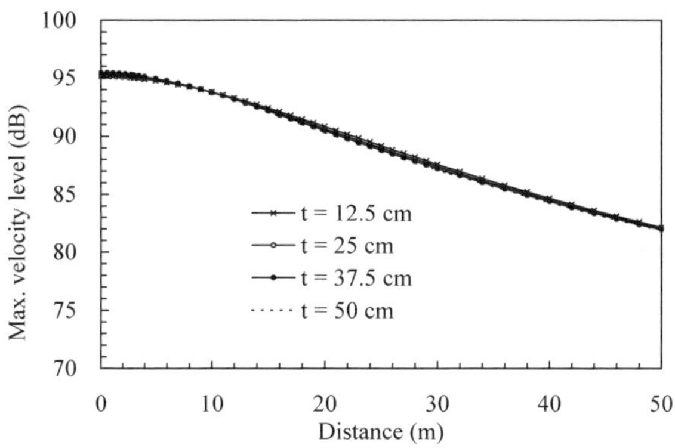

(b)

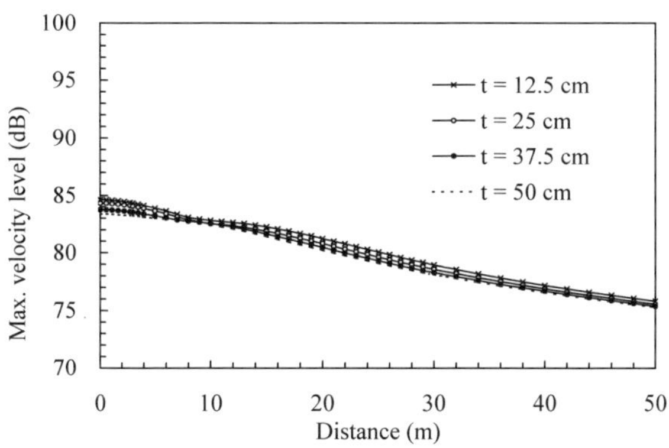

(c)

Fig. 10.24 Effect of tunnel lining thickness on ground vibration attenuation induced by an underground moving train with $f_0 = 0$ Hz (speed $c = 30$ m/s): (a) $\dot{u}$; (b) $\dot{v}$; (c) $\dot{w}$.

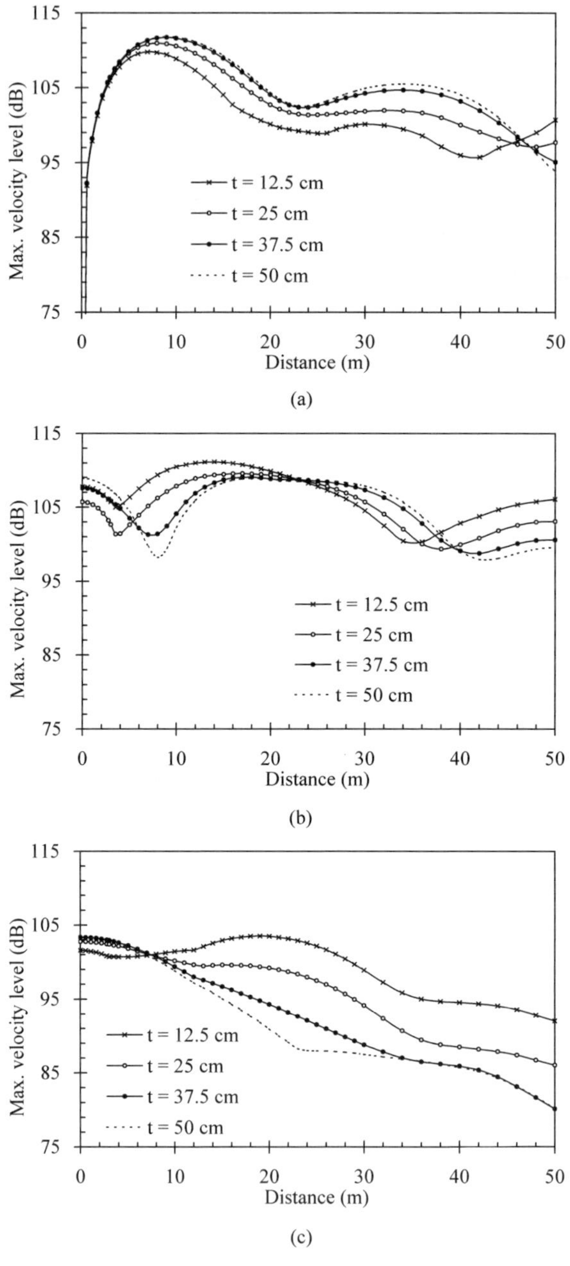

Fig. 10.25 Effect of tunnel lining thickness on ground vibration attenuation induced by an underground moving train with $f_0 = 5$ Hz (speed $c = 30$ m/s): (a) $\dot{u}$; (b) $\dot{v}$; (c) $\dot{w}$.

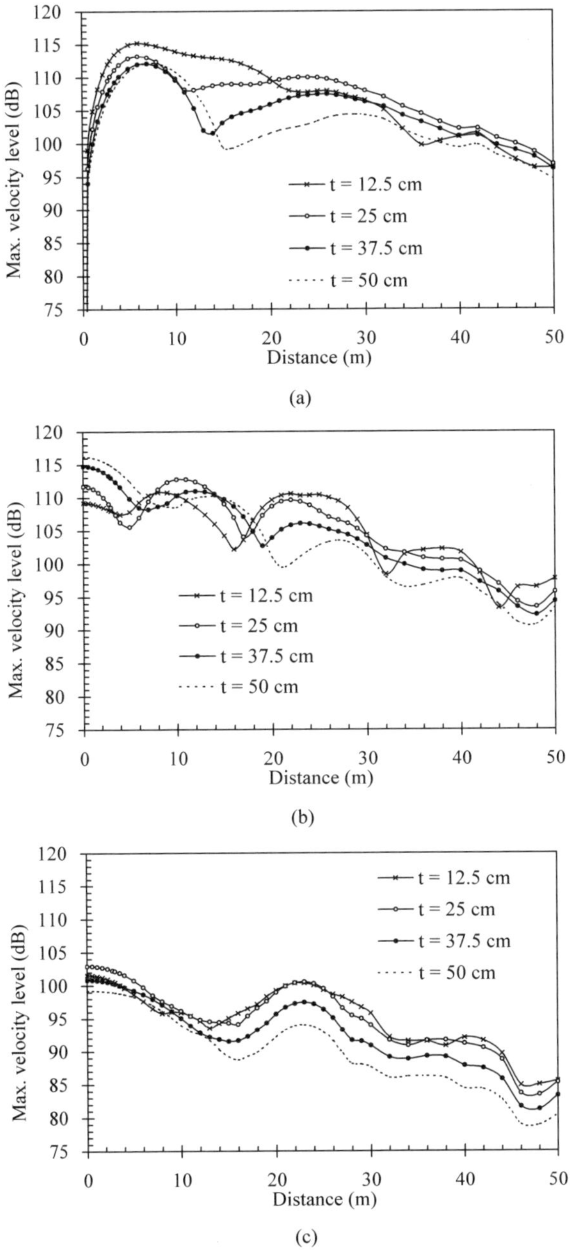

Fig. 10.26 Effect of tunnel lining thickness on ground vibration attenuation induced by an underground moving train with f_0=15 Hz (speed c = 30 m/s): (a) $\dot{u}$; (b) $\dot{v}$; (c) $\dot{w}$.

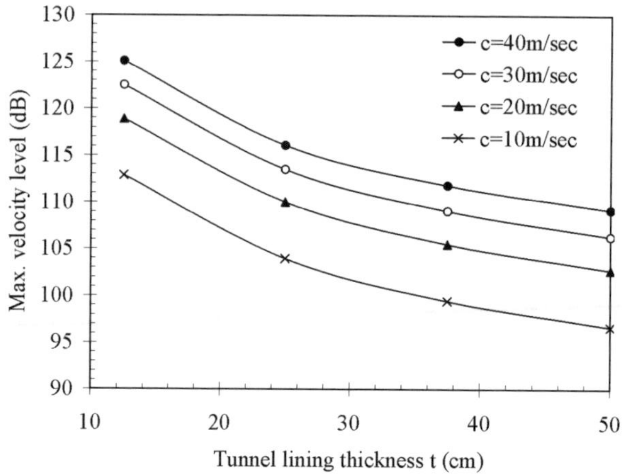

Fig. 10.27 Effect of tunnel lining thickness on tunnel vibration $\dot{v}$ at invert (point O) induced by a moving train with different speeds.

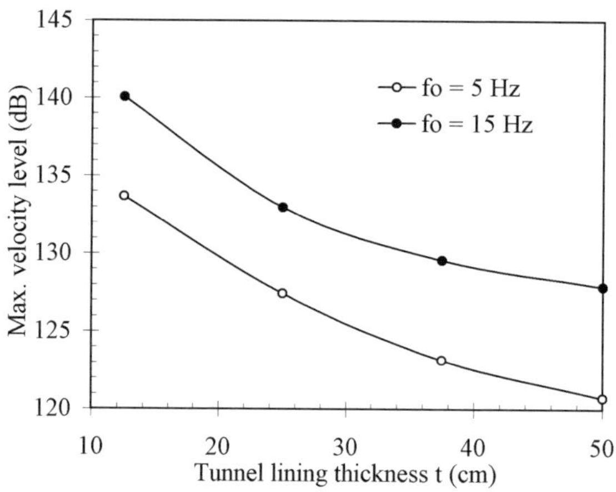

Fig. 10.28 Effect of tunnel lining thickness on tunnel vibration $\dot{v}$ at the invert (point O) induced by a moving train with speed c = 30 m/s.

10.6.6 *Effect of tunnel depth*

For the train speed c = 30 m/s, the effect of tunnel depths on the velocity levels of the three directions along the surface of the profile was plotted in Figs. 10.29-10.31 for the frequency f_0 = 0, 5, and 15 Hz, respectively. Five different depths were considered for the tunnel, h = 5, 10, 15, 20,

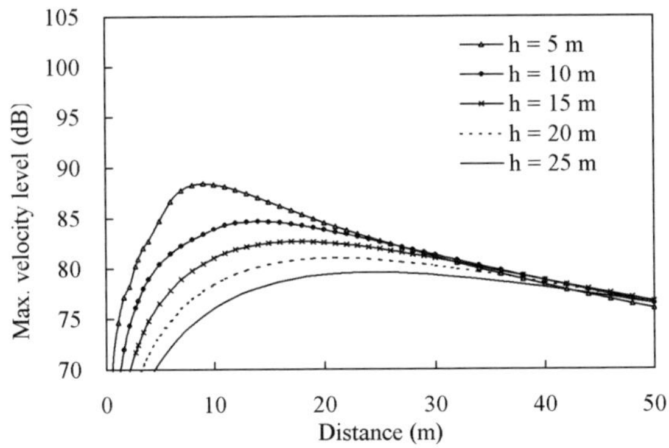

(a)

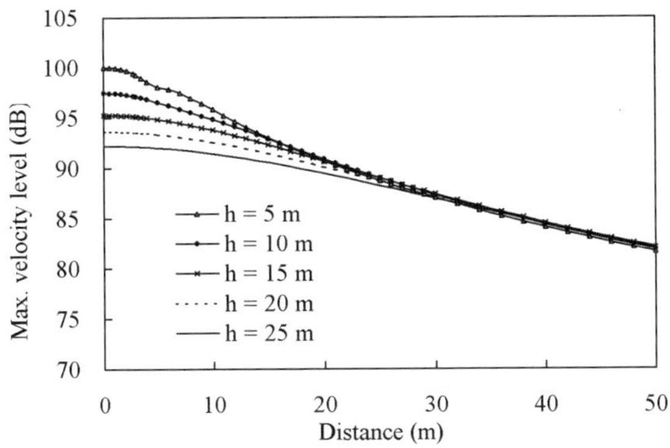

(b)

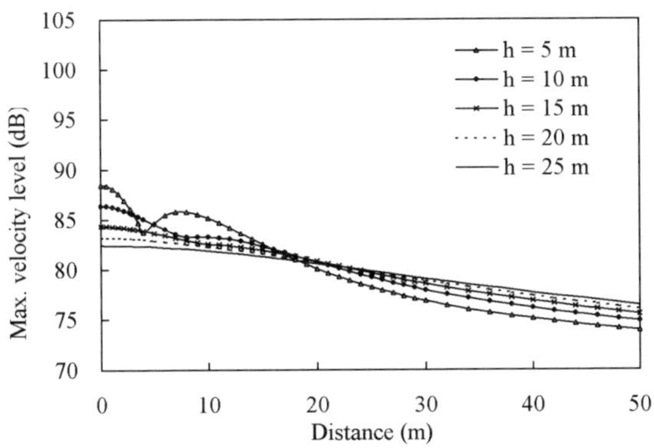

(c)

Fig. 10.29 Effect of tunnel depth on ground vibration attenuation induced by an underground moving train with $f_0 = 0$ Hz (speed $c = 30$ m/s): (a) $\dot{u}$; (b) $\dot{v}$; (c) $\dot{w}$.

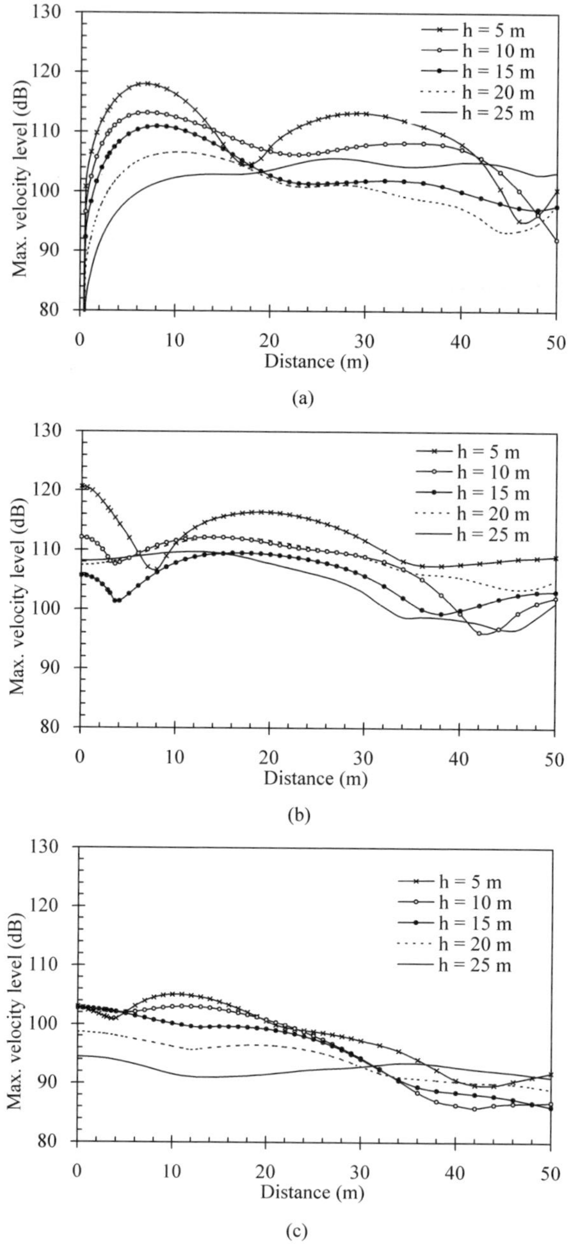

Fig. 10.30 Effect of tunnel depth on ground vibration attenuation induced by an underground moving train with $f_0 = 5$ Hz (speed $c = 30$ m/s): (a) $\dot{u}$; (b) $\dot{v}$; (c) $\dot{w}$.

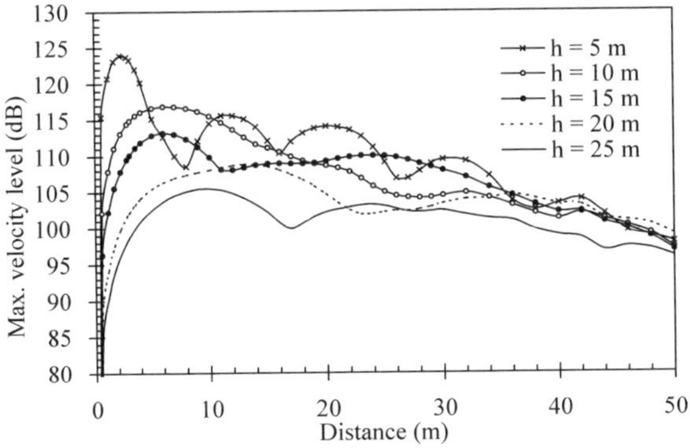

(a)

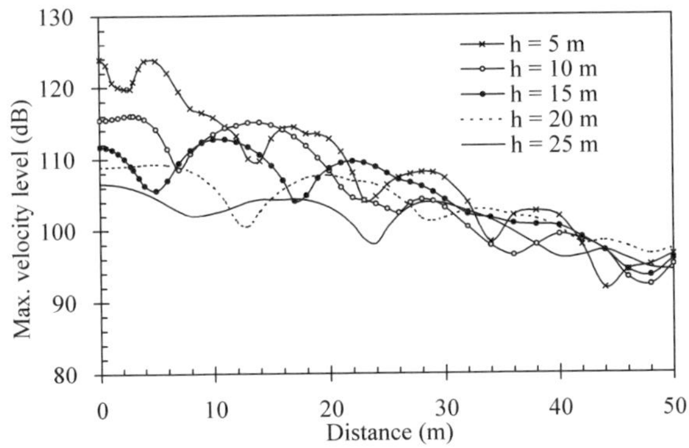

(b)

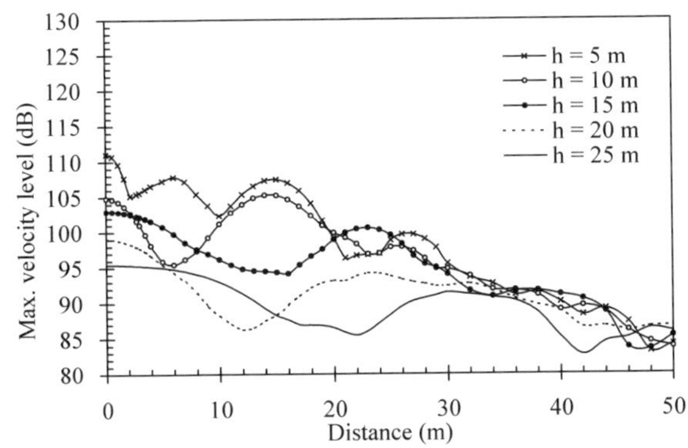

(c)

Fig. 10.31 Effect of tunnel depth on ground vibration attenuation induced by an underground moving train with $f_0 = 15$ Hz (speed $c = 30$ m/s): (a) $\dot{u}$; (b) $\dot{v}$; (c) $\dot{w}$.

and 25 m. For the case with no self oscillation, i.e., with $f_0 = 0$ as given in Fig. 10.29, the decay rate of ground surface responses for a deeper tunnel is generally lower than that for a shallower tunnel. On the other hand, an increase in the tunnel depth h is accompanied by a decrease in the velocity level in the vicinity of the source. Because of the above two factors, it turns out that as *the distance increases, the influence of tunnel depth becomes less noticeable or even a deeper tunnel may lead to higher vibration level*, as revealed by Fig. 10.29(c) for the horizontal velocity $\dot{w}$.

For the case with non-zero self oscillation frequencies ($f_0 \neq 0$), roughly similar trend can be observed for the case with $f_0 = 15$ Hz, as shown in Fig. 10.31. But for the case with $f_0 = 5$ Hz, the influence of tunnel depth is not as obvious as the other cases. Such a phenomenon could be attributed to the fact that the wave length for $f_0 = 5$ Hz is longer than that of $f_0 = 15$ Hz, which requires a longer observation range, than the range of 50 m considered in Fig. 10.30, for the phenomenon to be perceptible.

10.7 Concluding Remarks

In this chapter, the 2.5D finite/infinite element approach was adopted to investigate the various dynamic effects of the underground moving trains, as well as the soil and tunnel properties, on the vibrations of the ground surface. The following conclusions are drawn from the parametric studies, which remain strictly valid only for the conditions assumed in analysis.

(1) The velocity levels of both $\dot{u}$ and $\dot{v}$ increase as the carriage number increases. In contrast, the velocity level of $\dot{w}$ along the load-moving direction need not increase as the carriage number increases.

(2) For a tunnel embedded in a uniform half-space or in a soil stratum lying on a bedrock subjected to quasi-static moving loads (i.e., with $f_0 = 0$), the ground surface velocity increases with the increase in train speed. On the other hand, for the moving loads with nonzero self oscillation frequencies (i.e., with $f_0 \neq 0$), the influence of train speed becomes insignificant.

(3) The velocity level induced by moving loads with self oscillation (i.e., with $f_0 \neq 0$) is appreciably higher than that for the quasi-static case with $f_0 = 0$. This implies the importance of taking the effect of self oscillation frequencies f_0 of the moving loads into account.
(4) The effect of stratum depth is highly dependent on the self oscillation frequency of the moving loads. For the case when the stratum is shallower enough for the cut-off frequency to be influential, a decrease in the stratum depth is accompanied by a reduction of the ground surface vibration.
(5) For a tunnel built on a stiffer soil, the ground surface vibration can be greatly reduced.
(6) The influence of damping ratios on the ground surface responses is too small to be noted for the moving load with no self oscillation (i.e., with $f_0 = 0$). However, an increasing damping ratio may result in an apparent reduction of the responses for the case with non-zero self oscillation (i.e., with $f_0 \neq 0$).
(7) For the case with non-zero self oscillation (i.e., with $f_0 \neq 0$), the velocity level induced by higher frequencies attenuate faster than those by the lower ones.
(8) The effect of the tunnel wall thickness on the surface attenuation is insignificant for both cases with $f_0 = 0$ and $f_0 \neq 0$. However, for the response on the tunnel invert, the increase in wall thickness is accompanied by a substantial decrease in response.
(9) The velocity level of the ground surface shows a trend of decreasing with the increase in the tunnel depth, but this effect becomes less significant as the distance increases.

Appendix

Steady-State Response in Finite Integrals by Eason (1965)

For the sub-critical speed case, $c < c_S$, by setting $z - ct = r\cos\theta$ and $x = r\sin\theta$, the displacements for a uniform elastic half-space subjected to a moving point load $P\delta(x)\delta(y)\delta(z-ct)$ applied on the free surface can be written in a integral form as follows (Eason, 1965):

$$U_y = \frac{4\pi^2\mu}{P_y}u = \int_0^\pi \frac{rH\sin\phi}{K}\left[(Q_1 - Q_3)\cos(\theta-\phi)\right]d\phi \tag{A.1}$$

$$V_y = \int_0^\pi \frac{yH}{K}(Q_1 + Q_2)\,d\phi \tag{A.2}$$

$$W_y = \int_0^\pi \frac{rH\cos\phi}{K}\left[(Q_1 - Q_3)\cos(\theta-\phi)\right]d\phi \tag{A.3}$$

$$U_z = y\int_0^\pi \frac{H}{K}(Q_4 - Q_1)\cos\phi\sin\phi\, d\phi \tag{A.4}$$

$$V_z = r\int_0^\pi \frac{H}{K}(Q_4 + Q_1)\cos\phi\cos(\theta-\phi)\,d\phi \tag{A.5}$$

$$W_z = y\int_0^\pi \left[\frac{H}{K}(Q_4 - Q_1)\cos^2\phi + \frac{2}{R_2^2}\right]d\phi \tag{A.6}$$

$$U_x = y\int_0^\pi \left[\frac{H}{K}(Q_4 - Q_1)\sin^2\phi + \frac{2}{R_2^2}\right]d\phi \tag{A.7}$$

$$V_x = r\int_0^{\pi} \frac{H}{K}(Q_4 + Q_1)\sin\phi\cos(\theta - \phi)\,d\phi \tag{A.8}$$

$$W_x = y\int_0^{\pi} \frac{H}{K}(Q_4 - Q_1)\cos\phi\sin\phi d\phi \tag{A.9}$$

with related parameters given as follows:

$$Q_1 = \frac{\gamma_1\gamma_2 y^2(\beta^2 - 1)}{R_1^2 R_2^2} \tag{A.10}$$

$$Q_2 = \frac{\gamma_1\left[1 - \alpha_2^2(1 + \frac{1}{4}\beta^2)\cos^2\phi + \frac{1}{4}\alpha_2^4\cos^4\phi\right]}{R_1^2\left[\gamma_1(1 - \frac{1}{2}\alpha_2^2\cos^2\phi) + \gamma_2\right]} \tag{A.11}$$

$$Q_3 = \frac{1 - \alpha_2^2(1 - \frac{1}{4}\beta^2)\cos^2\phi}{R_1^2\left[1 - \frac{1}{2}\alpha_2^2\cos^2\phi + \gamma_1\gamma_2\right]} \tag{A.12}$$

$$Q_4 = \frac{1 - \alpha_2^2(1 - \frac{1}{4}\beta^2)\cos^2\phi}{R_2^2\left[1 - \frac{1}{2}\alpha_2^2\cos^2\phi + \gamma_1\gamma_2\right]} \tag{A.13}$$

$$H = (1 - \tfrac{1}{2}\alpha_2^2\cos^2\phi)^2 + \gamma_1\gamma_2 \tag{A.14}$$

$$\begin{aligned} K = {} & \beta^2 - 1 - \alpha_2^2(\tfrac{3}{2}\beta^2 - 1)\cos^2\phi \\ & + \tfrac{1}{2}\beta^2\alpha_2^4\cos^4\phi - \tfrac{1}{16}\beta^2\alpha_2^6\cos^6\phi \end{aligned} \tag{A.15}$$

$$\begin{aligned} \alpha_1 &= c / c_P \\ \alpha_2 &= c / c_S \\ \beta^2 &= c_P^2 / c_S^2 \end{aligned} \tag{A.16}$$

$$\begin{aligned} \gamma_1 &= (1 - \alpha_1^2\cos^2\phi)^{1/2} \\ \gamma_2 &= (1 - \alpha_2^2\cos^2\phi)^{1/2} \end{aligned} \tag{A.17}$$

$$\begin{aligned} R_1^2 &= r^2\cos^2(\theta - \phi) + \gamma_1^2 y^2 \\ R_2^2 &= r^2\cos^2(\theta - \phi) + \gamma_2^2 y^2 \end{aligned} \tag{A.18}$$

Bibliography

Abascal, R., and Domínguez, J. (1986). Vibration of footings on zoned viscoelastic soils, *J. Eng. Mech.*, ASCE, **112**(5), 433-447.

Aboudi, J. (1973). Elastic waves in half-space with thin barrier, *J. Eng. Mech. Div.*, **99**(EM1), 69-83.

Achenbach, J. D. (1976). *Wave Propagation in Elastic Solids*, North-Holland Publishing, New York, N.Y.

Ahmad, S., and Al-Hussaini, T. M. (1991). Simplified design for vibration screening by open and in-filled trenches, *J. Geot. Eng.*, ASCE, **117**(1), 67-88.

Ahmad, S., and Al-Hussaini, T. M., and Fishman, K. L. (1996). Investigation on active isolation of machine foundations by open trenches, *J. Geot. Eng.*, ASCE, **122**(6), 454-461.

Alabi, B. (1992). A parametric study on some aspects of ground-borne vibrations due to rail traffic, *J. Sound & Vibr.*, **153**(1), 77-87.

Al-Hussaini, T. M., and Ahmad, S. (1991). Design of wave barriers for reduction of horizontal ground vibration, *J. Geot. Eng.*, ASCE, **117**(4), 616-636.

Al-Hussaini, T. M., and Ahmad, S. (1996). Active isolation of machine foundations by in-filled trench barriers, *J. Geot. Eng.*, ASCE, **122**(4), 288-294.

Andersen, L., and Jones, C. J. C. (2006). Coupled boundary and finite element analysis of vibration from railway tunnels - a comparison of two- and three-dimensional models, *J. Sound & Vibr.*, **293**, 611-625.

Antes, H., and von Estorff, O. (1994). Dynamic response of 2D and 3D block foundations on a halfspace with inclusions, *Soil Dyn. & Earthq. Eng.*, **13**, 305-311.

Astley, R. J. (1983). Wave envelope and infinite elements for acoustic radiation, *Int. J. Num. Meth. Fluids*, **3**, 507-526.

Balendra, T., Chua, K. H., Lo, K. W., and Lee, S. L. (1989). Steady-state vibration of subway-soil-building system, *J. Eng. Mech.*, ASCE, **115**(1), 145-162.

Balendra, T., Koh, C. G., and Ho, Y. C. (1991). Dynamic response of buildings due to trains in underground tunnels, *Earthq. Eng. & Struct. Dyn.*, **20**, 275-291.

Beer, G., and Meek, J. L. (1981). Infinite domain elements, *Int. J. Num. Meth. Eng.*, **17**, 43-52.

Beskos, D. E. (1987). Boundary element methods in dynamic analysis, *Appl. Mech. Rev.*, **40**, 1-23.

Beskos, D. E. (1997). Boundary element methods in dynamic analysis: Part II (1986-1996), *Appl. Mech. Rev.*, **50**(3), 149-197 (1997).

Beskos, D. E., Dasgupta, B., and Vardoulakis, I. G. (1986). Vibration isolation using open or filled trenches, Part 1: 2-D Homogeneous Soil, *Comp. Mech.*, **1**(1), 43-63.

Beskos, D.E, Leung, K. L., Vardoulakis, I. G. (1990). Vibration isolation studies in non-homogeneous soils, *Boundary Elements in Mechanical and Electrical Engineering*, Eds.: Brebbia, C. A., and Chaudouet-Miranda, A., Springer-Verlag, Berlin, 205-217.

Bettess, P. (1977). Infinite elements, *Int. J. Num. Meth. Eng.*, **11**, 53-64.

Bettess, P., and Zienkiewicz, O. C. (1977). Diffraction and refraction of surface waves using finite and infinite elements, *Int. J. Num. Meth. Eng.*, **11**, 1271-1290.

Boroomand, B., and Kaynia, A. M. (1991), "Vibration isolation by an array of piles", *Soil Dyn. & Earthq. Eng.*, *V*, 684-691.

British Standards Institution (1992). *Evaluation of human exposure to vibration in buildings (1 Hz to 80 Hz)*. BS 6472, British Standards Institution, London.

Chatterjee, P., Degrande, G., Jacobs, S., Charlier, J., Bouvet, P., and Brassenx, D. (2003). Experimental results of free field and structural vibrations due to underground railway traffic, *10th Int. Cong. on Sound & Vibr.*, 7-10 July, Stockholm, Sweden.

Chen, Y. H., and Huang, Y. H. (2000). Dynamic stiffness of infinite Timoshenko beam on viscoelastic foundation in moving co-ordinate, *Int. J. Num. Meth. Eng.*, **48**, 1-18.

Chow, Y. K., and Smith, I. M. (1981). Static and periodic infinite solid elements, *Int. J. Num. Meth. Eng.*, **17**, 503-526.

Chua, K. H., Lo, K. W., and Balendra, T. (1995). Building response due to subway train traffic, *J. of Geot. Eng.*, ASCE, **121**(11), 747-754.

Clouteau, D., Arnst, M., Al-Hussaini, T. M., and Degrande, G. (2005). Freefield vibrations due to dynamic loading on a tunnel embedded in a stratified medium, *J. Sound & Vibr.*, **283**, 173-199.

Cole, J., and Huth, J. (1958). Stresses produced in a half plane by moving loads, *J. Appl. Mech.*, **25**, 433-436.

Cook, R. D., Malkus, D. S., and Plesha, M. E. (1988).*Concepts and Applications of Finite Element Analysis*, John Wiley & Sons, New York, N.Y.

Dawn, T. M., and Stanworth, C. G. (1979). Ground vibrations from passing trains, *J. Sound & Vibr.*, **66**(3), 355-362.

de Barros, F. C. P., and Luco, J. E. (1994). Response of a layered viscoelastic half-space to a moving point load, *Wave Motion*, **19**, 189-210.

Degrande, G., Clouteau, D., Othman, R., Arnst, M., Chebli, H., Klein, R., Chatterjee, P., and Janssens, B. (2006b). A numerical model for ground-borne vibrations from underground railway traffic based on a periodic finite element-boundary element formulation, *J. Sound & Vibr.*, **293**, 645-666.

Degrande, G., and Lombaert, G. (2000). High-speed train induced freefield vibrations: In situ measurements and numerical modelling. *Proc. Int. Workshop Wave 2000*, N. Chouw and G. Schmid, eds., Balkema, Rotterdam, The Netherlands, 29-41.

Degrande, G., Schevenels, M., Chatterjee, P., Van de Velde, W., Hölscher, P, Hopman, V., Wang, A., and Dadkah, N. (2006a). Vibration due to a test train at variable speeds in a deep bored tunnel embedded in London clay, *J. Sound & Vibr.*, **293**, 626-644.

Dieterman, H. A., and Metrikine, A. V. (1996). The equivalent stiffness of a half-space interacting with a beam: critical velocities of a moving load along the beam, *European J. Mech. A/Solids*, **15**(1), 67-90.

Dieterman, H. A., and Metrikine, A. V. (1997). Steady-state displacements of a beam on an elastic half-space due to a uniformly moving constant load, *European J. Mech. A/Solids*, **16**(2), 295-306.

Dobry, R., and Gazetas, G. (1988). Simple methods for dynamic stiffness and damping of floating pile groups, *Geotechnique*, **38**(4), 557-574.

Duffy, D. G. (1990). The response of an infinite railroad track to a moving, vibration mass, *J. Appl. Mech.*, ASME, **57**, 66-73.

Eason, G. (1965). The stresses produced in a semi-infinite solid by a moving surface Force, *Int. J. Eng. Sci.*, **2**, 581-609.

Emad, K., and Manolis, G. D. (1985). Shallow trenches and propagation of surface waves. *J. Eng. Mech.*, ASCE, **111**(2), 279-282.

Estorff, O. V., Kausel, E. (1989). Coupling of boundary and finite element for soil-structure interaction problems. *Earthq. Eng. & Struct. Dyn.*, **18**, 1065-1075.

Esveld, C. (1989). *Modern Railway Track*, MRT-Productions, West Germany.

Ewing, W. M., Jardetzky, W. S., and Press, F. (1957). *Elastic Waves in Layered Media*, McGraw-Hill Book Co., New York.

Forrest, J. A., and Hunt, H. E. M. (2006a). A three-dimensional tunnel model for calculation of train-induced ground vibration, *J. Sound & Vibr.*, **294**, 678-705.

Forrest, J. A., and Hunt, H. E. M. (2006b). Ground vibration generated by trains in underground tunnels, *J. Sound & Vibr.*, **294**, 706-736.

Frýba, L. (1972). *Vibration of Solids and Structures under Moving Loads*, Noordhoff International Publishing, Groningen, The Netherlands.

Fung, Y. C. (1965). *Foundations of Solid Mechanics*, Prentice-Hall, Englewood Cliffs, New Jersey.

Gakenheimer, D. C., and Miklowitz, J. (1969). Transient excitation of an elastic half space by a point load traveling on the surface, *J. Appl. Mech.*, **36**, 505-515.

Gardien, W., and Stuit, H. G. (2003). Modelling of soil vibrations from railway tunnels, *J. Sound & Vibr.*, 267, 605-619.

Graff, K. F. (1973). *Wave Motion in Elastic Solids*, Dover Publications, New York.

Griffin, M. J. (1996). *Handbook of Human Vibration*, Academic Press, London.

Grootenhuis, P. (1977). Floating track slab isolation for railways, *J. Sound & Vibr.*, **51**(3), 443-448.

Grundmann, H., and Lieb, M., and Trommer, E. (1999). The response of a layered half-space to traffic loads moving along its surface, *Archive Appl. Mech.*, **69**, 55-67.

Gupta, S., Hussein, M. F. M., Degrande, G., Hunt, H. E. M., and Clouteau, D. (2007). A comparison of two numerical models for the prediction of vibrations from underground railway traffic, *Soil Dyn. & Earthquake Eng.*, **27**(7), 608-624.

Gutowski, T. G., and Dym, C. L. (1976). Propagation of ground vibration: a review, *J. Sound & Vibr.*, **49**(2), 179-193.

Hall, L. (2003). Simulations and analyses of train-induced ground vibrations in finite element models, *Soil Dyn. & Earthq. Eng.*, **23**, 403-413.

Hanazato, T., Ugai, K., Mori, M., and Sakaguchi, R. (1991). Three-dimensional analysis of traffic-induced ground vibrations, *J. Geot. Eng.*, ASCE, **117**(8), 1133-1151.

Hao, H., and Ang, T. C. (1998). Analytical modeling of traffic-induced ground vibrations, *J. Eng. Mech.*, **124**(8), 921-928.

Heckl, M., Hauck, G., and Wettschureck, R. (1996). Structure-borne sound and vibration from rail traffic, *J. Sound & Vibr.*, **193**(1), 175-184.

Honjo, Y., and Pokharel, G. (1993). Parametric infinite elements for seepage analysis, *Int. J. Num. Anal. Meth. Geomechanics*, **17**, 45-66.

Howarth, H. V. C., and Griffin, M. J. (1991). The annoyance caused by simultaneous noise and vibration from railways, *J. Acoustical Soc. America*, **89**(5), 2317-2323.

Hung, H. H. (1995). *Vibration of Foundations and Soils Generated by High-Speed Trains*, M.Sc. Thesis, Department of Civil Engineering, National Taiwan University, Taipei, Taiwan (in Chinese).

Hung, H. H. (2000). *Ground Vibration Induced by High-Speed Trains and Vibration Isolation Countermeasures*, Ph.D. Thesis, Department of Civil Engineering, National Taiwan University, Taipei, Taiwan.

Hung, H. H., Kuo, J., and Yang, Y. B. (2001). Reduction of train-induced vibrations on adjacent buildings, *Struct. Eng. & Mech., an Int. J.*, **11**(5), 503-518.

Hung, H. H., and Yang, Y. B. (2001). Elastic waves in visco-elastic half-space generated by various vehicle loads, *Soil Dyn. & Earthq. Eng.*, **21**, 1-17.

Hung, H. H., Yang, Y. B., and Chang, D. W. (2004). Wave barriers for reduction of train-induced vibrations in soils, *J. Geot. & Geoenvironmental Eng.*, **130**(12), 1283-1291

Hunt, H. E. M. (1991). Stochastic modeling of traffic-induced ground vibration, *J. Sound & Vibr.*, **144**(1), 53-70.

Hunt, H. E. M. (1996). Modelling of rail vehicles and track for calculation of ground-vibration transmission into buildings, *J. Sound & Vibr.*, **193**(1), 185-194.

Hwang, R. N., and Lysmer, J. (1981). Response of buried structures to traveling waves, *J. Geot. Eng. Div.*, ASCE, **107**(GT2), 183-200.

International Organization for Standardization (1989). *Mechanical vibration and shock - Evaluation of human exposure to whole-body vibration – part 2: Continuous and shock-induced vibration in buildings (1 to 80 Hz).* ISO 2631-2, International Organization for Standardization, Switzerland.

International Organization for Standardization (1997). *Mechanical vibration and shock - Evaluation of human exposure to whole-body vibration – part 1: General requirements,* ISO 2631-1, International Organization for Standardization, Switzerland.

International Organization for Standardization (2003). *Mechanical vibration and shock - Evaluation of human exposure to whole-body vibration – part 2: Vibration in buildings (1 to 80 Hz).* ISO 2631-2, International Organization for Standardization, Switzerland.

Israil, A. S. M., and Ahmad, S. (1989). Dynamic vertical compliance of strip foundations in layered soils, *Earthq. Eng. & Struct. Dyn.*, **18**, 933-950.

Japanese Standards Association (1981). *Methods of Measurement for Vibration Level*, JIS Z 8735, Japanese Standards Association, Tokyo, Japan. (in Japanese).

Japanese Standards Association (1995). *Vibration Level Meters*, JIS C 1510, Japanese Standards Association, Tokyo, Japan.

Ju, S. H. (2002). Finite element analyses of wave propagations due to high-speed train across bridges, *Int. J. Num. Meth. Eng.*, **54**, 1391-1408.

Karasudhi, P., and Liu, Y. C. (1993). Vibration of elastic and viscoelastic multi-layered spaces, *Struct. Eng. and Mech., Int. J.*, **1**, 103-118.

Kausel E., and Roësset, J. M. (1977). Semianalytic hyperelement for layered strata. *J. Eng. Mech. Div.*, ASCE, **103**(EM4), 569-589.

Kausel, E., Roësset, J. M., Waas, G. (1975). Dynamic analysis of footings on layered media, *J. Eng. Mech. Div.*, ASCE, **101**(5), 679-693.

Kim, D. K., and Yun, C. B. (2000). Time domain soil-structure interaction in two-dimensional medium based on analytical frequency-dependent infinite elements, *Int. J. Num. Meth. Eng.*, **47**, 1241-1261.

Kim, S. M., and and Roësset, J. M. (1998). Moving loads on a plate in elastic foundation, *J. Eng. Mech.*, **124**(9), 1010-1017.

Knall, V. (1996). Railway noise and vibration: effects and criteria, *J. Sound & Vibr.*, **193**(1), 9-20.

Krylov, V. V. (1995). Generation of ground vibration by superfast trains, *Appl. Acoustics*, **44**, 149-164.

Krylov, V., and Ferguson, C. (1994). Generation of low frequency ground vibrations from railway trains, *Appl. Acoustics*, **42**, 199-213.

Kuppelwieser, H., and Ziegler, A. (1996). A tool for predicting vibration and structure-borne noise immissions caused by railways, *J. Sound & Vibr.,* **193**, 261-267.

Kurze, U. J. (1996). Tools for measuring, predicting and reducing the environmental impact from railway noise and vibration, *J. Sound & Vib.,* **193**(1), 237-251.

Kurzweil, L. G. (1979). Ground-borne noise and vibration from underground rail system, *J. Sound & Vibr.*, **66**(3), 363-370.

Laghrouche, O., and Le Houedec, D. (1994). Soil-railway interaction for active isolation of traffic vibration, *Adv. in Simulation & Interaction Techniques,* Eds.:

Papadrakakis, M., and Topping, B. H. V., Civil-Comp Press, Edinburgh, Scotland, 31-36.

Lamb, H. (1904). On the propagation of tremors over the surface of an elastic solid, *Phil. Trans. Roy. Soc. London, Ser. A*, CCIII **1**, 1-42.

Lang, J. (1988). Ground-born vibrations caused by trams, and control Measures, *J. Sound & Vibr.*, **120**(2), 407-412.

Lau, S. L., and Ji, Z. (1989). An efficient 3-D infinite element for water wave diffraction problems, *Int. J. Num. Meth. Eng.*, **28**, 1371-1387.

Lieb, M., and Sudret, B. (1998). A fast algorithm for soil dynamics calculations by wavelet decomposition, *Archive Appl. Mech.*, **68**, 147-157.

Luco, J. E., and Apsel, R. J. (1983). On the Green's functions for a layered half-space. part I, *Bull. Seismological of America*, **73**(4), 909-929.

Luco, J. E., and de Barros, F. C. P. (1994). Seismic response of a cylindrical shell embedded in a layered viscoelastic half-space. I: formulation, *Earthq. Eng. Struct. Dyn.*, **23**, 553-567.

Luco, J. E., and de Barros, F. C. P. (1995). Three-dimensional response of a layered cylindrical valley embedded in a layered half-space, *Earthq. Eng. & Struct. Dyn.*, **24**, 109-125.

Luco, J. E., and Westmann, R. A. (1972). Dynamic response of a rigid footing bonded to an elastic half space. *J. Appl. Mech.*, ASME, **39**, 527-534.

Lysmer, J., and Kuhlemeyer, R. L. (1969). Finite dynamic model for infinite media. *J. Eng. Mech. Div.*, ASCE, **95**(EM4), 859-877.

Lysmer, J., and Waas, G. (1972). Shear waves in plane infinite structures, *J. Eng. Mech. Div.*, ASCE, **98**(EM1), 85-105.

Madshus, C., Bessason, B., and Hårvik, L. (1996). Prediction model for low frequency vibration from high speed railways on soft ground, *J. Sound & Vibr.*, **193**(1), 195-203.

Medina, F., and Penzien, J. (1982). Infinite elements for elastodynamics, *Earthq. Eng. & Struct. Dyn.*, **10**, 699-709.

Medina, F., and Taylor, R. L. (1983). Finite element techniques for problems of unbounded domains, *Int. J. Num. Meth. Eng.*, **19**, 1209-1226.

Melke, J. (1988). Noise and vibration from underground railway lines: proposals for a prediction procedure, *J. Sound. Vibr.*, **120**(2), 391-406.

Melke, J., and Kraemer, S. (1983). Diagnostic methods in the control of railway noise and vibration, *J. Sound & Vibr.*, **87**(2), 377-386.

Metrikine, A. V., and Dieterman, H. A. (1997). The equivalent vertical stiffness of an elastic half-space interacting with a beam, including the shear stresses at the beam - half-space Interface, *European J. Mech. A/Solids*, **16**(2), 515-527.

Metrikine, A. V., Vostrukhov, A.V., and Vrouwenvelder, A. C. W. M. (2001). Drag experienced by a high-speed train due to excitation of ground vibrations, *Int. J. Solids Struct.*, **38**, 8511-8868.

Metrikine, A.V., and Vrouwenvelder, A. C. W. M. (2000a). Surface ground vibration due to moving train in a tunnel: two-dimensional model, *J. Sound & Vibr.*, **234**(1), 43-66.

Metrikine, A. V., and Vrouwenvelder, A. C. W. M. (2000b). Ground vibration induced by a high-speed train in a tunnel: two-dimensional model, *Wave 2000: Wave Propagation, Moving Load, Vibration Reduction*, Eds.: Chouw, N., and Schmid, G., 111-120, A. A. Balkema, Rotterdam, Netherlands.

Mohanan, V., Sharma, O., and Singal, S. P. (1989). A noise and vibration survey in an underground railway system, *Appl. Acoustics*, **28**, 263-275.

Nelson, J. T. (1996). Recent developments in ground-borne noise and vibration control, *J. Sound & Vibr.*, **193**(1), 367-376.

Newland, D. E., and Hunt, H. E. M. (1991). Isolation of buildings from ground vibration: a review of recent progress, *Proc. Instn. Mech. Engrs*, **205**, 39-52.

Ni, S. H., Feng, Z. Y., and Tsai, P. S. (1994). Analysis of the vibration response and screening effectiveness of strip foundations, *J. Chinese Inst. Civil and Hydraulic Eng.*, **6**(3), 269-277. (in Chinese)

Ni, S. H., and Hung, C. C. (1998). Screening effect of surface waves by rectangular strip in-filled trenches, *J. Chinese Inst. Civil and Hydraulic Eng.*, **10**(3), 457-465. (in Chinese)

Okumura, Y., and Kuno, K. (1991). Statistical analysis of field data of railway noise and vibration collected in an urban Area, *Appl. Acoustics*, **33**, 263-280.

Pan, C. S., and Xie, Z. G. (1990). Measurement and analysis of vibrations caused by passing trains in subway running tunnel, *China Civil Eng. J.*, **23**(2), 21-28 (in Chinese)

Pao, Y. H., and Mow, C. C. (1963). Scattering of plane compressional waves by a spherical obstacle, *J. Appl. Phys.*, **34**, 493-499.

Paolucci, R., Maffeis, A., Scandella, L., Stupazzini, M., and Vanini, M. (2003). Numerical prediction of low-frequency ground vibrations induced by high-speed trains at Ledsgaad, Sweden. *Soil Dyn. Earthq. Eng.*, **23**, 425–433.

Park, K. L., Watanabe, E., and Utsunomiya, T. (2004). Development of 3D elastodynamic infinite elements for soil-structure interaction problems, *Int. J. Struct. Stability & Dyn.*, **4**(3), 423-441.

Park, W. S., Yun, C. B., and Pyun, C. K. (1991). Infinite elements for evaluation of hydrodynamic forces on offshore structures, *Comp. & Struct.*, **40**, 837-847.

Park, W. S., Yun, C. B., and Pyun, C. K. (1992). Infinite elements for 3-dimensional wave-structure interaction problems, *Eng. Struct.*, **14**, 335-346.

Paulsen, R., and Kastka, J. (1995). Effects of combined noise and vibration on annoyance, *J. Sound & Vibr.*, **181**(2), 295-314.

Payton, R. G. (1967). Transient motion of an elastic half-space due to a moving surface line load, *Int. J. Eng. Sci.*, **5**, 49-79.

Piessens, R., de Doncker-Kapenga, E., Uberhube, C. W., and Kahaner, D. K. (1983). QUADPACK, Springer-Verlag, New York, N.Y.

Rajapakse, R. K. N. D., and Karasudhi, P. (1985). Elastostatic infinite elements for layered half space, *J. Eng. Mech.*, ASCE, **111**, 1144-1158.

Rajapakse, R. K. N. D., and Karasudhi, P. (1986). An efficient elastodynamic infinite element, *Int. J. Solids and Struct.*, **22**, 643-657.

Richart, F. E., Hall, J. R., and Woods, R. D. (1970). *Vibration of Soils and Foundations*, Prentice-Hall, Englewood Cliffs, N.J.

Rossi, F., and Nicolini, A. (2003). A simple model to predict train-induced vibration: theoretical formulation and experimental validation, *Environmental Impact Assessment Rev.*, **23**, 305-322.

Saini, S. S., Bettess, P., and Zienkiewicz, O. C. (1978). Coupled hydrodynamic response of concrete gravity dams using finite and infinite elements, *Earthq. Eng. & Struct. Dyn.*, **6**, 363-374.

Schmid, G., and Chouw, N., and Le, R. (1991). Shielding of structures from soil vibrations, *Soil Dyn. & Earthq. Eng. V, Int. Conf. Soil Dyn. & Earthq. Eng.*, Computational Mechanics Publications, Southampton, U.K., 651-662.

Seed, H. B., and Idriss, I. M. (1970). Soil modulus and damping factors for dynamic response analysis, *Report No. EERC 70-10*, University of California, Berkeley, California.

Segol, G., Lee, C. Y., and Abel, J. F. (1978). Amplitude reduction of surface waves by trenches, *J. Eng. Mech. Div.*, ASCE, **104**, 621-641.

Sheng, X., Jones, C. J. C., and Thompson, D. J. (2004). A theoretical model for ground vibration from trains generated by vertical track irregularities, *J. Sound & Vibr.*, **272**, 937-965.

Sheng, X. Jones C. J. C. and Thompson, D. J. (2005) Modelling ground vibration from tunnels using wavenumber finite and boundary element methods, *Proc. Royal Soc. A*, 461, 2043-2070.

Sheng, X., Jones, C. J. C., and Thompson, D. J. (2006). Prediction of ground vibration from trains using the wavenumber finite and boundary element method, *J. Sound & Vibr.*, **293**, 575-586.

Shyu R. J., Wang W. H., Cheng C. Y., and Hwang D. (2002). The characteristics of structural and ground vibration caused by the TRTS trains, *Metro's Impact on Urban Living*, Proceedings of *2002 World Metro Symposium*, Taipei, Taiwan, 610.

Sneddon, I. N. (1951). *Fourier Transforms*, McGraw-Hill, New York, N.Y.

Stamos, A. A., and Beskos, D. E. (1996). 3-D seismic response analysis of long lined tunnels in half-space, *Soil Dyn. & Earthq. Eng.*, **15**, 111-118.

Suiker, A. S. J., de Borst, R., and Esveld, C. (1998). Critical behaviour of a Timoshenko beam-half plane system under a moving load, *Archive Appl. Mech.*, **68**, 158-168.

Takahashi, D. (1985), "Wave propagation in ground-structure systems with line contact", *J. Sound & Vibr.*, **101**(4), 523-537.

Takahashi, D. (1986a), "Wave propagation in ground-structure systems, part I: analysis of the model with surface contact", *J. Sound & Vibr.*, **105**(1), 27-36.

Takahashi, D. (1986b), "Wave propagation in ground-structure systems, part II: parametric survey", *J. Sound & Vibr.*, **105**(1), 37-48.

Takemiya, H. (1997). Prediction of ground vibration induced by high-speed train operation, *18th Sino-Japan Technology Seminar*, Chinese Inst. of Engs., 1-10.

Takemiya, H. (1998a). Lineside ground vibrations induced by high-speed train passage, *Workshop on Effect of High-Speed Vibration on Structures and Equipment*, Dept. of Civil Eng., National Cheng Kung Univ., Tainan, Taiwan, R.O.C., 43-49.

Takemiya, H. (1998b). Paraseismic behavior of wave impeding block measured for ground vibration reduction, *Workshop on Effect of High-Speed Vibration on Structures and Equipment*, Dept. Civil Eng., Natl. Cheng Kung Univ., Tainan, Taiwan, R.O.C., 51-56.

Takemiya, H. (2004). Field vibration mitigation by honeycomb WIB for pile foundations of high-speed train viaduct, *Soil Dyn. & Earthq. Eng.*, **24**, 69-87.

Takemiya, H., and Fujiwara, A. (1994). Wave propagation/impediment in a stratum and wave impeding block (WIB) measured for SSI response reduction, *Soil Dyn. & Earthq. Eng.*, **13**, 49-61.

Thau, S. A., and Pao, Y. H. (1966). Diffractions of horizontal shear waves by a parabolic cylinder and dynamic stress concentrations, *J. Appl. Mech.*, ASME, **33**(4), 785-792.

Thiede, R., and Natke, H. G. (1991). The influence of thickness variation of subway walls on the vibration emission generated by subway traffic, *Soil Dyn. & Earthq. Eng. V: Int. Conf. Soil Dyn. & Earthq. Eng.*, Computational Mechanics Publications, Southampton, UK, 673-682.

Thomson, W. T., and Kobori, T. (1963). Dynamical ground compliance of rectangular foundations on an an elastic half-space, *J. Appl. Mech.*, ASME, **30**, 579-584.

Trochides, A. (1991). Ground-born vibrations in buildings near subways, *Appl. Acoustics*, **32**, 289-296.

Turunen-Rise, I. H., Brekke, A., Hårvik, L., Madshus, C. and Klæboe, R. (2003). Vibration in dwellings from road and rail traffic – Part I: a new Norwegian measurement standard and classification system, *Appl. Acoustics*, **64**, 71-87.

Ungless, R. F. (1973). *An Infinite Element*, M.A.Sc. Thesis, University of British Columbia.

Vadillo, E. G., Herreros, J., and Walker, J. G. (1996). Subjective reaction to structurally radiated sound form underground railways: field studies, *J. Sound & Vibr.*, **193**(1), 65-74.

Verhas, H. P. (1979). Prediction of the propagation of train-induced ground vibration, *J. Sound & Vibr.*, **66**(3), 371-376.

Ministry of the Environment (1976), *Vibration Regulation Law*. Tokyo, Japan.

Volberg, G. (1983). Propagation of ground vibrations near railway tracks, *J. Sound & Vibr.*, **87**(2), 371-376.

Wilson, G. P., Saurenman, H. J., and Nelson, J. T. (1983). Control of ground-borne noise and vibration, *J. Sound. Vibr.*, **87**(2), 339-350.

Wolf, J. P. (1985). *Dynamic Soil-Structure Interaction*, Prentice-Hall, Englewood Cliffs, New Jersey.

Wolf, J. P. (1988). *Soil-Structure Interaction Analysis in Time Domain*, Prentice-Hall, Englewood Cliffs, New Jersey.

Wolf, J. P., and Song, C. (1996). *Finite-Element Modeling of Unbounded Media*, John Wiley & Sons, New York, N.Y.

Woods, R. D. (1968). Screening of surface waves in soils, *J. Soil Mech. Found. Div.*, **4**, 951-979.

Wu, Y. S., Hsu, L. C., and Yang, Y. B. (2002). Ground vibrations induced by trains moving over a series of elevated bridges, *Proc. 10th Sound and Vibration Conference*, Tainan, Taiwan, 1-7 (in Chinese).

Wu, Y. S., and Yang, Y. B. (2004a). A semi-analytical approach for analyzing ground vibrations caused by trains moving over elevated bridges, *Soil Dyn. & Earthq. Eng.,* **24**, 949-962.

Wu, Y. S., and Yang, Y. B. (2004b). Train-induced ground and building vibrations considering bridge-foundation-soil interactions, *J. Chinese Inst. Civil & Hydraulic Eng.,* Special Issue, 845-859.

Xia, H., Zhang, N., and Cao, Y. M. (2005). Experimental study of train-induced vibration of environment and buildings, *J. Sound & Vib.*, **280**, 1017-1029.

Yang, S. C., and Yun, C. B. (1992). Axisymmetric infinite elements for soil-structure interaction analysis, *Eng. Struct.*, **14**, 361-370.

Yang, Y. B., and Hung, H. H. (1997). A parametric study of wave barriers for reduction of train-induced vibrations, *Int. J. Num. Meth. Eng.*, **40**, 3729-3747.

Yang, Y. B., and Hung, H. H. (2001). A 2.5D finite/infinite element approach for modelling visco-elastic bodies subjected to moving loads, *Int. J. Num. Meth. Eng.*, **51**(11), 1317-1336.

Yang, Y. B., and Hung, H. H. (2008). Soil Vibrations Caused by Underground Moving Trains, *J. of Geotech. & Geoenvironment. Eng.*, ASCE, to appear.

Yang, Y. B., Hung, H. H., and Chang, D. W. (2003). Train-induced wave propagation in layered soils using finite/infinite element simulation, *Soil Dyn. & Earthq. Eng.,* **23**(4), 263-278.

Yang, Y. B., Hung, H. H., and Hsu, L. C. (2008). Ground vibrations due to underground trains considering soil-tunnel interactions, *Interaction & Multiscale Mech., an Int. J.*, **1**(1), 157-175.

Yang, Y. B., Kuo, S. R., and Hung, H. H. (1996). Frequency-independent infinite element for analyzing semi-infinite problems, *Int. J. Num. Meth. Eng.*, **39**, 3553-3569.

Yang, Y. B., and Wu, Y. S. (2007). Semi-Analytical Approach for Analyzing Ground Vibrations Caused by Trains Moving over Elevated Bridges with Pile Foundations, chapter in *Linear and Non Linear Numerical Analysis of Foundations*, eds.: J. W. Bull, Taylor & Francis, London, U.K. 2007 (to appear).

Yang, Y. B., Yau, J. D., and Wu, Y. S. (2004). *Vehicle-Bridge Interaction Dynamics – with Applications to High-Speed Railways*, World Scientific, Singapore.

Yeh, C. S., Liao, W. I., Tsai, J. F., and Teng, T. J. (1997). *Train Induced Ground Motion and Its Mitigation by Trench and WIB*, report of NCREE-97-009 (in Chinese).

Yokota, A. (1996). Relationship of weighted acceleration levels between the ground surface and the floors of a wooden house, *J. Acoustical Soc. America*, **100**(4), 2685.

Yun, C. B., Kim, D. K., and Kim, J. M. (2000). Analytical frequency-dependent infinite elements for soil-structure interaction analysis in two-dimensional medium, *Eng. Struct.*, **22**(3), 258-271.

Yun, C. B., and Kim, J. M. (1995), "Axisymmetric elastodynamic infinite elements for multi-layered half-space," *Int. J.Num. Meth. Eng.*, **38**, 3723-3743.

Yun, C. B., Kim, J. M., and Hyun, C. H. (1995). Axisymmetric elastodynamic infinite elements for multi-layered half-space, *Int. J. Num. Meth. Eng.*, **38**, 3723-3743.

Zhang, C., and Zhao, C. (1987). Coupling method of finite and infinite elements for strip foundation wave problems, *Earthq. Eng. and Struct. Dyn.*, **15**, 839-851.

Zhao, C., and Valliappan, S. (1993). A dynamic infinite element for three-dimensional infinite-domain wave problems, *Int. J. Num. Meth. Eng.*, **36**, 2567-2580.

Author Index

Subject Index